Processes, Quality and Fertility of Soil

Processes, Quality and Fertility of Soil

Edited by Henry Wang

SYRAWOOD
PUBLISHING HOUSE
New York

Published by Syrawood Publishing House,
750 Third Avenue, 9th Floor,
New York, NY 10017, USA
www.syrawoodpublishinghouse.com

Processes, Quality and Fertility of Soil
Edited by Henry Wang

International Standard Book Number: 978-1-68286-384-8 (Hardback)

Cataloging-in-publication Data

Processes, quality and fertility of soil / edited by Henry Wang.
p. cm.
Includes bibliographical references and index.
ISBN 978-1-68286-384-8
1. Soil science. 2. Soils--Quality. 3. Soil fertility. 4. Fertilizers. I. Wang, Henry.
S591 .P76 2017
631.4--dc23

Printed in China.

TABLE OF CONTENTS

PREFACE

Soil is a major provider of nutrients to crops and undergoes various changes and processes. Soil quality and soil function are important branches of agricultural soil sciences. This book presents interdisciplinary view-points related to varied aspects of soil science such as soil nutrition, soil fertility as well as soil conservation. This book aims to shed light on some of the explored aspects of this field. It attempts to understand the multiple branches that fall under the discipline of soil science and how such concepts have practical applications. The extensive content of this book provides the readers with a thorough understanding of the subject. It will prove extremely helpful for experts in the field of soil ecology, agriculture, horticulture and soil surveying.

This book aims to highlight the current researches and provides a platform to further the scope of innovations in this area. This book is a product of the combined efforts of many researchers and scientists from different parts of the world. The objective of this book is to provide the readers with the latest information in the field.

I would like to express my sincere thanks to the authors for their dedicated efforts in the completion of this book. I acknowledge the efforts of the publisher for providing constant support. Lastly, I would like to thank my family for their support in all academic endeavors.

Editor

A Systematic Review of Biochar Research, with a Focus on Its Stability *in situ* and Its Promise as a Climate Mitigation Strategy

Noel P. Gurwick[1]*, Lisa A. Moore[2], Charlene Kelly[3], Patricia Elias[4]

1 Smithsonian Environmental Research Center. Edgewater, Maryland, United States of America, **2** Environmental Defense Fund, San Francisco, California, United States of America, **3** Western Carolina University, North Carolina, United States of America, **4** Union of Concerned Scientists, Washington, District of Columbia, United States of America

Abstract

Background: Claims about the environmental benefits of charring biomass and applying the resulting "biochar" to soil are impressive. If true, they could influence land management worldwide. Alleged benefits include increased crop yields, soil fertility, and water-holding capacity; the most widely discussed idea is that applying biochar to soil will mitigate climate change. This claim rests on the assumption that biochar persists for hundreds or thousands of years, thus storing carbon that would otherwise decompose. We conducted a systematic review to quantify research effort directed toward ten aspects of biochar and closely evaluated the literature concerning biochar's stability.

Findings: We identified 311 peer-reviewed research articles published through 2011. We found very few field studies that addressed biochar's influence on several ecosystem processes: one on soil nutrient loss, one on soil contaminants, six concerning non-CO_2 greenhouse gas (GHG) fluxes (some of which fail to support claims that biochar decreases non-CO_2 GHG fluxes), and 16–19 on plants and soil properties. Of 74 studies related to biochar stability, transport or fate in soil, only seven estimated biochar decomposition rates *in situ*, with mean residence times ranging from 8 to almost 4,000 years.

Conclusions: Our review shows there are not enough data to draw conclusions about how biochar production and application affect whole-system GHG budgets. Wide-ranging estimates of a key variable, biochar stability *in situ*, likely result from diverse environmental conditions, feedstocks, and study designs. There are even fewer data about the extent to which biochar stimulates decomposition of soil organic matter or affects non-CO_2 GHG emissions. Identifying conditions where biochar amendments yield favorable GHG budgets requires a systematic field research program. Finally, evaluating biochar's suitability as a climate mitigation strategy requires comparing its effects with alternative uses of biomass and considering GHG budgets over both long and short time scales.

Editor: Qinghua Sun, The Ohio State University, United States of America

Funding: The Tropical Forest and Climate Initiative at the Union of Concerned Scientists (UCS) briefly supported CK to perform an initial classification of some of the studies we reviewed. During the initial stages of this study, NPG was on staff at UCS, which paid his salary during that time. LAM is on staff at Environmental Defense Fund (EDF), which has covered her salary. EDF paid for NPG and LAM to attend a 5-day writing retreat that they used to make progress on this study. The material in this paper does not reflect any position of EDF. The funders had no role in study design, data collection and analysis, decision to publish, or preparation of the manuscript.

Competing Interests: The authors have declared that no competing interests exist.

* E-mail: Noel.Gurwick@Gmail.Com

Introduction

In the face of the dual challenge to mitigate global climate change and ensure food security for a growing global population, crop management strategies that build soil organic matter (SOM) have received considerable attention [1–4]. Because photosynthesis converts CO_2 to organic carbon, increases in plant carbon stocks reduce atmospheric CO_2 concentrations. When plant biomass does not rapidly decompose but instead is stored in wood or increased SOM, it keeps carbon out of the atmosphere. Increased SOM also enhances soil fertility and water holding capacity [4,5]. As a strategy for climate change mitigation, a key weakness of enhanced soil storage of carbon is the susceptibility of SOM to decomposition and re-release of CO_2 to the atmosphere [6,7].

A related carbon storage proposal that gained considerable attention at the beginning of the 21st century calls for heating biomass in the absence of oxygen (pyrolysis) and applying the resulting carbonized material to agricultural or forest soils [8–10]. The solid product of pyrolysis, called "biochar" in the context of climate change mitigation, is highly heterogeneous material with chemical composition that varies widely depending on feedstock and pyrolysis conditions [11]. The scale of biochar production and application being discussed is enormous. For example, Matovic [12] calculated that charring and burying 10% of global net primary productivity each year would offset the current annual increase in atmospheric CO_2.

This discussion has reached far beyond the academic literature, to highly visible media outlets [13], books about solutions to address climate change [14], and specific policy proposals for carbon offsets [15]. The California Energy Commission's Public Interest Energy Research Program has supported work to describe biochar projects that could qualify as greenhouse gas offsets, which would effectively trade carbon sequestration by biochar for emissions of fossil fuel derived CO_2 that would otherwise be limited by regulation [16]. With this funding, the Climate Trust, a US-based non-profit that specializes in "climate solutions for governments, utilities, and large businesses" has written that "at its maximum sustainable potential, biochar could" reduce annual global GHG emissions by 12% [16]; to arrive at this estimate, they assumed that 80% of biochar carbon persists in soil after 100 years. The International Biochar Initiative (IBI) promotes ubiquitous use of biochar as a soil amendment, advocates for inclusion of provisions favorable to biochar use in national and global climate mitigation policies, promotes biochar commercialization, and aspires to a global system that sequesters 2.2 Gt C/yr by 2050 [17]. Addressing international climate change policy, IBI has urged that the technical advisory body of the UN Framework Convention on Climate Change incorporate biochar within the work program on mitigation in the agricultural sector [18]. In April 2013, IBI also reported that a methodology for using biochar amendments as an offset protocol was soon to be submitted for consideration by the American Carbon Registry, which plays a key role in establishing protocols for carbon markets [19]. Finally, proponents of biochar use often suggest it has numerous benefits, as illustrated in IBI's description of biochar's benefits: "Biochar is uniquely positioned to aid in these critical overlapping arenas [of climate mitigation and adaptation and food security] by building soil carbon sinks and mitigating climate change while also enhancing soil quality and resilience to drought and certain diseases [18]."

Underlying the idea that incorporating biochar in soil effectively mitigates climate change is the argument that it resists degradation and will persist in soil for hundreds or thousands of years [14,20,21]. However, claims about biochar's stability in soil rest most frequently on observations of old charcoal in Amazonian and other soils [22]. The presence of ancient charcoal in soil shows only that some old charcoal persists; it could be a small fraction of a much larger stock of since-decayed charred biomass [11]. Even if it is the majority of the original stock, persistence of biochar carbon could well be context-dependent, remaining stable in some soils and climate regimes and not in others. The biochar systems that are being proposed for climate mitigation would apply highly variable biochars – produced from diverse feedstocks and under varying pyrolysis conditions – to diverse soils under a wide range of environmental conditions. What do empirical data indicate about biochar's stability *in situ*?

Although understanding biochar stability is critical to quantifying the impact of biochar amendments on net greenhouse gas (GHG) emissions to the atmosphere, it is not sufficient. Additional factors that need to be considered include emissions associated with growing, harvesting and transporting feedstock; and with biochar production and application to soil. Life cycle analysis (LCA) integrates these varied sources of emissions associated with biochar amendments to soil to understand the system-level impacts on terrestrial carbon stocks and atmospheric GHG concentrations [23]. The confidence we should place in LCA results is directly related to availability of data used to estimate model inputs.

Given the emphasis that biochar proponents have placed on biochar stability in soil, we focused this review on that aspect of biochar, but reported benefits of biochar also include increased soil fertility and water holding capacity, increased crop production, and remediation of contaminated soils. The literature provides some support for these claims [24–30], but there is no evidence that all these claims have been tested across a broad enough variety of landscapes, climates, and management systems to draw robust conclusions [31]. Furthermore some researchers have cautioned that biochar could have adverse effects by releasing toxic substances such as heavy metals into soil or reducing the efficacy of pesticides [32]. How frequently these unintended consequences of biochar amendment occur is unclear. To determine whether there is sufficient research to support adoption of biochar systems, there is a strong need to review the peer-reviewed literature with respect to the multiple benefits and risks biochar may deliver.

In this paper, we systematically review and organize the biochar literature towards the following three objectives: (1) quantitatively characterize the research effort that has been directed towards a variety of biochar attributes (*e.g.*, production processes, effects on soil or plant production, etc.); (2) provide a more focused characterization of the biochar literature concerning biochar stability; and (3) identify and review field studies that have calculated biochar stability. We also consider the broader implications of existing data for implementing biochar systems.

Methods

Searching

We searched for articles using the terms "biochar" and "bio-char" on the Web of Science (in the 'topic' field) and Agricola databases for papers published prior to January 1, 2012. Because the biochar literature is so diverse and the term "biochar" is relatively new, we supplemented our search with citations in recent studies. To further ensure that we had assembled a comprehensive list of studies, we asked researchers in a leading biochar research group to review and suggest additions to our database.

Screening

We excluded search results that were published in languages other than English or for which only an abstract was available, and then characterized each remaining search result as one of five publication types. "Primary research" papers appeared in the peer-reviewed literature and reported original data or results from observations, experiments, or models. "Methods" papers evaluated or described an investigative technique for studying biochar. "Review" papers summarized understanding of biochar but did not report new data. Our search also captured "other" publication types such as news stories, book chapters, extension newsletters, editorial notes, and letters-to-the-editor. Finally, we characterized as "incidental" any publication that did not concern biochar in any meaningful way; many publications, for example, listed but did not analyze or describe the products of pyrolysis, one of which is biochar. Throughout our screening process, we noted the number of publications identified during our search, the number of publications excluded in each stage, and the reasons for those exclusions, following the guidelines set forth in the Preferred Reporting Items for Systematic Reviews and Meta-Analyses (PRISMA) Statement (Figure S1) [33].

Objectives 1 and 2 – characterizing the literature

To achieve our first objective – a quantitative characterization of the biochar literature as a whole – we applied a second sorting process to the publications identified as primary research. Based on an initial reading of these papers, we identified ten topic areas and assigned each paper to all topic areas that applied. We also noted whether each study had a laboratory and/or *in situ* component. The ten topic areas were:

(1) *Stability, transport, or fate of biochar or soil carbon* – Studies that described changes in the physical and/or chemical properties of biochar (or soil organic matter mixed with biochar) over time; measured or inferred changes in a system's carbon balance after the addition of biochar; or described the amount or location of biochar in a soil or landscape at a single time point.
(2) *Model and/or life cycle analyses of biochar systems* – Studies that calculated the economic, energy and/or climate change mitigation potential of biochar production systems.
(3) *Influence of biochar on non-CO_2 trace gas emissions from soil* – Studies that reported rates of methane and/or nitrous oxide emissions from soils to which biochar had been added.
(4) *Soil nutrients* – Studies that reported nutrient levels in soils amended with biochar.
(5) *Plant responses* – Studies that reported responses, such as yield or nutrient status, of plants grown on soils amended with biochar.
(6) *Soil biology* – Studies that described or reported changes in soil microbes, fungi, earthworms, or other soil fauna on biochar or in soil amended with biochar.
(7) *Soil properties* – Studies that reported, for example, the pH, bulk density, water holding capacity, or cation exchange capacity of soils amended with biochar.
(8) *Nutrient loss (N, P, K)* – Studies that reported loss of nutrients, for example via leaching or gaseous emissions, from soils amended with biochar.
(9) *Biochar production and analysis* – Studies that described biochar production processes and/or characterized the physical or chemical properties of biochar.
(10) *Influence on contaminants* – Studies that described the effects of biochar on the levels or mobility of contaminants such as lead, arsenic, pesticides, and herbicides in water or soil.

To achieve our second objective – a quantitative characterization of the literature related to the stability of biochar – we focused on primary research articles that addressed the stability, transport, or fate of biochar or soil carbon (topic area 1, above). We sorted these studies in three stages. First, because estimating biochar stability by definition requires measuring a change in biochar amount or characteristics over time, we divided studies that took measurements at one time point from studies that tracked changes over time. We then divided the studies that tracked changes over time into field versus lab studies. Although laboratory incubations are valuable because they can test *relative* stability of different substrates and the mechanisms that control decomposition [11,34,35], we focused on field studies because they provide unique and essential data for deriving realistic estimates of biochar stability under real-world conditions. Finally, we divided the field studies into experiments that did or did not calculate a measure of biochar stability, such as mean residence time or turnover time.

Objective 3 – Reviewing scientific understanding of biochar stability

We looked closely at each study in this last group to address our third objective, which was to ascertain what can be said with confidence about the stability of biochar under field conditions. To allow a synthetic view of findings from these key studies, we noted for each study its location, ecosystem, type of biochar used, method of biochar application to the study site, and numerical estimate of biochar stability. Because converting among measures of stability (*e.g.*, half-life, turnover time, mean residence time) requires assumptions about the shape of the decay curve, we did not convert all these measurements into a "common currency" and instead report numbers as they appeared in each paper.

Results

Our initial search yielded 472 unique publications (Figure S1). We excluded 161 of these from further analysis either because full English text was unavailable (three publications) or because the publication was not primary research (158 publications, including 63 non-peer-reviewed ("other") publications, 53 literature reviews, 27 papers with only incidental mention of biochar, 14 articles describing new methods, and one paper captured by our search because it used the unrelated term "biocharacters").

Each of the remaining 311 peer-reviewed articles (Figure S2) described primary research in at least one of the ten topic areas (Figures 1 and S3). Most commonly, these studies tested pyrolysis techniques used to produce char and characterized the resulting materials. Full life cycle analysis of biochar systems and the effects of biochar on non-CO_2 trace gas and soil nutrient fluxes have received the least attention (Figure 1).

In situ research investigating the influence of biochar on nutrient loss and contaminants was almost non-existent (1 study each) (black bars in Figure 1). Field research was more common for studies of plant responses, soil nutrients, and soil properties (16–19 studies each, Figure 1). Of the 20 model/LCA studies, only four used field measurements, and of 74 original research papers that addressed biochar stability or fate, 29 included a field component.

Seventy-four studies addressed some aspect of the stability, transport, or fate of biochar in soil (Figures S2 and S3), with 60 measuring system change over time. Of those, 18 articles reported results from a field experiment. Seven of those field studies provided some quantitative estimate of biochar stability derived from measurements of biochar stocks or soil respiration (Table 1). Three of the seven studies experimentally added biochar to soil, and four used naturally-occurring biochar from fire at sites differing in fire frequency over a number of years.

The seven studies were diverse in numerous respects (Table 1). They included temperate and tropical ecosystems (*e.g.*, Eastern North America, Philippines, Colombia) and applied biochar produced from a variety of feedstocks (*e.g.*, mango wood, rice husks, char from forest fires). These studies also used different analytical approaches. For example, some measured soil CO_2 evolution for several years following char addition; others measured the amount of char in soil along a fire chronosequence at different times following the initiation of fire suppression, or after deliberate biochar addition at a site.

Four studies reported mean residence time (MRT) of biochar (ranging from ~8 to ~3,600 yrs), one reported turnover time (~300 yrs), one reported half-life ("considerably <50 years"), and one reported the percentage of original biochar lost over 130 years (22%) (Table 1).

Despite our efforts to ensure that we had a comprehensive set of studies, it is of course possible that we missed some relevant papers published before "biochar" became widely used. However, comparison with a recent meta-analysis of biochar's effects on nutrient cycling and plant productivity suggests we captured the vast majority of important publications. Biederman and Harpole [36] included "charcoal" and "black carbon" in their search terms, which we did not, and found 114 papers through 2012 and 84 papers through 2011 (our cutoff). By comparison, of the 311 peer-reviewed publications we found through 2011, 115 fell within topic areas 4–7 (which encompass the variables examined by Biederman and Harpole), and these 115 included the large

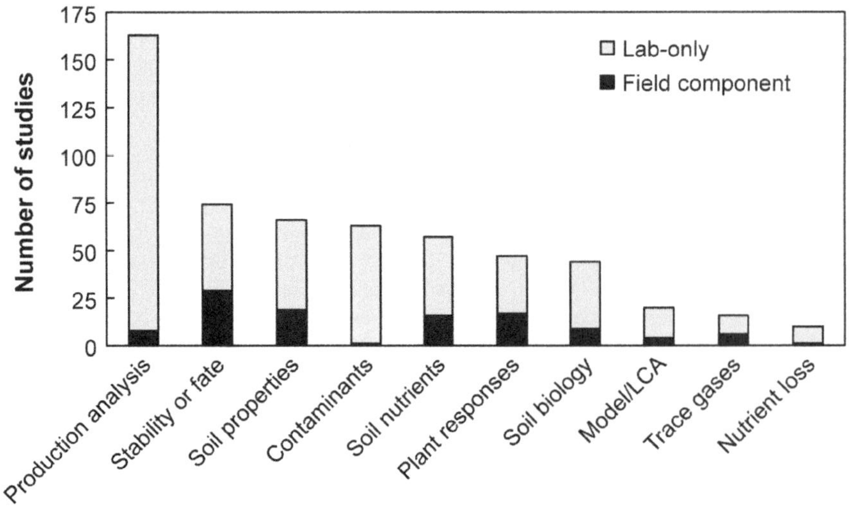

Figure 1. Number of primary research articles addressing each topic area. The dark shaded area indicates the studies that included a field component.

majority of the 84 papers included in their meta-analysis. Despite Biederman and Harpole's more expansive search terms, they found fewer, not more, references than we did in a comparable time frame, and even adding an additional year (2012) did not exceed the number of references we identified in their topic areas. The similar numbers of papers captured by the two studies may reflect our use of citations from recent studies and papers from the database of a leading biochar research laboratory, although these sources accounted for a relatively small proportion of the papers we reviewed. Although our criteria for excluding and categorizing studies differed from theirs, because the studies had different objectives, this comparison strongly suggests that adding additional search terms would not significantly have expanded the set of literature we analyzed, nor the primary patterns we found. Consistent with this conclusion, we conducted our study iteratively, and the patterns we discerned based on early analyses did not change as we added more papers and refined our exclusion and categorization criteria.

Discussion

Making decisions about biochar systems

At first glance, our review might suggest a robust literature on biochar (311 primary research articles; Figure 1), but information critical to evaluating biochar's influence on ecosystem properties and its fate in the environment is actually quite scarce and uneven

Table 1. Field experiments estimating biochar stability.

Study	Location (ecosystem)	Biochar source and application method	Study design	Biochar loss rate (years)
Major et al. [48]	Colombia (savanna)	Charred mango wood disked into soil	Measured soil respiration and leaching for 2 years after biochar addition	MRT 3,624
Haefele et al. [58]	Thailand and the Philippines (rice paddies)	Charred rice husks tilled into soil	Measured biochar C for 3 years after biochar addition	MRT >1,000
Knoblauch et al. [51]	Los Baños, Philippines (rice paddies)	Charred rice husks tilled into soil	Measured soil CO_2 and CH_4 emissions for 3 months immediately and 2 years after biochar addition	MRT "several hundred if not thousands"
Cheng et al. [68]	Eastern North America (various)	Collected from soils at historic charcoal furnaces	Compared C content of old charcoal to that of charcoal produced in reconstructed furnaces	22% of biochar C lost in 130
Hammes et al. [47]	Russia (steppe)	Naturally-occurring fire	Measured black carbon stocks at a 100-year fire suppression site	Turnover time 293
Bird et al. [50]	Zimbabwe (savanna)	Naturally-occurring fire	Measured charcoal and oxidation-resistant elemental carbon (OREC) abundance at a 50-year fire suppression site	Half-life "considerably <50 years" (charcoal) and <100 years (OREC)
Nguyen et al. [49]	Kenya (cropland)	Slash-and-burn conversion from forest to cropland	Measured black carbon stocks along a 100-year chronosequence	MRT 8.3

The locations, methods, and results of the seven experiments that measured or estimated biochar stability in a field setting. MRT is mean residence time. Assuming a steady decomposition rate, the results of Cheng et al. [68] imply a turnover time of 565 years. However, decomposition tends to slow over time, so turnover time is likely longer.

across topic areas. The literature is dominated by studies of biochar production and material properties. This understanding is essential for developing production facilities, and may in the future become relevant for understanding biochar's influence on ecosystem services, as relationships among biochar chemistry and physical processes and chemical reactions in the soil system become better understood. In two categories of clear interest from an ecosystem perspective – nutrient loss and contaminants – we found only one field study each. Other categories have more studies, but even in those cases critical questions have not been addressed. For example, Vaccari et al. [37] documented as much as a 2°C increase in soil temperatures for soils amended with biochar using rotary tillage, suggesting that biochar amendment to soil is likely to change albedo significantly. Decreased albedo would increase soil temperature, which has been shown to accelerate soil respiration and affect soil nutrient availability, ecosystem water dynamics, species composition, and growing season length [38–40]. But of 19 field studies that addressed soil physical properties, only this one addressed albedo. Interrogating the literature within the broad topic areas we used is likely to yield many instances such as this, where studies needed to answer more specific questions are scarce or absent. The database we produced (Figure S3), with studies grouped into descriptive categories, provides rapid access to the relevant literature for scientists and land managers with specific questions in mind.

Our systematic review of original research on biochar revealed two features of this relatively young field that complicate efforts to make science-based decisions about biochar systems. First, the biochar literature lacks standardized nomenclature and analytical methods, reflecting the varied traditions of bioenergy engineering, agronomy, climate science, soil science and geology – all disciplines that have begun using the term "biochar" instead of charcoal or black carbon. Bringing multiple perspectives to bear on challenging problems often yields insights that might be missed in the absence of an interdisciplinary approach [41,42], but the absence of standardized vocabulary and methods has complicated efforts to compare results among studies and distill the implications for policy and management. Second, laboratory studies dominate the biochar literature. Data from laboratory studies could be valuable for investigating mechanisms that control biochar behavior in the environment and its effects on ecosystem services [11,34,35]. However, whole ecosystem experiments have proven critical to testing policy-relevant hypotheses in numerous cases including the response of lakes to nutrient additions [43], the influence of acid precipitation on forests [44], and the response of terrestrial ecosystems to rising CO_2 levels [45,46]. In the absence of ecosystem-level experiments, the consequences of management actions are largely speculative, making it hard to justify specific policies. Plant-soil ecosystems are complex, with interactions and feedbacks that lab experiments do not fully mimic, and there is no reason to suppose that field experiments are any less essential in evaluating whole system responses to biochar additions than they have been for evaluating impacts of other anthropogenic activities.

How stable is biochar in the field?

Proponents of biochar use have claimed it is long-lasting [14,21], but very few data are available to evaluate the stability of biochar *in situ*. Only seven of the primary research papers we identified reported field investigations of biochar stability in soil, and their estimates of stability – although not easily and directly comparable – spanned three orders of magnitude, from years to millennia (Table 1). Moreover, only one of those studies quantified the uncertainty of the results: Hammes et al. [47] calculated that the turnover time of biochar in their study ranged from 182 to 541 years. Two studies calculated mean residence time via first-order decay models but did not discuss the uncertainty in the calculations [48,49]. Other researchers noted "comparatively large" uncertainties [50,51] and numerous investigators have cautioned that stability must be better understood [52–55].

Given the ubiquitous discussion of biochar as an extremely passive carbon pool, we asked whether we should discount the two studies reporting most rapid loss: <10 years mean residence time [49] and "considerably <50 years" half-life [50]. In both studies, the loss estimates include physical transport, so rates of microbial decomposition of biochar to CO_2 may be lower than the reported loss rates, but in neither case are there data to estimate how much of the loss resulted from transport vs. decomposition. Nguyen et al. [49] used two different methods to estimate biochar abundance in the top 10 cm of soil at 9 sites along a fire chronosequence in Kenya and observed a 70% decrease in biochar content during the first 30 years. The site was relatively flat, and the authors therefore suggested that surface flow did not dominate biochar loss from the site. However, they noted that illuviation to deeper soil horizons may have been significant. Yanai et al. [56] urged caution when inferring organic matter responses to disturbance from soils collected along chronosequences because the nature of disturbance (such as logging techniques – horse vs. tractor) can change through time, so time *of* disturbance is difficult to separate from time *since* disturbance (the independent variable of interest). Nguyen et al. collected soils from fields where forest was cleared using slash-and-burn followed by plowing to 0.1–0.12 m depth, at eight points in time over 100 years. It is possible that different amounts of biomass had accumulated at different points in time, or that clearing methods were not identical, but burning, plowing, and plow depth were consistent across the chronosequence. Bird et al. [50] measured charcoal content in the upper 5 cm of soil from plots in Zimbabwe; some plots have a fire return interval of 1 to 5 years, while in others fire had been suppressed since 1947. Like Nguyen et al. [49], Bird et al. [50] noted that some degradation processes may have altered large particles to small particles which then moved beneath the shallow sampling depth. On balance these studies do not appear any less credible than the other three [47,57,58] that estimated biochar stability by measuring changes in biochar content in soil (Table 1). These findings caution against assuming that biochar persists 100 years or more, a value used in some LCA models [59,60].

What might account for the wide variation in field-based estimates of biochar stability? These field experiments were conducted in a variety of ecosystems on several continents, leading to large variation in conditions such as temperature, moisture and microbial communities, all of which act on the biochar in each study (Table 1). The experiments also used different biochar feedstocks and pyrolysis conditions such as temperature, duration, and oxygen content, all of which affect biochar properties and hence stability [11]. Production methods included vegetation fires, historical kilns, carefully regulated commercial or laboratory reactor vessels, and simply piling biomass on top of a burning chamber and waiting for the pile to turn black.

This variation in experimental materials and conditions is a valuable feature of field-based studies of biochar. After all, biochar systems would be implemented in different ecosystems using a greater variety of biochars and methods than were reported in the seven field studies we identified. Similarly, the potential diversity of feedstocks and conditions that could be used is greater than represented in these field studies, as evidenced by the broader range of experimental conditions represented in the 311 primary research articles included in our review. For example, biochar feedstock could include animal waste [61,62], agricultural waste

[63,64], and natural vegetation [65,66]. Studies to date begin to establish the range of variation in biochar stability but do not go very far towards explaining it. As this young field begins to mature, field-based studies conducted across sites that vary systematically with respect to key variables such as temperature and moisture, and that span the full range of variation, combined with laboratory experiments, should help establish empirical understanding of why biochar stability ranges so widely and project how biochar might behave in a given setting [67].

Limitations on estimating whole-system GHG balance

Ultimately, the suitability of biochar as a climate mitigation tool will depend on more than just biochar stability. In the field, biochar additions may influence plant growth and associated carbon uptake; decomposition of native soil organic matter; and emissions of non-CO_2 greenhouse gases from soil, resulting from either changes in soil moisture or nutrient availability. Prior to applying biochar to soil, biomass harvest (or collection) and transport, and biochar production and field application influence GHG emissions. While it is neither necessary nor realistic to measure or estimate all aspects of a GHG budget with equal precision, and some terms may reasonably be judged *de minimus*, all elements do need to be at least considered in assessing the GHG and climate change mitigation potential of interventions like biochar addition to soil. In addition, using very different approaches to measure key terms of the GHG budget hampers confidence in comparisons of whole system GHG balances. For example, biochar stability estimated as mean residence time from changes in biochar stocks [47,49,50,58,68] is difficult to compare directly with stability estimated by soil CO_2 emissions [48,51]. Finally, choosing the best climate change mitigation strategy requires comparing total system emissions from a biochar-based system against emissions associated with alternative management strategies such as combusting biomass for energy or leaving biomass in the field.

Very few studies have reported the information needed to estimate how whole ecosystem GHG budgets respond to biochar additions, and rarely-studied components of the GHG budget (like non-CO_2 emissions) sometimes respond to biochar amendment in ways that aggravate rather than mitigate climate change. We found only one field study that attempted to construct a whole system carbon budget of a plant-soil system in response to biochar additions vs. leaving the biomass in place [51]. We found six field studies of the effects of biochar on non-CO_2 GHG emissions. Three found no significant effect of biochar on N_2O emissions [29,69,70], contrary to common assertions that biochar decreases N_2O flux [21,60]. Two found no effect on CH_4 emissions [69,70], and one measured an increase in CH_4 emissions [71]. Carbon additions to soil are well-known to stimulate decomposition of native soil carbon [72–75], and Wardle et al. [76] reported a "priming" effect of biochar, in which biochar amendment catalyzed decomposition of existing soil organic carbon (but see [77] and [78]). Whole system analysis of bioenergy crop production has included vigorous debate about GHG emissions associated with indirect land use change (ILUC) [79–81], and ILUC could affect evaluation of biochar systems compared to alternative biomass fates. For example, Roberts et al. [23] showed that ILUC can reduce or even reverse the climate mitigation potential of biochar systems.

Comprehensive comparisons of GHG emissions among management approaches are typically achieved through an LCA, and we found nine such studies in our review. These studies might appear to provide the kind of comparisons we need – taking into account the many potential sources of GHG emissions associated with implementing a biochar system – but in fact they face the data availability constraints discussed in this review. In addition, LCAs to date have considered only a very limited number of alternative biomass management options. Of the LCA studies included in our systematic review, only four considered management options besides pyrolysis as a basis for comparison [23,59,60,82]. Two of these analyses compared biomass combustion for energy production to biomass conversion to biochar, and found that in some circumstances biomass combustion avoided more GHG emissions than the biochar system [23,60]. Most of the LCAs we captured in our review cautioned that their default assumptions about biochar stability needed further testing and/or that their results were sensitive to their choice of biochar stability [52–55,59,60,82,83].

Evaluating the consequences of alternative fates for biomass takes on added weight when we consider that different management options lead to varied temporal patterns of carbon storage and release. Even in cases in which biochar additions to soil lead to GHG benefits over hundreds of years compared to alternative biomass management scenarios, approximately 50% of the biomass carbon will be released to the atmosphere as CO_2 over the short term, either vented during pyrolysis or emitted when the other pyrolysis products such as bio-oil and syngas are combusted for energy [60]. In contrast, the half-life of uncharred wood of northern temperate tree species ranges from 6.8 to 150 years [84,85], and wood decomposition proceeds gradually over that time period [85,86], avoiding a large, near-term pulse of CO_2 to the atmosphere. Similarly, most wood products are long-lived. For example, wood used for single-family housing in the US has a half-life of approximately 80 years [87], again avoiding large pulses of CO_2 to the atmosphere on time scales of weeks to years, although CO_2 release associated with harvest and wood processing need to be taken into account. Therefore, in many cases charring woody biomass would lead to greater radiative forcing in the short term than would leaving it to decompose or using it in long-lived wood products. Reducing GHG emissions in the short-term will lessen the unprecedented rate of climate change, facilitating to some extent the ability of ecosystems and human institutions to adapt.

We are not the first to caution that many claims about biochar are overly enthusiastic. For example, Kookana et al. [32] called for a more balanced evaluation of biochar's potential adverse environmental impacts, particularly a reduction in agrochemical effectiveness and the introduction of soil contaminants. Likewise, Jeffery et al. [31] called on the research community to "think 'outside of the pot'" and criticized a recent meta-analysis [36] for failing to acknowledge important limitations (including a reliance on very short-term studies to infer long-term carbon storage) and over-stating biochar's potential benefits relative to current scientific understanding. Our review looked specifically for field studies because they provide unique and essential data for deriving estimates of biochar stability under real-world conditions, and our findings suggest a need for caution with regard to claims about climate-relevant variables, including biochar stability and effects on non-CO_2 GHG emissions.

Conclusions

- The study of biochar behavior in soil is a very young field, as reflected in diverse, non-standardized terminology and methods, and uneven distribution of studies across topic areas.

- We need a systematic field research program that investigates stability of biochars representing a range of feedstocks and production methods, across climate and soil gradients.
- We lack the field studies that are needed to understand with confidence how biochar production and application affects whole-system GHG balance. Key variables include, for example, emissions associated with biochar production, transportation, and application to soils; the extent to which biochar amendment stimulates ("primes") decomposition of soil organic matter; the influence of biochar on non-CO_2 trace gas emissions; and the amount of energy captured during biochar production.
- The promise and limitations of biochar production and amendment to field soil should be evaluated against a range of biomass management options, including burning biomass for energy and leaving dead wood in place.
- Even with limited available data, it is evident that potential long-term benefits of biochar-based carbon sequestration come at a cost of short-term CO_2 pulses into the atmosphere and, consequently, near-term acceleration of climate change.
- Optimistic claims about biochar's benefits to the environment contrast sharply with the limited amount of research on biochar's behavior and effects. There is insufficient empirical evidence to support assertions that biochar amendment to soil mitigates climate change significantly, or that it provides overall environmental benefits when evaluated across a comprehensive set of metrics.

Supporting Information

Figure S1 PRISMA literature search flow chart for identifying primary research and studies to include in our qualitative synthesis of literature on biochar stability, transport, or fate.

Figure S2 Primary research publications. The numbers correspond to the "Reference Number" in the first column of Figure S3. The 74 citations in bold are publications that addressed the stability, transport, or fate of biochar or soil carbon.

Figure S3 The topic areas addressed by each primary research publication. The numbers in the "Reference Number" column correspond to the numbered list in Figure S2.

Acknowledgments

We thank Jeffrey Bird and Fernanda Santos for reviewing the list of biochar publications identified through our literature search and augmenting it with suggestions from their literature database. We also thank Doug Boucher and three anonymous colleagues for comments on an early draft of this manuscript.

Author Contributions

Conceived and designed the experiments: NPG LAM PE. Analyzed the data: NPG LAM CK. Wrote the paper: NPG LAM.

References

1. Lal R (2004) Soil Carbon Sequestration Impacts on Global Climate Change and Food Security. Science 304: 1623–1627.
2. Lal R, Follett RF, Stewart BA, Kimble JM (2007) Soil Carbon Sequestration to Mitigate Climate Change and Advance Food Security. Soil Science 172: 943–956 910.1097/ss.1090b1013e31815cc31498.
3. Paustian K, Six J, Elliott ET, Hunt HW (2000) Management options for reducing CO2 emissions from agricultural soils. Biogeochemistry 48: 147–163.
4. Smith P, Fang C, Dawson JJC, Moncrieff JB (2008) Impact of Global Warming on Soil Organic Carbon. In: Donald LS, editor. Advances in Agronomy: Academic Press. pp. 1–43.
5. Magdoff F, Van Es H (2009) Building soils for better crops: sustainable soil management 3rd edn. Sustainable Agriculture Network handbook series 10. Sustainable Agriculture Publications, Waldorf, MD.
6. Powlson DS, Gregory PJ, Whalley WR, Quinton JN, Hopkins DW, et al. (2011) Soil management in relation to sustainable agriculture and ecosystem services. Food Policy 36: S72–S87.
7. Olander LP, Murray BC, Profeta T, Galik C, Dawson M (2008) Designing Offsets Policy for the US: Principles, challenges, and options for encouraging domestic and international emissions reductions and sequestration from uncapped entities as part of a federal cap-and-trade for greenhouse gases: Nicholas Institute for Environmental Policy Solutions, Nicholas School of the Environment and Earth Sciences.
8. Laird DA, Brown RC, Amonette JE, Lehmann J (2009) Review of the pyrolysis platform for coproducing bio-oil and biochar. Biofuels Bioproducts & Biorefining-Biofpr 3: 547–562.
9. Sohi SP, Krull E, Lopez-Capel E, Bol R (2010) A review of biochar and its use and function in soil. In: Sparks DL, editor. Advances in Agronomy, Vol 105. pp. 47–82.
10. Larson RW. Using biochar for cost-effective CO2 sequestration in soils; 2009. Springer. pp. 2462–2467.
11. Spokas KA (2010) Review of the stability of biochar in soils: predictability of O: C molar ratios. Carbon Management 1: 289–303.
12. Matovic D (2011) Biochar as a viable carbon sequestration option: Global and Canadian perspective. Energy 36: 2011–2016.
13. Kanter J (2009) Debating the Climate Benefits of 'Biochar'. Green: The New York Times.
14. Lovins HL, Cohen B (2011) Biochar. The Way Out Kick-Starting Capitalizm to Save Our Economic Ass. New York: Hill and Wang. pp. 390.
15. De Gryze S, Cullen M, Durschinger L, Lehmann J, Bluhm D, et al. (2010) Evaluation of the opportunities for generating carbon offsets from soil sequestration of biochar. An issues paper commissioned by the Climate Action Reserve, final version.
16. Weisberg P, Delaney M, Hawkes J (2010) Carbon Market Investment Criteria for Biochar Projects. PortlandOR: The Climate Trust. 38 p.
17. International Biochar Initiative (2013) How much CO2 can biochar potentially remove from the atmosphere? : International Biochar Initiative website. Available: http://www.biochar-international.org/biochar/faqs#q10. Accessed 27 Aug 2013.
18. International Biochar Initiative (2013) International Policy. International Biochar Initiative website. Available: http://www.biochar-international.org/policy/international. Accessed 27 August 2013.
19. International Biochar Initiative (2013) Apr 2013 Update: Biochar Carbon Offset Protocol Submitted for Validation at American Carbon Registry. International Biochar Initiative website. Available: http://www.biochar-international.org/protocol. Accessed 27 Aug 2013.
20. Glaser B, Haumaier L, Guggenberger G, Zech W (2001) The 'Terra Preta' phenomenon: a model for sustainable agriculture in the humid tropics. Naturwissenschaften 88: 37–41.
21. International Biochar Initiative (2013) What Is Biochar? International Biochar Initiative website. Available: http://www.biochar-international.org/biochar. Accessed 27 Aug 2013.
22. Glaser B, Balashov E, Haumaier L, Guggenberger G, Zech W (2000) Black carbon in density fractions of anthropogenic soils of the Brazilian Amazon region. Organic Geochemistry 31: 669–678.
23. Roberts KG, Gloy BA, Joseph S, Scott NR, Lehmann J (2010) Life Cycle Assessment of Biochar Systems: Estimating the Energetic, Economic, and Climate Change Potential. Environmental Science & Technology 44: 827–833.
24. Asai H, Samson BK, Stephan HM, Songyikhangsuthor K, Homma K, et al. (2009) Biochar amendment techniques for upland rice production in Northern Laos 1. Soil physical properties, leaf SPAD and grain yield. Field Crops Research 111: 81–84.
25. Blackwell P, Krull E, Butler G, Herbert A, Solaiman Z (2010) Effect of banded biochar on dryland wheat production and fertiliser use in south-western Australia: an agronomic and economic perspective. Soil Research 48: 531–545.
26. Jeffery S, Verheijen FGA, van der Velde M, Bastos AC (2011) A quantitative review of the effects of biochar application to soils on crop productivity using meta-analysis. Agriculture Ecosystems & Environment 144: 175–187.
27. Kammann CI, Linsel S, Gossling JW, Koyro HW (2011) Influence of biochar on drought tolerance of Chenopodium quinoa Willd and on soil-plant relations. Plant and Soil 345: 195–210.
28. Karami N, Clemente R, Moreno-Jimenez E, Lepp NW, Beesley L (2011) Efficiency of green waste compost and biochar soil amendments for reducing lead and copper mobility and uptake to ryegrass. Journal of Hazardous Materials 191: 41–48.

29. Karhu K, Mattila T, Bergstrom I, Regina K (2011) Biochar addition to agricultural soil increased CH4 uptake and water holding capacity - Results from a short-term pilot field study. Agriculture Ecosystems & Environment 140: 309–313.
30. Streubel JD, Collins HP, Garcia-Perez M, Tarara J, Granatstein D, et al. (2011) Influence of Contrasting Biochar Types on Five Soils at Increasing Rates of Application. Soil Science Society of America Journal 75: 1402–1413.
31. Jeffery S, Verheijen FGA, Bastos AC, Van Der Velde M (2013) A comment on 'Biochar and its effects on plant productivity and nutrient cycling: a meta-analysis': on the importance of accurate reporting in supporting a fast-moving research field with policy implications. GCB Bioenergy: n/a-n/a.
32. Kookana RS, Sarmah AK, Van Zwieten L, Krull E, Singh B (2011) Biochar application to soil: Agronomic and environmental benefits and unintended consequences. In: Sparks DL, editor. Advances in Agronomy, Vol 112. pp. 103–143.
33. Moher D, Liberati A, Tetzlaff J, Altman DG, Group TP (2009) Preferred Reporting Items for Systematic Reviews and Meta-Analyses: The PRISMA Statement. PLoS Med 6: e1000097.
34. Anderson CR, Condron LM, Clough TJ, Fiers M, Stewart A, et al. (2011) Biochar induced soil microbial community change: Implications for biogeochemical cycling of carbon, nitrogen and phosphorus. Pedobiologia 54: 309–320.
35. Cheng C-H, Lehmann J, Thies JE, Burton SD, Engelhard MH (2006) Oxidation of black carbon by biotic and abiotic processes. Organic Geochemistry 37: 1477–1488.
36. Biederman LA, Harpole WS (2013) Biochar and its effects on plant productivity and nutrient cycling: a meta-analysis. GCB Bioenergy 5: 202–214.
37. Vaccari FP, Baronti S, Lugato E, Genesio L, Castaldi S, et al. (2011) Biochar as a strategy to sequester carbon and increase yield in durum wheat. European Journal of Agronomy 34: 231–238.
38. Knorr W, Prentice IC, House JI, Holland EA (2005) Long-term sensitivity of soil carbon turnover to warming. Nature 433: 298–301.
39. Luo Y (2007) Terrestrial Carbon–Cycle Feedback to Climate Warming. Annual Review of Ecology, Evolution, and Systematics 38: 683–712.
40. Melillo JM, Steudler PA, Aber JD, Newkirk K, Lux H, et al. (2002) Soil Warming and Carbon-Cycle Feedbacks to the Climate System. Science 298: 2173–2176.
41. Carpenter SR, Mooney HA, Agard J, Capistrano D, DeFries RS, et al. (2009) Science for managing ecosystem services: Beyond the Millennium Ecosystem Assessment. Proceedings of the National Academy of Sciences 106: 1305–1312.
42. Reid W, Chen D, Goldfarb L, Hackmann H, Lee Y, et al. (2010) Earth system science for global sustainability: Grand challenges. Science(Washington) 330: 916–917.
43. Schindler DW (1974) Eutrophication and Recovery in Experimental Lakes: Implications for Lake Management. Science 184: 897–898.
44. Likens GE, Bormann FH (1995) Biogeochemistry of a Forested Ecosystem. New York: Springer-Verlag. 159 p.
45. Dukes JS, Chiariello NR, Cleland EE, Moore LA, Shaw MR, et al. (2005) Responses of grassland production to single and multiple global environmental changes. PLoS Biology 3: e319.
46. Reich PB, Hungate BA, Luo Y (2006) Carbon-Nitrogen Interactions in Terrestrial Ecosystems in Response to Rising Atmospheric Carbon Dioxide. Annual Review of Ecology, Evolution, and Systematics 37: 611–636.
47. Hammes K, Torn MS, Lapenas AG, Schmidt MWI (2008) Centennial black carbon turnover observed in a Russian steppe soil. Biogeosciences 5: 1339–1350
48. Major J, Lehmann J, Rondon M, Goodale C (2010) Fate of soil-applied black carbon: downward migration, leaching and soil respiration. Global Change Biology 16: 1366–1379.
49. Nguyen BT, Lehmann J, Kinyangi J, Smernik R, Riha SJ, et al. (2009) Long-term black carbon dynamics in cultivated soil. Biogeochemistry 92: 163–176.
50. Bird MI, Moyo C, Veenendaal EM, Lloyd J, Frost P (1999) Stability of elemental carbon in a savanna soil. Global Biogeochem Cycles 13: 923–932.
51. Knoblauch C, Maarifat AA, Pfeiffer EM, Haefele SM (2011) Degradability of black carbon and its impact on trace gas fluxes and carbon turnover in paddy soils. Soil Biology & Biochemistry 43: 1768–1778.
52. Kameyama K, Shinogi Y, Miyamoto T, Agarie K (2010) Estimation of net carbon sequestration potential with farmland application of bagasse charcoal: life cycle inventory analysis through a pilot sugarcane bagasse carbonisation plant. Australian Journal of Soil Research 48: 586–592.
53. Ogawa M, Okimori Y, Takahashi F (2006) Carbon sequestration by carbonization of biomass and forestation: three case studies. Mitigation and Adaptation Strategies for Global Change 11: 421–436.
54. Pratt K, Moran D (2010) Evaluating the cost-effectiveness of global biochar mitigation potential. Biomass & Bioenergy 34: 1149–1158.
55. Cooper JM, Butler G, Leifert C (2011) Life cycle analysis of greenhouse gas emissions from organic and conventional food production systems, with and without bio-energy options. Njas-Wageningen Journal of Life Sciences 58: 185–192.
56. Yanai RD, Arthur MA, Siccama TG, Federer CA (2000) Challenges of measuring forest floor organic matter dynamics:: Repeated measures from a chronosequence. Forest Ecology and Management 138: 273–283.
57. Cheng C-h, Lehmann JC, Engelhard MH (2008) Natural Oxidation of Black Carbon in Soils: Changes in Molecular Form and Surface Charge along a Climosequence. Geochimica et Cosmochimica Acta 72: 1598–1610.
58. Haefele SM, Konboon Y, Wongboon W, Amarante S, Maarifat AA, et al. (2011) Effects and fate of biochar from rice residues in rice-based systems. Field Crops Research 121: 430–440.
59. Whitman T, Nicholson CF, Torres D, Lehmann J (2011) Climate Change Impact of Biochar Cook Stoves in Western Kenyan Farm Households: System Dynamics Model Analysis. Environmental Science & Technology 45: 3687–3694.
60. Woolf D, Amonette JE, Street-Perrott FA, Lehmann J, Joseph S (2010) Sustainable biochar to mitigate global climate change. Nature Communications 1.
61. Chan KY, Van Zwieten L, Meszaros I, Downie A, Joseph S (2008) Using poultry litter biochars as soil amendments. Australian Journal of Soil Research 46: 437–444.
62. Gaskin JW, Steiner C, Harris K, Das KC, Bibens B (2008) Effect of low-temperature pyrolysis conditions on biochhar for agricultural use. Transactions of the Asabe 51: 2061–2069.
63. Busscher WJ, Novak JM, Evans DE, Watts DW, Niandou MAS, et al. (2010) Influence of Pecan Biochar on Physical Properties of a Norfolk Loamy Sand. Soil Science 175: 10–14 10.1097/SS.1090b1013e3181cb1097f1046.
64. Spokas K, Reicosky D (2009) Impacts of sixteen different biochars on soil greenhouse gas production. Annals of Environmental Science 3.
65. Dai X, Boutton TW, Glaser B, Ansley RJ, Zech W (2005) Black carbon in a temperate mixed-grass savanna. Soil Biology and Biochemistry 37: 1879–1881.
66. Kane E, Hockaday W, Turetsky M, Masiello C, Valentine D, et al. (2010) Topographic controls on black carbon accumulation in Alaskan black spruce forest soils: implications for organic matter dynamics. Biogeochemistry 100: 39–56.
67. Fraser LH, Henry HAL, Carlyle CN, White SR, Beierkuhnlein C, et al. (2012) Coordinated distributed experiments: an emerging tool for testing global hypotheses in ecology and environmental science. Frontiers in Ecology and the Environment 11: 147–155.
68. Cheng C-H, Lehmann J, Engelhard MH (2008) Natural oxidation of black carbon in soils: Changes in molecular form and surface charge along a climosequence. Geochimica et Cosmochimica Acta 72: 1598–1610.
69. Castaldi S, Riondino M, Baronti S, Esposito FR, Marzaioli R, et al. (2011) Impact of biochar application to a Mediterranean wheat crop on soil microbial activity and greenhouse gas fluxes. Chemosphere 85: 1464–1471.
70. Scheer C, Grace PR, Rowlings DW, Kimber S, Van Zwieten L (2011) Effect of biochar amendment on the soil-atmosphere exchange of greenhouse gases from an intensive subtropical pasture in northern New South Wales, Australia. Plant and Soil 345: 47–58.
71. Zhang AF, Cui LQ, Pan GX, Li LQ, Hussain Q, et al. (2010) Effect of biochar amendment on yield and methane and nitrous oxide emissions from a rice paddy from Tai Lake plain, China. Agriculture Ecosystems & Environment 139: 469–475.
72. Cheng WX, Coleman DC (1990) Effect of Living Roots On Soil Organic-Matter Decomposition. Soil Biology & Biochemistry 22: 781–787.
73. Dijkstra FA, Cheng W (2007) Interactions between soil and tree roots accelerate long-term soil carbon decomposition. Ecology Letters 10: 1406.
74. Kuzyakov Y, Bol R (2006) Sources and mechanisms of priming effect induced in two grassland soils amended with slurry and sugar. Soil Biology & Biochemistry 38: 747–758.
75. Zhang W, Wang X, Wang S (2013) Addition of External Organic Carbon and Native Soil Organic Carbon Decomposition: A Meta-Analysis. PLoS ONE 8: e54779.
76. Wardle DA, Nilsson M-C, Zackrisson O (2008) Fire-Derived Charcoal Causes Loss of Forest Humus. Science 320: 629.
77. Lehmann J, Sohi S (2008) Comment on "Fire-Derived Charcoal Causes Loss of Forest Humus". Science 321: 1295.
78. Zimmerman AR, Gao B, Ahn MY (2011) Positive and negative carbon mineralization priming effects among a variety of biochar-amended soils. Soil Biology & Biochemistry 43: 1169–1179.
79. Kline KL, Dale VH (2008) Biofuels: effects on land and fire. Science (Washington) 321.
80. McKone TE, Nazaroff WW, Berck P, Auffhammer M, Lipman T, et al. (2011) Grand Challenges for Life-Cycle Assessment of Biofuels. Environmental Science & Technology 45: 1751–1756.
81. Searchinger T, Heimlich R, Houghton RA, Dong F, Elobeid A, et al. (2008) Use of US croplands for biofuels increases greenhouse gases through emissions from land-use change. Science 319: 1238–1240.
82. Hammond J, Shackley S, Sohi S, Brownsort P (2011) Prospective life cycle carbon abatement for pyrolysis biochar systems in the UK. Energy Policy 39: 2646–2655.
83. Gaunt JL, Lehmann J (2008) Energy Balance and Emissions Associated with Biochar Sequestration and Pyrolysis Bioenergy Production. Environmental Science & Technology 42: 4152–4158.

84. Freschet GT, Weedon JT, Aerts R, van Hal JR, Cornelissen JH (2012) Interspecific differences in wood decay rates: insights from a new short-term method to study long-term wood decomposition. Journal of Ecology 100: 161–170.
85. Schowalter T, Caldwell B, Carpenter S, Griffiths R, Harmon M, et al. (1992) Decomposition of fallen trees: effects of initial conditions and heterotroph colonization rates. In: Singh K, Singh J, editors. Tropical Ecosystems : Ecology and Management. New Delhi: Wiley Eastern Limited. pp. 373–383
86. Harmon ME, Franklin JF, Swanson FJ, Sollins P, Gregory S, et al. (1986) Ecology of coarse woody debris in temperate ecosystems. Advances in ecological research 15: 302.
87. Skog KE (2008) Sequestration of carbon in harvested wood products for the United States. Forest Products Journal 58: 56.

2

Straw Mulching Reduces the Harmful Effects of Extreme Hydrological and Temperature Conditions in Citrus Orchards

Yi Liu[1], Jing Wang[1], Dongbi Liu[2], Zhiguo Li[1], Guoshi Zhang[1,3], Yong Tao[1], Juan Xie[1], Junfeng Pan[1], Fang Chen[1,3]*

1 Laboratory of Aquatic Botany and Watershed Ecology, Wuhan Botanical Garden, Chinese Academy of Sciences China, Wuhan, China, **2** Institute of Plant Protection and Soil Fertilizer, Hubei Academy of Agricultural Sciences, Wuhan, China, **3** China Program, International Plant Nutrition Institute (IPNI), Wuhan, China

Abstract

Extreme weather conditions with negative impacts can strongly affect agricultural production. In the Danjiangkou reservoir area, citrus yields were greatly influenced by cold weather conditions and drought stress in 2011. Soil straw mulching (SM) practices have a major effect on soil water and thermal regimes. A two-year field experiment was conducted to evaluate whether the SM practices can help achieve favorable citrus fruit yields. Results showed that the annual total runoff was significantly ($P<0.05$) reduced with SM as compared to the control (CK). Correspondingly, mean soil water storage in the top 100 cm of the soil profile was increased in the SM as compared to the CK treatment. However, this result was significant only in the dry season (Jan to Mar), and not in the wet season (Jul to Sep) for both years. Interestingly, the SM treatment did not significantly increase citrus fruit yield in 2010 but did so in 2011, when the citrus crop was completely destroyed (zero fruit yield) in the CK treatment plot due to extremely low temperatures during the citrus overwintering stage. The mulch probably acted as an insulator, resulting in smaller fluctuations in soil temperature in the SM than in the CK treatment. The results suggested that the small effects on soil water and temperature changes created by surface mulch had limited impact on citrus fruit yield in a normal year (e.g., in 2010). However, SM practices can positively impact citrus fruit yield in extreme weather conditions.

Editor: Yi Liu, Wuhan Botanical Garden, Chinese Academy of Sciences China, China

Funding: This work was financially supported by the National Key Technology R&D Program (2012BAD15B01), the NSFC (31100386), the Open Funding Project (Y152741s04) of the Key Laboratory of Aquatic Botany and Watershed Ecology, Chinese Academy of Sciences and the Cooperated Program with the International Plant Nutrition Institute (IPNI-HB-33). The funders had no role in study design, data collection and analysis, decision to publish, or preparation of the manuscript.

Competing Interests: The authors have declared that no competing interests exist.

* E-mail: fchenipni@126.com

Introduction

One of the major challenges of climate change to agriculture and food security involves ever increasing extreme weather events world widely, such as droughts, heat waves, excessive cold, heavy and prolonged rainfalls, hailstorms, and so on [1]. They usually cause negative effects and sometimes fatal damages to crops, physiologically and/or physically [2]. However, the stresses could be relieved and neutralized by certain positive effects from field microclimate, beneficial soil water and thermal conditions under some farmland management practices [3]. Hence, there is an increasing need for understanding the response of soil water and temperature dynamics to changes in extreme weather conditions [4].

Soil water is considered to be one of the most important factors affecting plant growth and development [5,6]. Even a small change in soil water storage could greatly affect crop productivity [7]. Soil surface mulching, such as with plastic film [8], crop residue [9], or gravel and sand [10], has a large impact on many of the hydrological and biological processes of soil ecosystems, and the most prominent of these changes is the modification of the soil–plant–atmosphere continuum (SPAC) water cycling. For example, numerous reports indicate that soil straw mulching (SM) favorably influences the soil moisture regime by reducing evaporation from the soil surface [11,12], improving infiltration [13], and soil water retention [14]. SM has also led to improvements in crop yields in arid and semi-arid environments [5,12] and economized the use of irrigation water [14]. Thus, a better understanding of the impact of SM practices on soil hydrological processes is becoming critical, especially from the crop production perspective, because of the increasing shortage of water resources worldwide.

Soil temperature controls the rate of crop development, particularly when the meristem is within the soil [15]. Higher soil temperature accelerated the rates of leaf tip appearance and full leaf expansion, enabling the crop to attain maximum green leaf area index more rapidly [10]. SM has been reported to cause either a decrease, an increase or a negligible effect on soil temperature. For instance, SM during over-wintering period can improve soil thermal regime according to several studies [12,16,17]. However, Sarkar et al. [18] reported that SM could reduce soil temperature, while effective soil water conservation with SM may result in higher production. While Ghosh et al. [19] argued SM had little or no effect on soil temperature, and that its effects were almost entirely due to increased organic matter. These inconsistent results may depend on multiple factors including soil properties, climate, and species planted [15]. Lower soil temperature under SM has mostly been attributed to the reduced solar energy reaching the soil during hot periods, while increased soil temperature under SM has been attributed to the reduction in outgoing heat radiation from the soil during cold periods.

The Danjiangkou reservoir, established in the 1970s with a drainage area of 95,200 km^2, is a water source area for China's Middle Route of the South-to-North Water Transfer Project [20]. The staple crops of the region (wheat and rice) are generally grown in the flat land part of the Danjiangkou reservoir area. Citrus is one of the main types of fruit growing on sloping lands, and its high yields, averaging about 40 t hm^{-2}, are assumed to be a result of the beneficial thermal effects of the great lakes. After the wheat (or rice) harvest, farmers would generally burn the stalks, but this practice is now prohibited to restore and protect the Danjiangkou reservoir riparian ecosystem. Therefore, farmers have been using wheat (or rice) straw for mulching in citrus orchards. Notably, SM practices can effectively contribute to water conservation and decreased nutrient losses on sloping lands [21]. Although several studies have reported changes in water quality [22] and the role of surface mulching in soil nutrient losses [23], little is known about the impact of SM practices on reducing the harmful effects of extreme weather conditions. In the Danjiangkou reservoir area, citrus yields were strongly affected by cold and drought stress in 2011. The average temperature measured at the Danjiangkuo city meteorological station was 1.5°C during December 2010 to February 2011, which broke the record set in year 2000 and lowered the long-term (10 years) average temperature by 3°C for the corresponding period.

In the present study, we analyzed runoff, soil water content and storage, and seasonal variations in soil temperature under SM and non-mulching or control (CK) treatments. We specifically focused on the role of SM on soil water and temperature dynamics by comparing fruit yield under mulching and no mulching practices. This was done to test the hypothesis that the small effects on soil water conservation and thermoregulation created by surface SM practices can greatly impact citrus fruit yields in extreme weather conditions. The objectives of this study were: (i) to determine how productivity of citrus fruit was affected by soil water and temperature, (ii) to evaluate the influence of SM on the soil water and temperature in sloping citrus orchards in the Danjiangkou Reservoir area.

Materials and Methods

Site description

This study was conducted in 2010 and 2011 at the Xiaofuling experimental station (32°45′46″N, 111°9′26″E) in the Danjiangkou reservoir area (Fig. 1). The area has a subtropical zone climate with mean annual temperature of 15.7°C, and a monthly average temperature of 27.3°C in July and 4.2°C in January. Mean annual rainfall is approximately 834 mm, 80% of which concentrates between May and October. The study was carried out in a 10-year-old citrus orchard. The experimental site is owned by Wuhan Botanical Garden, Chinese Academy of Sciences China. The field studies did not involve endangered or protected species and no specific permits were required for the described field studies. The site is located about 300 m asl, with an average slope of 15°. The soil at the site is a cinnamon yellow soil as defined by the Chinese soil classification system [24] and textural composition is 14% clay, 23% silt, and 64% sand. At start of the field experiment in 2009, the soil had a pH of 6.5 and a bulk density of 1.45 g cm^{-3}. The amounts of organic matter, total nitrogen, available phosphorus, available potassium, and inorganic nitrogen were 9.1 g kg^{-1}, 0.88 g kg^{-1}, 16.0 mg kg^{-1}, 106.3 mg kg^{-1} and 101.8 mg kg^{-1}, respectively. These nutrient contents using routine analytical methods [25].

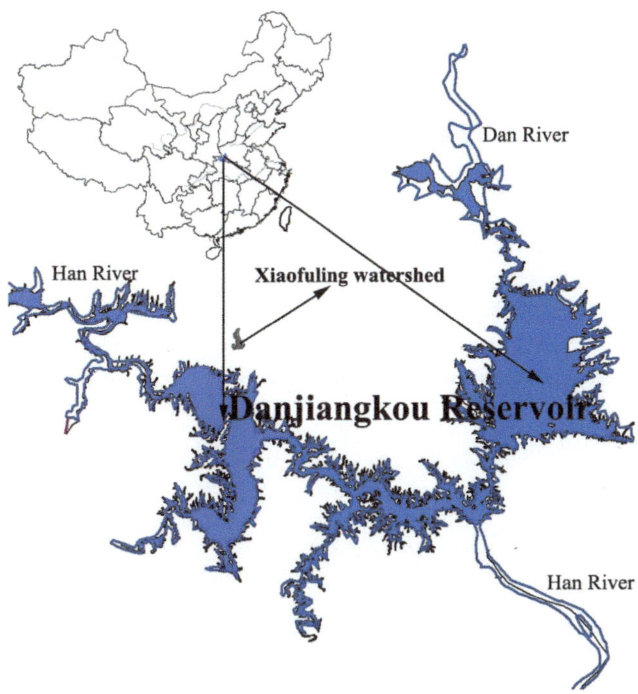

Figure 1. Location of the study area.

Experimental treatments and field management

The experiment was designed with two treatments, CK or without mulching and SM, (Fig. 2) with three replicates for each treatment. All plots, separated by concrete borders, were set up with a tank at the base of each plot to collect the runoff. The size of each experimental plot was 40.5 m^2 (4.5 m×9 m). The plantation consisted of trained citrus trees at 1.5 m×3 m spacing, with rows perpendicular to the slope. Hence, there are eight trees at each plot. In the plots mulched with straw, rice (or wheat) straw was uniformly applied at a rate of 6,000 kg hm^{-2}. Each citrus tree was fertilized in April by hand, with 0.5 kg N (urea), 0.3 kg P_2O_5 (superphosphate), and 0.4 kg K_2O (potassium chloride) in both years (2010 and 2011). Manual weeding was undertaken as required during the citrus growing season.

Sampling measurements and data calculation

To measure the runoff caused by rainfall, a standard recording rain gauge was sited about 100 m from the experimental plots. Rainfall was calculated at 1-day intervals. Runoff was collected at each plot after each rainfall event, and the depth of water in the runoff collection tank was recorded to calculate runoff volumes. The runoff were determined by the following formula: $R = S \times h \times 10 / Pa$, where R is runoff (mm), S is collection tank floorage (m^2), h is water depth (cm), Pa is plot area (m^2). The monthly soil water content was determined gravimetrically by oven drying (105°C for 24 h) the core samples that were taken at depth intervals of 20 cm down the 0–100 cm profile in each plot on the 26th (or 27th) day of each month (January 2010–December 2011). The soil water storage (W) in the profile was considered to be the total storage in all of the sampled layers in the plot, as was

Figure 2. Photographs showing plots under control or without mulching (CK) and straw mulching (SM) treatments.

calculated using the formula: $W = h \times \rho \times \theta \times 1000$, where h is soil depth (cm), ρ is soil bulk density (g cm^{-3}), θ is soil gravimetric water content (g g^{-1}). Soil bulk density was determined from the inner diameter of the core sampler, segment depth, and the oven-dry weight of the core samples at start of the field experiment in 2009 (see Table S1). Air temperature within the canopy (CA, height = 1 m) and soil temperature at 5 cm depth were also recorded at 0.5 h intervals by the StowAway TidbiT temperature recorder (Range: −20° to 70°C in air; −20° to 30°C in water; Accuracy: ±0.4° at 20°C). During the 24-h period, the values were averaged to calculate mean daily temperature.

At the time of commercial harvest, the citrus fruits were harvested gradually when they were ripe. The yield per plot (t hm^{-2}) was obtained by weighing the harvested fruit. A random sample of 25 fruits from each plot was collected to determine the average fruit weight. In addition, fruit size was measured by measuring the equatorial diameter with the help of Vernier caliper from each experimental plot. Using a standard juicer, 25 fruits were juiced. The juice was weighed and expressed as a percentage of the total fruit weight.

Statistical analyses

Statistically significant differences in mean runoff, soil water storage, and soil temperature between the CK and SM treatments were determined utilizing the Wilcoxon signed rank test (WSRT). A P value of <0.05 was considered statistically significant. Statistical analysis for fruit yield and quality was performed using Analysis of Variance (ANOVA). One-way analysis of variance was

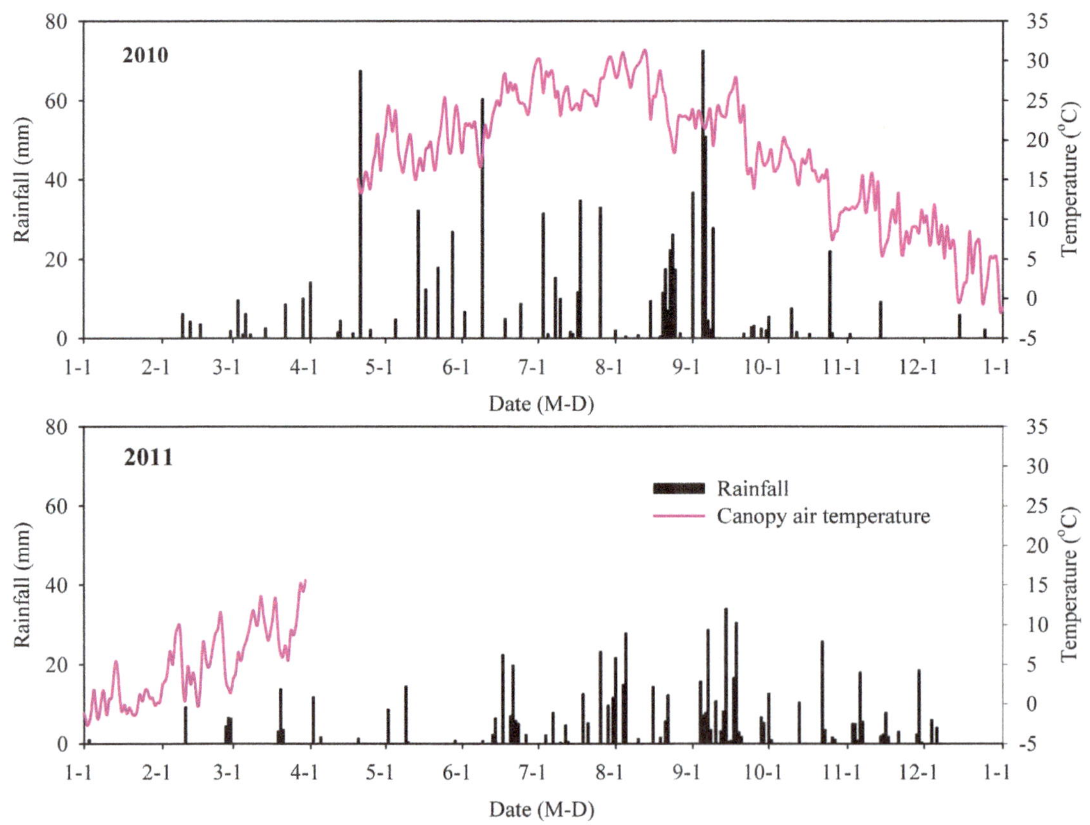

Figure 3. Canopy air temperature and rainfall data for 2010 and 2011 at Xiaofuling experimental station in Danjiangkou reservoir area in Hubei, China. Please note that the record of the air temperatures ended on March 31, 2011 due to the loss of the logger.

carried out to determine the differences between the measured parameters for different treatments. Least significant difference (LSD) at P = 0.05 was used to elucidate any significant differences.

Results

Rainfall and air temperature

The air temperature and rainfall of experimental field are shown in Fig. 3. Field data as well as summary statistics of air temperature and rainfall are provided as Table S2 and S3 in supporting information. The annual precipitation recorded at the experimental site was 849.3 mm in 2010 and 655.5 mm in 2011. 2011 was a dry year, which lowered precipitation by 21.5% compared with the long-term average. Fig. 3 also shows the air temperature within the canopy from April 20, 2010 to March 31, 2011. It should be noted that the record of the air temperatures ended on March 31, 2011 due to the loss of the logger. The daily air temperature varied from −2.5 to 31.2°C, averaging 15.0°C. The average minimum air temperature in the winter of 2010 (i.e., from December 2010 to February 2011) was −2.0°C. There were 66 days with minimum temperature <0°C during the study period, and not only the mean but also the maximum/minimum air temperatures were lower compared to the average of the last 10 years for the corresponding period.

Runoff

The surface mean runoff ranged from 1.0 to 37.7 mm in 2010 and from 0.4 to 15.8 mm in 2011 (see Fig. 4 or Table S4), and had three peak values with time (on April 21, June 9, and September 5) in 2010 and one peak value with time (on September 20) in 2011 indicating extreme rainfall events. Three rainstorm events were registered in 2010 as 67.7 mm (April 21, 2010), 60.4 mm (June 9, 2010) and 72.5 mm (September 5, 2010). While one continuous 7 day (From September 13 to 20) rainfall was registered in 2011 as 86.8 mm for accumulative precipitation (see Table S3). Annual total runoff volumes, calculated by adding the readings taken at the sampled points throughout the whole year, were much lower in the SM (107 mm in 2010, 78 mm in 2011) than in the CK (145 mm in 2010, 97 mm in 2011) plots (see Table S4). Lower surface runoff values observed in the SM plots were probably due to good ground coverage and slightly higher water infiltration than in the CK plots.

Soil water

The seasonal variations in water storage in the soil profile (0–100 cm) under CK and SM treatments are shown in Fig. 5. Field data as well as summary statistics of soil water storage are provided as Table S5 in supporting information. Mean soil water storage values, calculated by averaging the readings taken at the sampled points over one year, were much higher in the SM (ranged from 245 to 303 mm in 2010, and 254 to 291 mm in 2011, see Table S5) treatment than in the CK (ranged from 231 to 303 mm in 2010, and 237 to 290 mm in 2011, see Table S5) treatment. Soil water storage exhibited pronounced seasonal variations with minimum values at 231±2 mm during the dry season and maximum values at 303±26 mm during the wet season (Fig. 5). Largest differences in soil water storage between CK and SM occurred during the dry season (from January to March in 2011),

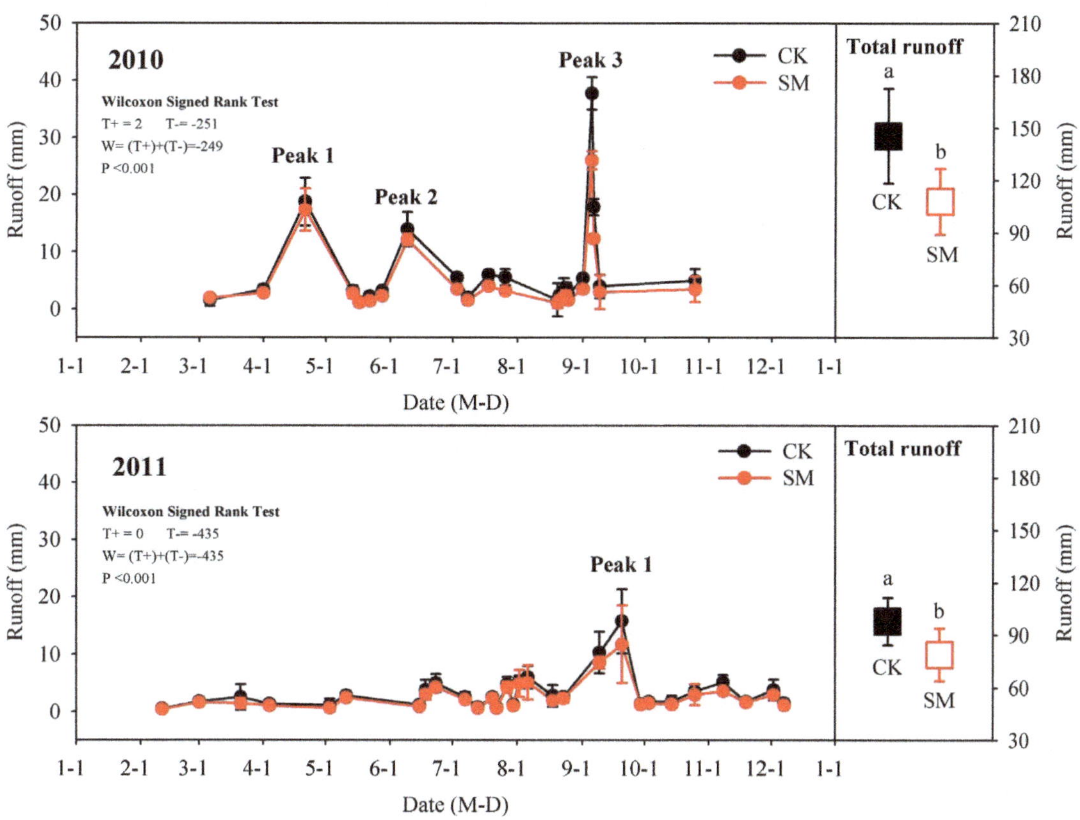

Figure 4. Seasonal variation in and annual total runoff under control (CK) and straw mulching (SM) treatments during 2010 and 2011. Error bars are twice the standard error of the mean (n = 3). Statistically significant differences are given after Wilcoxon signed rank test; Notations a and b indicate statistical significance at P<0.05 between CK and SM.

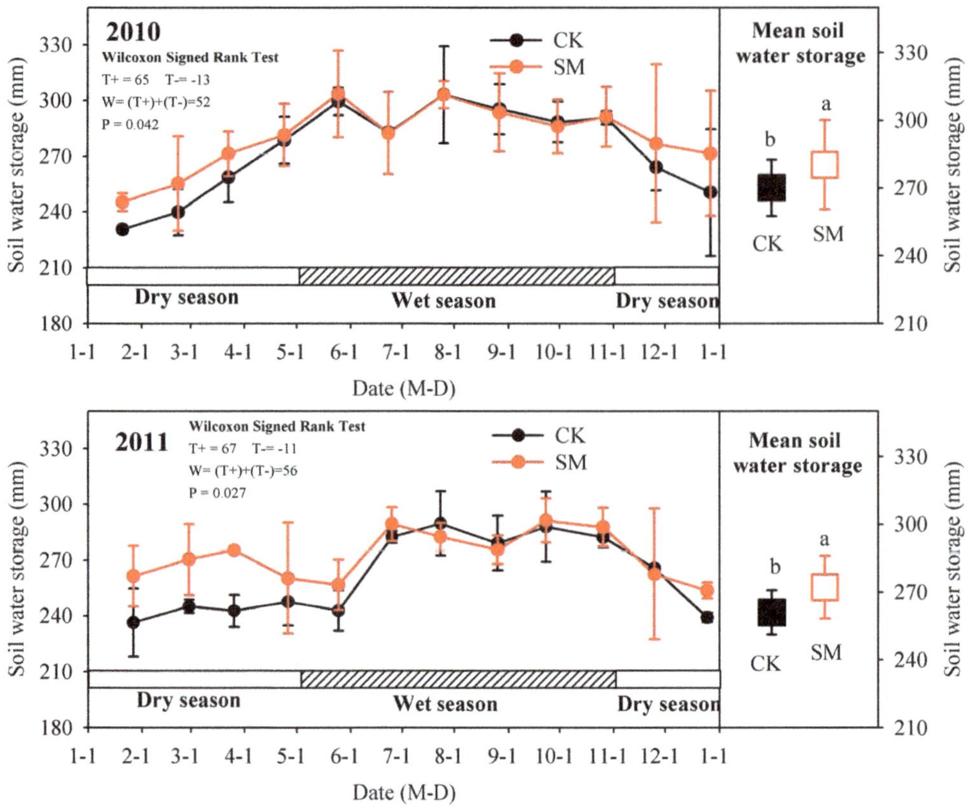

Figure 5. Seasonal variation in and mean soil water storage in the 0–100 cm soil profile under control (CK) and straw mulching (SM) treatments during 2010 and 2011. Error bars are twice the standard error of the mean (n = 3). Statistically significant differences are given after Wilcoxon signed rank test; Notations a and b indicate statistical significance at P<0.05 between CK and SM.

when soil under the SM treatment had about 10% (ranged from 5 to 13%) greater water storage than under the CK treatment. During the rainy season in the Danjiangkou reservoir area (July to September), citrus plants take up a great deal of water to maintain their luxuriant growth; the variation in seasonal soil water storage was therefore mainly affected by the amount of precipitation and citrus growth. As a result, no significant difference in soil water storage was observed between CK and SM treatments during the wet season.

The vertical distribution of soil water in a profile under both dry (26th February 2010 and 26th March 2011) and wet (27th August 2010 and 26th August 2011) seasons is shown in Fig. 6. Soil water distribution within the profile results from the combined effects of precipitation amount and movement of soil water. The soil water content was significantly higher in the SM treatment than in the CK treatment in the 0–40 cm soil layer during the dry season, indicating that SM reduced evaporation during the dry period because of the increased surface residue cover and/or the lack of

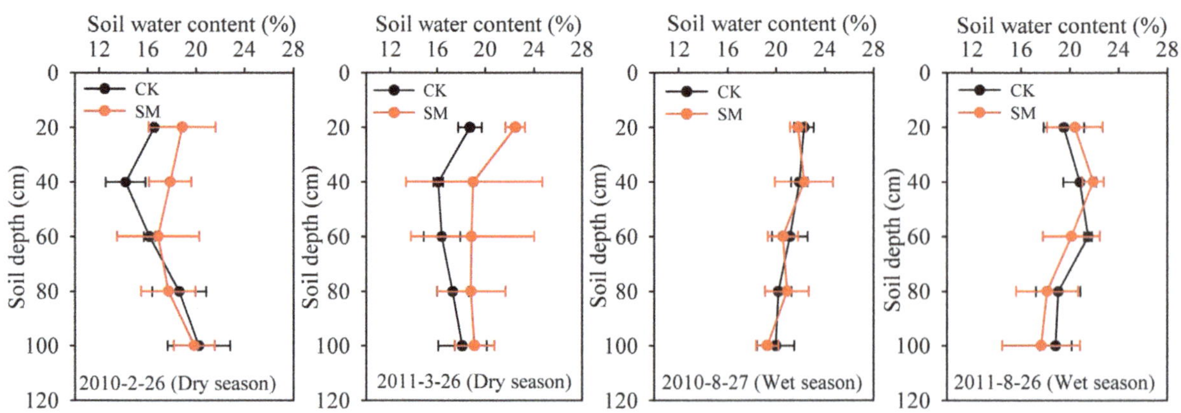

Figure 6. Differences in soil water content down the profile (0–100 cm) in dry (26th February 2010 and 26th March 2010) and wet (27th August 2010 and 26th August 2011) seasons between control (CK) and straw mulching (SM) treatment plots. Error bars are twice the standard error of the mean (n = 3).

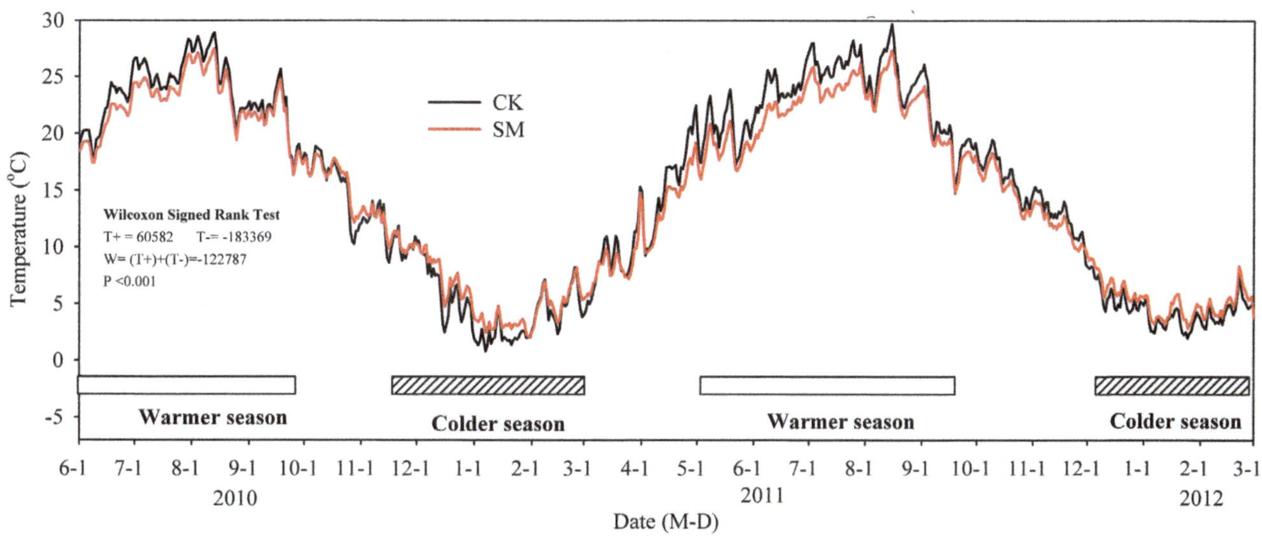

Figure 7. Seasonal variations in soil temperature under control (CK) and straw mulching (SM) treatment plots from June 2010 to March 2012. Statistically significant differences are given after Wilcoxon signed rank test; a P value of <0.05 was considered statistically significant.

soil disturbance. However, there was no significant difference in the soil water content in all soil layers between the SM and CK treatments during the wet season.

Soil temperature

The presence of SM altered soil temperature. Fig.7 shows the seasonal variations in soil temperature under CK and SM treatments. Monthly mean soil temperature as well as P values of WSRT are provided as Table 1. Our study demonstrated that soil temperatures under SM plots were higher during the colder seasons and lower during the warmer seasons when compared with the soil temperatures under CK plots. During the warmer period, reductions in soil temperature under the SM treatment were as high as 1.5°C (ranged from 0.3 to 2.9°C) as compared to the CK treatment (see Table S6). Because straw, that covered the soil surface, has a higher albedo and lower thermal conductivity than the bare soil, it helps to reduce the solar energy reaching the soil and, as a result, reduces temperature increases during warm conditions. Conversely, during the colder seasons, the presence of SM on the soil surface insulates the soil from the colder air temperatures. Therefore, heat loss from the soil is somewhat lower and soil temperatures are consequently higher under SM than under CK.

Four sets of typical diurnal trends (for summer, autumn, winter and spring) of CA and soil temperatures under both SM and CK conditions are presented in Fig. 8. In spring and autumn, when diurnal temperature range was large, soil temperature under mulch was lower during daytime, but higher at night. As daily radiation increased in summer, soil temperature under SM was always lower than that without mulch (CK). However, soil temperature was always higher in SM than in CK plots when the air temperature reached its minimum value in winter.

Table 1. Monthly mean soil temperature under the control (CK) and straw mulching (SM) treatments from June 2010 to March 2012.

Month-year	Soil temperature (°C)		WSRTP	Month-year	Soil temperature (°C)		WSRTP
	CK	SM			CK	SM	
Jun-2010	21.86±0.41	20.59±0.34	< 0.001	May-2011	20.33±0.32	18.55±0.24	< 0.001
Jul-2010	25.48±0.23	24.18±0.20	< 0.001	Jun-2011	23.95±0.24	22.02±0.24	< 0.001
Aug-2010	25.37±0.48	24.30±0.42	< 0.001	Jul-2011	26.32±0.19	24.40±0.16	< 0.001
Sep-2010	21.45±0.45	20.85±0.41	< 0.001	Aug-2011	25.10±0.37	23.82±0.30	< 0.001
Oct-2010	15.85±0.46	16.17±0.33	0.090	Sep-2011	20.08±0.47	19.04±0.41	< 0.001
Nov-2010	11.28±0.31	11.54±0.26	0.029	Oct-2011	16.32±0.33	15.42±0.31	< 0.001
Dec-2010	6.33±0.43	7.48±0.32	< 0.001	Nov-2011	12.75±0.30	11.82±0.25	< 0.001
Jan-2011	2.10±0.13	3.18±0.11	< 0.001	Dec-2011	5.80±0.24	6.69±0.23	< 0.001
Feb-2011	5.00±0.30	5.32±0.27	0.012	Jan-2012	3.47±0.16	4.27±0.16	< 0.001
Mar-2011	8.61±0.45	8.20±0.33	0.016	Feb-2012	4.41±0.20	5.14±0.21	< 0.001
Apr-2011	15.54±0.68	13.99±0.50	< 0.001	Mar-2012	8.05±0.57	6.37±0.52	< 0.001

Values are given as means ± standard error of means. WSRT: Wilcoxon Signed Rank Test, a P value of <0.05 was considered statistically significant.

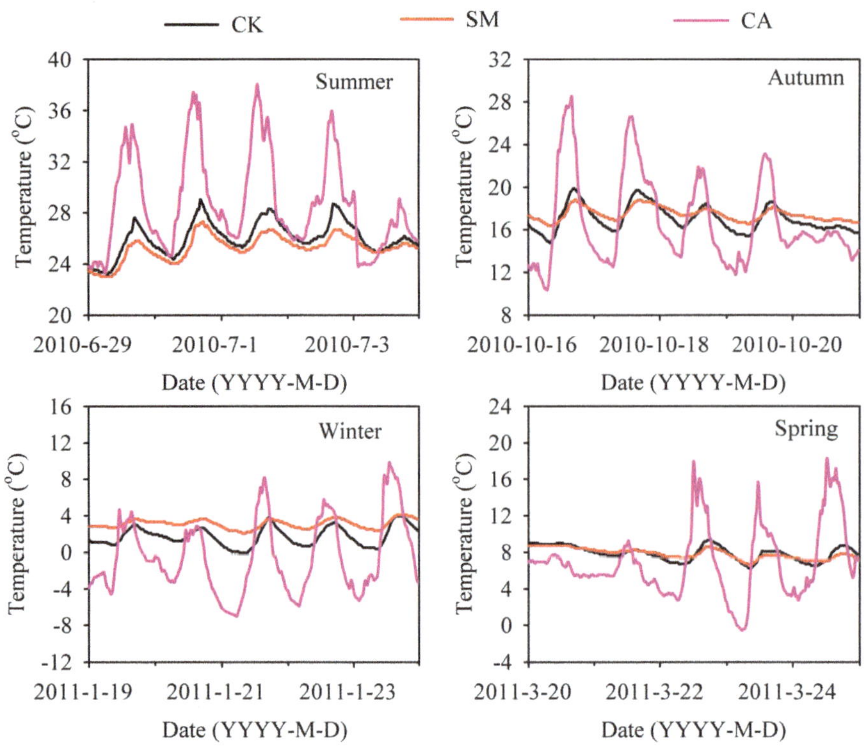

Figure 8. Typical diurnal air temperature trends for Summer, Autumn, Winter and Spring seasons within the canopy (CA) and soil temperatures under control (CK) and straw mulching (SM) treatment plots.

Fruit yield

Differences in citrus fruit yields were recorded in the SM and CK plots in different years (Table 2). There was no significant difference in fruit yield and fruit quality between CK and SM treatments in 2010, while in 2011, this difference in fruit yield was significant. Interestingly, the citrus yield was completely destroyed (zero fruit yield) in the CK treatment plot in 2011 due to extremely low temperatures during the citrus overwintering stage. According to visual observation, almost 50 to 60% of the leaves of citrus trees died or etiolated as a result of a cold and dry weather conditions that prevailed. In SM plots, despite the large reduction in fruit production, some fruit yield was recorded. The citrus fruit yield was 10.3±9.3 t hm^{-2} for SM in 2011.

Discussion

Field observations indicated that the SM treatment significantly decreased the annual runoff as compared to the CK treatment. For example, Adekalu, et al. [13] found that elephant grass (*Pennisetum purpureum*) can be a good alternative for rice straw to effectively reduce runoff and increase infiltration on sloping lands in southwestern Nigeria. Lal [26] found mulching tilled soil with 4–6 t hm^{-2} of rice straw to be effective in reducing soil loss and runoff on slopes ranging from 1% to 15%. Earlier studies have extensively discussed how soil surface mulching reduces the runoff by buffering the ground from raindrop action [21,27] and by modifying soil physical properties through the addition of litter and organic soil matter [28]. Furthermore, the absorption capability of straw provides additional pathways for water infiltration [21]. This trapped water in the straw is gradually released over several days, resulting in decreasing the velocity of surface flow and increasing infiltration. In general, surface runoff decreases with an increase of SM [13]. During the rainy season, a straw cover of 50% is necessary to significantly reduce runoff [29]. Results from the present study also provide indirect support for this conclusion, since the straw cover reached 100% in the SM plots, which was effective in controlling runoff water loss.

Table 2. Yield and fruit quality under the control (CK) and straw mulching (SM) treatments in 2010 and 2011.

Treatment	Yield (t hm^{-2})		Single fruit weight (g)		Equatorial diameter (mm)		Juice content (%)	
	2010	2011	2010	2011	2010	2011	2010	2011
CK	39.5±6.7 a	- -	172±32 a	- -	72.3±12.5 a	- -	42.5±4.0 a	- -
SM	43.2±5.6 a	10.3±9.3	168±7 a	179±42	70.8±3.8 a	76.0±10.9	41.2±5.1 a	40.9±3.6

Values are given as means ± standard error of means (n = 3). Values followed by different letters within a column are significantly different ($P<0.05$). Please note that the citrus crop was completely destroyed (zero citrus fruit yield) in the CK treatment plot in 2011 due to extremely low temperatures during the citrus overwintering stage.

Reduced runoff means an improvement in the soil water status in the root zone and a reduction in soil loss, which in turn leads to reduced land degradation and crop water stress [30]. Moreover, due to the evaporation reducing property of the surface-placed straw layer, mulching (SM treatment) increased soil water storage on an average by 10 mm as compared to the CK treatment (Fig 5). The effect was particularly pronounced during dry periods, when no rain occurred. Similar results have also been reported earlier [13]. Adekalu, et al. [13] stated that the large pores of crop residues permit rapid infiltration of water into the soil but retard evaporation. Water moves back to the atmosphere across the straw pores almost entirely in the vapor phase. The straw tends to act as a one-way water valve for the soil, thus water remains in the soil longer and benefits growing plants. Consequently, more soil water content in the SM treatment plot mainly resulted from higher infiltration, lesser runoff, and lower evaporation than in the CK treatment plot.

Under SM, soil temperature has been reported to have increased [31] as well as decreased [18]. And this can be mainly attributed to differences in climatic conditions. This effect can be explained with two basic mechanisms as observed in our field experiment. In the SM treatment, the mulch layer reduced soil radiation absorption during daytime, while at nighttime it reduced the outgoing heat radiation from the soil. Moreover, the mulch layer contains a significant amount of pore space. The majority of this pore space is likely to be filled with air, and air is known to be a very good insulator. The air space in the mulched layer prevents energy conduction. Therefore, in our study the SM treatment had lower thermal conductivity than the non-mulch control (CK), and acted as an insulator during the warmer period and helped to retain soil heat during the colder period, resulting in smaller fluctuations in soil temperatures (Fig. 7 and 8). Chen, et al. [12] and Olasantan [15] also observed similar results.

Previous research has shown that SM is an effective method to improve crop yield and soil water utilization [32]. Li, et al. [32] found that SM increased soil water content and maize (*Zea mays L.*) yields. Kar and Kumar [33] reported that potato tuber yield was higher in the SM treatments than with bare flat planting in eastern India. However, application of SM is restricted in some places because it is liable to lower the soil surface temperature and lead to a reduction in yield [34,35]. As pointed out by Doring, et al. [34], higher yields under mulch have mostly been attributed to increase soil water under arid and semiarid conditions [17,35], while reduced yields under SM have also been reported and have been attributed to below-optimum soil temperature, reduced soil nitrate levels, and mulching too early [35]. In our study, the results indicate that the effect of SM on citrus fruit yield can be positive in extreme weather conditions. This was possibly because SM reduced the outgoing heat radiation from the soil and, thus, increased the soil temperature compared to the no mulching or CK treatment. Higher soil water content during the dry season in the SM treatment may also have attributed to some fruit yield vis-à-vis the complete fruiting failure in the CK treatment.

Conclusions

From the comparison of runoff, soil water and temperature under both SM and CK conditions we conclude: (1) the surface runoff from the sloping citrus orchards were lower when the soil was mulching with straw than when it was unmulching. (2) Soil water storage in the top 100 cm of the soil profile was increased in the SM as compared to the CK treatment. However, this result was significant only in the dry season (Jan to Mar), and not in the wet season (Jul to Sep) for both years. (3) The mulch probably acted as an insulator, resulting in smaller fluctuations in soil temperature in the SM than in the CK treatment. The results of our study suggested that the small effects on soil water and temperature changes created by surface mulch had limited impact on citrus fruit yield in a normal year (e.g., in 2010). However, SM practices can positively impact citrus fruit yield in extreme weather conditions.

Supporting Information

Table S1 Soil bulk density at start of the field experiment in 2009.

Table S2 Field data and summary statistics of canopy air temperature for 2010 and 2011 at Xiaofuling experimental station.

Table S3 Field data and summary statistics of rainfall for 2010 and 2011 at Xiaofuling experimental station.

Table S4 Field data and summary statistics of runoff under control (CK) and straw mulching (SM) treatments during 2010 and 2011.

Table S5 Field data and summary statistics of soil water storage in the 0–100 cm soil profile under control (CK) and straw mulching (SM) treatments during 2010 and 2011.

Table S6 Field data and summary statistics of soil temperature under control (CK) and straw mulching (SM) treatment plots from May 2010 to March 2012.

Author Contributions

Conceived and designed the experiments: YL FC. Performed the experiments: JW ZGL. Analyzed the data: DBL GSZ JFP. Contributed reagents/materials/analysis tools: DBL YT JX. Wrote the paper: YL.

References

1. van der Velde M, Tubiello FN, Vrieling A, Bouraoui F (2012) Impacts of extreme weather on wheat and maize in France: evaluating regional crop simulations against observed data. Climatic Change 113: 751–765.
2. Porter JR, Semenov MA (2005) Crop responses to climatic variation. Philosophical Transactions of the Royal Society B-Biological Sciences 360: 2021–2035.
3. Farkas C, Birkas M, Varallyay G (2009) Soil tillage systems to reduce the harmful effect of extreme weather and hydrological situations. Biologia 64: 624–628.
4. Carrer M, Motta R, Nola P (2012) Significant Mean and Extreme Climate Sensitivity of Norway Spruce and Silver Fir at Mid-Elevation Mesic Sites in the Alps. Plos One 7: e50755.
5. Liu Y, Yang SJ, Li SQ, Chen XP, Chen F (2010) Growth and development of maize (*Zea mays L.*) in response to different field water management practices: Resource capture and use efficiency. Agricultural and Forest Meteorology 150: 606–613.
6. Silvente S, Sobolev AP, Lara M (2012) Metabolite Adjustments in Drought Tolerant and Sensitive Soybean Genotypes in Response to Water Stress. Plos One 7: e38554.

7. Liu Y, Li SQ, Chen F, Yang SJ, Chen XP (2010) Soil water dynamics and water use efficiency in spring maize (*Zea mays L.*) fields subjected to different water management practices on the Loess Plateau, China. Agricultural Water Management 97: 769–775.
8. Zhao H, Xiong YC, Li FM, Wang RY, Qiang SC, et al. (2012) Plastic film mulch for half growing-season maximized WUE and yield of potato via moisture-temperature improvement in a semi-arid.
9. Lou YL, Liang WJ, Xu MG, He XH, Wang YD, et al. (2011) Straw coverage alleviates seasonal variability of the topsoil microbial biomass and activity. Catena 86: 117–120.
10. Wang YJ, Xie ZK, Malhi SS, Vera CL, Zhang YB, et al. (2011) Effects of gravel-sand mulch, plastic mulch and ridge and furrow rainfall harvesting system combinations on water use efficiency, soil temperature and watermelon yield in a semi-arid Loess Plateau of northwestern China. Agricultural Water Management 101: 88–92.
11. Novak MD, Chen WJ, Orchansky AL, Ketler R (2000) Turbulent exchange processes within and above a straw mulch. Part II: Thermal and moisture regimes. Agricultural and Forest Meteorology 102: 155–171.
12. Chen SY, Zhang XY, Pei D, Sun HY, Chen SL (2007) Effects of straw mulching on soil temperature, evaporation and yield of winter wheat: field experiments on the North China Plain. Annals of Applied Biology 150: 261–268.
13. Adekalu KO, Olorunfemi IA, Osunbitan JA (2007) Grass mulching effect on infiltration, surface runoff and soil loss of three agricultural soils in Nigeria. Bioresource Technology 98: 912–917.
14. Balwinder-Singh, Humphreys E, Eberbach PL, Katupitiya A, Yadvinder-Singh, et al. (2011) Growth, yield and water productivity of zero till wheat as affected by rice straw mulch and irrigation schedule. Field Crops Research 121: 209–225.
15. Olasantan FO (1999) Effect of time of mulching on soil temperature and moisture regime and emergence, growth and yield of white yam in western Nigeria. Soil & Tillage Research 50: 215–221.
16. Huang YL, Chen LD, Fu BJ, Huang ZL, Gong E (2005) The wheat yields and water-use efficiency in the Loess Plateau: straw mulch and irrigation effects. Agricultural Water Management 72: 209–222.
17. Zhang SL, Lovdahl L, Grip H, Tong YN, Yang XY, et al. (2009) Effects of mulching and catch cropping on soil temperature, soil moisture and wheat yield on the Loess Plateau of China. Soil & Tillage Research 102: 78–86.
18. Sarkar S, Paramanick M, Goswami SB (2007) Soil temperature, water use and yield of yellow sarson (*Brassica napus* L. var. glauca) in relation to tillage intensity and mulch management under rainfed lowland ecosystem in eastern India. Soil & Tillage Research 93: 94–101.
19. Ghosh PK, Dayal D, Bandyopadhyay KK, Mohanty K (2006) Evaluation of straw and polythene mulch for enhancing productivity of irrigated summer groundnut. Field Crops Research 99: 76–86
20. Zhang QF (2009) The South-to-North Water Transfer Project of China: Environmental Implications and Monitoring Strategy1. Journal of the American Water Resources Association 45: 1238–1247.
21. Li XH, Zhang ZY, Yang J, Zhang GH, Wang B (2011) Effects of Bahia Grass Cover and Mulch on Runoff and Sediment Yield of Sloping Red Soil in Southern China. Pedosphere 21: 238–243.
22. Liu WZ, Liu GH, Zhang QF (2011) Influence of Vegetation Characteristics on Soil Denitrification in Shoreline Wetlands of the Danjiangkou Reservoir in China. Clean-Soil Air Water 39: 109–115.
23. Liu Y, Tao Y, Wan KY, Zhang GS, Liu DB, et al. (2012) Runoff and nutrient losses in citrus orchards on sloping land subjected to different surface mulching practices in the Danjiangkou Reservoir area of China. Agricultural Water Management 110: 34–40.
24. National Soil Survey Office (1992) Soil survey technique in China. Beijing: Agricultural Press (in Chinese).
25. Bao S.D (2000) Agro-chemical Analysis of Soil. Beijing: Agricultural Press (in Chinese).
26. Lal R (1979) Soil erosion on alfisols in Western Nigeria II: Effect of mulch rates. Geoderma 16: 377–382.
27. Wakindiki IIC, Danga BO (2011) Effect of straw mulch application on nutrient concentration in runoff and sediment in a humid region in Kenya. African Journal of Agricultural Research 6: 725–731.
28. She DL, Shao MA, Timm LC, Reichardt K (2009) Temporal changes of an alfalfa succession and related soil physical properties on the Loess Plateau, China. Pesquisa Agropecuaria Brasileira 44: 189–196.
29. Gutierrez J, Hernandez II (1996) Runoff and interrill erosion as affected by grass cover in a semi-arid rangeland of northern Mexico. Journal of Arid Environments 34: 287–295.
30. Fang Y, Xun F, Bai WM, Zhang WH, Li LH (2012) Long-Term Nitrogen Addition Leads to Loss of Species Richness Due to Litter Accumulation and Soil Acidification in a Temperate Steppe. Plos One 7: e47369.
31. Ramakrishna A, Tam HM, Wani SP, Long TD (2006) Effect of mulch on soil temperature, moisture, weed infestation and yield of groundnut in northern Vietnam. Field Crops Research 95: 115–125.
32. Li R, Hou XQ, Jia ZK, Han QF, Yang BP (2012) Effects of rainfall harvesting and mulching technologies on soil water, temperature, and maize yield in Loess Plateau region of China. Soil Research 50: 105–113.
33. Kar G, Kumar A (2007) Effects of irrigation and straw mulch on water use and tuber yield of potato in eastern India. Agricultural Water Management 94: 109–116.
34. Doring TF, Brandt M, Hess J, Finckh MR, Saucke H (2005) Effects of straw mulch on soil nitrate dynamics, weeds, yield and soil erosion in organically grown potatoes. Field Crops Research 94: 238–249.
35. Edwards L, Burney JR, Richter G, MacRae AH (2000) Evaluation of compost and straw mulching on soil-loss characteristics in erosion plots of potatoes in Prince Edward Island, Canada. Agriculture Ecosystems & Environment 81: 217–222.

3

Biochar Decelerates Soil Organic Nitrogen Cycling but Stimulates Soil Nitrification in a Temperate Arable Field Trial

Judith Prommer[1], Wolfgang Wanek[1]*, Florian Hofhansl[1], Daniela Trojan[2], Pierre Offre[2], Tim Urich[2], Christa Schleper[2], Stefan Sassmann[3], Barbara Kitzler[4], Gerhard Soja[5], Rebecca Clare Hood-Nowotny[1]*

1 Department of Microbiology and Ecosystem Science, University of Vienna, Vienna, Austria, **2** Department of Ecogenomics and Systems Biology, University of Vienna, Vienna, Austria, **3** Core Facility of Cell Imaging and Ultrastructure Research, University of Vienna, Vienna, Austria, **4** Institute of Forest Ecology and Soil, Federal Research and Training Centre for Forests, Natural Hazards and Landscape, Vienna, Austria, **5** Department of Health and Environment, Austrian Institute of Technology, Tulln, Austria

Abstract

Biochar production and subsequent soil incorporation could provide carbon farming solutions to global climate change and escalating food demand. There is evidence that biochar amendment causes fundamental changes in soil nutrient cycles, often resulting in marked increases in crop production, particularly in acidic and in infertile soils with low soil organic matter contents, although comparable outcomes in temperate soils are variable. We offer insight into the mechanisms underlying these findings by focusing attention on the soil nitrogen (N) cycle, specifically on hitherto unmeasured processes of organic N cycling in arable soils. We here investigated the impacts of biochar addition on soil organic and inorganic N pools and on gross transformation rates of both pools in a biochar field trial on arable land (Chernozem) in Traismauer, Lower Austria. We found that biochar increased total soil organic carbon but decreased the extractable organic C pool and soil nitrate. While gross rates of organic N transformation processes were reduced by 50–80%, gross N mineralization of organic N was not affected. In contrast, biochar promoted soil ammonia-oxidizer populations (bacterial and archaeal nitrifiers) and accelerated gross nitrification rates more than two-fold. Our findings indicate a de-coupling of the soil organic and inorganic N cycles, with a build-up of organic N, and deceleration of inorganic N release from this pool. The results therefore suggest that addition of inorganic fertilizer-N in combination with biochar could compensate for the reduction in organic N mineralization, with plants and microbes drawing on fertilizer-N for growth, in turn fuelling the belowground build-up of organic N. We conclude that combined addition of biochar with fertilizer-N may increase soil organic N in turn enhancing soil carbon sequestration and thereby could play a fundamental role in future soil management strategies.

Editor: Shibu Jose, University of Missouri, United States of America

Funding: The authors highly appreciate the funding through FWF Project V138-B17 (to RCHN) as well as the FFG Project 825438 (to GS). The funders had no role in study design, data collection and analysis, decision to publish, or preparation of the manuscript.

Competing Interests: The authors have declared that no competing interests exist.

* E-mail: rebecca.clare.hood-nowotny@univie.ac.at (RCHN); wolfgang.wanek@univie.ac.at (WW)

Introduction

Carbon farming opportunities in temperate agricultural systems are limited to no-till farming and conservation agriculture strategies. In these systems careful carbon management is required to avoid problems of nutrient, particularly N, immobilization by the soil microbial community, which can lead to substantial crop yield depressions. Biochar application to temperate agricultural soils offers the prospect for substantial carbon sequestration without the concomitant yield losses or greenhouse gas emissions associated with other strategies [1–4]. The pyrolysis derived aromatic-macromolecular structure of biochar [5] makes it more recalcitrant to microbial decomposition, which means it does not immobilize significant quantities of valuable nutrients when added to soil, rendering biochar a viable long-term carbon sink option for temperate agriculture [6–7]. Promoting novel carbon farming options in a wider agronomic context requires facing risk-benefit assessments by the soil custodians, namely farmers and other environmentally concerned stakeholders. As a result, an in-depth understanding of the impact of biochar on the soil N cycle, the most economically relevant biogeochemical cycle to farmers in mid to high latitudes, will be required to facilitate decision-making and allow for improved biochar management and subsequent adoption as a carbon sequestering strategy.

There is increasing evidence suggesting that biochar addition to soil impacts a number of processes of the soil N cycle [8] and that the impacts of biochar addition are strongly dependent on the biochar feedstock and the pyrolysis conditions, in addition to the respective soil properties and local environmental and climatic conditions [9]. For instance, biochar was shown to decrease nitrate leaching [10–11], to increase soil N immobilization [12–13], to increase ammonification, the conversion of organic N into ammonium [14–15], to increase nitrification, the oxidation of ammonium to nitrate [16–17], to promote volatilization, the loss of ammonia to the atmosphere [18], and to increase plant N uptake [19–21]. The mechanisms posited to explain the differences in N cycling processes when biochar is added to soil are mostly direct ones, resulting from the unique properties of the biochar such as its highly porous structure, the large surface area and its ion-exchange capacity [22], with associated beneficial effects on nutrient retention and water holding capacity. Some N cycle processes like NH_3 volatilization can be stimulated by inherent biochar properties such as the potential of alkaline biochar to increase soil pH [23]. Moreover biochars can have high nutrient contents and increase soil nutrient availability directly or through the effects of priming which may increase the bio-availability of soil nutrients [24–25]. In addition, indirect mechanisms associated with changes in the soil microbial community can be triggered by the properties specific to biochar and thereby have strong implications for soil microbial N processing [16,26]. To date the majority of biochar studies were carried out as short-term pot experiments, many of these focusing on poor, acidic and nutrient depleted soils where the addition of biochar has the potential to cause largest effects including fundamental changes in soil nutrient cycles, often resulting in pronounced increases in crop production [27–28]. Long-term field trials examining the effects of biochar on intensively managed and fertile agricultural soils in temperate regions are therefore urgently needed to advance our predictive capabilities of biochar - soil N cycle interactions [8,20].

The impacts of biochar on soil N transformation processes are extremely important, given the dominant role N nutrition plays in regulating crop production in temperate regions, as it has to be replenished either as inorganic N fertilizer or it is cycled back to the soil in a complex organic form such as crop residues. In the past two decades there has been a major paradigm shift in our conceptualization of the soil N cycle [29]. It places greater emphasis on the organic turnover processes in the soil N cycle such as the depolymerization of organic matter, which is now regarded as the rate limiting step of dissolved organic N production and subsequently of soil N mineralization and nitrification. As a result of the biochar characteristics described above biochar addition to soil could have major impacts on the organic N turnover component of the soil N cycle. The multitude of simultaneously occurring soil N processes (Figure 1) make measuring and understanding soil N transformation rates difficult and thus requires isotopic tools [30–31]. Using an array of parallel established and novel ^{15}N isotope pool dilution techniques [32–33], it should be possible to study each key transformation process independently. These techniques in consortium allow determining the impact of biochar on multiple soil processes in a synchronized manner and with a high degree of resolution and reliability. In the context of biochar addition this issue of complexity is of particular relevance as biochar addition could have conflicting impacts on different control points of the soil N cycle.

Our major objective therefore was to investigate the effects of biochar on key inorganic and organic N transformation processes, on soil C and N pools and on the soil microbial community. A hardwood-based biochar (pyrolysis at 500°C for two hours) was added in March 2011 to an arable field on calcareous Chernozem in Traismauer, Austria. The field trial consisted of four treatments replicated four times in circular plots of 6.5 m, comprising NPK fertilized plots as controls combined with biochar addition rates of 0, 24 and 72 t dry mass ha^{-1}. As agricultural systems rely on fertilization to grant crop production, fertilized plots were taken as controls (NPK, no biochar) and compared to the two biochar addition rates i.e. 24 t BC ha^{-1} with NPK amendment (BC1N) and 72 t BC ha^{-1} with NPK amendment (BC3N). A treatment with 72 t BC ha^{-1} lacking NPK amendment was also setup (BC3) to check for nutrient immobilization effects of BC-alone treatments. In this study we focused on the control (NPK) and the highest biochar treatment (BC3N). Soil samples were taken

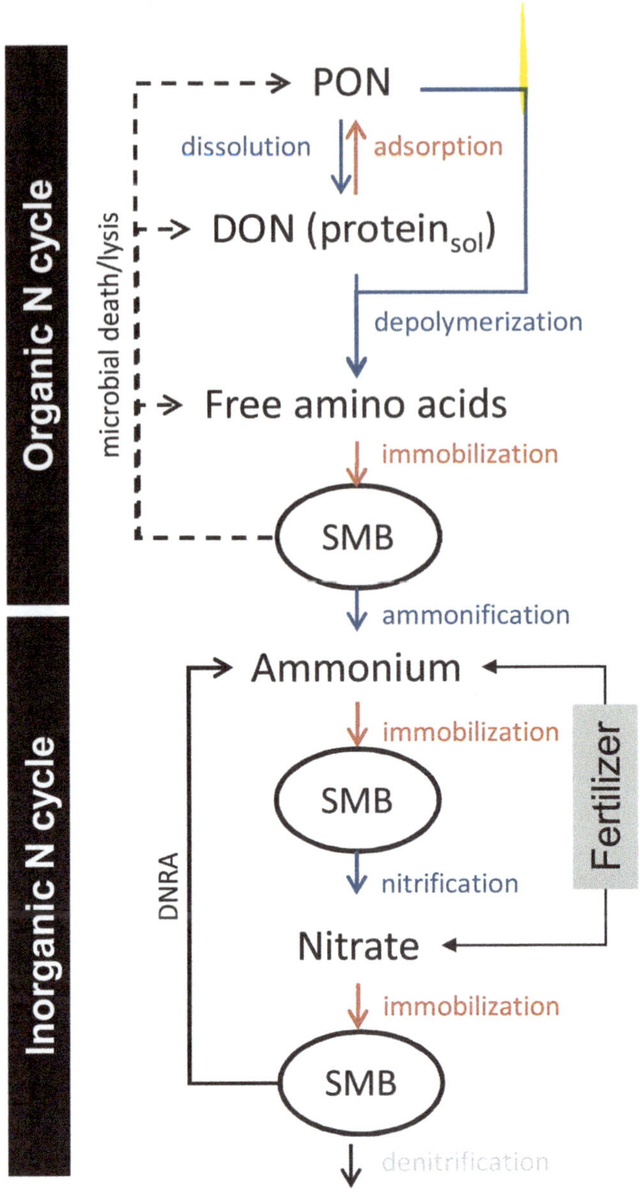

Figure 1. Overview of the agricultural soil N cycle. The organic N and inorganic N cycles are highlighted. DNRA...dissimilatory nitrate reduction to ammonium, DON...dissolved organic N, PON...particulate organic N such as detritus and soil organic matter, SMB...soil microbial biomass. Not shown - heterotrophic nitrification which results from microbial uptake of e.g. free amino acids, oxidation of reduced N compounds and release of nitrate into the soil.

four months after biochar application and sampling was repeated several times during a period of fourteen months. We applied a novel array of isotope pool dilution assays to measure gross rates of protein, amino acid, ammonium and nitrate dynamics (Figure 1) and molecular techniques to investigate nitrifier (ammonium oxidizer) population sizes. In addition we used scanning electron microscopy to examine the physical structure of the biochar in order to understand its possible role in shaping the soil microbial community.

Results

Biochar Porosity

The cellular structure of wood was well conserved in the biochar as seen in Figure S1. The mean pore coverage of xylem cross sections was 47.6±5.9% (mean±SD) for the biochar applied in this field trial. Overall pore size number distribution showed a maximum at approximately 0.7 μm diameter (related to simple xylem pits), and a second and third peak which are closely related to bordered xylem pits and parenchyma cells (around 3 μm) and xylem vessels (15–25 μm; Fig. 2, Fig. S2). Beech has a diffuse-porous wood structure [34] and early-wood and late-wood had distinctly different mean vessel diameters (24±12 and 15±9 μm, respectively), resulting in a decrease in mean pore coverage from 61.8±6.1% for early-wood to 40.2±4.1% for late-wood. Secondary cortex cells showed a wide range of diameters, with a peak ranging from 3 to 5 μm (data not shown). On an area basis the contribution of small pores decreased and the major contribution to total porosity shifted towards larger pores with diameters >10 μm i.e. xylem vessels (Fig. 2b).

Effect of Biochar on General Soil Properties

The effect of biochar on soil pH and cation exchange capacity was minimal. Soil pH decreased by 0.1 units from 7.5 to 7.4 in biochar-treated soils (Table 1). Soil cation-exchange capacity decreased slightly from an initial 22.5 to 20.8 cmolc kg^{-1} in biochar-treated soils but in both cases was dominated by Ca^{2+} (~90%) and Mg^{2+} (~7%), with a trend towards an increase of K^+ saturation from 1.6 to 4.8% (measured on one composite soil sample per treatment, Table 1).

Bulk Soil C and N Contents and Soil Water Relations

Biochar amendment resulted in increased soil organic C and there was a trend of enhanced total soil N concentration with biochar addition. Soil bulk densities and gravimetric soil water contents declined while water-filled pore space remained unaltered by biochar addition (two-way mixed ANOVA for effects of treatment and time, Table 2, Figure 3). Time affected most of these and the following soil parameters (Table 2), but emphasis here is given on treatment effects and treatment × time interactions; the latter might indicate a decreasing or increasing biochar-effect on soil properties or fluxes. However, all time × treatment interactions effects investigated in this study were either non-significant or not related to increasing or decreasing biochar-effects over time. Direct time effects were most likely caused by seasonal changes, agricultural management and subsequent plant growth effects. Soil organic C was significantly higher in biochar-treated soils (47.4 mg C g^{-1}) than in non-biochar controls (18.7 mg C g^{-1}; $p=0.001$) (Table 2). Total soil N responded analogously, with a slight but non-significant increase due to biochar-treatment (1.47 to 1.64 mg N g^{-1}; $p=0.062$). Concurrent with increases in soil organic C the soil bulk density decreased from 1.17 to 1.04 g cm^{-3} following biochar amendment ($p<0.001$). Along the same lines biochar amendment caused an increase in soil porosity from 56.0 to 60.6% ($p<0.001$) and in gravimetric soil water content from 0.140 to 0.163 g H_2O g^{-1} dry soil ($p<0.001$). As a result of increased soil water content and soil porosity with biochar addition, water-filled pore space (and air-filled pore space) did not change between controls (29.8%) and biochar-treated soils (28.9%, $p=0.229$). Interestingly, by comparing all four treatments we found the biochar-effect to be amount-dependent i.e. the magnitude of the effects increased or decreased with the amount of biochar added in the field (Table S1). Soil organic C increased with biochar rate, bulk density decreased and soil porosity and soil water content increased with biochar rate (Table S1).

Extractable C and N Concentrations

Dissolved organic C in soil extracts was slightly but significantly lower in the biochar treatment compared to the controls (111 vs. 123 μg C g^{-1} respectively; $p=0.010$) (Figure 3, Table 2). Concentrations of extractable ammonium were low compared to extractable nitrate and not significantly different between the control and biochar treatments, with 1.66 and 1.55 μg NH_4^+-N g^{-1} respectively ($p=0.383$). In contrast soil nitrate concentrations in the biochar (BC3N) treated soils were significantly lower than those from the control soils ($p=0.001$), with 8.40 and 14.4 μg NO_3^-N g^{-1} respectively. Dissolved organic N concentrations in the soil extracts decreased, though not significantly, from 9.62 in biochar-treated soils to 8.35 μg N g^{-1} in control treatments ($p=0.187$). The ratio of dissolved organic C: dissolved organic N varied around 16:1 and was not affected by biochar addition ($p=0.909$). Nitrate dominated the extractable N pool, accounting for between 67% (BC3N) and 76% (controls). The decrease in soil nitrate therefore translated into a significant decline of total dissolved N from 25.0 to 17.0 μg N g^{-1} in the biochar treatment ($p=0.001$). Free amino acids were low in concentration (around 0.7 μg N g^{-1}) and were not affected by treatment ($p=0.110$). Measured only once (July 2011, four months after the start of the field experiment) we found no significant differences in the concentrations of different size classes of dissolved organic N between the two treatments, with extractable high-molecular weight organic N concentrations around 3.3 μg N g^{-1} and low-molecular weight organic N concentrations of around 6 μg N g^{-1} (Table 2). Free amino acids accounted for approximately 12% of the low molecular weight organic N pool and comprised 8% of dissolved organic N in both treatments. Around 40% of the dissolved organic N and all the high-molecular weight organic N fraction (>3 kDa; 105–107%) was protein, although this measurement is based only on the July 2011 sampling. The negative effect of biochar on concentrations of dissolved organic C, total dissolved N and nitrate was consistent even when all four treatments were analyzed (Table S1), with decreases in dissolved organic C, nitrate and free amino acids with increasing biochar addition rate (Table S1).

Soil N Transformation Rates

The full set of gross N transformation processes was measured once, and one-way ANOVA results are presented in Table 3 and Figure 4; nitrate fluxes were measured on three occasions throughout the seasons (Table 3, Figure 4, see circles). Measured rates of protein influx into the extractable organic N pool were significantly ($p=0.016$) higher in the NPK treatment (80.0 μg N g^{-1} d^{-1}) and almost double the rates in the biochar treatment (44.2 μg N g^{-1} d^{-1}). Biochar showed a trend ($p=0.097$) of depressing protein efflux rates out of the extractable organic N pool, with rates in the biochar treatment (53 μg N g^{-1} d^{-1}) being half that of the control treatment (115 μg N g^{-1} d^{-1}). The decline in amino acid transformation rates was even more pronounced. Amino acid influx rates into the organic N pool were significantly ($p=0.0004$) lower in biochar treated soils (5.4 μg N g^{-1} d^{-1}), more than fivefold lower than those in the control treatment (28.5 μg N g^{-1} d^{-1}) and so too were amino acid consumption rates. Amino acid consumption rates declined more than six-fold from 50.3 μg N g^{-1} d^{-1} in the control treatment to 7.9 μg N g^{-1} d^{-1} in the biochar treatment ($p=0.039$). Gross mineralization rates were not significantly different between the two treatments ($p=0.616$) nor were ammonium consumption rates ($p=0.526$;

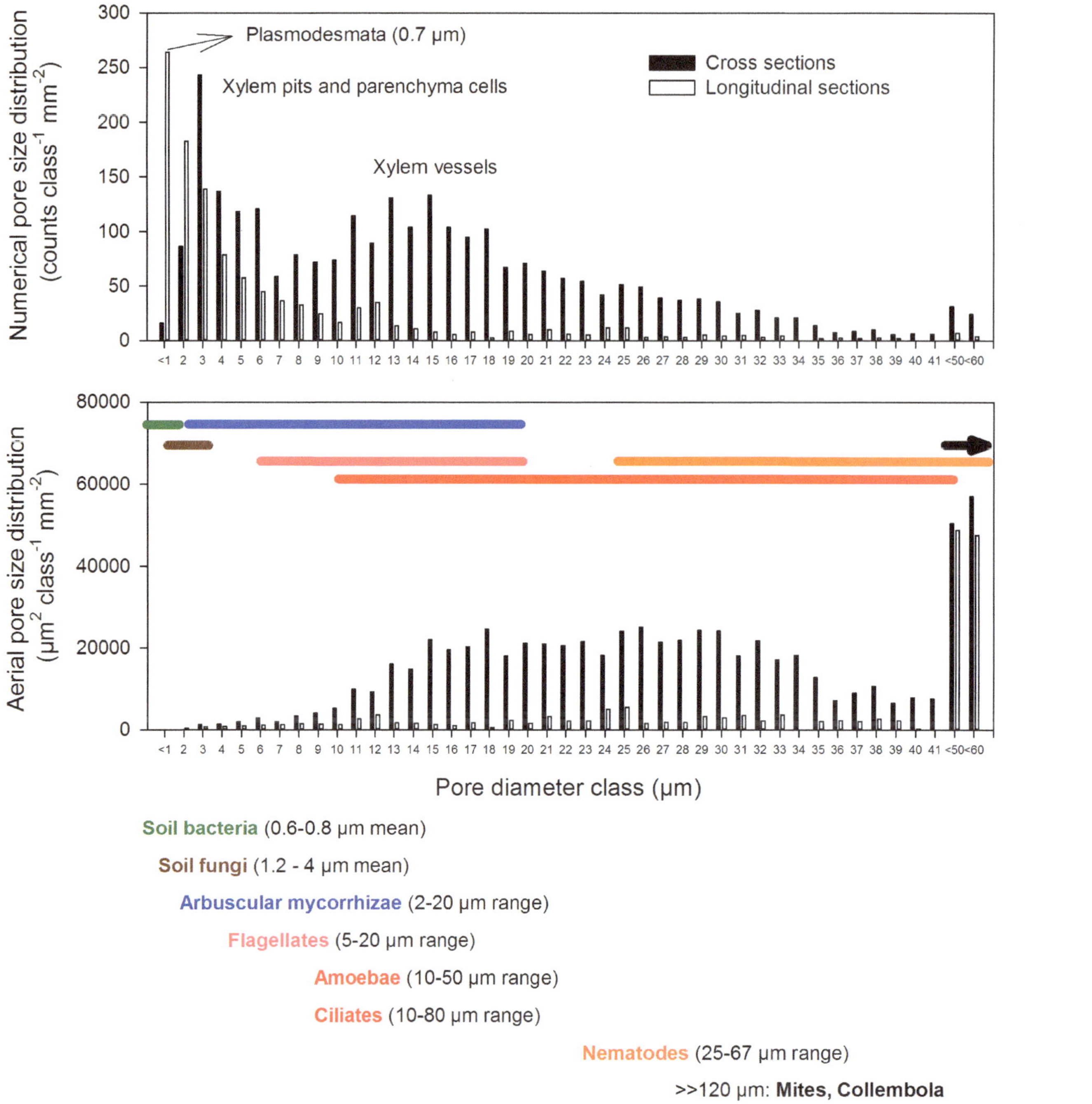

Figure 2. Pore size distribution of beech wood biochar as determined by scanning electron microscopy. In total, mean diameters of 7715 pores were measured in 6 longitudinal and 7 cross sections of charcoal pieces, each image covering 0.35 mm^2. Pore size distributions are presented (1, top panel) as counts mm^{-2} per 1 µm size class (top) and (2, middle panel) on an area basis, i.e. µm^2 mm^{-2} per 1 µm size class. Typical mean diameters (soil bacteria, soil fungi) and diameter ranges (arbuscular mycorrhizal fungi; protozoan grazers such as flagellates, amoebae, ciliates; and mesofaunal grazers and predators such as nematodes, mites and collembola) are presented color-coded (bottom panel) and with range lines in the middle panel.

Figure 3), and both processes ranged between 1.5 and 2.1 µg N g^{-1} d^{-1}. In contrast, gross nitrification and gross nitrate immobilization were strongly enhanced by biochar application. Gross nitrification rates in the biochar treatment (10.2 µg N g^{-1} d^{-1}) were fourfold higher than the rates in the control treatment (2.4 µg N g^{-1} d^{-1}), and significantly different ($p<0.001$). A similar trend ($p=0.0573$) was evident for nitrate consumption rates which almost doubled from 6.8 to 11.3 µg N g^{-1} d^{-1} in the biochar treatment.

Rates of nitrification and nitrate consumption were re-measured twice. Over the course of time nitrification rates were consistently higher (threefold) in the biochar treated compared to the control plots ($p<0.001$, Table 3). The same pattern was apparent for nitrate consumption rates which were consistently negatively affected by biochar amendment ($p=0.001$). On the sampling date

Table 1. Soil properties in control (NPK) and biochar-treated (BC3N) soils of the Traismauer biochar field trial.

Parameter	Unit	Control (NPK)	Biochar (BC3N)
pH ($CaCl_2$)		7.5	7.4
$CaCO_3$	%	15.8	15.2
Humus	%	2.4	18.1
Total N	%	0.148	0.203
P (CAL)	mg kg^{-1}	49	84
P_{tot} (acid digest)	g kg^{-1}	5.46	5.54
Sand	%	18.3	n.d.[1]
Silt	%	57.2	n.d.
Clay	%	24.5	n.d.
CEC	cmol kg^{-1}	22.5	20.8
Ca (CEC)	cmol kg^{-1}	20.7	18.2
Mg (CEC)	cmol kg^{-1}	1.46	1.53
K (CEC)	cmol kg^{-1}	0.36	0.99
Na (CEC)	cmol kg^{-1}	<0.04	<0.04
Al (CEC)	cmol kg^{-1}	<0.06	<0.06
Fe (CEC)	cmol kg^{-1}	<0.01	<0.01
Mn (CEC)	cmol kg^{-1}	<0.01	<0.01
H (CEC)	cmol kg^{-1}	0.002	0.002
Fe (EDTA)	mg kg^{-1}	40	67
Mn (EDTA)	mg kg^{-1}	107	128
Cu (EDTA)	mg kg^{-1}	7.2	7.1
Zn (EDTA)	mg kg^{-1}	2.3	7.5

All parameters were measured on composite samples collected in July 2011 from the two treatments (n = 1). n.d. – not determined.
[1]Soil texture is not shown for biochar treated soils. At high contents humus biases the particle size distribution and is therefore conventionally decomposed by prolonged H_2O_2 (or similar oxidant) treatment prior to soil texture analysis; biochar/charcoal is largely resistant against this oxidant treatment.

where we had gross nitrification rates measured for all four treatments nitrification rates were significantly higher in all the biochar treatments ($p<0.001$, Table S1) with rates increasing significantly with increasing rates of biochar application.

Microbial Community Size and Abundance of Archaeal and Bacterial Ammonia-oxidizers

Soil DNA concentration was significantly higher in the biochar treatment (18.0 µg DNA g^{-1} soil) compared to the controls (12.4 µg DNA g^{-1}; $p = 0.042$) (Figure 5, Table 2). Ammonia-oxidizing archaea (AOA) outnumbered bacterial ammonia-oxidizers (AOB) by a factor of 3.2 to 3.6, in both treatments ($p = 0.402$). At the same time archaeal amoA gene copy numbers (AOA) and bacterial amoA gene copy numbers (AOB) expressed on a dry soil mass basis increased through biochar application (AOA: $P = 0.045$, AOB: $p = 0.020$). AOA increased 1.5-fold from 1.68×10^7 to 2.47×10^7 amoA copy numbers g^{-1} soil and AOB 1.7-fold from 5.01×10^6 to 8.45×10^6 amoA copy numbers g^{-1} soil through biochar amendment. As a result of increasing soil DNA concentration and amoA copy numbers per gram soil we did not find a significant effect of biochar on the abundance of AOA ($p = 0.637$) and AOB ($p = 0.436$) expressed on a DNA basis, i.e. the relative share of ammonia oxidizers in the microbial community did not change according to treatment. Across all four treatments we did not find an effect of biochar rate on either soil DNA concentration or any measure of AOA or AOB abundance (Table S1). However, archaeal and bacterial amoA gene copy numbers were strongly positively related ($R^2 = 0.83$, $p = 0.0114$; data not shown). Moreover, we found a trend towards a positive relationship between both, AOA or AOB numbers on a dry soil mass basis and gross nitrification rates (AOA: $R^2 = 0.65$, $p = 0.051$, AOB: $R^2 = 0.60$, $p = 0.072$; Figure 6). The "weakness" of this relationship derived mainly from the control treatment where nitrification rates were relatively constant between the two measurement dates while amoA copy numbers varied more strongly. Nitrification rates were significantly positively correlated with the total number of amoA copy numbers on a soil dry mass basis (AOA plus AOB; $R^2 = 0.66$, $p = 0.049$, data not shown).

Discussion

Soil Organic N Cycling

Biochar addition had a strong negative effect on soil organic N cycling rates as it reduced protein production and protein consumption and concurrently decreased free amino acid production and consumption (Figure 4). This is important given that protein depolymerization plays the predominant role in organic N recycling with approximately 50% of soil organic matter and >90% of plant detritus being composed of protein [35–36], and with proteins representing around 40% of dissolved organic N in this study. Proteins are therefore considered the most important substrate for microbial N mobilization and N use [33], and thus any change in their dynamics strongly impacts whole-system N cycling. The influx into the K_2SO_4 extractable protein pool (representing a labile part of the total protein pool in soils) is mainly fuelled by microbial turnover and protein desorption processes while efflux from this pool relates to protein depolymerization or protein adsorption (Figure 1).The concomitant reductions in free amino acid turnover were most likely a direct consequence of the reduced protein-N cycling in biochar amended soils. Amino acid production is driven by depolymerization of accessible proteins through extracellular proteases and peptidases which release oligopeptides and free amino acids that are then taken up by microbes (free amino acid consumption) [33]. Though oligopeptides also represent a significant source of N for microbes and plants [37] we did not measure their dynamics, which would have been expected to change accordingly. We offer two possible explanations for the observed negative effects of biochar on organic N (protein) dynamics related to the surface properties of biochar and its unique pore structure.

(1) The observed decline in protein production rates might result from protein adsorption to biochar surfaces. The reactive surface of biochar has been shown to adsorb numerous low and high molecular weight organic molecules [38], including proteins and enzymes [39]. Hence, the reduced motility and accessibility of proteins in the biochar treated soils could explain the decrease in protein influx rates observed with resultant implications for amino acid dynamics. Very strong H-bonding between proteins and polyphenols are well known [40], and the abundant phenolic groups on low temperature biochars [41] provide ample ligands for strong sorption of proteins and peptides [9].
(2) Biochar could also provide structural protection, by the entrapment of organic matter within its inherent pore network. The "mesopore protection hypothesis" by Zimmerman et al. [42] suggests that small organic molecules can be occluded within mineral mesopores (2–50 nm diameter),

Table 2. Results of ANOVA of soil physicochemical data for the factors treatment and time in control (NPK) and biochar (BC3N) treatments.

Pools	Unit	NPK	BC3N	Treatment		Time		Interaction	
		mean	mean	F	P	F	P	F	P
C_{org}	mg C g^{-1} DW	18.66	47.43	(1, 6) 43.7	0.001	(1, 6) 0.0	0.951	2.5	0.165
N_{tot}	mg N g^{-1} DW	1.47	1.64	(1, 6) 5.3	0.062	(2, 12) 37.5	<0.001	0.3	0.778
BD	g cm^{-3}	1.17	1.04	(1, 6) 48.4	<0.001	(1, 6) 55.3	<0.001	0.3	0.595
Grav. WC	g H_2O g^{-1} DW	0.14	0.17	(1, 6) 46.7	<0.001	(6, 36) 218.7	<0.001	6.2	<0.001
Porosity	%	56.03	60.61	(1, 6) 48.6	<0.001	(1, 6) 55.3	<0.001	0.3	0.594
WFPS	%	29.79	28.92	(1, 6) 1.8	0.229	(6, 36) 215.4	<0.001	2.3	0.059
AFPS	%	70.21	71.08	(1, 6) 1.8	0.229	(6, 36) 215.4	<0.001	2.3	0.059
DOC	µg C g^{-1} DW	122.56	111.21	(1, 6) 13.6	0.010	(5, 30) 205	<0.001	2.9	0.028
DOC:DON		16.10	15.92	(1, 6) 0.0	0.909	(1.6, 9.9) 14.4	0.002	0.8	0.454
DON	µg N g^{-1} DW	9.62	8.35	(1, 6) 2.2	0.187	(2.3, 13.9) 28.5	<0.001	0.3	0.782
NO_3^-	µg N g^{-1} DW	14.43	8.40	(1, 6) 39.6	0.001	(6, 36) 95.9	<0.001	8.6	<0.001
NH_4^+	µg N g^{-1} DW	1.66	1.55	(1, 6) 0.9	0.383	(2.6, 15.5) 34.8	<0.001	1.0	0.417
FAA	µg N g^{-1} DW	0.77	0.70	(1, 6) 3.5	0.110	(1.6, 9.8) 3.2	0.091	0.9	0.429
TDN	µg N g^{-1} DW	24.95	17.03	(1, 6) 33.3	0.001	(5, 30) 97.2	<0.001	1.3	0.285
HMW-N_{org}	µg N g^{-1} DW	3.24	3.47	(1, 23) 1.1	0.306				
LMW-N_{org}	µg N g^{-1} DW	6.78	5.68	(1, 23) 0.7	0.401				
Protein	µg N g^{-1} DW	3.46	3.65	(1, 23) 1.4	0.253				
DNA	µg g^{-1} DW	12.40	18.04	(1, 6) 6.6	0.042	(1.1, 6.9) 0.1	0.774	2.6	0.154
AOA/DNA	amoA copy no $µg^{-1}$ DNA	1.41E6	1.36E6	(1, 6) 0.2	0.637	(2, 12) 8.2	0.006	3.3	0.070
AOA/soil	amoA copy no g^{-1} soil DW	1.68E7	2.47E7	(1, 6) 6.3	0.045	(1.1, 6.6) 3.5	0.103	0.9	0.378
AOB/DNA	amoA copy no $µg^{-1}$ DNA	419075	461583	(1, 6) 0.7	0.436	(1.2, 7) 7.2	0.028	1.6	0.246
AOB/soil	amoA copy no g^{-1} soil DW	5.01E6	8.45E6	(1, 6) 9.9	0.020	(2, 12) 4.5	0.035	1.3	0.319
AOA:AOB		3.64	3.19	(1, 6) 0.8	0.402	(2, 12) 0.4	0.681	0.0	0.970

Data were analyzed by two-way mixed ANOVA. Data for protein, HMW-N_{org} and LMW-N_{org} were available only for one sampling date and therefore were analyzed by one-way ANOVA. Abbreviations: C_{org}, soil organic C; N_{tot}, total soil N; BD, soil bulk density; Grav. WC, gravimetric soil water content; WFPS, water filled pore space; AFPS, aire filled pore space; DOC, dissolved organic C; DON, dissolved organic N; NO_3^-, nitrate; NH_4^+, ammonium; FAA, total free amino acids; TDN, total dissolved N; HMW-N_{org}, high molecular weight organic N; LMW-N_{org}, low molecular weight organic N; DNA, soil DNA content; AOA/DNA, archaeal amoA copy numbers on DNA basis; AOA/soil, archaeal amoA copy numbers on dry soil basis; AOB/DNA, bacterial amoA copy numbers on DNA basis; AOB/soil, bacterial amoA copy numbers on dry soil basis.

physically protecting organic molecules such as proteins (<10 nm) and amino acids (<2 nm) from breakdown by excluding decomposers and their extracellular enzymes. Mesopore surface area and volume are mainly determined by gas sorption analysis but are not targeted by scanning electron microscopy, and therefore have not been measured in this study. Nevertheless, low-temperature wood biochars produced at <650°C, just as the Romchar used in this study, have been shown to have a low mesopore volume and surface area as measured by gas sorption [39]. Physical protection of organic matter in mesopores by occlusion from decomposing microbes may thus have played a minor role in this study.

Just like the decrease in protein production, the decline in free amino acid production in biochar treated soils could also be triggered by protease and peptidase adsorption to or occlusion in biochar. Enzyme adsorption to solid support materials (e.g. soil matrix or biochar) has been reported to decrease soil enzymatic activity [43], and adsorbed substrates such as proteins and oligopeptides are also less accessible and bioavailable for further breakdown to free amino acids. Decreased aminopeptidase activity has been observed in experiments with fast pyrolysis biochar of switchgrass and was ascribed to sorption reactions between substrate and biochar [44]. Another study found accelerated N-mineralization of the recalcitrant soil organic N pool in arable soils treated with maize biochar [14]. The inconsistent effects of biochar on soil enzymes may be the result of several factors such as biochar feedstock, pyrolysis conditions, soil type as well as the respective enzyme and/or substrate [44]. For example, contrasting results obtained in our experiment and that of Nelissen et al. (2012) [14] could be explained by structural differences between the biochars used in these two studies, leading to variable enzyme diffusion impedance. Indeed, the Romchar employed here maintained its native wood-like structure according to the scanning electron microscopy images (Figure S1) while non-woody materials such as maize biochar have an amorphous structure with little evidence of retained pores [45].

Soil N Mineralization

Given that biochar application significantly decelerated the production and consumption of low molecular weight organic N in our field experiment, we would have expected a decrease in soil N mineralization rates. Somewhat surprisingly biochar had no significant effect on gross N mineralization and gross ammonium immobilization rates in our study, in line with a laboratory incubation of cultivated black Chernozem soil with wheat biochar

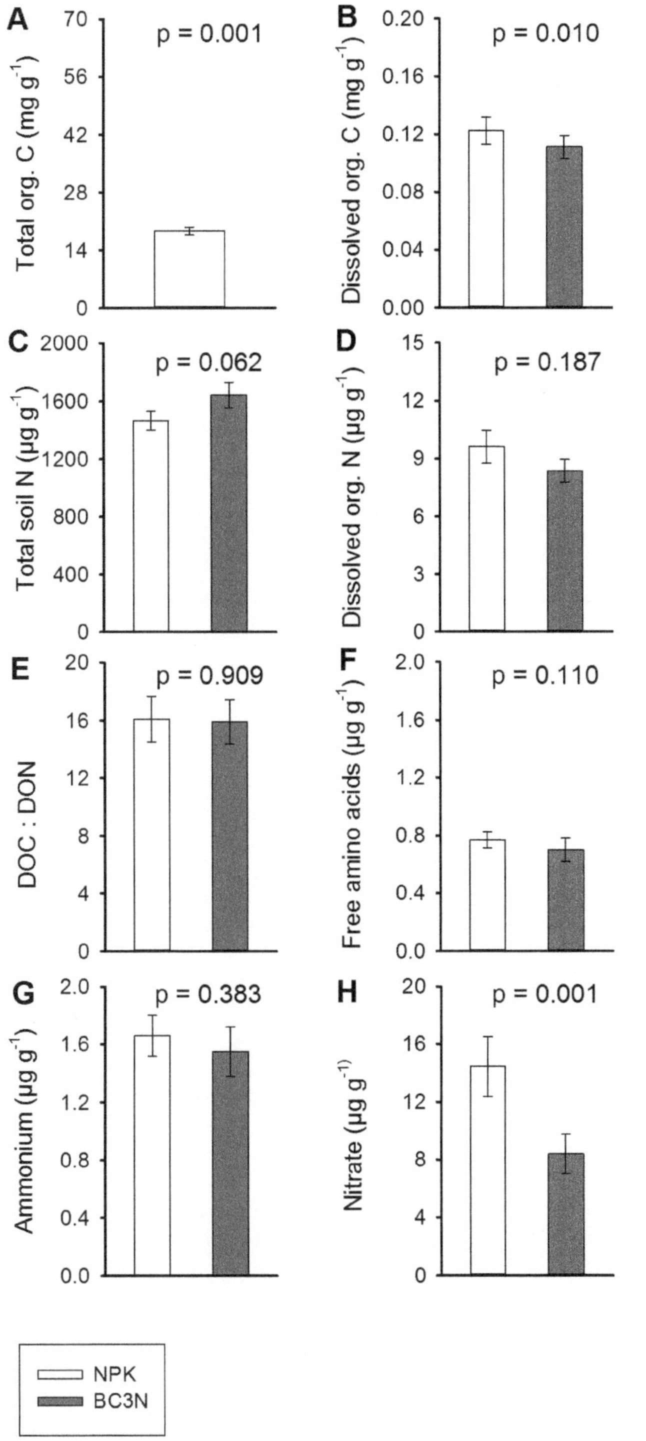

Figure 3. Soil C and N pool sizes for control (NPK) and biochar (BC3N) treatments. Bars represent means ±1 SE (n = 4) from seven measurements between July 2011 and September 2012 for extractable compounds, and from two measurements (September 2011 and 2012) for soil organic C and total soil N. Open bars, control treatment (NPK); grey bars, biochar treatment (BC3N). Units in mg C g^{-1} dry soil and µg N g^{-1} dry soil. P values are from two-way mixed ANOVA (see Table 2).

[46], but in contrast to observed accelerated rates of gross N mineralization of the recalcitrant soil organic N pool in an arable soil treated with maize biochar [14]. Interestingly mineralization (respiratory use) of amino acids and peptides, measured as the uptake of ^{14}C-labelled organic N and subsequent $^{14}CO_2$ release, was not affected by application of a wood-based biochar [15]. Divergent, positive, neutral or negative effects of biochar on soil N mineralization may be the upshot of different biochar types along with impacts on the C and N status of the soil microbial community. Carbon-rich biochar, poor in nitrogen, such as beech wood biochar (C:N ratio ~200:1) might promote N limitation, leading to the retention of catabolically produced ammonium in the N-limited microflora, which therefore results in a decrease of the amount of ammonium released to the soil [29]. Conversely, nitrogen-rich biochars with low C:N ratios such as maize-biochars [14] or manure-biochars [47] promote microbial C limitation, causing the excess of N to be mineralized and therefore N mineralization rates to increase. In our case the ammonium released (N mineralization) relative to organic N taken up by microbes (free amino acid consumption) was higher in the biochar treated soils than in controls (Figure 4), indicating that biochar addition did not lead to microbial N limitation as a consequence of the increased organic C content in soil. This suggests that the majority of organic C in the biochar was not metabolized or assimilated by the soil microbial community, and confirms that biochar is an efficient tool for long-term C sequestration in soil, with half-lifes >100 years [48]. The absence of microbial N limitation in the biochar treatment might be directly related to the field applications of fertilizer and might also be related to the slightly lower concentrations of dissolved organic C in the biochar treated plots.

Soil Nitrification

Apart from the negative effect that biochar had on soil organic N cycling, the most striking result was a strong and persistent (1.5 years) stimulation of gross nitrification. Several studies have examined the impact of biochar on soil net nitrification [8] but only few have reported gross nitrification rates, which allow the simultaneous analysis of both, productive and consumptive processes for nitrate. Biochar addition increased gross nitrification rates in two forest soils and one arable soil [14,26,49] but had no effect in other experiments with grassland and arable soils [46,49]. Several hypotheses have been raised to explain biochar-related (positive) effects on nitrification rates.

One of the dominant theories to explain the increase in nitrification rates following biochar amendment is the liming effect [14,26]. Biochar commonly has a neutral to alkaline pH [50] and therefore may increase the pH of acidic soils with a concurrent positive response of soil nitrifiers whose pH optimum is slightly acidic to neutral pH [51–52]. However the pH of the soil used in this study was slightly alkaline and did not increase but rather decreased in response to the addition of biochar.

Equally we can disregard the tenet that biochar addition leads to an increase in nitrification due to adsorption of nitrification inhibitors. In contrast to forest and grassland soils [49,53–54], intensively managed agricultural soils usually contain low amounts of naturally occurring nitrification inhibitors [55] and their effects might have been of minor importance in this experiment.

Nitrification rates have been shown to be strongly controlled by the ammonium supplied by soil N mineralization in a wide variety of ecosystem types including woodland, grassland and agricultural soils [56]. In this study, biochar amendment neither had an effect on gross ammonium production rates nor on the pool size of extractable ammonium, which obviates a simple substrate-driven explanation of increased nitrification rates in this field experiment. Whereas nitrification rates in the control soils were of the same order of magnitude as the mineralization rates, biochar treated

Table 3. Results of ANOVA of soil N transformation rates for the factor treatment in control (NPK) and biochar (BC3N) treatments.

Processes	Unit	NPK	BC3N	Treatment		Time		Interaction	
		mean	mean	F	P	F	P	F	P
One date									
Prot_GP	mg N kg^{-1} d^{-1}	79.99	44.15	(1, 11) 8.4	0.016				
Prot_GC	mg N kg^{-1} d^{-1}	115.06	52.37	(1, 11) 3.4	0.097				
AA_GP	mg N kg^{-1} d^{-1}	28.54	5.36	(1, 11) 26.7	0.000				
AA_GC	mg N kg^{-1} d^{-1}	50.25	7.91	(1, 11) 5.6	0.039				
N min	mg N kg^{-1} d^{-1}	1.88	2.10	(1, 11) 0.3	0.616				
NH_4^+_GC	mg N kg^{-1} d^{-1}	1.86	1.52	(1, 11) 0.4	0.526				
Nitrification	mg N kg^{-1} d^{-1}	2.42	10.16	(1, 11) 50.6	<0.001				
NO_3^-_GC	mg N kg^{-1} d^{-1}	6.83	11.27	(1, 11) 4.6	0.057				
Three dates									
Nitrification	mg N kg^{-1} d^{-1}	3.38	9.09	(1, 6) 82.2	<0.001	(2, 12) 0.4	0.698	3.6	0.061
NO_3^-_GC	mg N kg^{-1} d^{-1}	4.92	11.01	(1, 6) 35.2	0.001	(2, 12) 1.7	0.225	0.6	0.571

All transformation rates were measured and analyzed by one-way ANOVA for July 2011 data ("one date"). Nitrification and nitrate consumption was measured three times for NPK and BC3N treatments ("three dates") and analyzed by two-way mixed ANOVA. Abbreviations: GP, gross production (influx); GC, gross consumption (efflux); Prot, protein; AA, amino acid; N min, N mineralization; NH_4^+, ammonium; NO_3^-, nitrate.

soils exhibited nitrification rates four to five times higher than mineralization rates. This dichotomy between ammonium supply and nitrification rates could point to increased importance of heterotrophic nitrification fuelled by organic N compounds in biochar-treated soils. However heterotrophic nitrification mainly occurs in grassland and forest soils and especially in acidic soils [57–58], although this process has been recently detected in acidic arable and neutral soils [59–60]. This nitrification pathway will have to be investigated in follow-up incubation experiments, using acetylene as an inhibitor of autotrophic (archaeal and bacterial) nitrification [61] which does not affect heterotrophic nitrification [62–63].

Increased nitrification rates in soils treated with biochar could be attributed to the alteration of the soil water status, including its distribution and associated changes in soil oxygenation, since nitrification is dependent on soil O_2 availability. The soil air/water balance is essential for nitrification, which has been found to peak at 60% water-filled pore space [64]. Altered soil moisture conditions have been reported in various biochar studies [17,65–66]. However, in this study air-filled pore space was highly variable and on average greater than 70%, independent of treatments; therefore higher nitrification rates (Figure 4) could not be explained by improved oxygen availability.

While none of the proposed drivers of soil nitrification could explain the observed pattern of enhanced nitrification in biochar treated soils, the size of the ammonia-oxidizing community increased with biochar amendment (Fig. 5). Recently, the oxidation of ammonia to nitrite, the first and rate limiting step of nitrification, was shown to be performed by prokaryotes in both, the bacterial and archaeal domains [67]. Both ammonia oxidizing archaea (AOA) and ammonia oxidizing bacteria (AOB) abundances were significantly higher in the biochar treated soils, AOA outnumbering AOB in both treatments. There are numerous studies showing that AOA are often significantly more abundant than AOB in soils of different origins [68–69]. However the relative importance and contribution of these different groups to the nitrification process in natural environments is still unclear and to our knowledge there are no published studies available on the interaction of biochar and AOA, at least in temperate ecosystems. AOB abundances have been found to increase after natural forest fire events most likely as a consequence of the liming effect of charcoal and its sorption of nitrification inhibitors [26]. In a pot experiment the AOB community structure was unaffected by the sole addition of *Eucalyptus* wood biochar but changed when biochar was added together with an inorganic or organic N source [70], a possible result of substrate-limited nitrification as the ammonium concentrations declined in sole biochar treatments. The stimulation of both, AOA and AOB abundances through biochar addition in this field experiment was only significant when amoA gene copy numbers were expressed on a soil mass basis but not when expressed on a soil DNA basis. The microbial community size, indicated by total soil DNA content, increased with biochar addition revealing a general positive effect of biochar on the whole microbial community rather than a specific stimulation of the autotrophic nitrifiers. This unspecific stimulation of microbes and concomitant increase of nitrifiers is corroborated by the lack of biochar effects on more specific soil nitrification controls such as soil pH, ammonia and oxygen availability, and presence of nitrification inhibitors. The biochar effect was also unspecific in terms of stimulation of AOA and AOB; both AOA and AOB abundances increased with biochar and were strongly positively related across all soil samples independent of treatment. In addition ratios of AOA: AOB were unaffected by the biochar amendment.

AOA and AOB abundances were broadly ($P<0.1$) correlated with gross nitrification rates, suggesting that bacterial and archaeal abundance increased with increasing biochar addition and providing a functional link to the process data (Fig. 6). This and the total DNA data add weight to an underlying hypothesis, i.e. that the structural nature of the biochar provides a protective habitat for bacterial and archaeal growth. It is conceivable, that biochar promotes bacterial and archaeal community size by enhancing the colonizable surfaces in the soil and providing habitable protected pores. The unique pore-size distribution typically found in biochar with pores ranging from 1 nm to 100 µm [71] is believed to provide various beneficial functions to

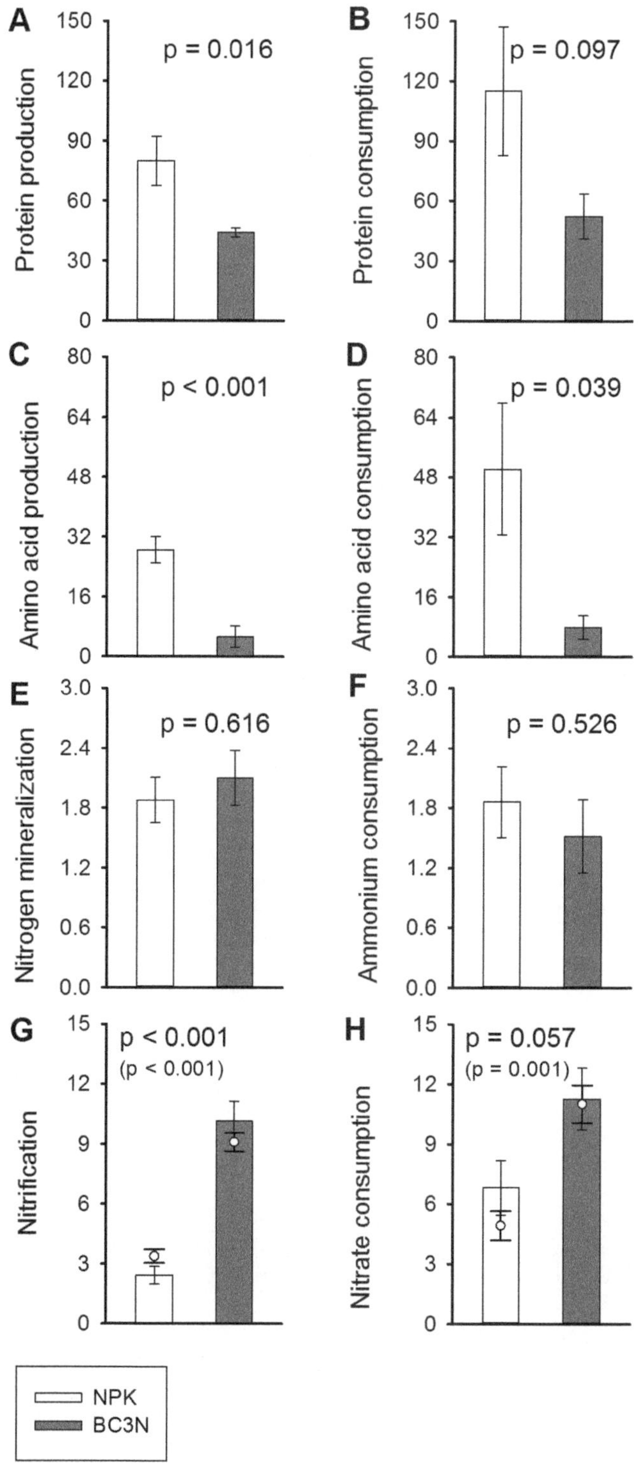

Figure 4. Soil N transformation rates for control (NPK) and biochar (BC3N) treatments. Bars represent means ±1 SE (n = 4) of measurements from July 2011. Open bars, control treatment (NPK); grey bars, biochar treatment (BC3N). Open circles represent means ±1 SE (n = 4) of nitrification rates from three measurement dates. Left panels represent gross influx rates into the target pools (production rates), right panels gross efflux rates (consumption rates). Units in μg N g^{-1} dry soil d^{-1}, equivalent to mg N kg^{-1} d^{-1}. P values are from one-way ANOVA (all rates) or two-way mixed ANOVA (nitrification for three time points, in brackets).

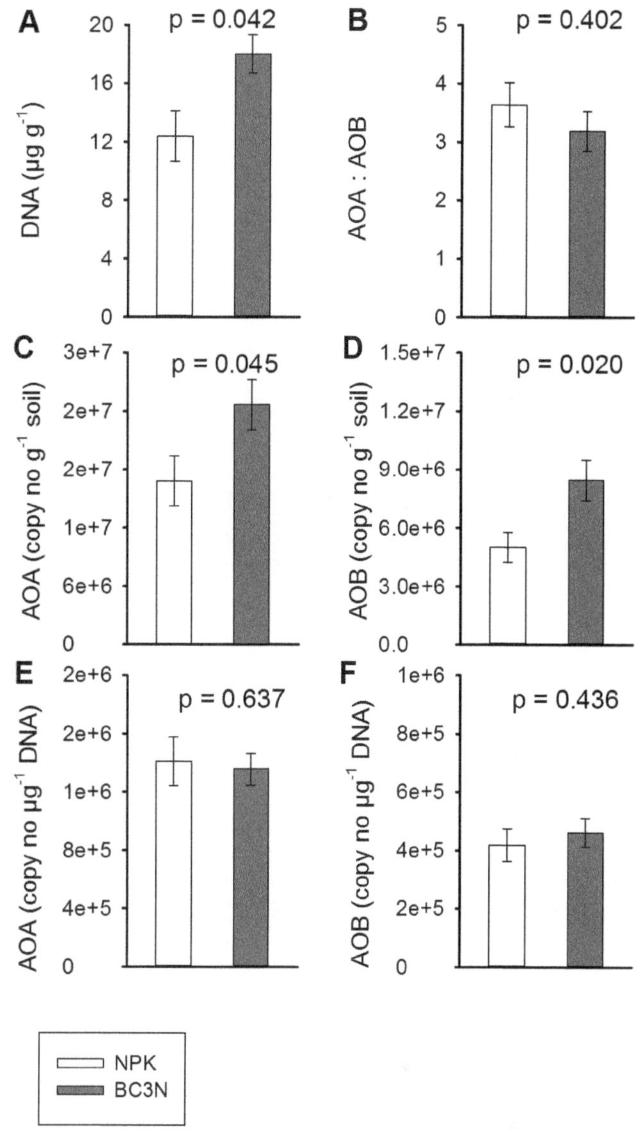

Figure 5. Soil microbial community size and ammonia-oxidizing community structure. Bars represents means ±1 SE (n = 4). Open bars, control treatment (NPK); grey bars, biochar treatment (BC3N). DNA, soil DNA content. AOA, archaeal amoA copy numbers; AOB, bacterial amoA copy numbers. AOA and AOB copy numbers are presented on soil dry mass and DNA basis. P values from two-way ANOVA (see Table 2).

soil bacteria and archaea [72]. Soil prokaryotes typically have mean cell sizes of 0.6 to 0.8 μm in diameter [73–74] with a length: diameter ratio of 1.5:1 [75] and were shown to inhabit pores greater than 0.6 μm [76]. Soil fungi have slightly larger mean diameters, of between 1.2 to 4 μm [74,77] but mycorrhizal hyphae vary enormously in diameter ranging between 2 to 20 μm [78]. As a consequence of the high proportion of colonizable pores, addition of *Eucalyptus* charcoal to a coarse textured soil increased habitable pore space surface area by 11% at a biochar rate of 25 t ha^{-1} [70] and habitable pore volume by 6.5% at 50 t ha^{-1} [79]. Indeed we found an increase of pore volume of 8% at biochar rates of 72 t ha^{-1}. More importantly it has been shown that interior surfaces and pores are rapidly colonized, microbiota showing no differences in densities between outer and internal biochar surfaces 3 years after application [79]. It has been

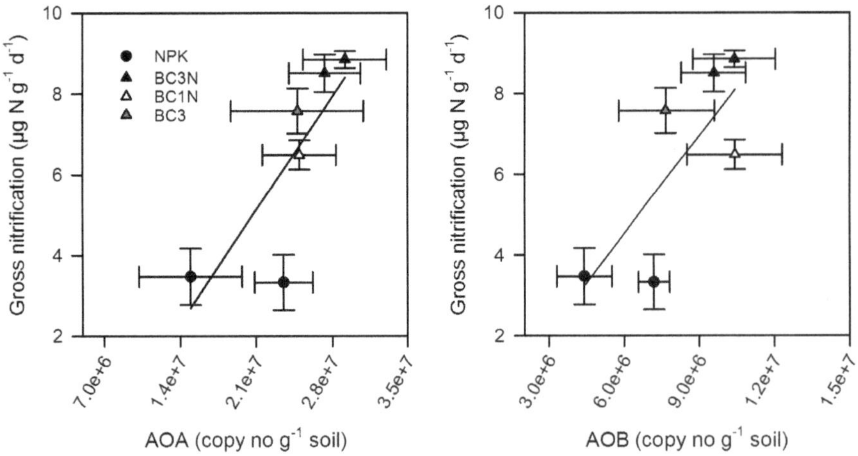

Figure 6. Relationship of archaeal (AOA) and bacterial amoA abundance (AOB) with gross soil nitrification rates. The relationships between archaeal amoA copy numbers (AOA) and bacterial amoA copy numbers (AOB) with gross nitrification rates were marginally significant for AOA ($p = 0.051$) and AOB ($p = 0.072$). Symbols represent the four treatments measured in September 2012. NPK and BC3N were also measured in June 2012. Values are means ±1 SE ($n = 4$).

suggested that soil microbes are protected from being grazed upon their natural predators by colonizing pore habitats in biochar that are inaccessible for soil microfauna (Fig. 2). By this mechanism biochar pores can act as a safe microhabitat for various soil microbes through the exclusion of larger predators [80–81] such as nematodes, mites and collembolans [73] but also microfaunal protozoan grazers and predators, i.e. flagellates, amoebae and ciliates [82–83]. Our scanning electron microscopy data suggest that there was an abundance of colonizable pores, particularly in the 0.3 to 40 µm range with maxima around 0.7, 3 and 15 µm. The abundant sack like structures of parenchyma cells and the long cave like structures of xylem vessels (Figure S1) could therefore offer physical protection for soil prokaryotes and fungi from grazers such as nematodes and amoeba, which are just physically too big to enter the cave or pore. This in turn could reduce microfaunal grazing, which leads to a decrease in mineralization of organic N via the soil "microbial loop" [84]. This scenario provides a possible explanation for the observed increases in nitrification rate, with bacteria, fungi, AOA and AOB taking refuge in the biochar. In the biochar pores AOA and AOB have ample access to carbon dioxide and available ammonia, given the high pH of this soil; this allows them to proliferate, but protected from grazing, thus not contributing significantly to N mineralization. Interestingly this scenario would suggest possible top down control of nitrification, with predators having been shown to consume up to 60% of bacterial production in some soils [85]. In addition to the provision of protected habitat provided by biochar co-location of nutrients and water on biochar surfaces and in pores may increase substrate availability, and thus may further stimulate microbial colonization [72] and nitrification.

Conclusions and Outlook

The main findings of this field study are that application of wood biochar led to marked decreases in soil organic N transformation rates while soil nitrification was significantly stimulated in this type of agricultural soil. This field study therefore indicates a de-coupling of the intrinsic organic and inorganic soil N cycles in the short- to medium-term (four to 18 months) as a consequence of biochar addition. Assuming these effects persist, the deceleration of soil organic N cycling ultimately leads to the dichotomy of two intertwined agronomic processes; the build-up of soil organic N and the slowing-down of inorganic N release from soil organic N essential for crop growth. Addition of fertilizer-N in combination with biochar effectively decouples these processes, with plants and microbes utilizing fertilizer-N for growth, in turn fuelling the belowground build-up of soil organic N. Since soil organic N and particularly refractory soil organic matter are peptide dominated materials [86], the accumulation of soil organic N is expected to enhance soil carbon sequestration due to increases in non-biochar derived soil organic matter. However, it remains an open question whether adsorptive capacity of biochar will saturate and pore-filling of biochar slows or curtails this process, though recent investigations with aged and fresh biochar suggested that the effects of biochar on organic compound sorption may be long lasting [87].

Stimulation of soil nitrification can have positive effects on plant N availability and crop growth but concurrently increases the risk of soil N losses through nitrate leaching and gaseous losses by denitrification and nitrifier-denitrification. The decrease in soil nitrate concentration and increase in microbial nitrate utilization however point to a relatively conservative N cycling in biochar-treated arable soils. It will therefore be of great importance to study the responses of soil organic N, nitrification, soil N losses and plant growth over longer time periods and in various agricultural settings, to better understand the long-term effects of biochar amendments on soil N cycling.

Given the central role of soil organic N cycling and nitrification in controlling global terrestrial N cycles, the implications of this work are of major importance. If it is more broadly proven that addition of biochar combined with fertilizer-N leads to a greater than sum increase in non-biochar derived soil organic matter, as a consequence of the mechanisms described here, it may prove to be a fundamental factor in future soil management: as building soil organic matter through carbon farming is the key step required to kick start the next black/green revolution.

Materials and Methods

Description of Field Trial

The field experiment was established in Traismauer, Lower Austria, Austria (48° 19′ 886″ N, 15° 44′ 308″ E; 268 m a. s. l) in

March 2011 by the Austrian Institute of Technology (AIT) Tulln, Austria. The temperate climate shows strong continental influence (Pannonian climate), with mean annual precipitation of 550 mm and mean annual temperature being 10°C. The arable soil was a deep sandy to loamy silt (18.3% sand, 57.2% silt and 24.5% clay), classified as a calcareous Chernozem on loess. Soil pH of non-biochar treated soil was 7.5 in 10 mM $CaCl_2$, the carbonate content was 15.8% and humus content 2.4% (Table 1). Cation exchange capacity of the soil was 22.5 cmol kg^{-1}, dominated by Ca^{2+} (~90%) and Mg^{2+} (~7%), with a small contribution by K^+ (2%). The site was used for agriculture for >100 years, with spring barley grown in 2011 and sunflower in 2012. Hardwood-derived biochar (80% beech, rest derived from other hardwoods) was produced by pyrolysis at 500°C for two hours and was purchased from S.C. Romchar S.R.L. (Harghita, Romania). This biochar had a pH of 8.6 in water and 7.5 in 10 mM $CaCl_2$ and comprised the following elemental concentrations: 80.3% C, 9.9% O, 0.4% N, and 1.6%H contents (Table 4). Volatiles made up 11.1% and ashes 15.2% of biochar dry mass (both determined at 550°C, Table 4). Biochar was applied once to the field at a dry mass rate of 0, 24 and 72 t ha^{-1} in March 2011 and incorporated to a depth of 15 cm. NPK fertilizer was applied annually in spring providing an annual N input of 120 (2011) and 75 kg ha^{-1} (2012), with corresponding annual rates of P input of 26 (2011) and 31 kg ha^{-1} yr^{-1} (2012) and K input of 50 (2011) and 100 kg ha^{-1} yr^{-1} (2012). The four treatments were: control i.e. NPK (no biochar application, NPK added), BC1N (24 t biochar ha^{-1}, NPK added), BC3N (72 t biochar ha^{-1}, NPK added) and BC3 (72 t biochar ha^{-1}, no NPK added). The four treatments were arranged according to a Latin square and replicated four times in circular plots of 6.5 m diameter.

The field is privately owned and not protected in any way. The local winegrower association Traisen valley (Weinbauverband Traisental) gave us the written permission to conduct the field experiment at the premises owned by Rudolf Hofmann who also agreed in his role as landowner to set up the experimental plots on his land. No governmental permits were required for the described field study, and the study did not involve endangered or protected species.

Table 4. Element contents of beech wood biochar pyrolyzed for two hours at 500°C.

Element	Content (% d.w.)
C	80.30
O	9.88
Ca	3.75
H	1.60
Si	1.58
Fe	0.79
K	0.73
N	0.40
Al	0.243
Mg	0.198
Mn	0.091
P	0.076
Si	0.053
Cl	0.010
Zn	0.009
Cr	0.005
Co	0.002
Ashes (550°C)	15.2
Volatiles (550°C)	11.1

Soil sampling was performed several times in 2011 and 2012 (see Table S2 for sampling dates and for the treatments sampled and the pools and processes measured at each date). Soil samples were taken from the topsoil layer (0–10 cm) using a soil corer with 8 cm diameter. From each treatment plot three soil samples were collected randomly and pooled to form one composite sample. The fresh samples were then weighed for bulk density calculation, sieved to pass 2 mm and stored at 4°C for a maximum of 2–3 days prior to analyses.

Soil Physicochemical Analyses

General soil properties. Soil water content (SWC, in g H_2O g^{-1} dry soil) was determined gravimetrically after drying fresh soil at 105°C for 24 hours in a forced-air drying oven. Bulk density (BD, in g cm^{-3}) was calculated from the fresh weight divided by soil core volume times dry weight: fresh weight ratio. The percentage of water-filled pore space (%WFPS) was calculated as follows:

$$\%WFPS = \frac{SWC \times BD \times 100}{1 - \frac{BD}{2.65}}$$

where 2.65 g cm^{-3} represents the mean density of rock.

Soil total C and N contents were measured on oven-dried (105°C) and homogenized soil samples. Samples were ground to a fine powder with a ball mill (MM2000, Retsch, Germany) and aliquots of 10 mg weighed into tin capsules. Total soil C and N contents were determined by an elemental analyzer (EA 1110; CE Instruments, Italy). For quantification of the soil organic carbon content samples were pretreated with excess acid (2 M HCl) for 60 min and then dried at 105°C to remove carbonate from the calcareous soil. Other basic soil properties, such as soil pH, soil organic matter content, carbonate content, soil texture, total P content, cation-exchange capacity, base saturation, and exchangeable cations (see Table 1) were only measured for control (NPK) and biochar-treated soils (BC3N) at the Institute of Sustainable Plant Production, Federal Agency for Food Security (AGES) in Vienna, Austria, according to standard ÖNORM procedures (www.austrian-standards.at/en).

Biochar characterization. The biochar (beech wood charcoal) was characterized for element, volatiles and ash contents by the German Biomass Research Center (DBFZ, Leipzig, Germany) and results are presented in Table 4. Biochar pH was measured in an aqueous slurry (5% (w/v)) using a pH electrode. For measurements of the porosity (pores >300 nm diameter) and the pore size distribution of the biochar, charcoal samples were mounted on 0.5″ aluminum specimen stubs equipped with SEM-Carbon foils (Agar scientific, UK). Since the specimens were conductive no additional coating was performed after air drying at 40°C for 24 hours. Micrographs were taken on a Philips XL 20 scanning electron microscope (SEM; Philips, Eindhoven, The Netherlands) at 350× and 2000× magnification with an acceleration voltage of 15 kV. For pore size and pore coverage measurements the software Scandium Version 5.0 (Soft Imaging Systems GmbH, Olympus, Vienna, Austria) and the image analyzing software GSA Image Analyzer (GSA, Gesellschaft für Softwareentwicklung und Analytik mbH, Germany) were used. To estimate pore size distribution in total, 7715 pore measurements in

6 longitudinal and in 7 cross-sectional images covering 0.35 mm^2 and in 25 longitudinal images covering 0.0112 mm^2 were performed. Pores down to a diameter of 300 nm could be safely measured. Pore size distributions were converted to counts mm^{-2} or circular area in μm^2 mm^{-2} and summed up for each 1 µm size class.

Pool sizes of extractable organic C, inorganic N and organic N forms. For measurements of extractable C and N pools fresh soils (aliquots of 4 g) were extracted for 30 min with 30 ml 0.5 M K_2SO_4, subsequently filtered through ash-free cellulose filters and extracts kept frozen at −20°C for later analyses. Total dissolved N and dissolved organic carbon (measured as non-purgeable organic carbon) were measured by high temperature catalytic oxidation on a Shimadzu TOC-VCPH with TNM-1 unit and ASI-V autosampler (Shimadzu Austria, Korneuburg, Austria). Ammonium and nitrate were determined with colorimetric methods, NH_4^+ with a modified indophenol reaction and NO_3^- with the VCl_3/Griess assay, as described in detail by Hood et al. [88]. Amino acid concentrations were quantified by a modified fluorometric OPAME procedure based on Jones et al. [89], optimized for free amino acid measurement in protein hydrolysates. The modified OPAME reagent was made up of 10 mg o-phthaldialdehyde dissolved in 1 ml of HPLC grade methanol to which 20 µl of 3-mercaptopropionic acid was added. This reagent was mixed with 40 ml potassium borate buffer (0.2 M; pH 9.5) and left to stand overnight. Sample aliquots (50 µl) were mixed with 200 µl reagent in the microtiter plate and measured after 10 min at an excitation wavelength of 340 nm and emission wavelength of 450 nm with a fluorescence microplate reader (Tecan Infinite M200, Tecan Austria GmbH, Grödig, Austria). As NH_4^+ in the soil extracts also gives a fluorescence signal, NH_4^+ concentration had to be determined [88] and its fluorescence contribution was determined via a NH_4^+ concentration series and subtracted from the total fluorescence of the samples. Total free amino acid-N concentration was then calculated using glycine as a standard and assuming that each mol of amino acid contains 1.37 mol N. Proteins were quantified in K_2SO_4 extracts by acid hydrolysis and subsequent amino acid analysis (modification of [89–90]; see amino acid determination above). Protein hydrolysis to free amino acids was performed in 6 M HCl and heating was performed for 15 hours at 100°C in a drying oven with Bovine serum albumin (BSA containing 15.8% N) as internal standard. Samples were then dried in an N_2 stream and redissolved in deionized water. Determination of the NH_4^+ concentration to correct the amino acid fluorescence was done as in the case of amino acid determination, with the exception that the sodium nitroprusside solution (the color reagent) contained 1.5 M NaOH instead of the original 0.3 M NaOH to neutralize residual acid in the hydrolysates.

Isotope Pool Dilution Assays

Gross rates of N transformation processes were determined using the ^{15}N isotope pool dilution (IPD) technique [31,91]. This method entails the addition of an isotopic tracer to a target pool and measuring the isotopic composition and concentration over time which allows estimating gross rates of influx to (production) and efflux from (consumption, immobilization) the target pool. In order to avoid the stimulation of consumption rates due to excessive addition of ^{15}N labeled substrate to the respective target pool, each pool size was determined prior to labeling the soil samples. Thereafter to each target pool 10% of the pool size was added in the form of ^{15}N tracer (98 at% ^{15}N). Before starting the IPD assays, all soils were adjusted to 40% water holding capacity and then the ^{15}N tracer solutions were added to bring the soils to 70% water holding capacity. After addition of the tracer to the soils the vials were shaken vigorously to homogenously distribute the label. The incubations were performed at room temperature (~20°C) and terminated by adding 0.5 M K_2SO_4 and subsequent extraction for 30 min on a rotary shaker followed by filtration through ash-free filter papers into new vials. Extracts were immediately frozen and kept for further analyses. An overview of the soil preparation pipeline for these isotope pool dilution assays is presented in Figure S2.

Gross N mineralization and NH_4^+ immobilization. Each 2 g soil sub sample was weighed into disposable HDPE vials (20 ml scintillation vials) and labeled with $^{15}NH_4Cl$ (98 at% ^{15}N purchased from Isotec-Sigma Aldrich, Vienna, Austria). The incubations were terminated after 4, 24 and 48 hours by extraction with 15 ml 0.5 M K_2SO_4. Prior to isotopic analyses ammonium concentrations in the labeled extracts were measured photometrically. Ammonium was isolated from the extracts by microdiffusion [92]. In brief, this method implies the conversion of ammonium into ammonia gas which is then trapped on an acidified cellulose filter disc enclosed in a semi-permeable Teflon membrane (acid trap), where the ammonia gas dissolves and redissociates to ammonium. The interconversion of ammonium and ammonia is a consequence of the pH-dependent equilibrium between these two N forms. To each 10 ml of soil extract one acid trap and 100 mg MgO were added to raise the pH to >9.5, the vials were closed immediately and left to stand at room temperature for three days. In cases where ammonium concentrations exceeded the optimal range of N (10–25 µg NH_4^+-N 10 ml^{-1}) for elemental analyzer-isotope ratio mass spectrometry (EA-IRMS) measurement, extracts were diluted with 0.5 M K_2SO_4 prior to microdiffusion. After microdiffusion acid traps were transferred to 2 ml HDPE screw-cap vials and dried in an evacuated desiccator over concentrated sulphuric acid for two days. The cellulose discs were retrieved and transferred into tin capsules and subjected to EA-IRMS. Conversely extracts with ammonium-N concentrations less than 2 µg NH_4^+-N 10 ml^{-1} were prepared to be measured with purge-and-trap isotope ratio mass spectrometry (PT-IRMS) [93]. For this purpose acid traps were made from glass fiber filters. Acid traps to be measured with PT-IRMS were opened and the filter discs transferred to 1.5 ml HDPE snap-cap vials containing 1 ml deionized water and shaken on a Vortex to redissolve the trapped ammonium. The alkaline persulfate oxidation with subsequent VCl_3/azide reactions were then applied to convert NH_4^+ via NO_3^- to N_2O for analysis by PT-IRMS (see below) [93].

Gross nitrification and NO_3^- immobilization. Soil labeling was done with $K^{15}NO_3$ (98 at% ^{15}N purchased from Isotec-Sigma Aldrich) and carried out as a mentioned above for gross N mineralization. As in the case of gross N mineralization a reduction-microdiffusion method was used to isolate nitrate-N after conversion by Devarda's alloy to ammonium-N from the extract into acid traps [92,94] and nitrate concentrations were determined before starting the microdiffusions. In a first step ammonium was eliminated from the extracts by adding one acid trap and MgO (100 mg) to 10 ml sample, then the vials were closed and microdiffusion proceeded for three days at room temperature. After retrieving this acid trap, a new one was added together with 50 mg Devarda's alloy (45% aluminum, 50% copper and 5% zinc), a reducing catalyst converting nitrate to ammonium, to the soil extracts. After five days of microdiffusion at room temperature to collect the ammonia produced from nitrate the second set of acid traps was treated as mentioned above for gross N mineralization and isotope ratios measured via EA-IRMS (see below) to estimate gross nitrification and gross nitrate immobilization.

The organic N-IPD assays described below are novel and were applied here for the first time to measure gross protein and free amino acid fluxes in soils (but see Wanek et al. [32] for the first amino acid IPD by gas chromatography-mass spectrometry).

Gross amino acid production and immobilization. The free amino acid pool was labeled by addition of a ^{15}N-labeled algal amino acid mixture (U-^{15}N-98%, Cambridge Isotope Laboratories Europe, Radeberg, Germany). For this, each 4 g fresh soil were weighed into 50 ml HDPE centrifuge tubes and the ^{15}N tracer was added. Samples were extracted for half an hour with 30 ml 0.5 M K_2SO_4 after 5 and 30 minutes incubation time. An ultrafiltration method was used to separate low-molecular weight organic N from high molecular weight organic N. Amicon Ultra-2 prelaunch centrifugal filter devices (Millipore Corporation) with a nominal molecular weight limit (NMWL) of 3 kDa were taken to collect the low molecular weight organic N fraction (mainly amino acids and oligopeptides) together with the inorganic N forms (NO_3^- and NH_4^+) in the permeate and the high molecular weight organic N forms (mainly proteins) in the retentate. Ultrafiltration was performed on a Beckman centrifuge (Beckman J2–21, Beckmann Coulter, Krefeld, Germany). From the labeled amino acid extract aliquots of 1.7 ml were centrifuged at 4000*g* for one hour at room temperature with a swinging bucket rotor (Beckman JS-13.1). In order to get rid of inorganic N from low molecular weight organic N, each 1 ml of permeate was filled into a 2 ml cryo-vial and 20 mg Devarda's and 20 mg MgO were added together with an acid trap. The closed cryo-vials were kept at 37°C for 3 days; thereafter the acid traps were discarded. Via alkaline persulfate oxidation and VCl_3/azide reaction the low molecular weight organic N was then converted to NO_3^- and finally to N_2O for analysis by PT-IRMS (see below). Moreover free amino acid concentrations were determined in each extract using the modified OPAME procedure as detailed above, to correct the isotope enrichment of free amino acids for the contributions of other low molecular weight organic N forms using a simple two-source isotope pool mixing model [95]. Free amino acid concentrations and the corrected isotope enrichments were used for later pool dilution calculations.

Gross protein production and consumption. Soil labeling was performed as in the case of the amino acid-IPD assays but with the ^{15}N tracer being a purified algal crude protein (98 at%^{15}N purchased from Isotec-Sigma Aldrich) and with incubation times of 5 min, 30 min, 4 and 24 hours. After ultrafiltration (similar to amino acid-IPD but centrifugation parameters altered) of each 1.7 ml of the labeled extract at 7500 *g* for one hour at room temperature with a fixed angle rotor (Beckman JA-20), the permeate was put aside. The high molecular weight organic N containing retentate was washed twice with each 1 ml 50 mM K_2SO_4 via 30 min centrifugation at room temperature and 7500 *g*, to remove inorganic N and low molecular weight organic N. Thereafter the retentate was recovered by 2 min centrifugation at 1000 *g* at room temperature and diluted to 0.5 ml with 50 mM K_2SO_4. The alkaline persulfate oxidation and subsequent VCl_3/azide reactions were then used to convert the high molecular weight organic N via NO_3^- to N_2O for analysis by PT-IRMS (see below).

Alkaline persulfate oxidation (APO) of N_{org} (NH_4^+) to NO_3^-. The APO reaction was used to convert ammonium and organic N forms into nitrate [93]. At elevated temperature persulfate decomposes forming persulfate radicals that efficiently oxidize organic and inorganic compounds, in the case of organic N and ammonium producing nitrate. The persulfate reagent was prepared by dissolving 44 g $Na_2S_2O_8$, 16.8 g NaOH and 30 g H_3BO_3 in deionized water with an end volume of 1 L [96]. For digestion aliquots of 750 µl of sample were pipetted into 1.5 ml HPLC glass vials and each 750 µl of persulfate solution were added. Vials were closed immediately (to avoid losses of ammonium due to the alkaline pH) and autoclaved for one hour at 120°C. Digested samples were stored at 4°C until further analyses.

VCl_3/azide conversion of NO_3^- to N_2O. The isotope ratios of nitrate after persulfate digestion were measured by conversion of nitrate to N_2O by a coupled VCl_3/azide reaction in which in a first step, VCl_3 is used to reduce nitrate to nitrite, which in a second step is further reduced to N_2O by sodium azide [93]. For this, aliquots of 1 ml sample were pipetted into 12-ml screw cap exetainers with butyl rubber septum (IVA-VC309; IVA Analysetechnik, Merbusch, Germany) which were closed tightly and purged with helium for 10 min to get rid of air-N_2O in the headspace. Thereafter 160 µl sodium azide and finally 1 ml VCl_3 were added using gas-tight syringes by injection through the septa. Exetainers were put on a rotary shaker and kept at 37°C for 18 hours. The reaction was stopped by neutralizing and deactivating the sodium azide through injecting each 160 µl 6 M NaOH into the exetainers and N_2O was analyzed by PT-IRMS.

PT-IRMS. The exetainer vials were then loaded into a 96-slot autosampler with a double-hole needle (GC-PAL, CTC Analytics, Zwingen, Switzerland) connected via a Gasbench II headspace analyzer (Thermo Fisher, Bremen, Germany) and a PreCon cryotrapping device (Thermo Fisher) to a gas isotope ratio mass spectrometer (IRMS, Delta V Advantage, Thermo Fisher) [93]. Through the double needle of the autosampler an auxiliary helium flow purged the contents of the exetainer vials through a water and CO_2 scrubber into a sequence of two cryo-traps held at liquid N_2 temperature, and the N_2O was then released and residual CO_2 and air (O_2, N_2) separated by gas chromatography. The purified sample N_2O was then transferred into the mass spectrometer via an open split with repeated pulses of standard N_2O gas for internal calibration (Air Liquide, Vienna, Austria). For calibration of at%^{15}N respective standards varying in ^{15}N:^{14}N ratios were run throughout the sample preparation methods. For further details on PT-IRMS analyses see Lachouani et al. [93].

EA-IRMS. Acid traps and homogenized dry soils were transferred into tin capsules and directly analyzed for N content and at%^{15}N by EA-IRMS. The system consisted of an elemental analyzer (EA 1110, CE Instruments, Milan, Italy) connected via a ConFlo III interface (Thermo Fisher) to the IRMS (Delta^{PLUS}, Thermo Fisher). External calibration was performed using laboratory standards for isotope composition and N concentration [93].

Isotope pool dilution calculations. Gross production (GP) and gross consumption (GC) rates, i.e. influx into and efflux from the target pools that was labeled with ^{15}N in the course of IPD experiments was then calculated based on the isotope pool dilution equations developed by others [97]:

$$GP = \frac{C_{t_2} - C_{t_1}}{t_2 - t_1} \times \frac{\ln\left(\frac{APE_{t_1}}{APE_{t_2}}\right)}{\ln\left(\frac{C_{t_2}}{C_{t_1}}\right)}$$

$$GC = \frac{C_{t_2} - C_{t_1}}{t_2 - t_1} \times \left(1 + \frac{\ln\left(\frac{APE_{t_2}}{APE_{t_1}}\right)}{\ln\left(\frac{C_{t_2}}{C_{t_1}}\right)} \right)$$

where C_{t1}, C_{t2}, APE_{t1} and APE_{t2} are the concentrations of N (μg N g^{-1} soil dry weight) and atom percent excess of ^{15}N (APE, in %) in the respective target N pool at t_1 and t_2 (days), i.e. the times of IPD sample termination. APE was calculated as the difference between the at%^{15}N of the given N pool at time t and the background ^{15}N abundance (at%^{15}N) measured in separate non-labeled samples. Gross rates are therefore expressed in μg N g^{-1} soil dry weight d^{-1}.

Extraction of nucleic acids and quantitative PCR analyses. Nucleic acids were extracted from soil samples (0.5 g fresh weight) using a modification of the method described by Griffiths et al. 2000 [98]. Samples were loaded in 2 mL Lysing Matrix E tubes (MP Biomedicals, Illkirch, France) and cells were disrupted by bead-beating with 0.5 mL of an SDS-based lysis buffer (1% SDS, 0.7 M NaCl, 0.1 M $Na_2(SO_3)$, 0.05 M EDTA pH8 and 0.1 M Tris/HCl pH 7.5) and 0.5 mL of a phenol:-chloroform:isoamyl-alcohol (25:24:1) mixture (pH 7.5). Precipitation of nucleic acids was performed for 1 h at room temperature with 20 μg glycogen (MBI Fermentas, Vienna, Austria), 0.5 mL isopropanol and 40 μL 5 M NaCl. Precipitated nucleic acids were washed twice in 500 μL ice-cold 70% ethanol, re-suspended in 50 μL DEPC-treated water and purified by filtration through PVPP mini-columns centrifuged at 1000×*g* for 2 min at 10°C. Purity and concentration of purified nucleic acids was checked spectrophotometrically using a NanoDrop spectrophotometer (Thermo scientific) and their integrity was determined by agarose gel electrophoresis. A SybrGreen I quantitative PCR approach was used to quantify thaumarchaeal and bacterial *amoA* gene abundance. Thaumarchaeal *amoA* genes were amplified with PCR primers Arch-amoA-104F/Arch-amoA-616R [99] and bacterial *amoA* genes with primers amoA-1F/amoA-2R [100]. Duplicate dilution series (10^2–10^7 copies) of PCR products obtained with M13F and M13R vector primers from Topo® TA cloning® vectors (Invitrogen) containing either a 632 bp fragment of *Nitrososphaera viennensis* EN76 [101] *amoA* gene, obtained with PCR primers Arch-amoA-7F/Arch-amoA-638R [99], or a 491 bp fragment of *Nitrosospira multiformis* ATCC 25196 *amoA1* gene (Nmul_A0799/Nmul_A2325), obtained with primers amoA-1F/ amoA-2R, were used as standards for the thaumarchaeal and bacterial *amoA* assays, respectively. Thaumarchaeal *amoA* standards were amplified with an efficiency of 93% and r^2 value of 0.997 and bacterial *amoA* standards with an efficiency of 93% and r^2 value of 0.991. All reactions (20 μL) contained 10 μL of 2× Quantifast™ SYBR® Green PCR Master Mix (Qiagen, Vienna, Austria), 0.2 mg.mL^{-1} BSA, 1 (Archaea) or 0.8 (Bacteria) μM of each primer and for the analysis of soil samples 10 ng of purified nucleic acid extracts. Amplification conditions were 95°C for 15 min, followed by 40 cycles of 95°C for 15 s, 60°C for 30 s (combined annealing and extension step) and plate read at 78°C for thaumarchaeal *amoA* genes and 35 cycles of 95°C for 15 s, 62°C for 30 s and plate read at 82°C for bacterial *amoA* genes, before a final extension step at 60°C for 10 min. Reactions were carried out in a Mastercycler® RealPlex^2 thermocycler (Eppendorf Austria GmbH, Vienna, Austria) and the specificity of the amplification was assessed by melting curve analysis and agarose gel electrophoresis.

Statistical analyses. Mixed between-within subjects ANOVAs (split-plot ANOVA) followed by Tukey HSD post hoc tests were performed in SPSS version 16.0 for Windows (SPSS Inc., USA) to test for the effects of treatment and sampling time on biological and physicochemical soil parameters. If normality (Shapiro-Wilk test) and variance homogeneity (Levene's test) were violated, data were log, square root or reciprocal transformed to meet these assumptions. Mauchly's test was used to test the assumption of sphericity and if significant, the degrees of freedom of the F-distribution were corrected using the Greenhouse-Geisser procedure as the estimated epsilon was always less than 0.75. Biochar effects on pools and processes that were only measured once were tested by oneway ANOVA. Regressions and correlations were calculated with Statgraphics Centurion XVI (StatPoint Technologies Inc., USA). Simple least squares linear regressions were calculated to estimate the amount dependency of biochar effects, with added biochar amount as independent variable. Correlations between microbial community structure parameters, or with nitrification rates were performed based on the Pearson product-moment correlation procedure.

Supporting Information

Figure S1 Structure of beech wood biochar shown by SEM analysis. Cross section across early to late wood transect in beech wood charcoal (top left), longitudinal section through beech charcoal showing open xylem vessels and xylem parenchyma cells (top right), close-up of xylem pits (bottom left) and close-up of a plasmodesmata channel (bottom right).

Figure S2 Scheme of the workflow for soil sample preparations to measure organic and inorganic N transformation rates using ^{15}N isotope pool dilution approaches.

Table S1 Results of ANOVA of soil physicochemical factors and gross nitrification rates for the factors treatment and sampling date for all four treatments.

Table S2 Soil sampling dates, treatments sampled and measurements done on these samples.

Acknowledgments

The setting up and maintenance of the field experiment by the AIT Tulln team is greatly acknowledged. All isotope analyses were run at the SILVER laboratory, Department of Microbiology and Ecosystem Science, University of Vienna, with technical assistance of Margarete Watzka. We further thank the local farmer and landowner, Rudolf Hofmann, Traismauer, Austria for allowing us to setup this biochar project.

Author Contributions

Conceived and designed the experiments: WW RCHN GS. Performed the experiments: JP DT PO FH SS. Analyzed the data: JP WW RCHN DT PO FH SS. Contributed reagents/materials/analysis tools: TU CS BK GS. Wrote the paper: JP WW RCHN PO.

References

1. Bruun EW, Muller-Stover D, Ambus P, Hauggaard-Nielsen H (2011) Application of biochar to soil and N2O emissions: potential effects of blending fast-pyrolysis biochar with anaerobically digested slurry. European Journal of Soil Science 62: 581–589.
2. van Zwieten L, Kimber S, Morris S, Downie A, Berger E, et al. (2010) Influence of biochars on flux of N2O and CO2 from Ferrosol. Australian Journal of Soil Research 48: 555–568.
3. Cayuela ML, Sanchez-Monedero MA, Roig A, Hanley K, Enders A, et al. (2013) Biochar and denitrification in soils: when, how much and why does biochar reduce N2O emissions? Scientific Reports 3.
4. Wang JY, Pan XJ, Liu YL, Zhang XL, Xiong ZQ (2012) Effects of biochar amendment in two soils on greenhouse gas emissions and crop production. Plant and Soil 360: 287–298.
5. Schmidt MWI, Noack AG (2000) Black carbon in soils and sediments: Analysis, distribution, implications, and current challenges. Global Biogeochemical Cycles 14: 777–793.
6. Baldock JA, Smernik RJ (2002) Chemical composition and bioavailability of thermally, altered *Pinus resinosa* (Red Pine) wood. Organic Geochemistry 33: 1093–1109.
7. Kuzyakov Y, Subbotina I, Chen HQ, Bogomolova I, Xu XL (2009) Black carbon decomposition and incorporation into soil microbial biomass estimated by C-14 labeling. Soil Biology & Biochemistry 41: 210–219.
8. Clough TJ, Condron LM, Kammann C, Müller C (2013) A review of biochar and soil nitrogen dynamics. Agronomy 3: 275–293.
9. Joseph SD, Camps-Arbestain M, Lin Y, Munroe P, Chia CH, et al. (2010) An investigation into the reactions of biochar in soil. Australian Journal of Soil Research 48: 501–515.
10. Chen Y, Shinogi Y, Taira M (2010) Influence of biochar use on sugarcane growth, soil parameters, and groundwater quality. Australian Journal of Soil Research 48: 526–530.
11. Ventura M, Sorrenti G, Panzacchi P, George E, Tonon G (2013) Biochar Reduces Short-Term Nitrate Leaching from A Horizon in an Apple Orchard. Journal of Environmental Quality 42: 76–82.
12. Bruun EW, Ambus P, Egsgaard H, Hauggaard-Nielsen H (2012) Effects of slow and fast pyrolysis biochar on soil C and N turnover dynamics. Soil Biology & Biochemistry 46: 73–79.
13. Zheng H, Wang Z, Deng X, Herbert S, Xing BS (2013) Impacts of adding biochar on nitrogen retention and bioavailability in agricultural soil. Geoderma 206: 32–39.
14. Nelissen V, Rütting T, Huygens D, Staelens J, Ruysschaert G, et al. (2012) Maize biochars accelerate short-term soil nitrogen dynamics in a loamy sand soil. Soil Biology & Biochemistry 55: 20–27.
15. Dempster DN, Jones DL, Murphy DV (2012) Organic nitrogen mineralisation in two contrasting agro-ecosystems is unchanged by biochar addition. Soil Biology & Biochemistry 48: 47–50.
16. Anderson CR, Condron LM, Clough TJ, Fiers M, Stewart A, et al. (2011) Biochar induced soil microbial community change: Implications for biogeochemical cycling of carbon, nitrogen and phosphorus. Pedobiologia 54: 309–320.
17. Dempster DN, Jones DL, Murphy DV (2012) Clay and biochar amendments decreased inorganic but not dissolved organic nitrogen leaching in soil. Soil Research 50: 216–221.
18. Taghizadeh-Toosi A, Clough TJ, Sherlock RR, Condron LM (2012) A wood based low-temperature biochar captures NH3-N generated from ruminant urine-N, retaining its bioavailability. Plant and Soil 353: 73–84.
19. Vaccari FP, Baronti S, Lugato E, Genesio L, Castaldi S, et al. (2011) Biochar as a strategy to sequester carbon and increase yield in durum wheat. European Journal of Agronomy 34: 231–238.
20. Jones DL, Rousk J, Edwards-Jones G, DeLuca TH, Murphy DV (2012) Biochar-mediated changes in soil quality and plant growth in a three year field trial. Soil Biology & Biochemistry 45: 113–124.
21. Ma L, Hou Z, Lv N, Ye J, Su S, et al. (2012) Effects of biochar application on wheat growth and nitrogen balance. Xinjiang Agricultural Sciences 49: 589–594.
22. Glaser B, Haumaier L, Guggenberger G, Zech W (2001) The 'Terra Preta' phenomenon: a model for sustainable agriculture in the humid tropics. Naturwissenschaften 88: 37–41.
23. DeLuca TH, MacKenzie MD, Gundale MJ (2009) Biochar Effects on Soil Nutrient Transformations. In: Lehmann J, Joseph S, editors. Biochar for Environmental Management: Science and Technology. London: Earthscan. 251–270.
24. Luo Y, Durenkamp M, De Nobili M, Lin Q, Brookes PC (2011) Short term soil priming effects and the mineralisation of biochar following its incorporation to soils of different pH. Soil Biology & Biochemistry 43: 2304–2314.
25. Fontaine S, Mariotti A, Abbadie L (2003) The priming effect of organic matter: a question of microbial competition? Soil Biology & Biochemistry 35: 837–843.
26. Ball PN, MacKenzie MD, DeLuca TH, Holben WE (2010) Wildfire and Charcoal Enhance Nitrification and Ammonium-Oxidizing Bacterial Abundance in Dry Montane Forest Soils. Journal of Environmental Quality 39: 1243–1253.
27. Liu X, Zhang A, Ji C, Joseph S, Bian R, et al. (2013) Biochar's effect on crop productivity and the dependence on experimental conditions–a meta-analysis of literature data. Plant and Soil DOI 10.1007/s11104-013-1806-x.
28. Jeffery S, Verheijen FGA, van der Velde M, Bastos AC (2011) A quantitative review of the effects of biochar application to soils on crop productivity using meta-analysis. Agriculture Ecosystems & Environment 144: 175–187.
29. Schimel JP, Bennett J (2004) Nitrogen mineralization: Challenges of a changing paradigm. Ecology 85: 591–602.
30. Barraclough D (1991) The Use of Mean Pool Abundances to Interpret N-15 Tracer Experiments.1. Theory. Plant and Soil 131: 89–96.
31. Murphy DV, Recous S, Stockdale EA, Fillery IRP, Jensen LS, et al. (2003) Gross nitrogen fluxes in soil: Theory, measurement and application of N-15 pool dilution techniques. Advances in Agronomy, Vol 79 79: 69–118.
32. Wanek W, Mooshammer M, Blochl A, Hanreich A, Richter A (2010) Determination of gross rates of amino acid production and immobilization in decomposing leaf litter by a novel N-15 isotope pool dilution technique. Soil Biology & Biochemistry 42: 1293–1302.
33. Mooshammer M, Wanek W, Schnecker J, Wild B, Leitner S, et al. (2012) Stoichiometric controls of nitrogen and phosphorus cycling in decomposing beech litter. Ecology 93: 770–782.
34. Gryc V, Vavrcik H, Gomola S (2008) Selected properties of European beech (*Fagus sylvatica* L.). Journal of Forest Science 54: 418–425.
35. Jones DL, Healey JR, Willett VB, Farrar JF, Hodge A (2005) Dissolved organic nitrogen uptake by plants - an important N uptake pathway? Soil Biology & Biochemistry 37: 413–423.
36. Kelleher BP, Simpson MJ, Simpson AJ (2006) Assessing the fate and transformation of plant residues in the terrestrial environment using HR-MAS NMR spectroscopy. Geochimica et Cosmochimica Acta 70: 4080–4094.
37. Farrell M, Hill PW, Farrar J, DeLuca TH, Roberts P, et al. (2013) Oligopeptides represent a preferred source of organic N uptake: A global phenomenon? Ecosystems 16: 133–145.
38. Kasozi GN, Zimmerman AR, Nkedi-Kizza P, Gao B (2010) Catechol and Humic Acid Sorption onto a Range of Laboratory-Produced Black Carbons (Biochars). Environmental Science & Technology 44: 6189–6195.
39. Bergeron SP, Bradley RL, Munson A, Parsons W (2013) Physico-chemical and functional characteristics of soil charcoal produced at five different temperatures. Soil Biology & Biochemistry 58: 140–146.
40. Haslam E (1974) Polyphenol-Protein Interactions. Biochemical Journal 139: 285–288.
41. Mukherjee A, Zimmerman AR, Harris W (2011) Surface chemistry variations among a series of laboratory-produced biochars. Geoderma 163: 247–255.
42. Zimmerman AR, Chorover J, Goyne KW, Brantley SL (2004) Protection of Mesopore-Adsorbed Organic Matter from Enzymatic Degradation. Environmental Science and Technology 38: 4542–4548.
43. Burns RG, DeForest JL, Marxsen J, Sinsabaugh RL, Stromberger ME, et al. (2013) Soil enzymes in a changing environment: Current knowledge and future directions. Soil Biology & Biochemistry 58: 216–234.
44. Bailey VL, Fansler SJ, Smith JL, Bolton H Jr (2011) Reconciling apparent variability in effects of biochar amendment on soil enzyme activities by assay optimization. Soil Biology & Biochemistry 43: 296–301.
45. Mullen CA, Boateng AA, Goldberg NM, Lima IM, Laird DA, et al. (2010) Bio-oil and bio-char production from corn cobs and stover by fast pyrolysis. Biomass & Bioenergy 34: 67–74.
46. Cheng Y, Cai Z-c, Chang SX, Wang J, Zhang J-b (2012) Wheat straw and its biochar have contrasting effects on inorganic N retention and N2O production in a cultivated Black Chernozem. Biology And Fertility Of Soils 48: 941–946.
47. Wang T, Arbestain MC, Hedley M, Bishop P (2012) Chemical and bioassay characterisation of nitrogen availability in biochar produced from dairy manure and biosolids. Organic Geochemistry 51: 45–54.
48. Singh BP, Cowie A, Smernik RJ (2012) Biochar carbon stability in a clayey soil as a function of feedstock and pyrolysis temperature. Environmental Science & Technology 46: 11770–11778.
49. DeLuca TH, MacKenzie MD, Gundale MJ, Holben WE (2006) Wildfire-produced charcoal directly influences nitrogen cycling in ponderosa pine forests. Soil Science Society of America Journal 70: 448–453.
50. Singh B, Singh BP, Cowie AL (2010) Characterisation and evaluation of biochars for their application as a soil amendment. Australian Journal of Soil Research 48: 516–525.
51. Prosser JI, Nicol GW (2012) Archaeal and bacterial ammonia-oxidisers in soil: the quest for niche specialisation and differentiation. Trends in Microbiology 20: 523–531.
52. Stienstra AW, Gunnewiek PK, Laanbroek HJ (1994) Repression of Nitrification in Soils under A Climax Grassland Vegetation. Fems Microbiology Ecology 14: 45–52.
53. Berglund LM, DeLuca TH, Zackrisson O (2004) Activated carbon amendments to soil alters nitrification rates in Scots pine forests. Soil Biology and Biochemistry 36: 2067–2073.
54. Gundale MJ, DeLuca TH (2006) Temperature and source material influence ecological attributes of ponderosa pine and Douglas-fir charcoal. Forest Ecology and Management 231: 86–93.

55. Clough TJ, Condron LM (2010) Biochar and the Nitrogen Cycle: Introduction. Journal of Environmental Quality 39: 1218–1223.
56. Booth MS, Stark JM, Rastetter E (2005) Controls on nitrogen cycling in terrestrial ecosystems: A synthetic analysis of literature data. Ecological Monographs 75: 139–157.
57. Nakajima Y, Ishizuka S, Tsuruta H, Iswandi A, Murdiyarso D (2005) Microbial processes responsible for nitrous oxide production from acid soils in different land-use patterns in Pasirmayang, central Sumatra, Indonesia. Nutrient Cycling in Agroecosystems 71: 33–42.
58. Wrage N, Velthof GL, van Beusichem ML, Oenema O (2001) Role of nitrifier denitrification in the production of nitrous oxide. Soil Biology & Biochemistry 33: 1723–1732.
59. Wang L, Cai Z (2008) Nitrous oxide production at different soil moisture contents in an arable soil in China. Soil Science And Plant Nutrition 54: 786–793.
60. Bateman EJ, Baggs EM (2005) Contributions of nitrification and denitrification to N2O emissions from soils at different water-filled pore space. Biology And Fertility Of Soils 41: 379–388.
61. Offre P, Prosser JI, Nicol GW (2009) Growth of ammonia-oxidizing archaea in soil microcosms is inhibited by acetylene. Fems Microbiology Ecology 70: 99–108.
62. Daum M, Zimmer W, Papen H, Kloos K, Nawrath K, et al. (1998) Physiological and molecular biological characterization of ammonia oxidation of the heterotrophic nitrifier Pseudomonas putida. Current Microbiology 37: 281–288.
63. Hynes RK, Knowles R (1982) Effect of Acetylene on Autotrophic and Heterotrophic Nitrification. Canadian Journal of Microbiology 28: 334–340.
64. Linn DM, Doran JW (1984) Effect of Water-filled Pore-space on Carbon-dioxide and Nitrous-oxide Production in Tilled and Nontilled Soils. Soil Science Society of America Journal 48: 1267–1272.
65. Busscher WJ, Novak JM, Evans DE, Watts DW, Niandou MAS, et al. (2010) Influence of Pecan Biochar on Physical Properties of a Norfolk Loamy Sand. Soil Science 175: 10–14.
66. Case SDC, McNamara NP, Reay DS, Whitaker J (2012) The effect of biochar addition on N2O and CO2 emissions from a sandy loam soil - The role of soil aeration. Soil Biology & Biochemistry 51: 125–134.
67. Schleper C (2010) Ammonia oxidation: different niches for bacteria and archaea? ISME Journal 4: 1092–1094.
68. He J, Shen J, Zhang L, Zhu Y, Zheng Y, et al. (2007) Quantitative analyses of the abundance and composition of ammonia-oxidizing bacteria and ammonia-oxidizing archaea of a Chinese upland red soil under long-term fertilization practices. Environmental Microbiology 9: 2364–2374.
69. Leininger S, Urich T, Schloter M, Schwark L, Qi J, et al. (2006) Archaea predominate among ammonia-oxidizing prokaryotes in soils. Nature 442: 806–809.
70. Dempster DN, Gleeson DB, Solaiman ZM, Jones DL, Murphy DV (2012) Decreased soil microbial biomass and nitrogen mineralisation with Eucalyptus biochar addition to a coarse textured soil. Plant and Soil 354: 311–324.
71. Downie A, Crosky A, Munroe P (2009) Physical properties of biochar. In: Lehmann J, Joseph S, editors. Biochar for Environmental Management: Science and Technology. London, UK: Earthscan.
72. Lehmann J, Rillig MC, Thies J, Masiello CA, Hockaday WC, et al. (2011) Biochar effects on soil biota - A review. Soil Biology & Biochemistry 43: 1812–1836.
73. Mulder C, Cohen JE, Setala H, Bloem J, Breure AM (2005) Bacterial traits, organism mass, and numerical abundance in the detrital soil food web of Dutch agricultural grasslands. Ecology Letters 8: 80–90.
74. Frey SD, Elliott ET, Paustian K (1999) Bacterial and fungal abundance and biomass in conventional and no-tillage agroecosystems along two climatic gradients. Soil Biology & Biochemistry 31: 573–585.
75. Mulder C, Van Wijnen HJ, Van Wezel AP (2005) Numerical abundance and biodiversity of below-ground taxocenes along a pH gradient across the Netherlands. Journal of Biogeography 32: 1775–1790.
76. Strong DT, Sale PWG, Helyar KR (1998) The influence of the soil matrix on nitrogen mineralisation and nitrification. II. The pore system as a framework for mapping the organisation of the soil matrix. Australian Journal of Soil Research 36: 855–872.
77. Baath E, Soderstrom B (1979) Significance of Hyphal Diameter in Calculation of Fungal Bio-volume. Oikos 33: 11–14.
78. Smith SE, Read DJ (2000) Mycorrhizal Symbiosis. Amsterdam, Netherlands: Academic Press.
79. Quilliam RS, Glanville HC, Wade SC, Jones DL (2013) Life in the 'charosphere' - Does biochar in agricultural soil provide a significant habitat for microorganisms? Soil Biology & Biochemistry 65: 287–293.
80. Pietikainen J, Kiikkila O, Fritze H (2000) Charcoal as a habitat for microbes and its effect on the microbial community of the underlying humus. Oikos 89: 231–242.
81. Warnock DD, Lehmann J, Kuyper TW, Rillig MC (2007) Mycorrhizal responses to biochar in soil - concepts and mechanisms. Plant and Soil 300: 9–20.
82. Finlay BJ, Black HIJ, Brown S, Clarke KJ, Esteban GF, et al. (2000) Estimating the growth potential of the soil protozoan community. Protist 151: 69–80.
83. Foissner W (1999) Soil protozoa as bioindicators: pros and cons, methods, diversity, representative examples. Agriculture Ecosystems & Environment 74: 95–112.
84. Coleman DC (1994) The Microbial Loop Concept as used in Terrestrial Soil Ecology Studies. Microbial Ecology 28: 245–250.
85. Moore JC, McCann K, Setala H, De Ruiter PC (2003) Top-down is bottom-up: Does predation in the rhizosphere regulate aboveground dynamics? Ecology 84: 846–857.
86. Knicker H (2011) Soil organic N - An under-rated player for C sequestration in soils? Soil Biology & Biochemistry 43: 1118–1129.
87. Jones DL, Edwards-Jones G, Murphy DV (2011) Biochar mediated alterations in herbicide breakdown and leaching in soil. Soil Biology & Biochemistry 43: 804–813.
88. Hood-Nowotny R, Hinko-Najera Umana N, Inselbacher E, Oswald-Lachouani P, Wanek W (2010) Alternative methods for measuring inorganic, organic, and total dissolved nitrogen in soil. Soil Science Society of America Journal 74: 1018–1027.
89. Jones DL, Owen AG, Farrar JF (2002) Simple method to enable the high resolution determination of total free amino acids in soil solutions and soil extracts. Soil Biology & Biochemistry 34: 1893–1902.
90. Roberts P, Jones DL (2008) Critical evaluation of methods for determining total protein in soil solution. Soil Biology & Biochemistry 40: 1485–1495.
91. Di HJ, Cameron KC, McLaren RG (2000) Isotopic dilution methods to determine the gross transformation rates of nitrogen, phosphorus, and sulfur in soil: a review of the theory, methodologies, and limitations. Australian Journal of Soil Research 38: 213–230.
92. Sorensen P, Jensen ES (1991) Sequential Diffusion of Ammonium and Nitrate from Soil Extracts to a Polytetrafluoroethylene Trap for N-15 Determination. Analytica Chimica Acta 252: 201–203.
93. Lachouani P, Frank AH, Wanek W (2010) A suite of sensitive chemical methods to determine the delta N-15 of ammonium, nitrate and total dissolved N in soil extracts. Rapid Communications in Mass Spectrometry 24: 3615–3623.
94. Khan SA, Mulvaney RL, Hoeft RG (2000) Direct-diffusion methods for inorganic-nitrogen analysis of soil. Soil Science Society of America Journal 64: 1083–1089.
95. Phillips DL, Gregg JW (2001) Uncertainty in source partitioning using stable isotopes. Oecologia 127: 171–179.
96. Doyle A, Weintraub MN, Schimel JP (2004) Persulfate digestion and simultaneous colorimetric analysis of carbon and nitrogen in soil extracts. Soil Science Society of America Journal 68: 669–676.
97. Kirkham D, Bartholomew WV (1954) Equations for following nutrient transformations in soil, utilizing tracer data. Soil Science Society of America Proceedings 18: 33–34.
98. Griffiths RI, Whiteley AS, O'Donnell AG, Bailey MJ (2000) Rapid method for coextraction of DNA and RNA from natural environments for analysis of ribosomal DNA- and rRNA-based microbial community composition. Applied and Environmental Microbiology 66: 5488–5491.
99. Alves RJE, Wanek W, Zappe A, Richter A, Svenning MM, et al. (2013) Nitrification rates in Arctic aoils are associated with functionally distinct populations of ammonia-ixidizing archaea. ISME Journal 7: 1620–1631.
100. Rotthauwe JH, Witzel KP, Liesack W (1997) The ammonia monooxygenase structural gene amoA as a functional marker: Molecular fine-scale analysis of natural ammonia-oxidizing populations. Applied and Environmental Microbiology 63: 4704–4712.
101. Tourna M, Stieglmeier M, Spang A, Konneke M, Schintlmeister A, et al. (2011) Nitrososphaera viennensis, an ammonia oxidizing archaeon from soil. Proceedings of the National Academy of Sciences of the United States of America 108: 8420–8425.

4

Dynamics and Sources of Soil Organic C Following Afforestation of Croplands with Poplar in a Semi-Arid Region in Northeast China

Ya-Lin Hu[1]*, Li-Le Hu[2], De-Hui Zeng[1]*

1 State Key Laboratory of Forest and Soil Ecology, Institute of Applied Ecology, Chinese Academy of Sciences, Shenyang, People's Republic of China, **2** Chinese Research Academy of Environmental Sciences, Beijing, People's Republic of China

Abstract

Afforestation of former croplands has been proposed as a promising way to mitigate rising atmospheric CO_2 concentration in view of the commitment to the Kyoto Protocol. Central to this C sequestration is the dynamics of soil organic C (SOC) storage and stability with the development of afforested plantations. Our previous study showed that SOC storage was not changed after afforestation except for the 0–10 cm layer in a semi-arid region of Keerqin Sandy Lands, northeast China. In this study, soil organic C was further separated into light and heavy fractions using the density fractionation method, and their organic C concentration and ^{13}C signature were analyzed to investigate the turnover of old vs. new SOC in the afforested soils. Surface layer (0–10 cm) soil samples were collected from 14 paired plots of poplar (*Populus* × *xiaozhuanica* W. Y. Hsu & Liang) plantations with different stand basal areas (the sum of the cross-sectional area of all live trees in a stand), ranging from 0.2 to 32.6 m^2 ha^{-1}, and reference maize (*Zea mays* L.) croplands at the same sites as our previous study. Soil Δ_C stocks (Δ_C refers to the difference in SOC content between a poplar plantation and the paired cropland) in bulk soil and light fraction were positively correlated with stand basal area ($R^2 = 0.48$, $p < 0.01$ and $R^2 = 0.40$, $p = 0.02$, respectively), but not for the heavy fraction. SOC_{crop} (SOC derived from crops) contents in the light and heavy fractions in poplar plantations were significantly lower as compared with SOC contents in croplands, but tree-derived C in bulk soil, light and heavy fraction pools increased gradually with increasing stand basal area after afforestation. Our study indicated that cropland afforestation could sequester new C derived from trees into surface mineral soil, but did not enhance the stability of SOC due to a fast turnover of SOC in this semi-arid region.

Editor: Ben Bond-Lamberty, DOE Pacific Northwest National Laboratory, United States of America

Funding: This work was supported by the National Natural Science Foundation of China (No. 31000297) (http://www.nsfc.gov.cn). The funders had no role in study design, data collection and analysis, decision to publish, or preparation of the manuscript.

Competing Interests: The authors have declared that no competing interests exist.

* E-mail: huyl@iae.ac.cn (YH); zengdh@iae.ac.cn (DZ)

Introduction

Afforestation of degraded croplands has occurred globally within the framework of the Kyoto Protocol [1], and has the potential to mitigate the rising atmospheric CO_2 concentration caused by anthropogenic emissions [2,3]. Consequently, changes of soil organic C (SOC) stocks after cropland afforestation have been well studied [3–8], considering the fact that soil has the greatest potential C sink capacity and the longest time period [9–11]. However, most previous studies have suggested that there were large variations in the direction and magnitude of SOC stock changes following afforestation of croplands, which are related to previous land-use, climate, soil texture, tree species, stand age, and management practices [3,4,8]. In addition, the changes of SOC stocks are often not detected by conventional methodologies within a short-time frame for most experiments due to the small changes in soil C when compared with the size of SOC reservoir [6,9]. In order to accurately assess small changes in SOC stocks and stability following cropland afforestation, it is necessary to investigate the dynamics in different SOC pools characterized by different physical and chemical properties, microbial degradability and turnover time [9].

Soil organic matter (SOM) can be divided into discrete fractions with different stability and ecological functions [12,13]. Density fractionation physically separates SOC into light fraction and heavy fraction (LF and HF, respectively), which has been increasingly used to assess SOC dynamics induced by land use change and management practices [8,14–17]. However, soil fractionation and C analysis only provide information on net changes of SOC, but not for soil C balance between the loss of old C and the input of new C. The ^{13}C natural abundance technique offers an elegant approach to quantify the relative contribution of new vs. old SOC, for example, where C_3 plants ($\delta^{13}C$ ca. −28‰) grow on soils derived from C_4 crops ($\delta^{13}C$ ca. −12‰) [9]. Several studies have elucidated the dynamics in SOC following cropland afforestation using soil fractionation techniques combined with the stable C isotope techniques [9,18–20], but this approach has not been used in semi-arid regions.

The SOC stocks are determined by the balance between the inputs of C derived from litterfall and rhizodeposition and the losses of C mainly through soil organic matter decomposition [1,21]. Therefore, the dynamics of SOC stocks following afforestation are not only correlated with stand age, but also with tree density that influences soil microclimatic conditions and the amount of litterfall [1,22,23]. Stand basal area (BA, the sum of the cross-sectional area of all live trees in a stand) can integrates information of both stand age and tree density, which is more feasible to evaluate the dynamics and sources of SOC following afforestation considering the important roles of trees in SOC inputs from litter and root exudates.

Understanding the changes in SOC stocks and stability following afforestation in semi-arid regions is important principally because of the vast area involved with 2.4 billion ha or ~17.7% of total global land surface area [24], and the different changes in soil C stocks and stability after afforestation compared with humid regions [25,26]. Poplar (*Populus*) species is one of the most widely grown trees on croplands in the Keerqin Sandy Lands, a semi-arid region in northeastern China, in two large scale afforestation programs i.e., the Three-North Shelter Forest Program and the Grain for Green Project. The net changes in above- and below-ground C stocks after afforestation have been well investigated in the Keerqin Sandy Lands [27–29]. In this present study, the objectives are to further evaluate the dynamics and sources of SOC in different soil C pools along BA following afforestation of croplands with poplars in this semi-arid region. We hypothesized that: (1) SOC stocks in bulk soil and light fraction would increase following cropland afforestation due to the enhanced inputs of litterfall with increasing BA, and (2) the sources of SOC would gradually convert from crop-derived sources to tree-derived sources because of the decomposition of old soil C and the accumulation of new soil C. To test these hypotheses, we selected 14 paired stands of poplar plantations with different BAs afforested on croplands and adjacent maize (*Zea mays* L.) croplands as controls in the southeastern region of Keerqin Sandy Lands, and then analyzed SOC concentrations and $\delta^{13}C$ values in bulk soil, LF and HF.

Materials and Methods

2.1 Ethics Statement

This study was carried out on collective-owned lands, and the owners of the lands gave us permission to conduct the study on these stands. The field studies did not involve endangered or protected species.

2.2 Site Description and Experimental Design

This study was carried out in the southeastern region of the Keerqin Sandy Lands (42°30′–42°55′N, 122°19′–122°30′E), a typical semi-arid region in northeast China. The climate is temperate continental monsoon. Mean annual temperature is about 5.7°C, ranging from −23.2°C in January to 32.4°C in July (1954–2004). Mean annual precipitation is about 450 mm (ranging from 224 mm to 661 mm during 1954–2004), with more than 60% occurring from June to August, and mean annual potential evaporation ranges from 1300 to 1800 mm with an average length of frost-free season of about 150 days. The soil is a sandy soil with 90.9% sand, 5.0% silt and 4.1% clay, and classified into the Entisol order, Semiaripsamment group (according to the United States Soil Classification System) and developed from sandy parent material through the action of wind [30]. Before croplands are established in this region, the dominant species of the native vegetation include *Agropyron cristatum*, *Arundinella hirta*, *Cleistogenes chinensis*, *Lespedeza davurica* and *Artemisia capillaris* var. *simplex* [31].

Since 1978, a large area of marginal croplands has been afforested with trees in this region under the Three-North Shelter Forest Program and the Grain for Green Project, in order to control windy erosion and desertification. By now, the wind erosion is effectively reduced. In June 2011, we selected 14 poplar (*Populus* × *xiaozhuanica* W. Y. Hsu & Liang, a hybrid of *P. nigra* var. *italica* and *P. simonii*) plantations (ranging from 2 to 20 years old) afforested on maize croplands and 14 adjacent maize cropland stands as control in Kezuohouqi and Zhangwu counties on the basis of a paired-plot experimental design (Table 1). Most of the paired cropland and plantation stands are conterminous except for several cases, and the distance of sampling point in each paired poplar plantation and cropland is less than 500 m. The topography and soil conditions are similar in each paired cropland and poplar plantation, and the slope of each stand is very gentle and less than 5°. All the poplar plantation stands were planted on maize croplands that had cultivated for at least 20 years before afforestation, and the paired croplands were continually planted to maize. Usually, croplands are fertilized with urea fertilizer each year, while fertilizer is no longer used after afforestation.

2.3 Stand Investigation and Soil Sampling

One 20 × 20 m plot was established in each stand. For poplar plantations, the diameter at breast height (DBH) and tree height were measured for all live trees in each plot. DBH was measured at breast level (1.3 m above ground) using a caliper. For measurement of tree height, we used a long pole to extend vertically to the top of the tree and then measured the length of the pole. The BA (m^2 ha^{-1}) was calculated from measurements of the DBH (cm) of all trees in a known area (A, ha), which was expressed as:

$$BA = \frac{\pi}{4 \times 10000} \times \frac{\sum DBH^2}{A} \quad (1)$$

Considering the changes of soil organic C stocks were only observed in 0–10 cm layer in our previous study [28] at the same sites, soil samples in the surface layer were only collected in this present study. Four soil cores were sampled randomly using an auger (2.5 cm in diameter) at the surface mineral soil layers (0–10 cm), and thoroughly mixed to form a homogenized sample for each stand (i.e., a total of 28 samples including 14 samples from poplar plantations and 14 samples from croplands). Soil samples were dried at room temperature (20 °C) and then passed through a mesh sieve with a size of 2 mm. Soil bulk density (ρ) was determined at three randomized sampling points in each plot for calculation of SOC content. For measurement of soil bulk density, a metal corer (volume is 100 cm^3) was driven into the soil at the desired depth, and then soil samples were oven dried at 115°C for 24 h and weighed. Soil bulk density was calculated as:

$$\rho = \frac{M}{V} \quad (2)$$

where M is dry mass of soil and V is volume of soil (i.e., 100 cm^3).

2.4 Soil Density Fractionation

A soil sample was physically separated into two pools by the modified density fractionation method of Sohi et al. [32]. Briefly, 10 g of air-dried soil (<2 mm) were placed in a centrifuge tube with 40 mL sodium iodide (NaI) solution at a density of 1.7 g cm^{-3}. The tubes were shaken up and down by hand for ten times. The release of light fraction was accelerated by sonication at 58 Watts for 180 s using a sonicator (Bilon96, Bilon Instruments Co., Ltd, China). After sonication was finished, the tubes were

Table 1. Stand location and characteristics.

Polar plantation stand							Cropland stand		
Plot	**Location**	**Elevation (m)**	**Tree height (m)**	**DBH† (cm)**	**Density (Trees ha^{-1})**	**Stand basal area (m^2 ha^{-1})**	**Plot**	**Location**	**Elevation (m)**
F1	42°54′11″N, 122°23′30″E	252	3.35	2.92	1100	0.85	C1‡	42°54′04″N, 122°23′32″E	255
F2	42°37′06″N, 122°20′54″E	166	4.38	4.62	1075	1.85	C2	42°37′04″N, 122°20′52″E	166
F3	42°54′00″N, 122°23′25″E	252	4.80	6.35	850	2.92	C3	42°53′59″N, 122°23′23″E	251
F4	42°53′04″N, 122°24′08″E	250	5.26	5.73	1275	3.65	C4	42°52′58″N, 122°24′09″E	250
F5	42°54′02″N, 122°25′32″E	247	6.85	7.39	1300	6.11	C5	42°54′00″N, 122°25′34″E	246
F6	42°53′18″N, 122°24′11″E	251	8.77	9.21	1100	8.20	C6	42°53′17″N, 122°24′07″E	252
F7	42°59′25″N, 122°20′25″E	244	8.95	11.58	825	9.24	C7	42°59′26″N, 122°20′20″E	244
F8	42°37′58″N, 122°22′24″E	186	15.80	13.37	750	10.69	C8	42°38′01″N, 122°22′24″E	185
F9	42°54′13″N, 122°23′36″E	252	15.20	15.10	700	13.15	C9	42°54′17″N, 122°23′41″E	251
F10	42°53′59″N, 122°25′10″E	249	18.70	10.54	1575	14.62	C10	42°53′56″N, 122°25′07″E	248
F11	42°55′42″N, 122°24′39″E	246	17.76	13.16	1025	15.69	C11	42°55′42″N, 122°24′37″E	246
F12	42°37′06″N, 122°20′58″E	165	30.72	22.00	625	24.79	C12	42°37′04″N, 122°20′52″E	165
F13	42°54′25″N, 122°23′28″E	250	20.68	17.11	1150	28.51	C13	42°54′23″N, 122°23′28″E	250
F14	42°37′05″N, 122°21′06″E	164	29.87	22.90	775	32.60	C14	42°37′06″N, 122°21′05″E	165

†DBH: Diameter at breast height (1.3 m); an average value for all living trees in each plot.
‡All croplands were planted to maize.

centrifuged at 6000 rpm for 20 min. The floating material was aspirated together with the NaI solution from the surface of tubes, and then filtered using Whatman GF/A filter papers. This procedure was repeated three times. The material collected on the filter paper (light fraction, LF) and the residue remaining in the centrifuge tube (heavy fraction, HF) were rinsed thoroughly with deionized water and collected. The samples of LF and HF were dried at 60°C for 48 h.

2.5 SOC Concentration and ^{13}C Analysis

The samples of bulk soil, LF and HF were ground to a fine powder with a ball mill and analyzed for SOC concentration and $\delta^{13}C$. SOC concentration was determined using the Walkey and Black $K_2Cr_2O_7$–H_2SO_4 oxidation method [33]. The isotope ratio $^{13}C/^{12}C$ was determined using isotope ratio mass spectrometer (Finnigan DELTAplusXP, Thermo Fisher Scientific, USA), and the ^{13}C abundance was expressed in delta-units ($\delta^{13}C$, ‰) according to the following equation:

$$\delta^{13}C = [(R_{sam}/R_{std}) - 1] \times 1000 \tag{3}$$

where R_{sam} is the $^{13}C/^{12}C$ ratio of soil sample and R_{std} is the $^{13}C/^{12}C$ ratio of the international Pee Dee formation belemnite carbonate standard (PDB).

2.6 Data Calculation and Statistical Analysis

Soil organic C content of bulk soil, LF and HF was calculated as:

$$SOC_{cont} = c \times \rho \times d \times r \tag{4}$$

where SOC_{cont} was soil organic C content of bulk soil, LF or HF; c was SOC concentration of bulk soil, LF or HF; ρ was soil bulk density; d was soil depth (i.e. 10 cm); and r was the dry mass ratio of LF or HF to bulk soil.

We estimated the sources of SOC in the poplar plantations based on an isotope mass balance and $\delta^{13}C$ values, and the fractional tree-derived SOC ($F_{tree\text{-}C}$) was calculated using a two-component mixing equation [34]:

$$F_{Tree-C}(\%) = \frac{\delta^{13}C_{poplar} - \delta^{13}C_{crop}}{\delta^{13}C_{tree} - \delta^{13}C_{crop}} \times 100\% \tag{5}$$

where $\delta^{13}C_{poplar}$ and $\delta^{13}C_{crop}$ values are actual measured values in bulk soil, LF or HF in poplar plantation and its paired cropland, respectively. The $\delta^{13}C_{tree}$ value is the measured $\delta^{13}C$ of poplar leaf litter (−29.63‰). Subsequently, the mass of tree-derived SOC (SOC_{tree}) and crop-derived SOC (SOC_{crop}) were calculated as:

$$SOC_{tree} = F_{Tree-C} \times SOC_{cont} \tag{6}$$

$$SOC_{crop} = SOC_{cont} - SOC_{tree} \tag{7}$$

All statistical analyses were done using the open source statistical software R version 2.14.1. Paired t tests were used to examine the changes of SOC concentration and SOC content in bulk soil, LF and HF between poplar plantations and the paired croplands. To test the dynamics and sources of SOC after afforestation, the relationships of BA and SOC content, and BA and $\delta^{13}C$ of bulk soil, LF and HF in poplar plantation stands were tested with a linear regression analysis. The significance level was set at $\alpha = 0.05$ for all the statistical analyses unless otherwise noted.

Results

3.1 SOC in Bulk Soil, Light Fraction and Heavy Fraction

Across all popular plantation stands, the average SOC concentration and SOC content in bulk soil were 64% and 54% higher than that across croplands, respectively (all $p<0.001$)

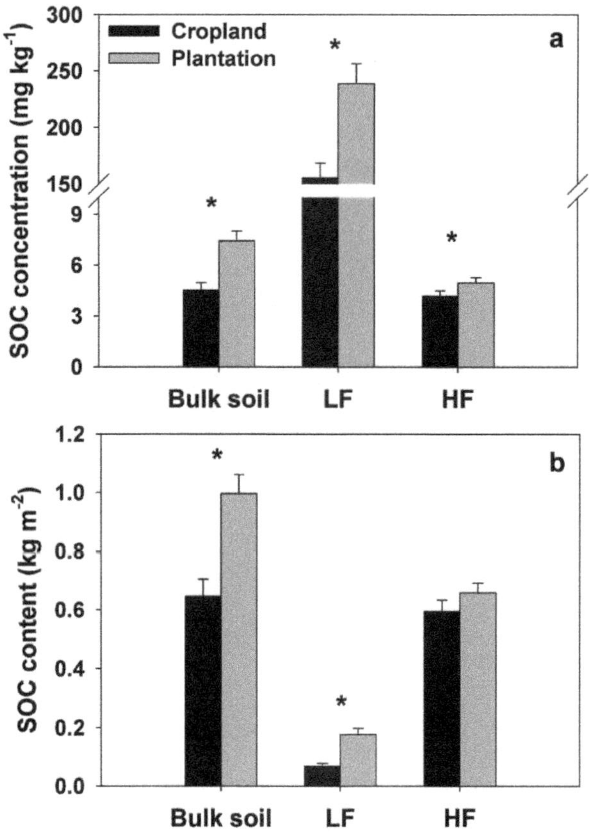

Figure 1. Averages of SOC concentration (a) and content (b) of bulk soil, LF and HF in croplands and poplar plantations. The asterisk above the bars indicates a significant difference between cropland and poplar plantation at $\alpha = 0.05$ level. The vertical error bars are standard errors of the means ($n = 14$).

(Fig. 1). SOC concentrations in bulk soil ranged from 4.67 to 12.50 g kg^{-1} in poplar plantations, and from 2.39 to 7.28 g kg^{-1} in croplands. SOC contents had a range from 0.67 to 1.54 kg C m^{-2} in poplar plantations, and a range from 0.38 to 1.06 kg C m^{-2} in cropland stands.

SOC concentrations in LF and HF in poplar plantations (an average of 238 g kg^{-1} in LF and 4.97 g kg^{-1} in HF, respectively) were also significantly higher than that in croplands (an average of 155 g kg^{-1} in LF and 4.19 g kg^{-1} in HF, respectively) (Fig. 1a). SOC was mainly stored in the HF pools in both poplar plantations and croplands. Average values of SOC content in LF and HF were 0.18 and 0.66 kg m^{-2} in poplar plantations, and 0.07 and 0.59 kg m^{-2} in croplands, respectively. SOC content in the LF of poplar plantations was significantly higher than that in croplands ($p<0.001$), but not for HF ($p = 0.12$) (Fig. 1b).

3.2 Changes of SOC in Bulk Soil, LF and HF with BA

Soil Δ_C stocks (Δ_C refers to the difference in SOC content between a poplar plantation and the paired cropland) in bulk soil had a linear increase trend with increasing BA (Fig. 2). Similarly, there was a significant positive correlation between soil Δ_C stocks in LF and BA, but not for HF.

3.3 Soil $\delta^{13}C$ in Poplar Plantations

Soil $\delta^{13}C$ values in bulk soil and LF were negatively correlated with BA (Fig. 3). Soil $\delta^{13}C$ in bulk soil was depleted from −21‰ (BA was 0.85 m^2 ha^{-1}) to −27‰ (BA was 28.51 m^2 ha^{-1}), and soil $\delta^{13}C$ in LF was depleted from −22‰ (BA was 0.85 m^2 ha^{-1}) to −30‰ (BA was 28.51 m^2 ha^{-1}). However, the relationship between soil $\delta^{13}C$ in HF and BA was not significant ($p = 0.18$). Soil $\delta^{13}C$ in HF (average value of −24‰) were significantly higher than that in LF (an average value of −28‰) in poplar plantations ($p<0.001$).

SOC_{crop} (SOC derived from crops) content in bulk soil of poplar plantations was slightly lower as compared with SOC content in croplands ($p = 0.17$), while SOC_{crop} contents in LF and HF were 51% and 27% lower, respectively (all $p<0.05$) (Fig. 4). SOC_{tree} (SOC derived from poplar trees) contents in bulk soil, LF and HF were all increased significantly with increasing BA (Fig. 5). SOC_{tree} contents in bulk soil ranged from 0.08 kg m^{-2} in the poplar stand (BA was 0.85 m^2 ha^{-1}) to 0.93 kg m^{-2} in the poplar stand (BA was 32.6 m^2 ha^{-1}). The percentages of SOC_{tree} to total SOC content in poplar plantations were on average 40% in bulk soil, 77% in LF, and 33% in HF.

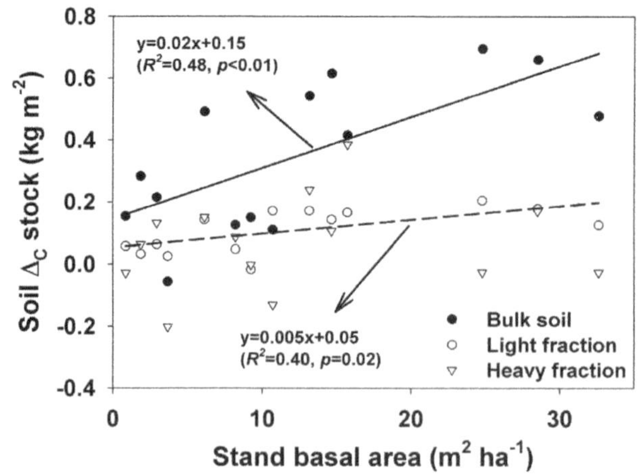

Figure 2. Dynamics of soil Δ_C stocks in bulk soil, LF and HF along stand basal area in poplar plantation stands. Solid line is regression for the bulk soil, and dash line is regression for light fraction.

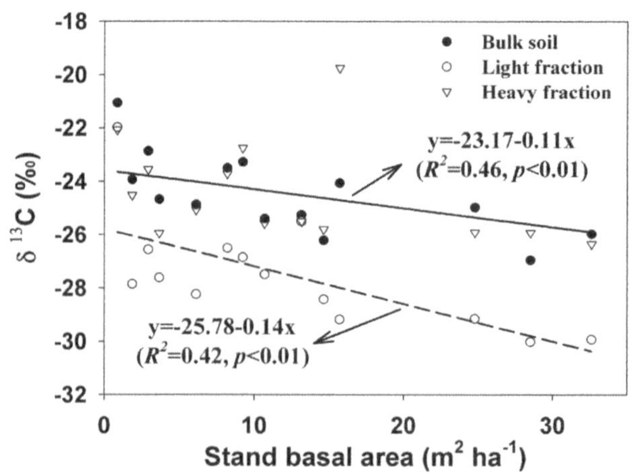

Figure 3. Dynamics of soil $\delta^{13}C$ in bulk soil, LF and HF along stand basal area in poplar plantations. Solid line is regression for bulk soil, and dash line is regression for light fraction.

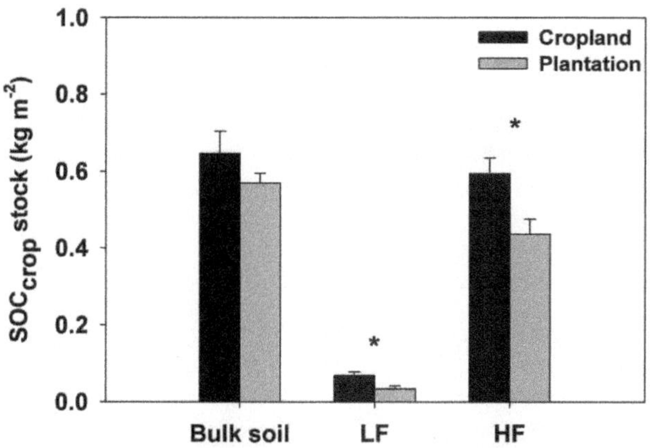

Figure 4. Averages of SOC_{crop} contents in bulk soil, LF and HF in croplands and poplar plantations. The asterisk above the vertical bars indicates a significant difference between cropland and poplar plantation at $\alpha = 0.05$ level. The vertical error bars are standard errors of the means ($n = 14$).

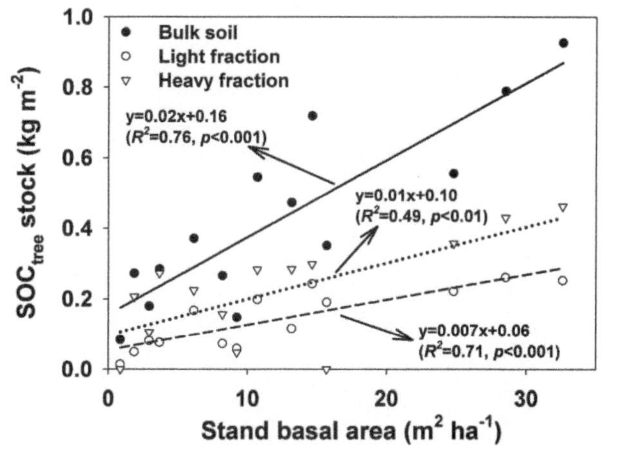

Figure 5. Dynamics of SOC_{tree} contents in bulk soil, LF and HF along stand basal area in poplar plantations. Solid line is regression for bulk soil, and dash line is regression for light fraction, and dot line is regression for heavy fraction.

Discussion

4.1 Changes of SOC in Bulk Soil, LF and HF

In this study, the afforested soil had higher SOC content than croplands (an average of 0.35 kg C m^{-2}) in the surface soils, implying that afforestation with hybrid poplar on croplands could sequester 12.8 Mg CO_2 ha^{-1} into surface mineral soils in this semi-arid region. Similarly, Mao et al. [29] observed that SOC sequestration in 1 m depth was 1.9 kg C m^{-2} in a 20-year-old poplar plantation that was afforested on the marginal agricultural land. Li et al. [35] reported that soil sequestered 0.39 kg C m^{-2} in 0–15 cm depth in a 25-year-old Mongolian pine (*Pinus sylvestris* var. *mongolica*) afforested on active sand dune in the Keerqin Sandy Lands. Increased SOC stocks induced by afforestation with poplar trees on croplands were also reported in other regions [36–38]. When land use change from crop to forest, an average increase of 18% in SOC stock was observed, and afforested broadleaved tree species were more effective in sequestering CO_2 into soils [3].

Furthermore, this present study showed that soil Δ_C stocks increased with increasing BA, suggesting the potential capacity of storing soil C would enhance with the increase of BA after afforestation. It is consistent with results in the literature: for example, Vesterdal et al. [39] found that SOC contents in 0–5 cm increased with stand age after afforestation with Norway spruce (*Picea abies* (L.) Karst) on former arable lands. Sartori et al. [38] also observeed that SOC stocks in 0–10 cm layer had an increasing trend along a choronosequence of poplar plantations in the Columbia Plateau, Oregon, USA. Mao et al. [29] and Arevalo et al. [36] all found that SOC stocks increased with stand age though there was a loss of SOC in the early stage after poplar plantations were established on croplands. It has been proposed that the increases of soil C inputs from litterfall with increasing stand age or tree density, and the lack of tillage disturbance cause the increase in soil C stocks after cropland afforestation [8,19].

The light fraction of SOM is commonly referred to as plant-like SOM with high C concentration, and the heavy fraction contains more decomposed SOM with lower C concentration [12]. Consistently, we found that SOC concentrations (ranging from 99 to 270 g kg^{-1} in croplands and 138 to 351 g kg^{-1} in poplar plantations, respectively) in the LF were obviously higher than in the HF (ranging from 2.72 to 5.89 g kg^{-1} in croplands and 2.89 to 7.01 g kg^{-1} in poplar plantations, respectively) in this present study. However, SOC content in the HF accounted for 89% of SOC in croplands and 79% of SOC in poplar plantations (Fig. 1b), indicating that SOC stocks were mainly distributed in the HF in both croplands and poplar plantations. In general, SOC contents in the LF accounted for 17–47% of total SOC contents in the surface mineral soil in a temperate zone [12], and the larger mass of the heavy fraction in soils was also observed by the other studies [8,15–17,40].

Changes of SOC stock in LF are usually more sensitive to land use change and management practices [14–17]. In this study, we observed that SOC concentration and content in the LF were all significantly increased after afforestation (Fig. 1). Similarly, the increases of SOC concentration in the LF following cropland afforestation were also observed by Laik et al. [41]. Furthermore, Li et al. [35] found that afforestation with Mongolian pine on active sand dunes resulted in an increase of light fraction C concentration in Keerqin Sandy Lands. There was a significant positive relationship between soil Δ_C stocks in LF and BA, indicating the gradual increases of SOC sequestration into the LF with increasing BA. It is consistent with the results of Marin-Spiotta et al. [42] who found that the mass and SOC concentrations in the LF increased along a chronosequence of natural reforestation of abandoned tropical pastures. The increases of soil C in the LF could be associated with the enhanced C input with increasing BA [14], considering the significant positive relationship between SOC_{tree} in the LF and BA (Fig. 5).

Soil organic C content in the HF plays an important role in the stability of SOC associated with its slow decomposition due to physical protection. However, we did not observe the increase of SOC stock in HF after afforestation in this semi-arid region. It implies that SOC stability might not be enhanced after afforestation with poplar on croplands. Huang et al. [14] also reported that there were no significant changes in SOC stock in the heavy fraction after afforestation on grasslands. However, Clark et al. [18] observed that the stability of SOC increased when native forests were allowed to invade abandoned agricultural fields in western New England. SOC stability is controlled by soil texture rather than land use management [5]. The enhancement of SOC stability after afforestation more likely occurred in soils with more clay and silt, and under climate conditions with more precipitation and warmer temperature [4,18,26].

4.2 Turnover of SOC

The SOC stocks are determined by the balance between the input of C derived from plant litter and the loss of C mainly through soil organic matter decomposition [1,21]. Our results showed that SOC derived from crops (i.e., old SOC) in the LF and HF in poplar plantations was all significantly lower than that in croplands, though the difference of SOC_{crop} in bulk soil was not significant (Fig. 4), implying that cropland afforestation led to the loss of old SOC. However, the new SOC derived from poplar trees in bulk soil, LF and HF all gradually increased with increasing BA (Fig. 5). Consistently, a net loss of old SOC and a gain of new SOC after cropland or grass afforestation were also observed in several other studies [14,19,20]. Our results imply that the increased C inputs from trees (litter and roots) following afforestation are the major causes enhancing soil C sequestration rather than the inhibition of soil old C decomposition.

Though there was a positive relationship between SOC_{tree} in the HF of the poplar plantations and BA, we could not infer that the long-term stability of SOC derived from trees would be enhanced considering the fast turnover of SOC_{crop} in the HF in poplar plantation stands in this semi-arid region. Galdo et al. [9] and Dondini et al. [19] suggested that afforestation could enhance SOC stability because they did not find the loss of old soil SOC due to the formation of soil aggregates. Furthermore, Paul et al. [5] suggested that the presence of soil C stabilization processes did not necessarily mean that recently incorporated soil C will also be effectively stabilized. In our study, a fast replacement of the old SOC by new SOC might be associated with the weak physical protection in the sandy soil [13], and the frequent drying-rewetting in the semi-arid region [43].

Conclusions

Our results confirm that cropland afforestation with poplars has the potential to sequester C rapidly into soils considering the gradual increases of SOC following afforestation because of the substantial replenishment of old soil C derived from crops by new C derived from trees. However, soil C sequestration was mainly caused by the increase in soil LF, but no significant changes of SOC content were observed in soil passive C pool due to a loss of old soil C in HF in this semi-arid region. It implies that the stability of soil organic C is not enhanced after a short-term afforestation on croplands.

Acknowledgments

We are grateful to Dr. Rong Mao for reviewing this manuscript, and Gui-Gang Lin and Bo Li for their considerable help in field work and laboratory analyses. We also thank two anonymous reviewers and Ben Bond-Lamberty for their helpful comments that substantially improved an earlier version of this manuscript.

Author Contributions

Conceived and designed the experiments: YH DZ. Performed the experiments: YH. Analyzed the data: YH. Contributed reagents/materials/analysis tools: YH LH. Wrote the paper: YH DZ LH.

References

1. Laganière J, Angers DA, Paré D (2010) Carbon accumulation in agricultural soils after afforestation: a meta-analysis. Glob Chang Biol 16: 439–453.
2. Nilsson S, Schopfhauser W (1995) The carbon-sequestration potential of a global afforestation program. Clim Chang 30: 267–293.
3. Guo LB, Gifford RM (2002) Soil carbon stocks and land use change: a meta analysis. Glob Chang Biol 8: 345–360.
4. Paul KI, Polglase PJ, Nyakuengama JG, Khanna PK (2002) Change in soil carbon following afforestation. For Ecol Manage 168: 241–257.
5. Paul S, Flessa H, Veldkamp E, Lpez-Ulloa M (2008) Stabilization of recent soil carbon in the humid tropics following land use changes: evidence from aggregate fractionation and stable isotope analyses. Biogeochemistry 87: 247–263.
6. Morris SJ, Bohm S, Haile-Mariam S, Paul EA (2007) Evaluation of carbon accrual in afforested agricultural soils. Glob Chang Biol 13: 1145–1156.
7. Li D, Niu S, Luo Y (2012) Global patterns of the dynamics of soil carbon and nitrogen stocks following afforestation: a meta-analysis. New Phytol 195: 172–181.
8. Tang G, Li K (2013) Tree species controls on soil carbon sequestration and carbon stability following 20 years of afforestation in a valley-type savanna. For Ecol Manage 291: 13–19.
9. Galdo ID, Six J, Peressott A, Cotrufo MF (2003) Assessing the impact of land-use change on soil C sequestration in agricultural soils by means of organic matter fractionation and stable C isotopes. Glob Chang Biol 9: 1204–1213.
10. Lal R (2004) Soil carbon sequestration impacts on global climate change and food security. Science 304: 1623–1627.
11. Shi S, Zhang W, Zhang P, Yu Y, Ding F (2013) A synthesis of change in deep soil organic carbon stores with afforestation of agricultural soils. For Ecol Manage 296: 53–63.
12. Christensen BT (1992) Physical fractionation of soil and organic matter in primary particle size and density separates. Adv Soil Sci 20: 1–90.
13. Lützow M, Kögel-Knabner I, Ekschmitt K, Flessa H, Guggenberger G, et al (2007) SOM fractionation methods: relevance to functional pools and to stabilization mechanisms. Soil Biol Biochem 39: 2183–2207.
14. Huang Z, Davis MR, Condron LM, Clinton PW (2011) Soil carbon pools, plant biomarkers and mean carbon residence time after afforestation of grassland with three tree species. Soil Biol Biochem 43: 1341–1349.
15. Janzen HH, Campbell CA, Brandt SA, Lafond GP, Townley-Smith L (1992) Light-fraction organic matter in soils from long-term crop rotations. Soil Sci Soc Am J 56: 1799–1806.
16. John B, Yamashita T, Ludwig B, Flessa H (2005) Storage of organic carbon in aggregate and density fractions of silty soils under different types of land use. Geoderma 128: 63–79.
17. Tan Z, Lal R, Owens L, Izaurralde RC (2007) Distribution of light and heavy fractions of soil organic carbon as related to land use and tillage practice. Soil Till Res 92: 53–59.
18. Clark JD, Planter AF, Johnson AH (2011) Soil organic matter quality in chronosequences of secondary northern hardwood forests in Western New England. Soil Sci Soc Am J 76: 684–693.
19. Dondini M, Groenigen KJ, Galdo ID, Jones MB (2009) Carbon sequestration under Miscanthus: a study of ^{13}C distribution in soil aggregates. Glob Change Biol Bioenergy 1: 321–330.
20. Wei X, Qiu L, Shao M, Zhang X, Gale WJ (2012) The accumulation of organic carbon in mineral soils by afforestation of abandoned farmland. PLoS ONE 7(3): e32054. doi:10.1371/journal.pone.0032054.
21. Arai H, Tokuchi N (2010) Factors contributing to greater soil organic carbon accumulation after afforestation in a Japanese coniferous plantation as determined by stable and radioactive isotopes. Geoderma 157: 243–251.
22. Helmisaari H, Derome J, Nöjd P, Kukkola M (2007) Fine root biomass in relation to site and stand characteristics in Norway spruce and Scots pine stands. Tree Physiol 27: 1493–1504.
23. Litton CM, Ryan MG, Knight DH (2004) Effects of tree density and stand age on carbon allocation patterns in postfire lodgepole pine. Ecol Appl 14: 460–475.
24. Lal R (2004) Carbon sequestration in dryland ecosystems. Environ Manage 33: 528–544.
25. Jackson RB, Banner JL, Jobbágy EG, Pockman WT, Wall DH (2002) Ecosystem carbon loss with woody plant invasion of grasslands. Nature 418: 623–626.
26. Richter DD, Markewitz D, Trumbore SE, Wells CG (1999) Rapid accumulation and turnover of soil carbon in a re-establishing forest. Nature 400: 56–58.
27. Hu YL, Zeng DH, Fan ZP, Chen GS, Zhao Q, et al (2008) Changes in ecosystem carbon stocks following grassland afforestation of semiarid sandy soil in the southeastern Keerqin Sandy Lands, China. J Arid Environ 72: 2193–2200.
28. Hu YL, Zeng DH, Chang SX, Mao R (2013) Dynamics of soil and root C stocks following afforestation of croplands with poplars in a semi-arid region in northeast China. Plant Soil 368: 619–627.
29. Mao R, Zeng DH, Hu YL, Li LJ, Yang D (2010) Soil organic carbon and nitrogen stocks in an age-sequence of poplar stands planted on marginal agricultural land in Northeast China. Plant Soil 332: 277–287.
30. Zhenghu D, Honglang X, Zhibao D, Gang W, Drake S (2007) Morphological, physical and chemical properties of aeolian sandy soils in northern China. J Arid Environ 68: 66–76.
31. Jiao SR (1989) The ecosystem structure and function of sandy-fixation forests in Zhanggutai. Shenyang: Liaoning Science & Technology Press. 9–10 p.
32. Sohi SP, Mahieu N, Arah JRM, Powlson DS, Madari B, et al (2001) A procedure for isolating soil organic matter fractions suitable for modeling. Soil Sci Soc Am J 65: 1121–1128.
33. Nelson DW, Sommers LE (1996) Total carbon, organic carbon and organic matter, in: Sparks DL (Eds.), Methods of soil analysis. Part 3. Chemical methods. Wisconsin: Soil Science Society of America. 961–1010.

34. Hernandez-Ramirez G, Sauer TJ, Cambardella CA, Brandle JR, James DE (2011) Carbon sources and dynamics in afforested and cultivated corn belt soils. Soil Sci Soc Am J 75: 216–225.
35. Li Y, Awada T, Zhou X, Shang W, Chen Y, et al (2012) Mongolian pine plantations enhance soil physico-chemical properties and carbon and nitrogen capacities in semi-arid degraded sandy land in China. Appl Soil Ecol 56: 1–9.
36. Arevalo CBM, Bhatti JS, Chang SX, Sidders D (2011) Land use change effects on ecosystem carbon balance: from agricultural to hybrid poplar plantation. Agric Ecosyst Environ 141: 342–349.
37. Jug A, Makeschin F, Rehfuess KE, Hofmann-Schielle C (1999) Short-rotation plantations of balsam poplars, aspen and willows on former arable land in the Federal Republic of Germany. III. Soil ecological effects. For Ecol Manage 121: 85–99.
38. Sartori F, Lal R, Ebinger MH, Eaton JA (2007) Changes in soil carbon and nutrient pools along a chronosequence of poplar plantations in the Columbia Plateau, Oregon, USA. Agric Ecosyst Environ 122: 325–339.
39. Vesterdal L, Ritter E, Gundersen P (2002) Change in soil organic carbon following afforestation of former arable land. For Ecol Manage 169: 137–147.
40. Bu X, Ruan H, Wang L, Ma W, Ding J, et al (2012) Soil organic matter in density fractions as related to vegetation changes along an altitude gradient in the Wuyi Mountains, southeastern China. Appl Soil Ecol 52: 42–47.
41. Laik R, Kumar K, Das DK, Chaturvedi OP (2009) Labile soil organic matter pools in a calciorthent after 18 years of afforestation by different plantations. Appl Soil Ecol 42: 71–78.
42. Marin-Spiotta E, Silver WL, Swanston CW, Ostertag R (2009) Soil organic matter dynamics during 80 years of reforestation of tropical pastures. Glob Chang Biol 13: 1584–1597.
43. Unger S, Máguas C, Pereira JS, David TS, Werner C (2010) The influence of precipitation pulses on soil respiration-assessing the "Birch effect" by stable carbon isotopes. Soil Biol Biochem 42: 1800–1810.

5

Assessment of the AquaCrop Model for Use in Simulation of Irrigated Winter Wheat Canopy Cover, Biomass, and Grain Yield in the North China Plain

Xiu-liang Jin[1,2,3], Hai-kuan Feng[2,3], Xin-kai Zhu[1], Zhen-hai Li[2,3], Sen-nan Song[1,2,3], Xiao-yu Song[2,3], Gui-jun Yang[2,3], Xin-gang Xu[2,3]*, Wen-shan Guo[1]*

1 Key Laboratory of Crop Genetics and Physiology of Jiangsu Province, Yangzhou University, Yangzhou, China, **2** Beijing Research Center for Information Technology in Agriculture, Beijing Academy of Agriculture and Forestry Sciences, Beijing, China, **3** National Engineering Research Center for Information Technology in Agriculture, Beijing, China

Abstract

Improving winter wheat water use efficiency in the North China Plain (NCP), China is essential in light of current irrigation water shortages. In this study, the AquaCrop model was used to calibrate, and validate winter wheat crop performance under various planting dates and irrigation application rates. All experiments were conducted at the Xiaotangshan experimental site in Beijing, China, during seasons of 2008/2009, 2009/2010, 2010/2011 and 2011/2012. This model was first calibrated using data from 2008/2009 and 2009/2010, and subsequently validated using data from 2010/2011 and 2011/2012. The results showed that the simulated canopy cover (CC), biomass yield (BY) and grain yield (GY) were consistent with the measured CC, BY and GY, with corresponding coefficients of determination (R^2) of 0.93, 0.91 and 0.93, respectively. In addition, relationships between BY, GY and transpiration (T), ($R^2 = 0.57$ and 0.71, respectively) was observed. These results suggest that frequent irrigation with a small amount of water significantly improved BY and GY. Collectively, these results indicate that the AquaCrop model can be used in the evaluation of various winter wheat irrigation strategies. The AquaCrop model predicted winter wheat CC, BY and GY with acceptable accuracy. Therefore, we concluded that AquaCrop is a useful decision-making tool for use in efforts to optimize wheat winter planting dates, and irrigation strategies.

Editor: Dafeng Hui, Tennessee State University, United States of America

Funding: The National Natural Science Foundation of China (Grant No. 41001244, 41271415), Beijing Nova Program (Grant No.2011036), the 8th Six Talents Peak Project of Jiangsu Province (2011-NY039) and the Key Projects in the National Science & Technology Pillar Program during the Twelfth Five-Year Plan Period (Grant No.2012BAH27B04). The authors are grateful to Mr. Weiguo Li, Mrs. Zhihong Ma and Hong Chang for data collection. The funders had no role in study design, data collection and analysis, decision to publish, or preparation of the manuscript.

Competing Interests: The authors have declared that no competing interests exist.

* E-mail: xxgpaper@126.com (XX); guows@yzu.edu.cn (WG)

Introduction

Winter wheat (*Triticum aestivum* L.) is an important staple food crop for the majority of the North China Plain (NCP) population [1]. However, increasing industrial, and domestic water use has resulted in a reduction in water available for irrigation of these crops. Thus there is a growing need for improvement to this region's agriculture water resources management, especially given increasing food demands of the region's increasing population.

It is widely known that well-timed irrigation can substantially increase water use efficiency (WUE) [2,3], providing an optimal growth environment throughout the season [4,5]. In fact, various studies have described several such irrigation strategies for use by farmers [6–16]. Since the mid-1960s, the relationship between water and crop yield has been described with both empirical and mechanistic models [17–20]. For example, De Wit [21] proposed that a linear relationship between yield and water consumption exists. In contrast, Downey [22], via deficit irrigation studies, suggested that there exists a nonlinear relationship between water and yield. Based on the above studies, the Minhas model [23], Rao model [18], Blank model [24], and the Stewart model [25] were developed. More recently, Wang and Sun [26] showed that a quadratic relationship between crop yield and crop water consumption did in fact exist. Their work was followed by Kang et al. [27] in which a multiple and synergistic model (developed under deficit irrigation conditions) was proposed. At present, the simulation of the soil-plant-atmosphere continuum remains an important part of such research, especially with regard to expansion of the application range of resulting models to a wider array of cropping systems.

Therefore, the Food and Agricultural Organization (FAO) developed the AquaCrop model in an effort to meet this need in 2009. This model was originated from the "yield response to water" data of Doorenbos and Kassam [28], and evolved to a normalized crop water productivity (NCWP) concept [29]. Compared with other models, AquaCrop is relatively simple to operate by those with little, or no research experience, and allows for simulation of crop performance in multiple scenarios. In addition to a high level of accuracy, this robust model requires a limited set of input parameters, most of which are relatively easy to acquire [29,30]. The AquaCrop model is also capable of predicting crop productivity, water requirements, and water use efficiency under water-limiting conditions [31]. To date, this

Table 1. Winter wheat cultivars and planting dates were selected in the 2008, 2009, 2010 and 2011.

Winter wheat cultivars	Planting dates
Nongda195, Jingdong8, Jing9428	Sep. 28th, Oct. 7th, and Oct. 20th, 2008
Nongda195, Jingdong13, Jing9428	Sep. 25th, Oct. 5th, and Oct. 15th, 2009
Nongda195, Yannong19, Jing9428	Sep. 25th, Oct. 5th, and Oct. 15th, 2010
Nongda211, Zhongmai175, Jingdong8, Jing9843	Sep. 25th, 2011

Note: There are three winter wheat cultivars and each has three planting dates in 2008, 2009 and 2010. In 2011, four cultivars are planted on the same date.

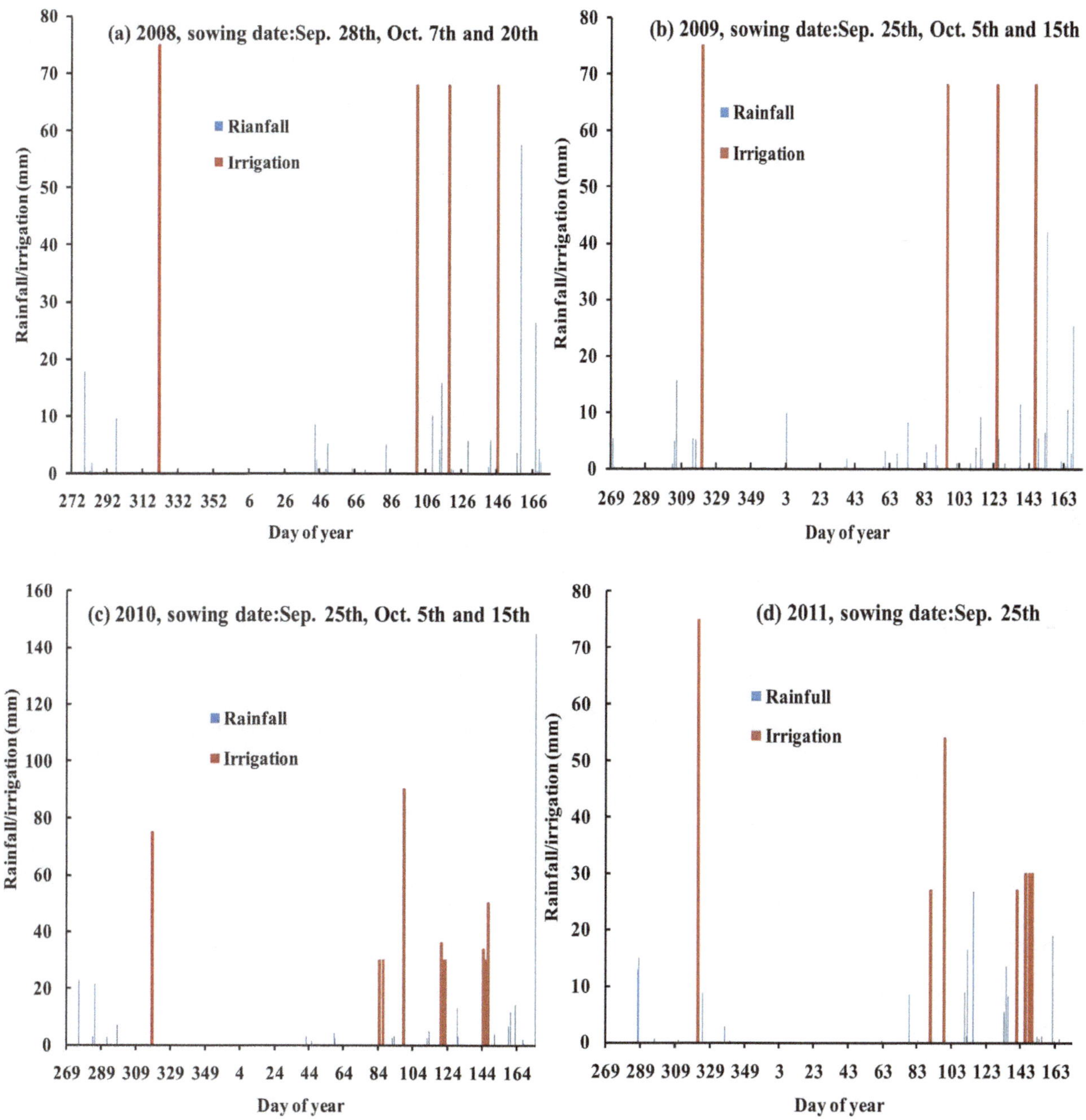

Figure 1. Daily rainfall, and supplemental irrigation for the Xiaotangshan site during the cropping seasons 2008/2009 (a), 2009/2010(b), 2010/2011 (c) and 2011/2012(d).

Table 2. Irrigation schedule during experimental period (2008/2009, 2009/2010, 2010/2011 and 2011/2012).

Day of year	Year	Month	Day	Irrigation amount (mm)
321	2008	11	16	75
100	2009	4	10	68
118	2009	4	28	68
146	2009	5	26	68
320	2009	11	15	75
95	2010	4	5	68
124	2010	5	4	68
146	2010	5	26	68
318	2010	11	13	75
84	2011	3	24	30
86	2011	3	26	30
98	2011	4	7	90
120	2011	4	29	36
121	2011	4	30	30
122	2011	5	2	30
144	2011	5	23	34
145	2011	5	24	30
146	2011	5	25	30
147	2011	5	26	54
321	2011	11	16	75
90	2012	3	30	27
98	2012	4	7	54
140	2012	5	19	27
145	2012	5	24	30
147	2012	5	26	30
149	2012	5	28	30

model has been successfully tested for cotton [32,33], maize [30,34–38], wheat [39–42], sugar beet [37], sunflower [37,43], groundnut [44], potato [45,46], quinoa [16], Teff [47], barley [48,49], green onion [50] and tomato [51] under a wide-range of environments. Previous studies have demonstrated that the AquaCrop model accurately simulates crop canopy cover (CC), biomass yield (BY) and grain yield (GY) under both regular, and deficit irrigation, and in low soil fertility conditions. In such unreliable water-limited environments as the NCP, the AquaCrop model is a potentially valuable tool for use in efforts to maximize this region's winter wheat yield. Therefore, the objective of this study was to validate this model in simulating the effects of planting date, and multiple irrigation scenarios on: (1) canopy cover, (2) biomass yield, (3) grain yield, and (4) water use efficiency of winter wheat in the NCP. These data will provide some guidelines for efforts to optimize irrigation management for winter wheat crops in this region.

Materials and Methods

Study Site

This field experiments were conducted in the 2008/2009, 2009/2010, 2010/2011 and 2011/2012 growing seasons at the Xiaotangshan experimental site (44.17° N, 116.433°E), Beijing, PR China. This area is representative of the overall soil and crop management practices in this region. The soil is fine-loamy, with a nitrate Nitrogen (NO_3–N) content of 3.16–14.82 mg kg^{-1}, an ammonium Nitrogen (NH_3–N) content of 10.20–12.32 mg kg^{-1}, an Olsen P of 3.14–21.18 mg kg^{-1}, an exchangeable K of 86.83–120.62 mg kg^{-1}, and an organic matter content of 15.84 20.24g kg^{-1} with in the uppermost 0–30 cm layer. Beijing is characterized by a typical continental climate, with maximum temperatures of 26.1°C in the summer and minimum temperatures of −4.7°C in the winter. Throughout all seasons, the temperature fluctuated daily with significant differences between night and day. During this experimental period, the average annual precipitation was 650 mm, and the frost-free period was on average 180 days.

Local winter wheat cultivars and planting dates are shown in Table 1. Each plot area is 100 m^2 in 2008, 2009 and 2010, and 300 m^2 in 2011. The experiment was designed as a 2-way factorial arrangement of treatments in a randomized complete block design, with three replications for each treatment. Plot management followed local standard practices (weed control, pest management and fertilizer application) for wheat production in this region. The Xiaotangshan Experimental Site belongs to the National Engineering Research Center for Information Technology in Agriculture. It gives some permission for us to study relative agriculture research within this area. We confirm that the field studies did not involve endangered or protected species.

Climate Data Collection and Analysis

Climate data for the experimental site was obtained from the local Xiaotangshan meteorological station. The daily reference evapotranspiration (ET_o) for the growing season from 2008 to 2012 was calculated based on the FAO Penman-Monteith method as described in Allen et al. [52], and the ET_o calculator (FAO, 2009) [53]. Daily maximum and minimum temperature, relative humidity, wind speed, rainfall, and total sunshine hours were recorded directly at the Xiaotangshan experimental site. The total rainfall, from sowing to harvest was 199, 208, 145 and 168 mm in

Table 3. Physical soil characteristics of the Xiaotangshan experimental site.

Site	Soil texture	Groundwater table	Depth (m)	Moisture content (vol%)			Ksat (mm day^{-1})	CN
				Sat	FC	WP		
Xiaotangshan	fine-loamy	3.5 m	0.0–0.1	51.1	27.3	8.8	240	75
			0.1–0.2	51.3	27.3	8.7	240	
			0.2–0.3	54.7	34.8	13.2	224	

Note: FC, field capacity; WP, wilting point; Sat, water content at saturation; CN, curve number; and K_{sat}, saturation hydraulic conductivity describes water movement through saturated media. The values of saturated hydraulic conductivity in soils vary within a wide range of several orders of magnitude, depending on the soil material.

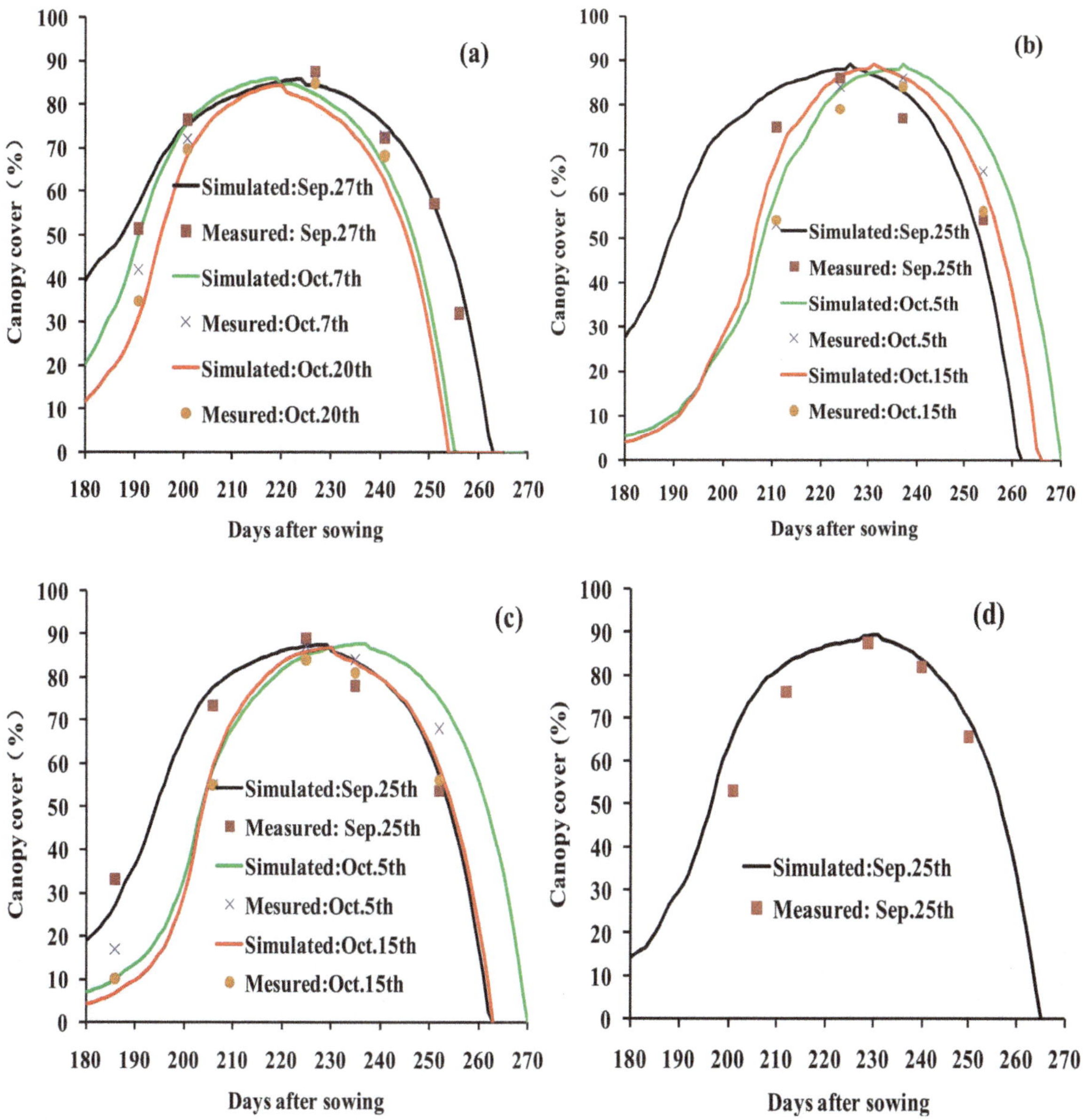

Figure 2. Simulated and measured canopy cover (CC) for winter wheat under different planting dates and irrigation strategies in the cropping season 2008/2009 (a), 2009/2010 (b), 2010/2011 (c) and 2011/2012 (d).

2008/2009, 2009/2010, 2010/2011 and 2011/2012, respectively. Supplemental irrigation was applied to treatments following cessation of any rain events (Fig. 1a–d and Table 2).

Soil Data of the Experimental Site

The soil at the Xiaotangshan experimental site represents the major soil type (fine-loamy) on which winter wheat is grown in NCP. The soil was at maximum field capacity (27.3% at 0.0–0.1 m, 27.3% at 0.1–0.2 m and 34.8% at 0.2–0.3 m) during sowing and early establishment. The physical soil characteristics were measured directly in the field and used for input into AquaCrop (Table 3).

Field Experiments and Crop Data Collection

In 2008/2009, 2009/2010, 2010/2011, and 2011/2012 aboveground biomass was determined 5–6 times from a 0.25 m^2 area by randomly cutting four representative plants from each plot. All plant samples were heated to 105°C, oven dried at 70°C to a constant weight, and final dry weight (DW) recorded.

Leaf area index (LAI) was estimated using the following two methods:

(1) By multiplying the plant population by the leaf area per plant as described in Kar et al. [54]. Area of the leaf was measured manually from 20 plants using a straightedge. Counting of plant populations was conducted manually from a 0.1 m^2 area. The LAI equation is as follows:

Table 4. Input data of crop parameters used in AquaCrop model.

Description	Value	Unit
Base temperature	0	°C
Upper temperature	26	°C
Canopy growth coefficient (CGC): Increase in CC per day	0.03	%/day
Canopy decline coefficient (CDC): Decrease in CC per day at senescence	0.09	%/day
Maximum canopy cover (CC_x)	90	%
Water productivity (NCWP)	15	g/cm^2
Reference harvest index (HI)	46	%
Upper threshold for canopy expansion (P_{upper})	0.20	% of TAW
Lower threshold for canopy expansion (P_{lower})	0.65	% of TAW
Leaf expansion stress coefficient curve shape	3.0	–
Upper threshold for stomatal closure (P_{upper})	0.65	% of TAW
Minimum effective rooting depth	0.3	m
Maximum effective rooting depth	1.2	m
Canopy senescence stress coefficient (P_{upper})	0.70	% of TAW
Shape factor describing root zone expansion	1.5	–
Crop coefficient when canopy is complete but prior to senescence	1.1	–
Senescence stress coefficient curve shape	3.0	–
Allowable maximum increase of specified HI	15	%
Minimum air temperature below which pollination starts to fail	5	°C
Maximum air temperature above which pollination starts to fail	35	°C
Water Productivity normalized for ET_o and CO_2 during yield formation	100	%
Time from sowing to emergence	7	Days
Time from sowing to flowering	232	Days
Time from sowing to start senescence	236	Days
Length of the flowering stage (days)	10	Days

$$LAI = 0.75 \times \rho \times \left(\frac{\sum_{i=1}^{m}\sum_{i=1}^{n}(L_{ij} \times B_{ij})}{m}\right) \quad (1)$$

Where ρ is plant density, m is the number of measured plants, L_{ij} is leaf length, B_{ij} is the maximum leaf width, and n is the number of leaves of the nth plant.

(2) The LAI-2000 Plant Canopy Analyzer (LI-COR Inc., Lincoln, NE, USA) was used in measuring for determination of LAI. The resulting values were similar to those obtained via the manual LAI calculations, thus the LAI-2000 data was used as the model input data.

Canopy cover was estimated from different irrigation treatments in 2008/2009, 2009/2010, 2010/2011, and 2011/2012 based on Hsiao et al. [30] using the following:

$$CC = 1 - \exp^{(-0.65 \times LAI)} \quad (2)$$

where CC is canopy cover, and LAI is the leaf area index.

Grain yield was measured following maturation from samples obtained from a 1.5 m^2 area in each plot, with three replications for each treatment. Collected grain was dried and weighed on an electronic scale (±0.01 g). As there were no significant differences

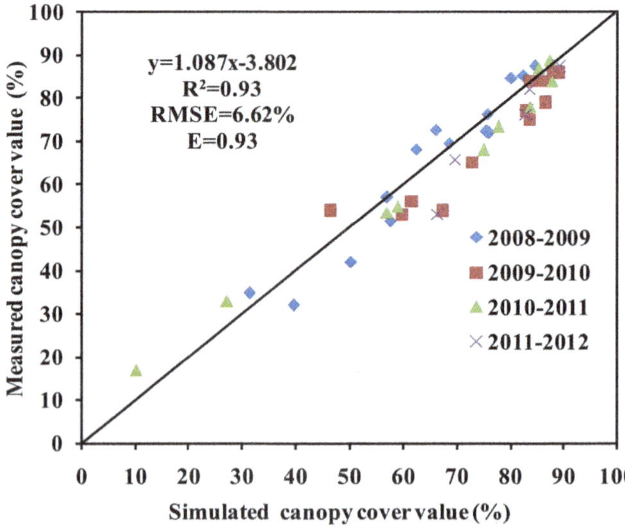

Figure 3. Relationship between the measured and simulated canopy cover (CC) in winter across 4 years. Note: x represents the simulated CC, y represents the measured CC. The intercept represents the relative error between the simulated CC and the measured CC. The slope represents the consistency between the simulated CC and the measured CC.

Table 5. Simulated and measured canopy cover (CC) for winter wheat under different planting dates in 2008/2009, 2009/2010, 2010/2011 and 2011/2012.

Year	Planting date	Slope	Intercept	R^2	RMSE (%)	E
2008–2009	28/9/2008	1.036	1.575	0.89	4.39	0.90
2008–2009	7/10/2008	1.251	−19.64	0.97	5.84	0.94
2008–2009	20/10/2008	1.018	2.032	0.98	4.25	0.96
2008–2009		1.060	−2.093	0.91	6.62	0.91
2009–2010	25/9/2009	1.201	−11.98	0.91	5.93	0.90
2009–2010	5/10/2009	1.082	−6.316	0.95	4.35	0.94
2009–2010	15/10/2009	0.930	8.443	0.92	7.02	0.91
2009–2010		1.120	−8.039	0.93	4.94	0.93
2010–2011[a]	25/9/2010	1.157	−12.31	0.97	5.73	0.94
2010–2011[a]	5/10/2010	1.180	−13.21	0.92	3.18	0.92
2010–2011[a]	15/10/2010	1.183	−7.109	0.96	3.78	0.95
2010–2011[a]		1.134	−7.996	0.96	7.19	0.94
2011–2012[a]	25/9/2011	0.743	24.35	0.96	7.15	0.93

Note: [a]Validation data set: R^2, determination coefficient; E, model efficiency; RMSE, root mean square of error.
The intercept represents the relative error between the simulated CC and the measured CC, The slope represents the consistency between the simulated CC and the measured CC.

between winter wheat varieties in many of the measured characteristics (e.g. phenological development, canopy cover, etc.), the average grain yield of the different varieties was considered in model simulations.

Water Use Efficiency (WUE)

Water use efficiency (WUE) is defined as the grain yield per unit amount of water consumed [55]. In this study grain water use efficiency (Grain-WUE), and biomass water use efficiency (Biomass-WUE) were calculated using Eq. (3) and Eq. (4), respectively, as in Araya et al. (2010b) [48]:

$$\text{Grain-WUE} = \left[\frac{\text{GY}}{\sum \text{T}}\right] \tag{3}$$

$$\text{Biomass-WUE} = \left[\frac{BY}{\sum T}\right] \tag{4}$$

Where GY is the grain yield kg ha^{-1} (measured), T is the transpiration as determined using AquaCrop model, and BY is the total final aboveground biomass yield in kg ha^{-1} (measured).

Description of AquaCrop Model

The AquaCrop model was proposed by the FAO in 2009, with a detailed description presented in Steduto et al. [29], and Raes et al. [31]. The model computes a daily water balance, and separates evapotranspiration into evaporation and transpiration components. Transpiration is correlated with canopy cover, which is proportional to the degree of soil cover, and evaporation is proportional to the area of soil not covered by vegetation. The crop's stomata conductance, canopy senescence, leaf growth, and yield response to water stress are modeled using four stress coefficients (stomata closure, leaf expansion, canopy senescence, and change in harvest index (HI)). The model subsequently estimates yield from the daily crop transpiration values.

In general, the normalized crop water productivity (NCWP) is considered constant for a given climate condition and crop (For crops not nutrient-limited, the model provides categories ranging from slight to severe deficiencies corresponding to lower water productivity (WP).) is applicable for using in different locations, seasons, and even future climates [29]. Depending on the crop, NCWP increases slightly with an increase in atmospheric CO_2 concentration [29]. NCWP is set between 13 and 20 g m^{-2} for C_3 crops. For example, NCWP is set at 15 g m^{-2} for the winter wheat according to the AquaCrop Manual (Annex I Section I.10 Wheat, Pages A39–A42) [56,57]. In our current study, we have not included any of the water stress study data; therefore NCWP remained at 15 g m^{-2} for the winter wheat. The crop's daily aboveground biomass is calculated using NCWP from the AquaCrop model [29,30]. Biomass yield (BY) is calculated by multiplying NCWP by the ratio of crop transpiration (T), and evapotranspiration (ET_o), following calculation of BY (its harvestable portion), and the grain yield (GY) is determined via harvest index (HI).

These changes are described by the following Eqs. (5) and (6):

$$BY = \sum (T/ET_O) \times NCWP \tag{5}$$

$$GY = BY \times HI \tag{6}$$

Where BY is biomass yield in kg ha^{-1}, T is crop transpiration in mm, ET_o is evapotranspiration in mm, $NCWP$ is the normalized crop water productivity in g m^{-2}, HI is harvest index, and GY is grain yield in kg ha^{-1}.

Data Analysis

Winter wheat canopy cover (CC), biomass yield (BY) and grain yield (GY) in AquaCrop were calibrated using the measured data sets of 2008/2009 and 2009/2010, and validated using the 2010/2011 and 2011/2012 measured data sets. The good fit regression equation between the simulated and observed values was corroborated using prediction error statistics. The coefficient of determination (R^2), root mean square error (RMSE), and model efficiency (E) were used as the error statistics to evaluate both calibration and validation results. The E and R^2 were used to access the predictive power of the model, and the RMSE indicated the error in model prediction. In this study, the prediction model output for CC, GY and BY during harvest was used for model evaluation. These statistical indices were used to compare measured and simulated values. Model performance was assessed using E (Nash and Sutcliffe, 1970) as follows:

$$E = 1 - \frac{\sum_{i=1}^{n} (S_i - O_i)^2}{\sum_{i=1}^{n} (O_i - \bar{O}_i)^2} \tag{7}$$

where S_i and O_i are predicted, and observed data, respectively. $\bar{O}_i$ is the mean value of O_i, and n is the number of observations.

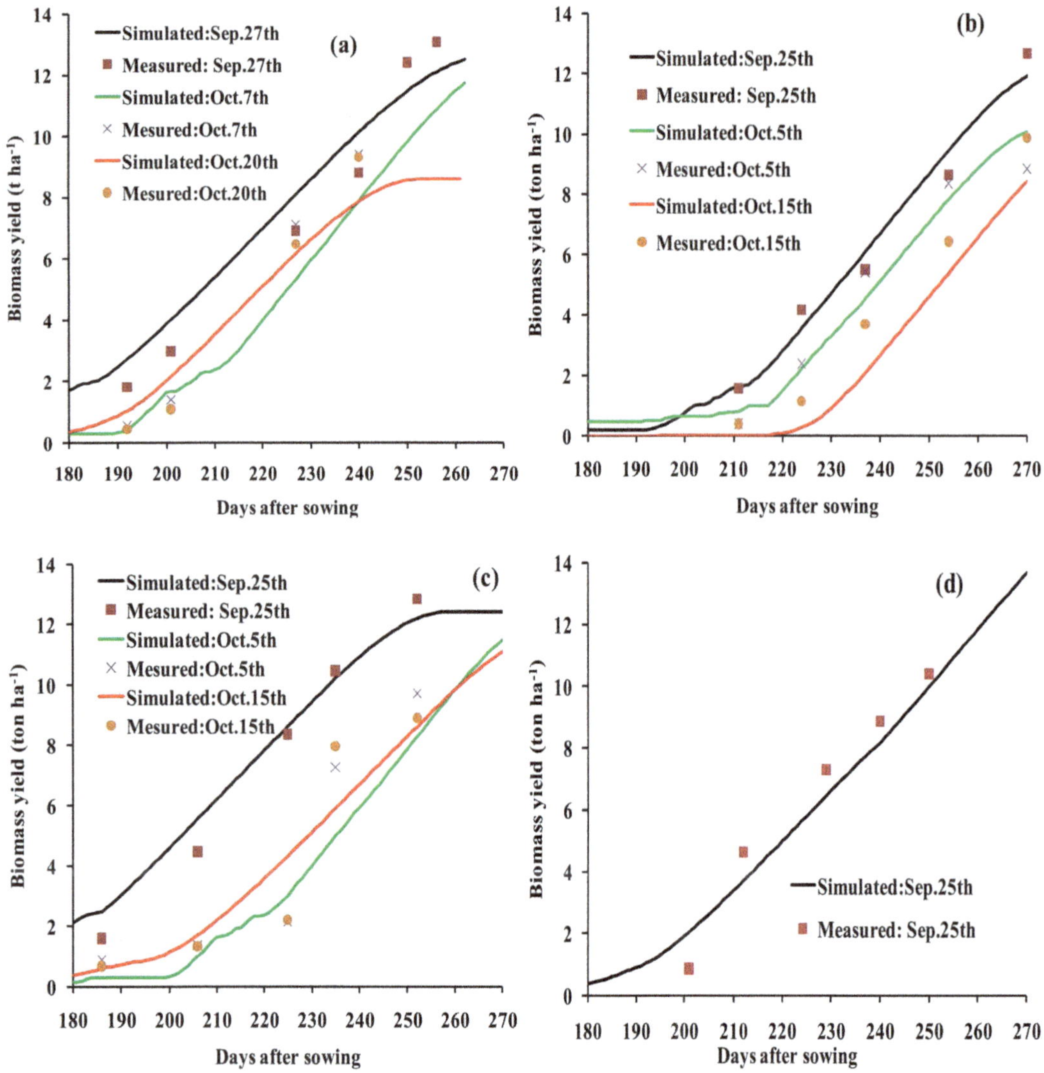

Figure 4. The simulated as compared with the measured aboveground biomass accumulation at different growth stages for winter wheat with different planting dates and irrigation strategies in the cropping season 2008/2009 (a), 2009/2010 (b), 2010/2011 (c) and 201/2012 (d).

$$RMSE=\sqrt{\frac{\sum_{i=1}^{n}(S_i-O_i)^2}{n}} \tag{8}$$

E and R^2 approaching one, and a RMSE near zero were indicators of improved model performance. Following model calibration, and validation, Grain-WUE and Biomass-WUE were calculated using Eqs. 3 and 4.

Result and Analysis

AquaCrop Model Calibration and Validation Results

The crop parameters used to calibrate the AquaCrop model are presented in Table 4. Key stress parameters (e.g. canopy growth, canopy senescence stress coefficient) (P_{upper}) were adjusted as needed to simulate CC. There was a strong linear relationship between the simulated and the measured CC ($R^2=0.93$, RMSE = 6.62% and E = 0.93) for winter wheat under different planting dates, and irrigation strategies in the cropping season 2008/2009, 2009/2010, 2010/2011 and 2011/2012 (Figs. 2a, 2b, 2c, 2d and 3). The R^2, RMSE and E of the simulated and measured CC were 0.91, 6.62% and 0.91, respectively in 2008/2009. And the R^2, RMSE and E values of CC were 0.93, 4.94% and 0.93 in 2009/2010, 0.96, 7.19%, 0.94 in 2010/2011, and 0.96, 7.15%, and 0.93 in 2011/2012, respectively (Table 5).

The simulated aboveground BY was similar to that measured (Figs. 4a, 4b, 4c and 4d). The stress coefficients were also adjusted, and readjusted as needed to simulated aboveground biomass. There was a strong relationship between measured and simulated BY across the four years (Fig. 5 and Table 6). The GY was also similar to the measured GY across all four years (R^2, RMSE and E values of 0.93, 0.52 ton ha^{-1} and 0.92, respectively) (Fig. 6).

The R^2, RMSE and E also showed good performance between the simulated and the measured values for CC ($R^2=0.89$–0.98, RMSE = 3.18–7.19% and E = 0.90–0.96) and BY ($R^2=0.92$–0.98, RMSE = 1.12–1.84 ton ha^{-1} and E = 0.92–0.96) (Tables 5 and 6). Higher R^2 and E values and lower RMSE values indicated good

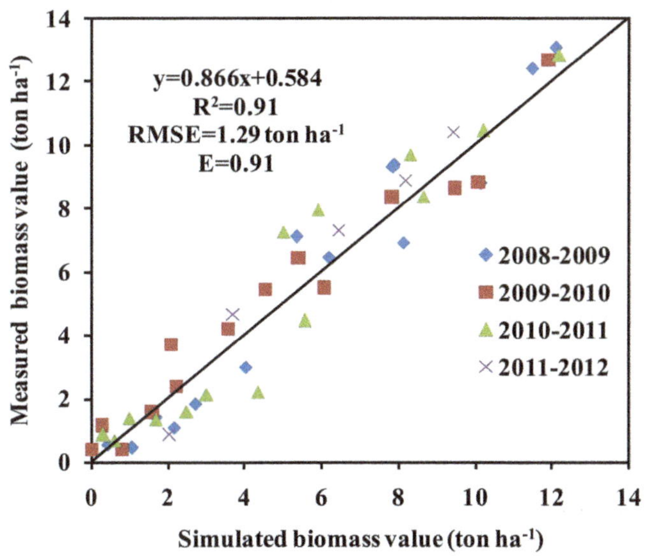

Figure 5. Relationship between the measured and simulated biomass yield (BY) in winter wheat across 4 years. Note: x represents the simulated BY, y represents the measured BY. The intercept represents the relative error between the simulated biomass yield and the measured BY. The slope represents the consistency between the simulated BY and the measured BY.

Table 6. Simulated and measured biomass yield (BY) for winter wheat under different planting dates in 2008/2009, 2009/2010, 2010/2011 and 2011–2012.

Year	Planting date	Slope	Intercept	R^2	RMSE (ton ha^{-1})	E
2008–2009	28/9/2008	0.772	1.286	0.98	1.34	0.95
2008–2009	7/10/2008	0.787	−0.270	0.98	1.54	0.96
2008–2009	20/10/2008	0.802	−0.094	0.98	1.12	0.96
2008–2009		0.778	0.361	0.94	1.21	0.93
2009–2010	25/9/2009	0.993	1.315	0.92	1.39	0.92
2009–2010	5/10/2009	0.960	1.163	0.95	1.22	0.94
2009–2010	15/10/2009	0.954	1.331	0.97	1.27	0.95
2009–2010		0.945	1.284	0.95	1.29	0.94
2010–2011[a]	25/9/2010	0.809	0.617	0.97	1.21	0.95
2010–2011[a]	5/10/2010	1.043	0.378	0.97	1.37	0.95
2010–2011[a]	15/10/2010	0.797	0.019	0.97	1.84	0.94
2010–2011[a]		0.820	0.162	0.97	1.25	0.96
2011–2012[a]	25/9/2011	0.953	0.158	0.96	0.93	0.95

Note: [a]Validation data set: R^2, determination coefficient; E, model efficiency; RMSE, root mean square of error.
The intercept represents the relative error between the simulated BY and the measured BY, The slope represents the consistency between the simulated BY and the measured BY.

model performance. The calibrated results were also consistent with the validated results for CC, BY and GY (Tables 5 and 6). These results suggest that the AquaCrop model is useful for simulating winter wheat CC, BY and GY under different planting dates, and irrigation strategies.

Biomass, Grain Yield and Water Use Efficiency

Winter wheat was planted on Sep. 25th (normal sowing), Sep. 28th (normal sowing), Oct. 5th (late sowing), Oct. 7th (late sowing), Oct. 15th (late sowing) and Oct. 20th (late sowing) during 2008/2011. Winter wheat that was planted on Sep. 25th and 28th had greater biomass and grain yield than did those planted on Oct. 5th, 7th, 15th, and 20th (Table 6). There was relatively more transpiration and biomass yield, yet lower grain yield in 2008 than in 2010, but there was relatively more transpiration (T), biomass yield, and even grain yield in 2008 than in 2009 (Table 7). The highest biomass and grain yield was obtained in crops planted on Sep. 28th 2008 and on Sep. 25th 2011, respectively; the lowest biomass and grain yield was obtained in crops planted on Oct. 15th 2009. A relatively higher grain and biomass yield per m^3 of water was obtained from crops planted in 2010 than was in 2008 or 2009. Therefore, the biomass and grain yield water use efficiency (biomass-WUE and yield-WUE) was higher in 2010 than both 2008 and 2009. The biomass yield WUE was higher in 2008 than in 2009, yet the grain yield WUE was lower in 2008 than in 2009. Frequent irrigation with a small amount of water obviously improved grain yield in 2010/2011 and 2011/2012 and increased the biomass-WUE and grain-WUE (Table 5). A good relationship did exist between GY, BY, and T (R^2 values of 0.57 and 0.71, respectively) (Fig. 7), thus suggesting that T might be used in estimating biomass and grain yield.

Discussion

In this study, the AquaCrop model successfully predicted CC, BY GY in winter wheat. The crop parameters are adjusted to simulate CC, BY and GY for winter wheat under different planting dates and irrigation strategies. These adjustments were made to obtain more stable and closer relationships between the simulated values and the measured values. The results showed that the model calibration data sets from 2008/2009 and 2009/2010 were very consistent with the model validation data sets from 2010/2011 and 2011/2012. Good relationships were obtained

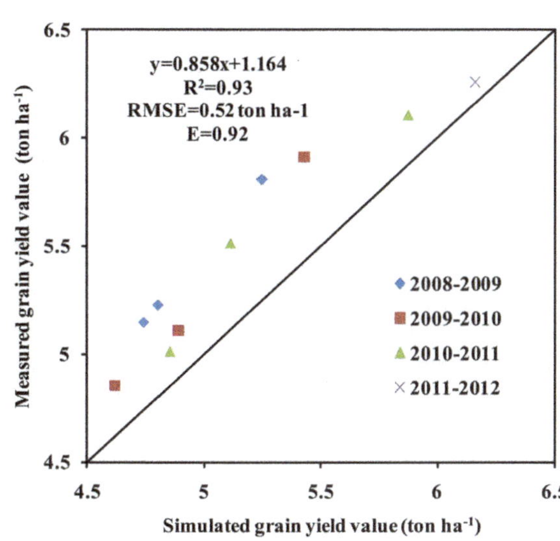

Figure 6. Relationship between the measured and simulated grain yield (GY) in winter wheat across 4 years. Note: x represents the simulated GY, y represents the measured GY. The intercept represents the relative error between the simulated GY and the measured GY. The slope represents the consistency between the simulated GY and the measured GY.

Table 7. Biomass-WUE and Grain-WUE in response to the seasonal transpiration over the different planting dates across years at Xiaotangshan experimental site.

Year	Planting date	Biomass (kg ha^{-1})	Grain (kg ha^{-1})	T (m^3 ha^{-1})	Biomass-WUE (kg m^{-3})	Grain-WUE (kg m^{-3})
2008–2009	28/9/2008	13072	5808	341.2	3.83	1.70
2008–2009	7/10/2008	11612	5227	259.1	4.48	2.02
2008–2009	20/10/2008	11144	5146	255.7	4.36	2.01
2009–2010	25/9/2009	12666	5909	322.3	3.93	1.83
2009–2010	5/10/2009	10212	5106	296.3	3.45	1.72
2009–2010	15/10/2009	9861	4852	207	4.76	2.34
2010–2011	25/9/2010	11812	6103	292.9	4.03	2.08
2010–2011	5/10/2010	11024	5512	251.1	4.39	2.20
2010–2011	15/10/2010	10062	5013	220.2	4.57	2.28
2011–2012	25/9/2011	12514	6257	306	4.09	2.04

Note: T, transpiration; WUE, water use efficiency.

between the simulated CC, BY and GY, and the measured CC, BY and GY across four years (Figs. 2,3,4,5 and 6, Tables 5 and 6). The best fitting model was obtained in 2009/2010, and the poorest fitting model was obtained in 2011/2012. These differences are very likely due to differences in rainfall and irrigation application (Fig. 1). Compared to 2008/2009, 2009/2010 and 2010/2011, the severe drought and inadequate irrigation occurred in 2011/2012. In the same year, CC, BY and GY were significantly different under different planting dates, likely due to accumulated temperature difference (Figs 3 and 4, Table 7). The results indicate that the AquaCrop model can be used to simulate CC, BY and GY for winter wheat under different planting dates and irrigation strategies. Our results (along with those in the Salemi et al. [39]), suggested that climate conditions, variety planted, and irrigation strategy could induce some differences in model simulations under different years. Heng et al. [34] demonstrated that the AquaCrop model is a good predictor of biomass and yield when irrigation is adequate; and this was corroborated by the results of this present study. In addition, the simulated BY was also consistent with the measured BY under different planting dates (Table 6 and Figs.4a–d). The CC results in this study were also similar to that observed in Salemi et al. [39], and Du et al. [40]. The results suggested that AquaCrop model could be used to simulate winter wheat CC, BY and GY under different planting dates and irrigation strategies. Winter wheat obtained higher biomass and grain yield when planted on Sep. 25th and Sep. 28th than on Oct. 5th, Oct. 7th, Oct. 15th and Oct. 20th (Table 7). This was likely due to the higher growing degree days (accumulated temperature) promoting CC growth, and BY, GY accumulation at the earlier planting date. The degree of CC affects the rate of transpiration and consequently BY and GY accumulation [32]. Therefore, the relatively higher BY and GY required relatively more temperature accumulation.

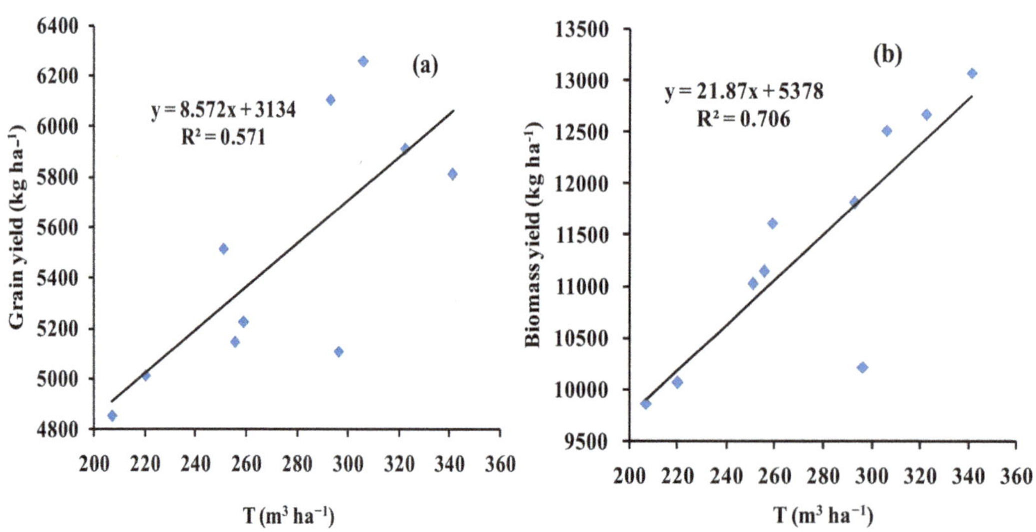

Figure 7. Relationships between the measured grain yield (GY), biomass yield (BY) and transpiration (T) in winter wheat. Note: (a) x represents transpiration, y represents the measured GY. The intercept represents the relative estimation error between transpiration and the measured GY. The slope represents the estimation consistency between transpiration and the measured GY. (b) x represents transpiration, y represents the measured BY. The intercept represents the relative estimation error between transpiration and the measured BY. The slope represents the estimation consistency between transpiration and the measured BY.

Biomass, and grain yield WUE decreased with increasing transpiration amount for all four years. This is consistent with that presented in Farahani et al. [32], but is not with that presented in Hedge [58]. In this present study, grain yield WUE ranged from 1.70 to 2.34 kg m^{-3}, reaching its maximum value on Oct. 15th 2009. Wang et al. [21] reported that grain yield WUE for winter wheat was between 0.7 and 1.3 kg m^{-3}, and Li et al. [50] reported it to be between 0.93 and 1.51 kg m^{-3} in WUE. These results are; however, not consistent with Wang et al. [26], and Li et al. [50], in which grain yield WUE was reported to be much greater. It indicated that winter wheat varieties developed faster, resulting in greater yield, leading to improvement in WUE. However, our results are consistent with that found in Fang et al. [59], in which the grain yield WUE ranged from 1.71 to 2.21 kg m^{-3}. The BY and GY in 2010/2011 and 2011/2012 were higher than in 2008/2009 and 2009/2010. This suggested that the frequent irrigation could be used to increase BY, GY, and promote biomass yield WUE and grain yield WUE for winter wheat at drought stages. This might be one of the reasons that the effects of drought on BY and GY were reduced by the frequent irrigation, thereby improving WUE.

This study demonstrated that the AquaCrop model could be used to evaluate different planting date and irrigation strategies for winter wheat in NCP. Many others have conducted AquaCrop model application studies under different crops and environment conditions [32,35,37,38,43,49]. These results indicated that AquaCrop model is stable and usable for different crops and environmental conditions. It is therefore plausible to use the AquaCrop model to improve irrigation management strategies that would maximize grain and biomass yield It is important to note that this study was limited to winter wheat in Xiaotangshan experimental site, Beijing, China. A subsequent study is focused on validating this model under deficit irrigation, thereby expanding the extent of this model's application in the future.

Conclusion

This paper demonstrated that the AquaCrop model adequately simulated the CC, BY, and GY of winter wheat under different planting dates and irrigation strategies. The simulated CC agreed well with the measured CC across all 4 years. The R^2, RMSE, E of CC winter wheat ranged from 0.89 to 0.98, 3.18% to 7.19% and 0.90 to 0.96, respectively. The measured and simulated BY were also closely related. The AquaCrop model calibrated the BY with the prediction error statistics of $0.92< R^2<0.98$, $1.12<$ RMSE <1.84 ton ha^{-1} and $0.92< E<0.96$. The simulated GY was also consistent with the measured GY with the R^2, RMSE and E values of 0.93, 0.52 ton ha^{-1} and 0.92, respectively. The results demonstrated that frequent irrigation obviously improved BY, GY, biomass WUE and grain WUE for winter wheat in 2010/2011. These results suggest that the AquaCrop model could be used to predict CC, BY and GY of winter wheat with a high degree of reliability under various planting dates and irrigation strategies situations in the North China Plain (NCP).

Acknowledgments

We are very grateful to the reviewers for their comments, Mr. Weiguo Li and Mrs. Hong Chang for data collection.

Author Contributions

Conceived and designed the experiments: XJ HF ZL. Performed the experiments: XJ HF ZL SS. Analyzed the data: XJ XZ XS GY XX WG. Contributed reagents/materials/analysis tools: XZ XS GY XX WG. Wrote the paper: XJ. Measure data: XJ HF ZL SS.

References

1. Zhang HC (2000) Thoughts on cultivation techniques for high quality of wheat in China and its processing. Jiangsu Agricultural Sciences. 5: 2–6.(in Chinese).
2. Molden D (2003) A water-productivity framework for understanding and action. In: Kijne, J.W., Barker, R., Molden, D. (Eds.), Water Productivity in Agriculture: Limits and Opportunities for Improvement. International Water Management Institute, Colombo, Sri Lanka, pp. 1–18.
3. Zwart SJ, Bastiaanssen WGM (2004) Review of measured crop water productivity values for irrigated wheat, rice, cotton and maize. Agriculture Water Management. 69: 115–133.
4. Anac MS, Ali Ul M, Tuzel IH, Anac D, Okur B, et al. (1999) Optimum irrigation schedules for cotton under deficit irrigation conditions. In: Kirda, C.,Moutonnet, P., Hera, C., Nielsen, D.R. (Eds.), Crop Yield Response to Deficit Irrigation. Kluwer Academic Publishers, Dordrecht, Boston, London, pp. 196–212.
5. Raes D, Geerts S, Kipkorir E, Wellens J, Sahli A (2006) Simulation of yield decline as a result of water stress with a robust soil water balance model. Agriculture Water Management. 81: 335–357.
6. Hill RW, Allen RG (1996) Simple irrigation calendars: a foundation for water management. In: Food and Agricultural Organization of the United Nations (FAO) (Ed.), Irrigation Scheduling: From Theory to Practice. Rome, Italy, pp. 69–74.
7. Raes D, Sahli A, Van Looij J, Ben Mechlia N, Persoons E (2000) Charts for guiding irrigation in real time. Irrigation Drainage System.14: 343–352.
8. De Nys E, Kipkorir EC, Sahli A, Vaes R, Raes D (2001) Design of farmer oriented irrigation charts by using a soil-water balance model. In: Proceedings of the 4th Inter Regional Conference on Environment-Water, ICID, pp. 235–243.
9. Legesse A (2006) Sustainable irrigation in the Tigray Highlands of Northern Ethiopia. Master dissertation of the Master in Water Resources engineering. K.U. Leuven University, Leuven, Belgium.
10. Farre F, Faci JM (2009) Deficit irrigation in maize for reducing agricultural water use in a Mediterranean environment. Agriculture Water Management. 96: 384–394.
11. Kipkorir EC (2002) Optimal planning of deficit irrigation for multiple crop systems according to user specified strategy. Dissertations de Agricultura No. 514. Faculty of Agricultural Sciences. K.U. Leuven University, Belgium.
12. Debaek P, Aboudrare A (2004) Adaptations of crop manage to water-limited environments. European Journal of Agronomy. 21: 433–446.
13. Ali MH, Talukder MSU (2008) Increasing water productivity in crop production. A synthesis. Agricultural Water Management. 95: 1201–1213.
14. Behera SK, Panda RK (2009) Integrated management of irrigation water and fertilizers for wheat crop using field experiments and simulation modeling. Agricultural Water Management. 96, 1532–1540.
15. Blum FA (2009) Effective use of water (EUW) and not water-use efficiency (WUE) is the target of crop yield improvement under drought stress. Field Crops Research. 112: 119–123.
16. Geerts S, Raes D, Gracia M, Miranda R, Cusicanqui JA, et al. (2009) Simulating yield response of Quinoa to water availability with AquaCrop. Agronomy Journal. 101: 499–508.
17. Jensen ME (1968) Water consumption by agricultural plants. In:Water Deficits in Plant Growth (T T Koslowski, ed.). Academic Press, New York, 1: 1–22.
18. Rao NH, Sarma PBS, Chander S (1988) Irrigation scheduling under limited water supply. Agriculture Water Management. 15: 165–175.
19. Rao NH, Sarma PBS, Chander S (1992) Real time adaptive irigation scheduling under a limited water supply. Agriculture Water Management. 20: 267–279.
20. Penning de Vries FWT, Jansen DM, Berge ten HFM, Bakema A (1989) Simulation of Ecophysiological process of growth in several annual crops. Simulation Monographys, Pudoc, Wagenningen, pp. 82–96.
21. De Wit CT (1970) Dynamic concepts in biology. In: Prediction and Management of Photosynthetic Productivity. Proceedings International Biological Program/Plant Production Technical Meeting. Wageningen, Netherlands: PUDOC, pp. 17–23.
22. Downey LA (1972) Water-yield relations for non-forage crops. Journal of Irrigation and Drainage Division, American Society of Civil Engineers. 98: 107–115.
23. Minhas BS, Parikh KS, Srinivasan TN (1974) Towards the structure of a production function for wheat yields with dated in Puts of irrigation water. Water Resource Research. 10: 383–393.
24. Blank H (1975) Optimal irrigation decisions with limited water (D). Ph. D. Thesis, Department of Civil Engineering, Colorado State University, Fort Collins, CO.
25. Stewart JI, Hagan RM, Pruitt WO (1976) Production function sand predicted irrigation programs for Principal crops as required for water resources planning and increased water use efficiency (R). Final Report, U.S. Department of Interior, Washington DC, 1976, pp. 80.

26. Wang JP, Sun XF (2001) A Combining Water-Crop Relation Model. Irrigation and Drainage. 2: 76–78.
27. Kang SZ, Cai HJ, Feng SY (2004) Technique innovation and research fields of modern agricultural and ecological water-saving in the future. Transactions of the Chinese Society of Agricultural Engineering. 20: 1–6. (in Chinese).
28. Doorenbos J, Kassam AH (1979) Yield response to water. FAO irrigation and drainage paper no. 33. FAO, Rome.
29. Steduto P, Hsiao TC, Raes D, Fereres E (2009) AquaCrop-The FAO crop model to simulate yield response to water. I. Concepts. Agronomy Journal. 101: 426–437.
30. Hsiao TC, Heng L, Steduto P, Roja-Lara B, Raes D, et al. (2009) AquaCrop-The FAO model to simulate yield response to water: parametrization and testing for maize. Agronomy Journal. 101: 448–459.
31. Raes D, Steduto P, Hsiao TC, Fereres E (2009a) AquaCrop-The FAO Crop Model to Simulate Yield Response to Water: Reference Manual Annexes., www.fao.org/nr/water/aquacrop.html.
32. Farahani HJ, Izzi G, Oweis TY (2009) Parameterization and evaluation of the AquaCrop model for full and deficit irrigated cotton. Agronomy journal.101: 469–476.
33. Garcia-Vila M, Fereres E, Mateos L, Orgaz F, Steduto P (2009) Deficit irrigation optimization of cotton with AquaCrop. Agronomy journal. 101: 477–487.
34. Heng LK, Hsiao TC, Evett S, Howell T, Steduto P (2009) Validating the FAO AquaCrop model for irrigated and water deficient field maize. Agronomy Journal 101: 488–498.
35. Vanuytrecht E, Raes D, Willems P (2011) Considering sink strength to model crop production under elevated atmospheric CO_2. Agricultural and Forest Meteorology 15: 1753–1762.
36. Zinyengere N, Mhizha T, Mashonjowa E, Chipindu B, Geerts S, et al. (2011) Using seasonal climate forecasts to improve maize production decision support in Zimbabwe. Agricultural and Forest Meteorology. 151: 1792–1799.
37. Stricevic R, Cosic M, Djurovic N, Pejic B, Maksimovic L (2011) Assessment of the FAO AquaCrop model in the simulation of rainfed and supplementally irrigated maize, sugar beet and sunflower. Agriculture Water Management. 98: 1615–1621.
38. Abedinpour M, Sarangi A, Rajput TBS, Singh M, Pathak H, et al. (2012) Performance evaluation of AquaCrop model for maize crop in a semi-arid environment. Agricultural Water Management. 110: 55–66.
39. Salemi H, Mohd-Soom MA, Lee TS, Mousavi SF, Ganji A, et al. (2011) Application of AquaCrop model in deficit irrigation management of Winter wheat in arid region. African Journal of Agricultural Research. 610: 2204–2215.
40. Du WY, He XK, Shamaila Z, Hu ZF, Zeng AJ, et al. (2011) Yield and biomass prediction testing of aquaCrop model for winter wheat. Transactions of the Chinese Society for Agricultural Machinery.42: 174–183. (in Chinese).
41. Wang XX, Wang QJ, Fan J, Fu QP (2013) Evaluation of the AquaCrop model for simulating the impact of water deficits and different irrigation regimes on the biomass and yield of winter wheat grown on China's Loess Plateau. Agricultural Water Management 129: 95–104.
42. Soddu A, Deidda R, Marrocu M, Meloni R, Paniconi C, et al. (2013) Climate variability and durum wheat adaptation using the AquaCrop model in southern Sardinia. Procedia Environmental Sciences 19: 830–835.
43. Todorovic M, Albrizio R, Zivotic L, Abi saab M, Stwckle C, et al. (2009) Assessment of AquaCrop, CropSyst, and WOFOST models in the simulation of sunflower growth under different water regimes. Agronomy Journal. 101: 509–521.
44. Karunaratne AS, Azam-ali SN, Izzi G, Steduto P (2011) Calibraion and validation of FAO-AquaCrop model for irrigated and water deficient bambara groundnut. Experimental Agriculture 47: 509–527.
45. Garcia-Vila M, Fereres E (2012) Combining the simulation crop model AquaCrop with an economic model for the optimization of irrigation management at farm level. European Journal Agronomy. 36: 21–31.
46. Yuan M, Zhang L, Gou F, Su Z, Spiertz JHJ, et al. (2013) Assessment of crop growth and water productivity for five C_3 species in semi-arid Inner Mongolia. Agricultural Water Management 122: 28–38.
47. Araya A, Keesstra SD, Stroosnijder L (2010a) Simulating yield response to water of Teff (Eragrostis tef) with FAO's AquaCrop model. Field Crops Research. 116: 196–204.
48. Araya A, Habtu S, Hadgu KM, Kebede A, Dejene T (2010b) Test of AquaCrop model in simulating biomass and yield of water deficient and irrigated barley (Hordeum vulgare). Agricultural Water Management. 97: 1838–1846.
49. Abrha B, Delbecque N, Raes D, Tsegay A, Todorvovic M, et al. (2012) Sowing strategies for barley (Hordeum vulgare L.) based on modeled yield response to water with AquaCrop. Experimental Agriculture. 48: 252-271.
50. Li ZZ, Xu Y, Lu XJ, Hu KL, Jiang LH, et al. (2011) Evaluation of the AquaCrop model for simulating biomass for Chinese green onion and soil water storage. Journal o f China Agricultural University.16: 59–66. (in Chinese).
51. Katerji N, Campi P, Mastrorilli M (2013) Productivity, evapotranspiration, and water use efficiency of corn and tomato crops simulated by AquaCrop under contrasting water stressconditions in the Mediterranean region. Agricultural Water Management 130: 14–26.
52. Allen RG, Periera LS, Raes D, Smith M (1998) Crop evapotranspiration. Guidelines for computing crop water requirement. FAO Irrigation and Drainage Paper No.56. FAO, Rome.
53. FAO (2009) ETo calculator version 3.1. In: Evapotranspiration from reference surface, FAO, Land and Water Division, Rome, Italy.
54. Kar G, Verma HN, Singh R (2006) Effect of winter crop and supplementary irrigation on crop yield, water us efficiency and profitability in rainfed rice based on cropping system of eastern India. Agriculture Water Management. 79: 280–292.
55. Oweis T, Hachum A (2006) Water harvesting and supplementary irrigation for improved water productivity of dry farming systems in West Asia and North Africa. Agriculture Water Management. 80: 57–73.
56. Raes D, Steduto P, Hsiao TC, Fereres E (2009b) Crop Water Productivity. Calculation Procedures and Calibration Guidance. AquaCrop version 3.0. FAO, Land and Water Development Division, Rome.
57. Raer D, Steduto P, Hsiao TC, Fereres E (2010) Reference Manual. Annexes-AquaCrop, Crop parameters. Annex I Section I.10 Wheat, A39–A42. FAO, Land and Water Development Division, Rome.
58. Hedge DM (1987) The effect of soil water potential, method of irrigation, canopy temperature, yield and water use of radish. Horticultural Science. 62: 507–511.
59. Fang QX, Chen YH, Li QQ, Yu SZ, Luo Y, et al. (2004) Effect of irrigation on water use efficiency of winter wheat. Transactions of the Chinese Society of Agricultural Engineering. 20: 34–39. (in Chinese).

6

Microbial Degradation of Acetamiprid by *Ochrobactrum* sp. D-12 Isolated from Contaminated Soil

Guangli Wang*, Xiao Chen, Wenlong Yue, Hui Zhang, Feng Li, Minghua Xiong

College of Life Sciences, Huaibei Normal University, Huaibei, People's Republic of China

Abstract

Neonicotinoid insecticides are one of the most important commercial insecticides used worldwide. The potential toxicity of the residues present in environment to humans has received considerable attention. In this study, a novel *Ochrobactrum* sp. strain D-12 capable of using acetamiprid as the sole carbon source as well as energy, nitrogen source for growth was isolated and identified from polluted agricultural soil. Strain D-12 was able to completely degrade acetamiprid with initial concentrations of 0–3000 mg·L^{-1} within 48 h. Haldane inhibition model was used to fit the special degradation rate at different initial concentrations, and the parameters q_{max}, K_s and K_i were determined to be 0.6394 (6 h)$^{-1}$, 50.96 mg·L^{-1} and 1879 mg·L^{-1}, respectively. The strain was found highly effective in degrading acetamiprid over a wide range of temperatures (25–35°C) and pH (6–8). The effects of co-substrates on the degradation efficiency of acetamiprid were investigated. The results indicated that exogenously supplied glucose and ammonium chloride could slightly enhance the biodegradation efficiency, but even more addition of glucose or ammonium chloride delayed the biodegradation. In addition, one metabolic intermediate identified as *N*-methyl-(6-chloro-3-pyridyl)methylamine formed during the degradation of acetamiprid mediated by strain D-12 was captured by LC-MS, allowing a degradation pathway for acetamiprid to be proposed. This study suggests the bacterium could be a promising candidate for remediation of environments affected by acetamiprid.

Editor: Stephen J. Johnson, University of Kansas, United States of America

Funding: The authors would like to thank the financial support from the National Natural Science Foundation (No. 31100083), P. R. China. The funders had no role in study design, data collection and analysis, decision to publish, or preparation of the manuscript.

Competing Interests: The authors have declared that no competing interests exist.

* E-mail: wanf-3344@163.com

Introduction

Neonicotinoid insecticides, which are one of the most important classes of commercial insecticides worldwide, are systemic in plants and animals and are used to manage crop pests and control fleas on cats and dogs [1–4]. Acetamiprid, a chloropyridinyl neonicotinoid, was considered to be a favorable choice for controlling those pests that are severely resistant to organophosphorus, urethane and synthetic pyrethroid pesticides, so it is regarded by EPA as an important substitute of organophosphorus pesticides [5]. Owing to its broad insecticidal spectrum and relatively low acute and chronic mammalian toxicity, acetamiprid is used widely in crop protection [24,25]. Though the half life of acetamiprid in field was reported to be about 2.8–14 days [6], the risk of its ambient pollution, principally in water, is still present [7]. Acetamiprid exposure has been proven to have adverse effect on greenhouse workers spraying acetamiprid [8], soil microorganisms [25] and beneficial insects [24]. In recent years, acetamiprid residues in crops are receiving considerable attention due to their potential toxicity to humans [9,10], and methods for the biotransformation of neonicotinoids are being actively researched. Microbial degradation is considered as an efficient "green" solution to eliminate environmental polluting chemicals [11,12]. Acetamiprid metabolism in microorganisms has been studied in *Stenotrophomonas* sp. THZ-XP, *Rhodotorula mucilaginosa* IM-2, *Stenotrophomonas maltophilia* CGMCC 1.1788, *Pseudomonas* sp. FH2, *Phanerochaete sordida* YK-624 and *Pigmentiphaga* sp. strain AAP-1 [4,13–19].

In this report, a highly effective acetamiprid-degrading strain, designated D-12, was isolated from polluted agricultural soil and identified as *Ochrobactrum* sp. Strain D-12 was able to use acetamiprid as the sole source of carbon and nitrogen for growth and completely degraded 300 mg·L^{-1} acetamiprid within 6 h. A kinetic model of acetamiprid degradation and transformation was proposed. Effects of several parameters including pH, temperature, initial substrate concentration and second carbon/nitrogen source as co-substrate on the biodegradation of acetamiprid by the isolated strain were investigated. Moreover, the pathway of acetamiprid biodegradation mediated by strain D-12 is proposed. This paper highlights a significant potential use of pure cultures of microbial cells for the cleanup of acetamiprid-contaminated soil.

Materials and Methods

Chemicals and Media

Analytical grade acetamiprid (purity, 99%), purchased from Wako Pure Chemical Industries (Osaka, Japan), was used as a standard. Acetamiprid samples (>97% purity) were purchased from Dongfeng Pesticides Factory (Shanghai, China). Chromatographic grade acetonitrile and acetic acid were purchased from Sigma-Aldrich (St. Louis, USA). Molecular biology reagents were purchased from TaKaRa Biotechnology Co., Ltd (Dalian). All other chemicals were of the highest grade that was commercially available. Luria-Bertani (LB) medium (10.0 g NaCl, 10.0 g peptone and 5.0 g yeast extract per litre water, pH 7.0) and mineral salts medium (MSM) (1.5 g K_2HPO_4, 0.5 g KH_2PO_4, 0.2 g $MgSO_4 \cdot 7H_2O$, 1.0 g NaCl per litre water, pH 7.0) were utilised in this study. For solid medium, agar powder was added at a concentration of 1.6%. When necessary, acetamiprid was added to the media at an appropriate concentration. All media used in this study were prepared using Milli-Q water (>18.2 MΩ) and sterilised by autoclaving at 121°C for 25 min.

Isolation and identification of bacteria

A soil sample was collected from an agricultural field, which had been exposed to acetamiprid for more than 10 years, in the city of Yancheng, China. Enrichment and isolation of degrading bacterial isolates were conducted as described in detail previously [18]. The ability of isolates to degrade acetamiprid was determined by HPLC following the protocol described below. One pure isolate designated D-12 showing the highest degradation activity was selected for further study.

Strain D-12 was characterized based on its morphological, physiological and biochemical properties [20] and genetic analysis based on 16S rRNA gene sequence. The cell morphology was examined by light microscopy (BH-2, Olympus, Japan) and transmission electron microscopy (H-7650, Hitachi High-Technologies Corp., Japan), using cells from an exponentially growing culture. The genomic DNA of D-12 strain was extracted by high-salt precipitation [21]. Pure cultures were phylogenetically characterized using 16S rRNA gene sequencing. Two PCR primers 27F/1492R were designed to amplify the 16S rRNA gene [22]. The 1389 bp 16S rRNA gene sequence was compared to sequences in GenBank using BLAST program. Multiple sequence alignment was carried out using Clustal X 1.8.3 with the default settings. For further phylogenetic analysis, MEGA version 4.0 software [23] was used. Distances were calculated using the Kimura two-parameter distance model. Unrooted trees were built using the Neighbor Joining method. The date set was bootstrapped 1000 times.

Acetamiprid degradation experiments

The isolated strain was grown on LB medium for 16 h at 30°C on a rotary shaker (160 rpm). Then cells were harvested and washed three times with a 0.02 mol·L^{-1} phosphate buffer (pH 7.0). The washed cells were re-suspended in the same buffer, resulting in a cell suspension with an OD_{600} of 1.0. In addition, a stock solution of acetamiprid (4,000 mg·L^{-1}) was prepared by dissolving the acetamiprid in MSM. After transferring appropriate volume of this stock solution to a 250 mL sterile flask, 100 mL of sterilized MSM was added. 1 mL of the prepared cell suspension was then inoculated into the medium for biodegradation assessment. All tests were conducted in triplicate.

To assess the effects of pH, temperature and initial substrate concentration on acetamiprid degradation, incubation temperature (15, 20, 25, 30, 35, 40 and 45°C), initial acetamiprid concentration (250, 500, 1000, 1500, 2000, 2500 and 3000 mg·L^{-1}) and media pH (4.0–10.0, in increments of 1.0 pH units) on biodegradation of acetamiprid by strain D-12 were studied.

To study the effect of second carbon or nitrogen sources as co-substrate on the degradation of acetamiprid, different dosages (100, 200, 300, 400, and 500 mg·L^{-1}) of glucose as a second carbon source as well as different dosages (300, 600 and 900 mg·L^{-1}) of ammonium chloride as a nitrogen source as well as different dosage of glucose (100, 200, 300, 400 and 500 mg·L^{-1}) as a second carbon source with the presence of 300 mg·L^{-1} of ammonium chloride as a nitrogen source was respectively added into flasks which contained 100 mL of MSM medium, 5 mL of strain D-12 culture and 300 mg·L^{-1} of acetamiprid. Acetamiprid added as sole carbon and nitrogen source was set as control. Each treatment was set in triplicate.

Chemical Analysis

Cell growth was monitored by measuring the optical density of culture samples at 600 nm (OD_{600}). Non-inoculated medium served as control. For acetamiprid extraction from liquid culture, 5 mL of sample collected from the medium was extracted with 10 mL of dichloromethane. After shaking for 10 min, the dichloromethane phase was dried over anhydrous Na_2SO_4, and the solvent was removed using a stream of nitrogen at room temperature. The residues were dissolved in 200 μL of acetonitrile. Samples in acetonitrile were then filtered through a 0.22 μm Millipore membrane filiter. An aliquot of the solution (20 μL) was injected into an HPLC system for detection.

The concentration of acetamiprid was determined by HPLC using a Zorbax C-18 ODS Spherex column (250 mm*4.6 mm). The mobile phase was 65% (volume) water and 35% (volume) acetonitrile as well as 0.01% acetic acid at a flow rate of 1 mL·min-1. The eluate was monitored by measuring the A242 with a Waters 2487 Dual Wavelength Absorbance Detector, and the injection volume was 20 μL. Recovery efficiency of the stated method was evaluated at the concentrations of 10, 30, 50 and 70 mg·L^{-1} acetamiprid that appended in MSM.

The metabolites produced during acetamiprid degradation were purified using thin-layer chromatography (TLC) by concentrating the extract on a pre-coated silica-gel TLC plate (silica G, 20×20 cm, 0.25 mm thickness) with chloroform-methanol solution (20:1 by volume). The collected metabolite was dissolved in acetonitrile and centrifuged at 10,000 g to remove the silica. The organic solvents were then removed and the residue was dried under vacuum condition.

The purified metabolites were analyzed by standard MS, ionized by electrospray with a positive polarity, and scanned in the normal mass. Characteristic fragment ions were detected with second-order MS.

The MS apparatus was an LC-MSD-Trap-SL system equipped with an electrospray ionization source and was operated in the positive polarity mode. The ES-MS interface was operated using a gas temperature of 35°C and a drying gas flow of 9.0 L min^{-1}. The nebulizer nitrogen gas pressure was 45 psi. Full scan signals were recorded within the m/z range from 50 m/z to 600 m/z. For LC-MS, the spray voltage was 7.0 kV. The sheath and auxiliary gases were nitrogen. The sheath and auxiliary gases were adjusted to 65 and 10 arbitrary units, respectively. Auto Gain Control mode was used to optimize the injection time.

Data Analysis

Results were also assessed by analysis of variance (ANOVA) and statistical analyses were performed on three replicates of data obtained from each treatment. The significance ($P<0.05$) of differences was treated statistically by single factor analysis of variance using SPSS software packages.

Ethics statement

The location that we collected samples was a very common field, so it did not need specific permission and involve any endangered or protected species. In order to avoid being influenced by plant diseases and insect pests, acetamiprid was used discontinued ten years ago.

Table 1. Recovery efficiency and regression equation determined by different concentrations of AAP for the analytical method evaluation.

Append concentration ($mg \cdot L^{-1}$)	Detected concentration	Recovery efficiency (%)	Regression equation
10	10.44±0.53	104	
30	28.12±1.45	94	y = 12.224x+1.0760
50	43.34±0.32	87	$R^2 = 0.9982$
70	67.35±2.15	91	

Regression equation was determined by HPLC using different concentrations of standard AAP; y, AAP concentration; x, the peak area of HPLC; R^2, correlation coefficient of the regression equation.

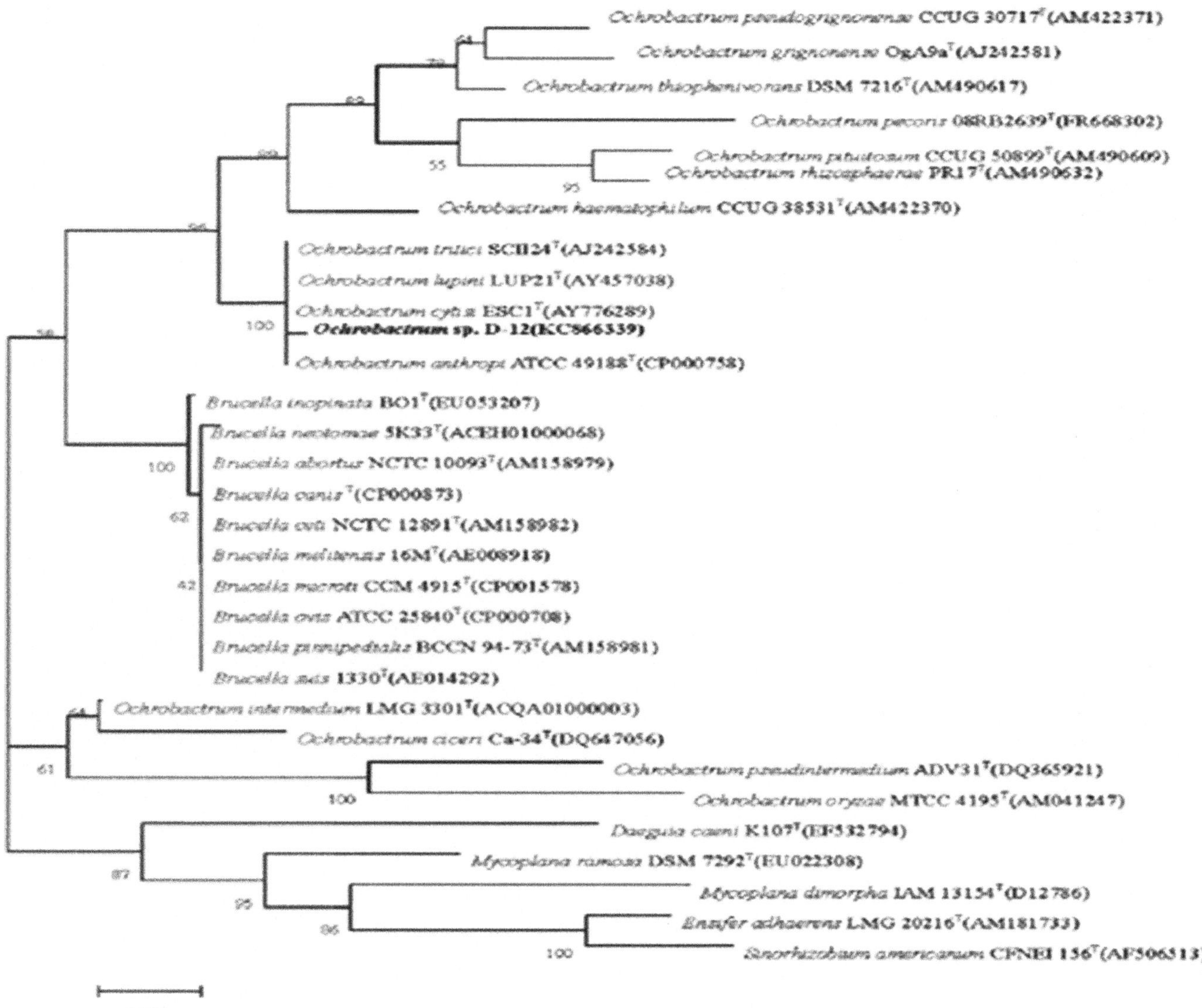

Figure 1. Phylogenetic tree constructed by the neighbor-joining method based on 16S rRNA gene sequences of D-12 and type strains of *Ochrobactrum* species. Bootstrap values, expressed as percentages of 1,000 replications, are given at branching points. Bar, 0.002 nucleotide substitutions per nucleotide position.

Table 2. Morphological and physio-biochemical characteristics of *Ochrobactrum* sp. D-12.

Characteristics	Results	Characteristics	Results
Colonies	**white to beige, mucoid with entire edges**	**Cells**	**2–3 mm in diameter within 24 h**
Gram-staining	–	β-galactosidase	+
Growth at 42°C	+	α-Glucosidase	+
Growth at pH 9	+	Lipase C14	+
Growth with 7% NaCl	+	Glucose	+
Nitrate reduction	+	Citrate (24 h)	–
indole	–	D-Mannose	+
Aesculin hydrolysis	+	L-Arabinose	–
oxidase	+	D-Turanose	+
catalase	+	L-Lyxose	+
Urease (48 h)	–	Sodium acetate	+

+, tested positive/utilized as substrate; 2, tested negative/not utilized as substrate.

Results and Discussion

Evaluation of the analytical method for acetamiprid determination

The recovery efficiency of acetamiprid in MSM are arranging from 87% to 104%. The evaluation results displayed that the analytical method applied in this study well satisfied the requirement of the pesticide analysis standard (Table 1; satisfy scopes are from 80% to 120%). The LOD (limit of detection) and LOQ (limit of quantification) of the method were 1.0×10^{-10} g and 0.002 $mg\cdot L^{-1}$, respectively.

Isolation and identification of the acetamiprid-degrading strain D-12

Acetamiprid is a member of the neonicotinoid group of insecticides commonly used against wide range of insect pests. Owing to its broad insecticidal spectrum and relatively low acute and chronic mammalian toxicity, acetamiprid is used widely in crop protection [24,25], therefore, it is probable that several bacteria have adapted to this acetamiprid-contaminated environment. A pure bacterial strain that could grow by using acetamiprid as the sole source of carbon and nitrogen was obtained from contaminated soil. The isolated strain was found to be aerobic, non-spore-forming, Gram-negative rods and motile with a polar flagellum (Fig. S1). Colonies of strain D-12 on LB agar were white to beige, mucoid with entire edges and 2–3 mm in diameter within 24 h at 30°C. Strain D-12 was positive for oxidase, catalase, β-galactosidase, α-Glucosidase, Lipase C14 and Nitrate reduction and negative for indole and urease. It could hydrolyze Aesculin. Strain D-12 could grow in LB in the presence of 7% NaCl (w/v), and the optimal growth was at 1–2% NaCl (w/v). Growth occurs at temperature of 4–42°C with an optimum at 28–30°C. Growth occurs at pH 6.0–9.0, with an optimum at pH 7.0–7.5. Strain D-12 could use glucose, D-Mannose, L-Arabinose, D-Turanose, L-Lyxose, Sodium acetate as sole carbon sources for growth but failed to utilize Citrate and L-Arabinose (Table 2). Phylogenetic analysis of the 16S rRNA gene sequences (Fig. 1) revealed strain D-12 clustered with members of the genus *Ochrobactrum*, and had 100% sequences similarity with *Ochrobactrum tritici* SCII24^{T} (AJ242584), *Ochrobactrum lupini* LUP21^{T} (AY457038), *Ochrobactrum anthropi* ATCC 49188^{T} (CP000758) and *Ochrobactrum cytisi* ESC1^{T} (AY776289). Members of this genus have become the focus of some studies because of its versatile biodegradability. A few of strains from *Ochrobactrum* genus have been reported to degrade various xenobiotics such as vinyl chloride [26], chlorothalonil [27,28], methyl parathion [29] dimethyl formamide [30] and nicotine [31]. However, this is the first report of *Ochrobactrum* sp. degrading acetamiprid. *Ochrobactrum* are ubiquitous and numerous in soil and able to survive under extremely harsh conditions. These features make them ideal candidates for bioremediation of contaminated environments.

The pathway of acetamiprid biodegradation by strain D-12

As shown in Fig. S2, the metabolite was detected in the HPLC analysis of a 10-day MM culture fluid inoculated with *Ochrobactrum* sp. strain D-12. To determine the structure of the metabolite produced during the degradation of acetamiprid, 15-day cultures of *Ochrobactrum* sp. strain D-12 in MM medium supplemented with 100 μM acetamiprid were subjected to TLC and HPLC. The purified metabolite with the characteristic second-order MS fragment ion peaks at m/z 64.02, 73.87, 82.80,104.57,122.20,142.02 and 157.01 (Fig. 2–3), was identified as *N*-methyl-(6-chloro-3-pyridyl)methylamine. Therefore, the degradation pathway of acetamiprid by isolated D-12 was proposed (Fig. S3). The metabolite idenfied is known to be less toxic to mammals and bees, which was previously identified in mice and honeybees as IM1-4 [32–34]. The hydrolytic mechanism is similar to the metabolic conversion of the compound in mammals and insects [13].

Effects of pH, temperature and initial concentrations on acetamiprid degradation by D-12

The effects of pH and temperature on the aerobic degradation of D-12 in the culture medium were tested after incubation 48 h. From the analysis,we found that pH has significant influence ($P<0.05$) on the experiment by SPSS software(Version 19.0). when the pH was between 6.0 and 8.0, more than 95% of 300 $mg\cdot L^{-1}$ acetamiprid could be degraded by D-12 within 16 h. At pH 7.0, acetamiprid was completely degraded in 14 h. However, when the pH was at 4.0, 5.0, 9.0 or 10.0, acetamiprid biodegradation was distinctly inhibited, implying that restricted growth of D-12

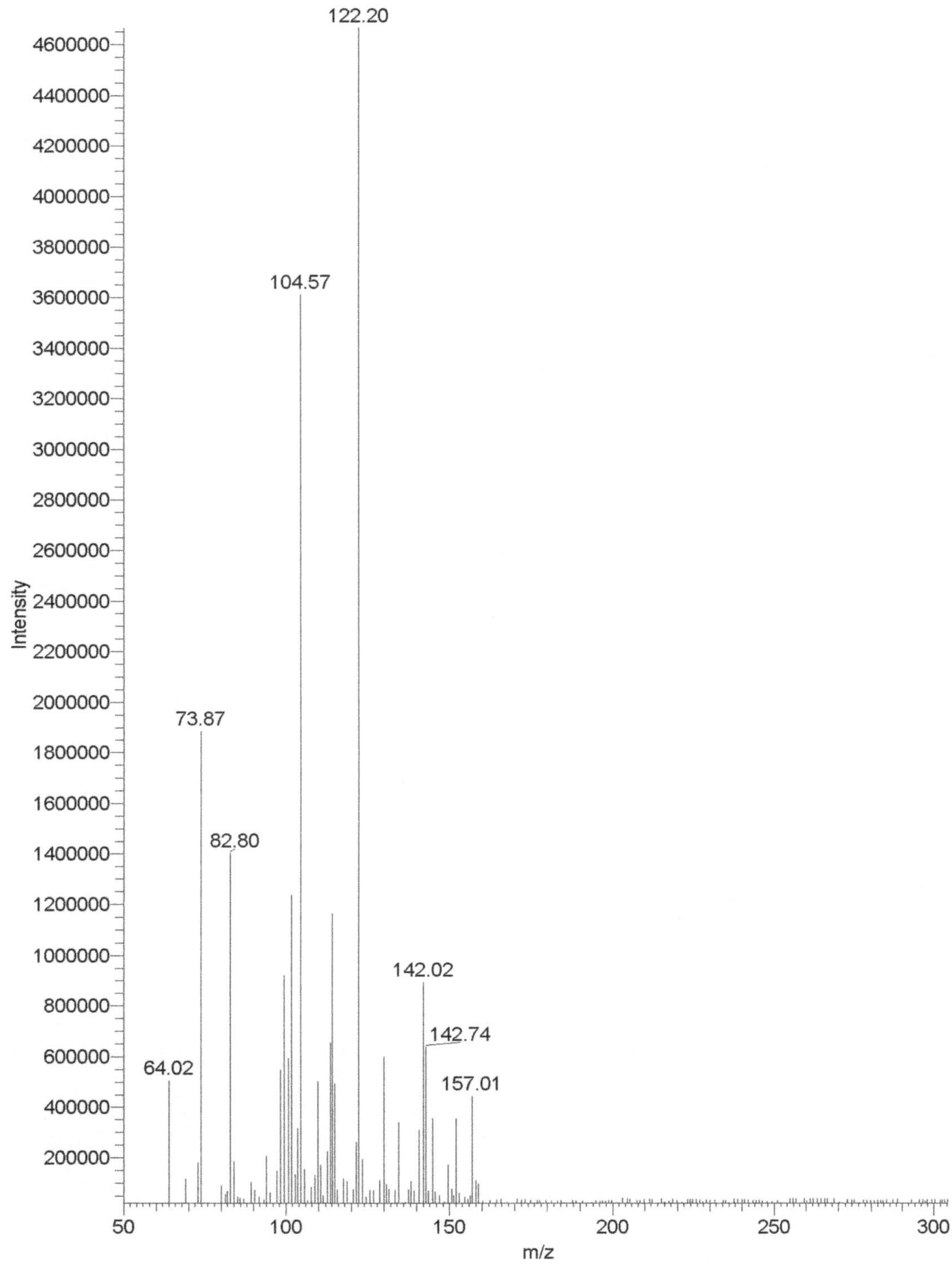

Figure 2. Mass spectrum of the metabolite.

occurred at these pH values. But the temperature has no significant ($P>0.05$).Experiment showed that the optimum temperature for the biodegradation of acetamiprid was 25–35°C. Acetamiprid biodegradation efficiency was the highest at this temperature range; however, acetamiprid biodegradation decreased when the temperature dropped to 20°C or rose to 40°C, indicating that lower and higher temperatures were not beneficial for the biodegradation of acetamiprid by D-12. Therefore, pH 7.0 and 30°C were chosen for all subsequent experiments. At the same time, control experiments were carried out under the same conditions without bacteria. No obvious degradation was detected in the control experiments. The recovery rates of acetamiprid after sample pretreatment were measured between 94% and 102%.

In order to determine the effect of initial acetamiprid concentrations on degrading efficiency, biodegradation of acetamiprid by *Ochrobactrum* sp. D-12 was conducted under acetami-

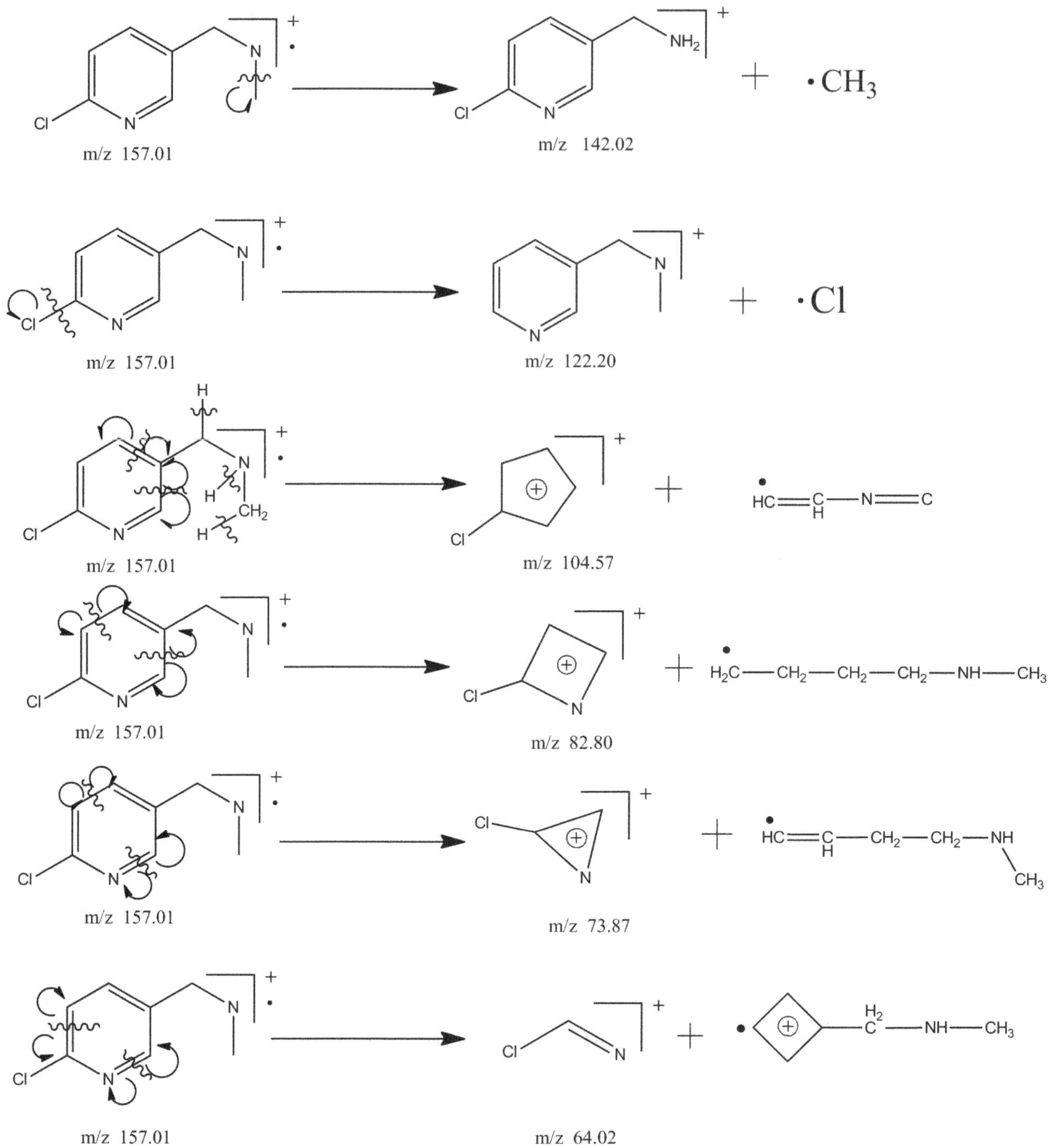

Figure 3.The detailed explanation for the MS fragment ion peaks.

prid concentrations ranging 250, 500, 1000, 1500, 2000, 2500 and 3000 mg·L^{-1}. Strain D-12 grew on acetamiprid up to the concentration, as high as 3000 mg·L^{-1}. As shown in Fig. 4, the lag phase was extended at higher acetamiprid concentration. At concentrations of 250, 500, 1000, 1500 and 2000 mg·L^{-1}, the degradation rates reached 70.80%, 66.80%, 64.00%, 60.27% and 55.60% after 12 hours of incubation, respectively. However, only 50.68% and 39.27% degradation were achieved at higher initial concentrations of 2500 and 3000 mg·L^{-1}, respectively.

The decrease in the specific acetamiprid degradation rate with an increase in the initial acetamiprid concentration suggests that acetamiprid may act as a partial inhibitor to strain D-12. Therefore, the substrate inhibition model Eq. (1) adapted from [35] was used to explain the degradation kinetics of acetamiprid by strain D-12.

$$q = \frac{q_{\max} S}{S + K_s + (S^2/K_i)} \quad (1)$$

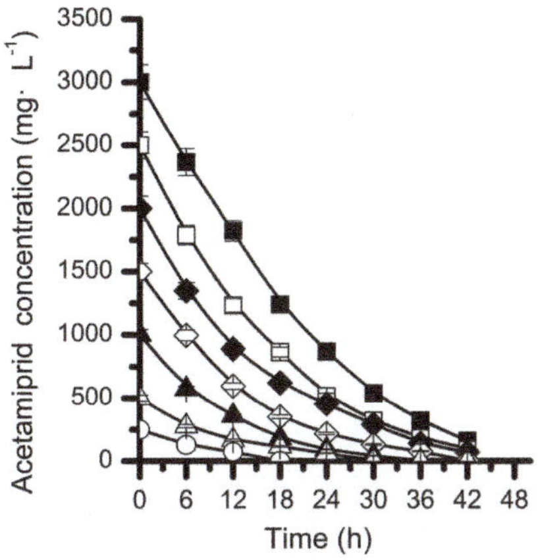

Figure 4. Degradation kinetics of acetamiprid at different initial concentrations by strain D-12. (■), 3000 mg·L^{-1}; (□), 2500 mg·L^{-1}; (◆), 2000 mg·L^{-1}; (◇), 1500 mg·L^{-1}; (▲), 1000 mg·L^{-1}; (Δ), 500 mg·L^{-1}; (○), 250 mg·L^{-1}; values are the means of three replicates with standard deviation.

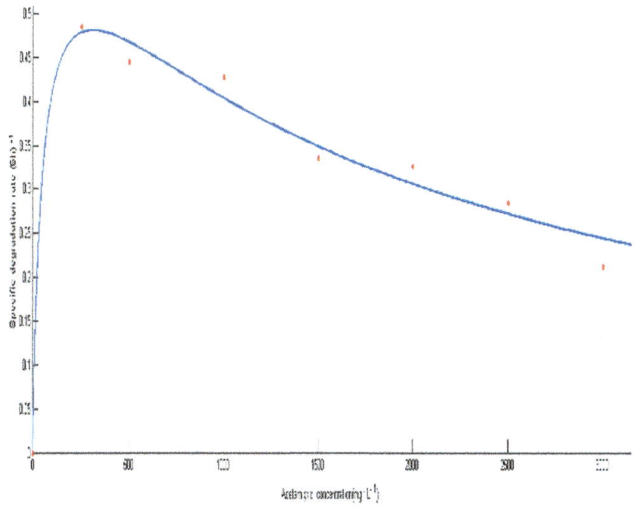

Figure 5. Relationship between specific degradation rate and initial acetamiprid concentration by strain D-12.

where q_{max} is the maximum specific acetamiprid degradation rate (6 h)$^{-1}$, K_i is the substrate inhibition constant (mg·L^{-1}), K_s is the half-saturation constant (mg·L^{-1}), S is the substrate concentration (mg·L^{-1}), and S_m is a critical inhibitor concentration of the substrate which decreases degradation.

The relationship between q and initial acetamiprid concentration is shown in Fig. 5. The kinetic parameters of strain D-12 estimated from non-linear regression analysis using matrix laboratory (MATLAB) software (Version 7.8) were q_{max} = 0.6394 (6 h)$^{-1}$, K_s = 50.96 mg·L^{-1}, and K_i = 1879 mg·L^{-1}, respectively. The S_m was established to be 312.8 mg·L^{-1}. The value of R^2 was 0.9827, which indicates that the experimental data were well correlated with the model. As indicated in Fig. 5, when S were lower than 312.8 mg·L^{-1}, q gradually increased. At higher concentrations, inhibition by acetamiprid became prominent.

It was noteworthy that this particular strain could tolerate and efficiently degrade acetamiprid up to the concentration, as high as 3000 mg·L^{-1}. However, the specific acetamiprid degradation rate decreased with an increase in the initial acetamiprid concentration (Fig. 5). These findings indicated that increased acetamiprid concentration had a marked effect on degradation performance of strain D-12, but did not lead to complete inhibition. These results proved that strain D-12 was responsible for acetamiprid degradation.

Effects of different co-substrates on the biodegradation of acetamiprid by strain D-12

Industrial wastes often contain a mixture of recalcitrant compounds as well as easily biodegradable compounds. Investigation of the biodegradation of refractory pollutants in the presence of accessible carbon and nitrogen sources might aid in reducing the toxic and growth-inhibiting effects of xenobiotics on cells, thereby increasing the transformation rates of xenobiotics [36–38]. However, the reverse consequences had been also reported [39–42]. Consequently, effect of the two substrates on acetamiprid biodegradation by strainD-12 was still investigated. Generally, for the bacteria, glucose and ammonium chloride are accessible carbon and nitrogen source, respectively. As shown in Fig. 6 the effect of glucose as a second carbon source or/and ammonium chloride as a second nitrogen source (co-substrate) on the biodegradation of acetamiprid by strain D-12 was investigated.

The acetamiprid degradation efficiency could reach as high as 73% within 12 h when only acetamiprid was added as a sole carbon source (control). Comparatively, after adding 100 mg·L^{-1} glucose as the second carbon source (Fig. 6a), the acetamiprid degradation efficiency was enhanced and 200 mg·L^{-1} acetamiprid was completely degraded within 10 h. However, when more than 300 mg·L^{-1} glucose was added, the acetamiprid degradation efficiency decreased slightly within the same period (Fig. 6a). It has been reported that the addition of glucose promotes the growth of the strain and thus stimulates the degradation of piperazine [43]. The same phenomenon could be observed in the literatures [36–38,44,45].

It was found that the addition of 300 mg·L^{-1} and 600 mg·L^{-1} ammonium chloride could evidently enhance the degradation process and acetamiprid was completely degraded in less 12 h than the control (Fig. 6b), but even more addition (above 900 mg·L^{-1}) of ammonium chloride would delay the biodegradation of acetamiprid by strain D-12. These results are consistent with those of Wang et al. (2012) who found that the addition of 500 mg·L^{-1} ammonium chloride could enhance the biodegradation of 600 mg·L^{-1} nitrobenzene, while the addition of greater amounts of ammonium chloride delayed biodegradation[4]. Luo et al. (2008) founded that the degradation of 100 mg·L^{-1} bensulphuron-methyl in the presence of 1000 mg·L^{-1} ammonium chloride was greater than in the sample without it. Moreover [37], Qiao and Wang (2010), demonstrated the opposite effect, who found that 100 mg·L^{-1} ammonium chloride mildly inhibits the biodegradation of 900 mg·L^{-1} pyridine [46]; Cai et al. (2013) illustrated that the addition of ammonium did not have an evident growth-promoting effect while slightly inhibiting the degradation of piperazine [43]. Based on these studies, appropriate amounts of an extra nitrogen source may inhibited the biodegradation of toxic compounds.

Furthermore, when both glucose and ammonium chloride were added as co-substrates, the degradation efficiency of acetamiprid was much higher compared to that with addition of either glucose or ammonium chloride as carbon or nitrogen source. As shown in Fig. 6a and Fig. 6c, when adding both 100 mg·L^{-1} glucose and 300 mg·L^{-1} ammonium chloride, the acetamiprid degradation

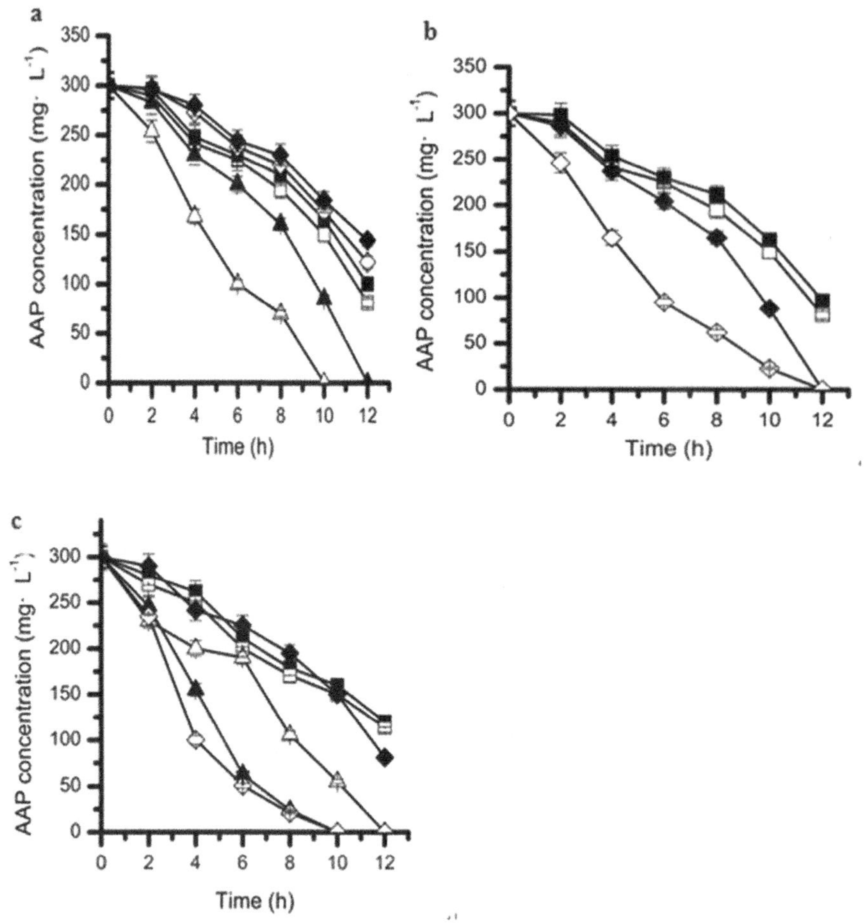

Figure 6. Effect of addition of (a) different concentrations of glucose. (◆), 500 mg·L^{-1}; (◇), 400 mg·L^{-1}; (■), 300 mg·L^{-1}; (□), control; (▲), 100 mg·L^{-1}; (Δ), 200 mg·L^{-1}. (b) different concentrations of ammonium chloride. (■), 900 mg·L^{-1}; (□), control; (◆), 600 mg·L^{-1}; (◇), 300 mg·L^{-1} (c) different concentrations of glucose with the presence of 300 mg·L^{-1} ammonium chloride as co-substrates on the biodegradation of acetamiprid by strain D-12. (◆), control; (■), 500 mg·L^{-1}; (□), 400 mg·L^{-1}; (Δ), 300 mg·L^{-1}; (▲), 200 mg·L^{-1}; (◇), 100 mg·L^{-1}. Error bars, mean ± SD of three replicates.

efficiency was enhanced by 20% or 50% than that with the addition of same dosage of glucose or ammonium chloride within 10 h, which implied that the performance of biodegradation is likely to be even better in the presence of both carbon and nitrogen sources.

Conclusions

Strain D-12 isolated in the present study appeared to be highly efficient in degrading acetamiprid in different contaminated soils and water resources, thus suggesting the isolate could be a significant potential use for the cleanup of acetamiprid-contaminated soil. Degradation of acetamiprid occurred at 25–35°C and pH 6–8. This is an important feature of a microorganism to be employed for bioremediation of variable environments. Another important feature which is worth mentioning is that the bacterium utilized acetamiprid as the sole carbon source as well as energy, nitrogen source for growth. Moreover, the strain harbors the metabolic pathway for the detoxification of acetamiprid. Finally, this is the first report about biodegradation of acetamiprid by a bacterial strain from the *Ochrobactrum* genus.

Supporting Information

Figure S1 Transmission electron micrograph of strain D-12. Bar, 1.0 μm.

Figure S2 The degradation products of acetamiprid in the culture extracts were detected by HPLC.

Figure S3 Proposed pathway for acetamiprid degradation by *Ochrobactrum* sp. strain D-12.

Acknowledgments

The authors would like to thank the academic editor and the two anonymous reviewers for their helpful comments and suggestions. We also gratefully acknowledge B.S. Hao Gao of Huaibei Normal University for excellent assistance in the kinetics analysis.

Author Contributions

Conceived and designed the experiments: GLW, WLY. Performed the experiments: XC, MHX. Analyzed the data: HZ, FL. Contributed reagents/materials/analysis tools: GLW. Wrote the manuscript: GLW.

References

1. Arther RG, Cunningham J, Dorn H, Everett R, Herr LG, et al. (1997) Efficacy of imidacloprid for removal and control of fleas (Ctenocephalides felis) on dogs. Am J Vet Res 58: 848–850.
2. Jacobs DE, Hutchinson MJ, Fox MT, Krieger KJ (1997) Comparison of flea control strategies using imidacloprid or lufenuron on cats in a controlled simulated home environment. Am J Vet Res 58: 1260–1262.
3. Yamamoto I, Casida JE (1999) Nicotinoid insecticides and the nicotinic acetylcholine receptor. Springer, Tokyo.
4. Wang J, Hirai H, Kawagishi H (2012) Biotransformation of acetamiprid by the white-rot fungus *Phanerochaete sordida* YK-624. Appl Microbiol Biotechnol 93: 831–835.
5. U.S. Environmental Protection Agency, http://www.epa.gov/opprd001/factsheets/ acetamiprid.pdf. (2002).
6. Singh DK, Kumar S (2008) Nitrate reductase, arginine deaminase, urease and dehydrogenase activities in natural soil (ridges with forest) and in cotton soil after acetamiprid treatments. Chemosphere 71: 412–418.
7. Seccia S, Fidente P, Barbini DA, Morrica P (2005) Multi residue determination of nicotinoid insecticide residues in drinking water by liquid chromatography with electrospray ionization mass spectrometry. Anal. Chim. Acta 553:21–26.
8. Marín A, Vidal JLM, Gonzalez FJE, Frenich AG, Glass CR, et al. (2004) Assessment of potential (inhalation and dermal) and actual exposure to acetamiprid by greenhouse applicators using liquid chromatography-tandem mass spectrometry. J Chromatogr B 804:269–275.
9. Pramanik SK, Bhattacharyya J, Dutta S, Dey PK, Bhattacharyya A, et al.(2006) Persistence of acetamiprid in/on mustard (Brassica juncea L.). Bull Environ Contam Toxicol 76: 356–360.
10. Sanyal D, Chakma D, Alam S (2008) Persistence of a neonicotinoid insecticide, acetamiprid on chili (*Capsicum annum* L.). Bull Environ Contam Toxicol 81: 365–368.
11. Arora PK, Sasikala C, Ramana CV (2012) Degradation of chlorinated nitroaromatic compounds. Appl Microbiol Biotechnol 93: 2265–2277.
12. Fan X, Liu X, Huang R, Liu Y (2012) Identification and characterization of a novel thermostable pyrethroid-hydrolyzing enzyme isolated through metagenomic approach. Microb Cell Fact 11: 33.
13. Tang HZ, Li J, Hu HY, Xu P (2012) A newly isolated strain of *Stenotrophomonas* sp. hydrolyzes acetamiprid, a synthetic insecticide. Process Biochem 47: 1820–1825.
14. Dai Y, Zhao Y, Zhang W, Yu C, Ji W, et al. (2010) Biotransformation of thianicotinyl neonicotinoid insecticides: diverse molecular substituents response to metabolism by bacterium *Stenotrophomonas maltophilia* CGMCC 1.1788. Bioresour Technol 101: 3838–43.
15. Dai YJ, Ji WW, Chen T, Zhang WJ, Liu ZH, et al. (2010) Metabolism of the neonicotinoid insecticides acetamiprid and thiacloprid by the yeast *Rhodotorula mucilaginosa* strain IM-2. J Agric Food Chem 58: 2419–25.
16. Zhao YJ, Dai YJ, Yu CG, Luo J, Xu WP, et al. (2009) Hydroxylation of thiacloprid by bacterium *Stenotrophomonas maltophilia* CGMCC 1.1788. Biodegradation 20: 761–768.
17. Chen T, Dai YJ, Ding JF, Yuan S, Ni JP, et al. (2008) N-demethylation of neonicotinoid insecticide acetamiprid by bacterium *Stenotrophomonas maltophilia* CGMCC 1.1788. Biodegradation 19: 651–658.
18. Yao XH, Min H (2006a) Isolation, characterization and phylogenetic analysis of a bacterial strain capable of degrading acetamiprid. J Environ Sci-China 18: 141–146.
19. Wang GL, Yue WL, Liu Y, Li F, Xiong MH, et al. (2013) Biodegradation of the neonicotinoid insecticide Acetamiprid by bacterium *Pigmentiphaga* sp. strain AAP-1 isolated from soil.Bioresour Technol 138: 359–368.
20. Holt JG, Krieg NR, Sneath PHA, Staley JT, Williams ST, et al. (1994) Bergey's Manual of Determinative Bacteriology, 9th ed. Williams and Wilkins: Baltimore.
21. Miller SA, Dykes DD, Polesky HF (1988) A simple salting out procedure for extracting DNA from human nucleated cells. Nucleic Acid Res 16: 12–15.
22. Hu JY, Fan Y, Lin YH, Zhang HB, Ong SL, et al. (2003) Microbial diversity and prevalence of virulent pathogens in biofilms developed in a water reclamation system. Res Microbiol 154: 623–629.
23. Tamura K, Dudley J, Nei M (2007) MEGA 4: molecular evolutionary genetics analysis (MEGA) software version 4.0. Mol Biol Evol 24: 1596–1599.
24. Fitzgerald J (2004) Laboratory bioassays and field evaluation of insecticides for the control of Anthonomus rubi, Lygus rugulipennis and Chaetosiphon fragaefolii, and effects on beneficial species in UK strawberry production, Crop Prot 23: 801–809.
25. Yao XH, Min H, Lü ZH, Yuan HP (2006b) Influence of acetamiprid on soil enzymatic activities and respiration. Eur J Soil Biol 42: 120–126.
26. Danko AS, Luo MZ, Bagwell CE, Brigmon RL, Freedman DL, et al. (2004) Involvement of Linear Plasmids in Aerobic Biodegradation of Vinyl Chloride. Appl Environ Microbiol 70: 6092–6097.
27. Liang B, Li R, Jiang D, Sun JQ, Qiu JG, et al. (2010) Hydrolytic dechlorination of chlorothalonil by *Ochrobactrum* sp. CTN-11 isolated from a chlorothalonil-contaminated soil. Curr Microbiol 61: 226–233.
28. Kim YM, Park K, Joo GJ, Jeong EM (2004) Glutathione-dependent biotransformation of the fungicide chlorothalonil. J Agric Food Chem 52: 4192–4196.
29. Qiu XH, Bai WQ, Zhong QZ, Li M, He MQ, et al. (2006) Isolation and characterization of a bacterial strain of the genus *Ochrobactrum* with methyl parathion mineralizing activity. J Appl Microbiol 101: 986–994.
30. Veeranagouda Y, Paul PVE, Gorla P (2006) Complete mineralisation of dimethylformamide by *Ochrobactrum* sp. DGVK1 isolated from the soil samples collected from the coalmine leftovers. Appl Microbiol Biotechnol 71: 369–375
31. Yuan YJ, Lu ZX, Huang LJ, Li Y, Lu FX, et al. (2007) Biodegradation of nicotine from tobacco waste extract by *Ochrobactrum intermedium* DN2. J Ind Microbiol Biotechnol 34: 567–570.
32. Brunet JL, Badiou A, Belzunces LP (2005) In vivo metabolic fate of [^{14}C]-acetamiprid in six biological compartments of the honeybee, Apis mellifera L. Pest Manage Sci 61: 742–748.
33. Tomizawa M, Lee DL, Casida JE (2000) Neonicotinoid insecticides: Molecular features conferring selectivity for insect versus mammalian nicotinic receptors. J Agric Food Chem 48 (12): 6016–6024.
34. Ford KA, Casida JE (2008) Comparative metabolism and pharmacokinetics of seven neonicotinoid insecticides in spinach. J Agric Food Chem 56: 10168–10175.
35. Chen S, Dong YH, Chang C, Deng Y, Zhang XF, et al. (2013) Characterization of a novel cyfluthrin-degrading bacterial strain *Brevibacterium aureum* and its biochemical degradation pathway. Bioresour Technol 132: 16–23.
36. Wang JL, Quan XC, Han LP, Qian Y, Werner H, et al. (2002) Kinetics of co-metabolism of quinoline and glucose by *Burkholderia pickettii*. Process Biochem 37: 83–836.
37. Luo W, Zhao YH, Ding HT, Lin XY, Zheng HB, et al. (2008) Co-metabolic degradation of bensulfuron-methyl in laboratory conditions. J Hazard Mater 158: 208–214.
38. Mohammad ZK, Pijush KM, Suhail S, Vinod T (2011) Degradation pathway, toxicity and kinetics of 2,4,6-trichlorophenolwith different co-substrate by aerobic granules in SBR. Bioresour Technol 102: 7016–7021.
39. Lin Q, Wanga JL (2010) Microbial degradation of pyridine by *Paracoccus* sp. isolated from contaminated soil. J Hazard Mater 176: 220–225.
40. Xiong RL, Chen LJ, Liu JJ (2009) Biodegradation of pyridine with co-substrates by *Paracoccus denitrificans* W12. Journal of Tsinghua University (Science and Tech- nology) 49: 826–829.
41. Quan XC, Wang JL, Han LP, Shi HC, Qian Y, et al. (2001). Biodegradation kinetics of a mixture containing quinoline and glucose by *Burkholderia pickettii* strain. Acta Scientiae Circumstantiae 21: 416–419.
42. Sharma A, Thakur IS (2008) Characterization of pentachlorophenol degrading bacterial consortium from chemostat. Bulletin of Environment Contamination and Toxicology 81: 12–18.
43. Cai S, Li X, Cai TM, He J (2013) Degradation of piperazine by *Paracoccus* sp. TOH isolated from activated sludge. Bioresour Technol 130: 536–542.
44. Wen QX, Chen ZQ, Zhao Y, Zhang HC, Feng YJ, et al. (2010) Biodegradation of polyacrylamide by bacteria isolated from activated sludge and oil-contaminated soil. J Hazard Mater 175: 955–959.
45. Ziagova M, Kyriakou G, Liakopoulou KM (2009) Co-metabolism of 2, 4-dichlorophenol and 4-Cl-m-cresol in the presence of glucose as an easily assimilated carbon source by *Staphylococcus xylosus*. J Hazard Mater 163: 383–390.
46. Qiao L, Wang JL (2010) Microbial degradation of pyridine by *Paracoccus* sp. isolated from contaminated soil. J Hazard Mater 176: 220–225.

Functional Traits Differ between Cereal Crop Progenitors and Other Wild Grasses Gathered in the Neolithic Fertile Crescent

Jennifer Cunniff[1*¤], Sarah Wilkinson[1], Michael Charles[2], Glynis Jones[3], Mark Rees[1], Colin P. Osborne[1]

1 Department of Animal and Plant Sciences, University of Sheffield, Sheffield, United Kingdom, **2** School of Archaeology, University of Oxford, Oxford, United Kingdom, **3** Department of Archaeology, University of Sheffield, Sheffield, United Kingdom

Abstract

The reasons why some plant species were selected as crops and others were abandoned during the Neolithic emergence of agriculture are poorly understood. We tested the hypothesis that the traits of Fertile Crescent crop progenitors were advantageous in the fertile, disturbed habitats surrounding early settlements and in cultivated fields. We screened functional traits related to competition and disturbance in a group of grass species that were increasingly exploited by early plant gatherers, and that were later domesticated (crop progenitors); and in a set of grass species for which there is archaeological evidence of gathering, but which were never domesticated (wild species). We hypothesised that crop progenitors would have greater seed mass, growth rate, height and yield than wild species, as these traits are indicative of greater competitive ability, and that crop progenitors would be more resilient to defoliation. Our results show that crop progenitors have larger seed mass than wild species, germinate faster and have greater seedling size. Increased seed size is weakly but positively correlated with a higher growth rate, which is primarily driven by greater biomass assimilation per unit leaf area. Crop progenitors also tend to have a taller stature, greater grain yield and higher resilience to defoliation. Collectively, the data are consistent with the hypothesis that adaptations to competition and disturbance gave crop progenitors a selective advantage in the areas surrounding early human settlements and in cultivated environments, leading to their adoption as crops through processes of unconscious selection.

Editor: Gary W. Crawford, University of Toronto Mississauga, Canada

Funding: This work was supported by a post graduate scholarship from the University of Sheffield to Jennifer Cunniff; a NERC standard grant (NE/H022716/1) awarded to Colin P Osborne, Glynis Jones and Michael Charles; and an ERC advanced research grant (269830-EOA) awarded to Glynis Jones. The funders had no role in study design, data collection and analysis, decision to publish, or preparation of the manuscript.

Competing Interests: The authors have declared that no competing interests exist.

* E-mail: jennifer.cunniff@rothamsted.ac.uk

¤ Current address: Department of Agroecology, Rothamsted Research, Harpenden, United Kingdom

Introduction

Grain assemblages from early settlements in the Fertile Crescent of Southwest Asia show that hunter-gatherers collected a large, diverse range of species before the origin of agriculture [1–8]. However, later assemblages in the region suggest a progressive specialization in diet, characterized by a decline in the occurrence of some wild plant species, and an increase of other species as they were brought into cultivation [5,8,9]. Domesticated crops emerged in the Fertile Crescent of Southwest Asia between ~11,000 and 8,000 BP [10–12], and are characterized by the evolution of a 'domestication syndrome', including the loss of natural dispersal mechanisms and an increase in seed mass [13,14]. However, it is not clear what drove the selection of some species as crops and what caused the abandonment of others.

The selection of crop species could have been an intentional selection process by early farmers in response to demographic [15] or social [16] pressures, or it could be the result of unconscious selection [17–20]. Unconscious selection arises from the interactions between humans and their food plants, and has the potential to determine the species that were domesticated and to drive the domestication process. Some authors have suggested a prolonged period of pre-domestication cultivation, during which certain species were cultivated but had not yet acquired the morphological changes associated with domestication [8,21–23]. We propose an ecological model, whereby selection could have occurred in two distinct settings prior to the establishment of full agricultural systems, each introducing plant species to novel anthropogenic environments. Initially, the gathering of wild plants for subsistence and the accidental spillage of seeds may have concentrated food plants in the area surrounding settlements. Selection then acted on this local species pool via disturbance, competition and high soil fertility. Cultivation subsequently modified the selection regime through the deliberate sowing of seeds in tilled soil and, at some point, by other management practices such as weeding, manuring and irrigation. This paper explores the potential role of these ecological selection processes in the domestication of crops. We set up an experiment to compare the crop progenitors and co-collected wild species to investigate whether there are functional traits of these crop progenitors that would have favoured their selection in these early agricultural environments.

High relative growth rate (RGR), defined as the rate of dry matter production per unit of dry matter under optimal conditions

[24,25], is generally considered a crucial adaptation to fertile, disturbed niches. RGR represents one of the fundamental axes of ecological variation, corresponding with rapid rates of resource acquisition, and trading off against allocation to storage, defence and survival [25–27]. High RGR has previously been shown to correlate negatively with seed mass, suggesting that small-seeded species have growth strategies (physiological, morphological or structural adaptations) to facilitate rapid growth [28,29]. However, because RGR typically declines as plants grow, comparisons of species with different seed sizes will inevitably confound size effects with true differences in growth strategy [30–32]. To separate the effects of size and growth strategy, Metcalf *et al.* [33] suggested making comparisons at a common size and, when this was done, a positive relationship between RGR and seed mass was found in short-lived species [34] raising the possibility that fast growth may be an important component of the domestication syndrome in large-seeded crops. Large seed size also correlates with a range of other traits that would be an advantage in early anthropogenic environments, including quicker and earlier germination [35], enhanced competitive ability [36] and survival following burial [37].

This paper tests the overarching idea that Fertile Crescent crop progenitors might be better adapted to disturbed, fertile environments than other wild grass species exploited by pre-agricultural societies by testing five hypotheses: (i) crop progenitors have larger seeds and seedlings, and faster germination than the other wild species; (ii) the RGR under fertile soil conditions is greater in crop progenitors than the other wild species, when compared at a common size; (iii) size and seed yield at maturity are greater in crop progenitors than the other wild species; (iv) resilience (survival and fecundity) of crop progenitors when defoliated is greater than that of wild species; and (v) the differences between wild species and crop progenitors outlined in (ii) to (v) can be explained by correlations of each trait with seed mass.

Materials and Methods

Three experiments were designed to test the hypotheses. Experiment 1 investigated whether seed mass had an effect on the timing and rate of germination. Experiment 2 investigated the relationship between seed mass and growth rate using a functional approach with repeated harvests to calculate RGR using conventional and size standardised methods. The RGR data have been presented previously as part of a meta-analysis investigating the association of seed size, plant size and growth rate [34]. Here, the data are further analysed to determine how variation in plant allometry and physiological processes contribute to interspecific variation in RGR. The third experiment tested how crop progenitors and wild species respond to simulated disturbance via complete removal of above ground biomass, comparing survival and yield.

Species selection

Nine species were selected for the experiments. All were grown from seed obtained from germplasm holdings, and different seed accessions were used for the three experiments (Table S1). To ensure that these accessions were reasonable representatives of the original progenitors, they were selected based on their region of origin, being as close to the centres of domestication as possible. While intraspecific variation and maternal effects are to be expected, previous experiments show that they are typically less important than interspecific variation. Species identity was ensured by the use of taxonomic and molecular markers to look for contamination in the germplasm holding centres [38]. The outer glumes were removed from all seeds before weighing. Since this study focuses on cereal cultivation, only grasses were chosen to ensure that comparisons between crop progenitors and wild species were not confounded by growth habit or phylogeny.

Three wild grasses, which became the major domesticates and staple foods in Southwest Asia were selected to represent the crop progenitors: *Hordeum spontaneum* Koch, *Triticum boeoticum* Boiss., and *Triticum dicoccoides* Koern. Six wild grass species that were never domesticated were also investigated: *Aegilops crassa* Boiss., *Aegilops speltoides* Coss., *Aegilops tauschii* Coss., *Eremopyrum bonaepartis* (Spreng.) Nevski, *Eremopyrum distans* (K.Koch) Nevski and *Taeniatherum caput-medusae* (L.) Nevski. These particular species were chosen because they are present in significant numbers in archaeobotanical assemblages, from numerous sedentary sites across Southwest Asia ranging in age from 23,000- 9,700 BP [1,2,4,5,7,8,10,39,40]. From the range of preserved plant remains, wild annual grasses were abundant at the majority of these early settlements and were chosen as a focus for the study.

For example, from the 90,000 plant remains collected at Ohalo II, an Upper Palaeolithic site in Israel, 19,000 were grass grains [4,5]. The abundance of wild grasses at this location suggests they were a staple food source. Furthermore, there is evidence for the processing of plant material at this site, with a large concentration of remains around a grinding stone, and the presence of starch grains indicating the pounding and grinding of grass seeds [41]. Four sites in northern Syria show evidence of the collection of a wide range of grasses, with *Aegilops* spp., *T. caput-medusae* and *Eremopyrum* spp. being identified at all locations and *H. spontaneum* and *T. boeoticum* highly abundant in the majority of samples collected [8]. At Neolithic Çatalhöyük, east Turkey, two out of eight archaeobotanical samples from storage structures in burnt houses were dominated by *T. caput-medusae* accompanied by a lesser, but still substantial, amount of *Eremopyrum* type grass [7]. The purity of the samples, including one consisting solely of *T. caput-medusae* and *Eremopyrum* spp., suggests that seeds were being gathered, processed and stored.

Experiment 1 – Seed mass and germination rate

(a) Plant material and growth conditions. Germination took place under propagator lids and a thin layer of wet compost, within a controlled environment chamber (BDR 16, Conviron, Winnipeg, Manitoba, Canada) under a 20/10°C day/night with an 8 hour photoperiod, PPFD of 300 µmol photons m^{-2} s^{-1}, and humidity of 70%/50% day/night. Seeds were uncovered after two days to observe radicle length and then observed daily for germination. Seedlings were harvested when the ligule of the first true leaf emerged. Roots were washed free of compost, and seedlings were oven dried to a constant weight at 80°C before weighing.

(b) Statistical analysis. Regressions lines were fitted to plots of germination *vs* seed size, and seedling mass *vs* seed size. R^2 values were calculated to estimate the goodness of fit and the significance of the relationship was tested using the *lm()* function in R (version 2.6.14, The R Foundation for Statistical Computing).

Experiment 2 – Seed mass and RGR

(a) Plant material and growth conditions. Seeds were germinated on moist, washed sand (Chelford 52; WBB Minerals, Sandbach, Cheshire) under propagator lids in a controlled environment room (BDW 40, Conviron). Conditions were 20/10°C (day/night) with an 8 hour photoperiod and PPFD of 300 µmol photons m^{-2} s^{-1}.

The general approach for growth analysis was based on studies by Grime and Hunt [25], Poorter and Remkes [42], Poorter [43],

Hunt and Cornelissen [44] and Hendry & Grime [45]. Three days after germination, 24 seedlings of a uniform size were selected for each species and planted into 1 litre pots containing washed sand. These were top-watered every two days with full-strength Long Ashton nutrient solution ([46] Tables 40, 41), and bottom-watered with distilled water on alternate days. The seedlings were returned to the controlled environment room at: 20/10°C (day/night) with a 16 hour photoperiod, maximum PPFD of 756 µmol photons $m^{-2}s^{-1}$ and RH of 70%. Although these light levels are lower than the peak value in full sunlight, the levels are comparable to or higher than the irradiance levels used in most previous growth experiments measuring RGR and its growth components (see Shipley *et al.* [47]). For reference, the daily quantum input (DQI) in this experiment was 43.6 mol m^{-2} day^{-1}.

(b) Harvesting schedule and measurements. The experiment was started 7 days after germination, when all plants had reached a fresh weight of approximately 100 mg (day 7). Harvests were carried out over a 3-week period on days 7, 10, 14, 17, 21 and 27. Each day, 4 plants of each species were harvested and divided into leaf blades, roots and leaf sheaths. Each plant was cleaned of the growth medium, and dried to a constant weight at 80°C for 2 days. The total leaf area of each plant was determined by scanning the freshly harvested leaf blades and using image analysis software (ImageJ 1.36b., Wayne Rasband, National Institute of Health, Bethesda, Maryland, USA).

(c) Statistical analysis. Regression lines were fitted to plots of seedling mass *vs* seed size, final dry mass *vs* seed size and final leaf area *vs* seed size. R^2 values were calculated to estimate the goodness of fit and the significance of the relationship was tested using the *lm()* function in R (version 2.6.14, The R Foundation for Statistical Computing).

(d) Estimating RGR at a common size. A species-specific functional growth analysis was performed by fitting growth functions to plots of ln-total plant mass against time (Fig. S1), which were then used to estimate RGR at a common reference size for each species. For full details of the fitting and RGR estimation see references: [30,32,34,48]. Fitting these curves and estimating RGR at a common size (sRGR) rather than calculating RGR via classical methods [44] accounts for size-dependent variation in growth rate bought about by changes in physiology, morphology and allocation. The 30th percentile (0.067 g) of the distribution of plant mass was used as the reference size because it encompasses all the species, resource limitation should still be minimal, and growth approaching its maximum rate. Plant mass in the data set ranged from 0.005 g to 5.816 g.

RGR is determined by the net assimilation rate (NAR), which is the absolute growth rate per unit leaf area, and the leaf area per unit of plant mass, termed the leaf area ratio (LAR). The LAR is further factorized into: (1) leaf mass ratio (LMR), the proportion of biomass invested in leaves, and (2) specific leaf area (SLA), the leaf area per unit leaf mass. To better understand why RGR varies among species, the components of sRGR were calculated at the reference size (sNAR, sLMR and sSLA) based on predictions from linear regressions of ln-leaf area and ln-leaf mass against ln-total plant mass. In all cases the species × total plant mass interaction was significant ($P<.05$), so that all regression models had species-specific intercepts and slopes [30,48].

Experiment 3 – Components of yield and response to defoliation

(a) Plant material and growth conditions. Seeds were germinated in 24-cell plug-trays containing a 1:1 sand: compost mix. Trays were placed in a controlled environment chamber (BDR 16, Conviron) under a 20/10°C day/night with an 8 hour photoperiod and a PPFD of 300 µmol photons m^{-2} s^{-1}. Once the seeds had germinated and the seedlings had reached the second leaf stage, an 8-week vernalization treatment was imposed to enable flowering. Temperatures were set at 4°C (day and night) with PPFD and photoperiod as during germination. At the end of vernalization in late April, plants were re-potted into 4 litre pots containing the same growth medium. Plants were divided between three rooms in a glasshouse and organised in a randomised block design (Tapton Experimental Gardens, Broomhill, Sheffield). They were grown until maturity in mid-July. Artificial lighting was supplied by sodium lamps at 16 h d^{-1} throughout the experiment to supplement and extend natural daylight. The PPFD, temperature and humidity were logged at 5 minute intervals (DL2e data logger, Delta-T Devices Ltd., Cambridge, UK) to give mean maximum daily PPFD values of 420±31.1 µmol photons m^{-2} s^{-1}, mean maximum daily temperatures of 24±0.6°C and mean minimum temperatures of 16±0.2°C. Mean maximum and minimum humidity values over a diurnal cycle were 62±0.8% and 43±1.6%, respectively. Over the full growing season, maximum daily temperatures ranged from 16°C to 35°C, maximum daily PPFD varied from 79 µmol photons m^{-2} s^{-1} to 830 µmol photons m^{-2} s^{-1}, and maximum daily RH from 52% to 73%. Plants were watered every other day at the beginning of the experiment; this was reduced after flowering as the water demands of the plant decreased.

(b) Defoliation treatment and data collection. A defoliation treatment was applied to all nine species during the vegetative stage, two weeks after plants were transferred to the glasshouse by completely removing all plant material at 2 cm above the soil surface. There were eight replicates in the defoliated and unperturbed control treatments, giving a total of 16 individuals (8 replicates ×2 treatments) per species.

Measurements were made of six characters: survival (whether the plants flowered and set seed), time of flowering (days from end of vernalization period to extrusion of inflorescence on the leading tillers), plant height (measured from soil surface to the collar of the leading spike), the number of tillers (flowering and not flowering), number of seeds and potential yield. Survival and flowering were recorded whenever plants were watered. Plants were harvested after grain filling, but before spikes had started to shatter. Archaeological records show that prior to crop domestication, plants were harvested when grains were immature and still attached to the seed head, since the rachis of wild grasses shatters when mature, thereby dispersing the seeds [13]. Plant height and number of tillers (flowering and not-flowering) were recorded. The number of seeds on each plant was estimated by finding a relationship between the length of a spike and the number of seeds on a spike for each species. Grain yield of each species was estimated by multiplying seed number by the average values of seed weight measured in the RGR analysis.

(c) Statistical analysis. Data were analysed using the statistical computing package R (version 2.6.14, The R Foundation for Statistical Computing). All characters were tested using a generalized linear model (*glm*) with the appropriate error distribution and link function. In all cases the putative minimal model was determined by using the *dropterm()* function, and the best fit was determined by comparing dispersion parameters and AIC values. Comparisons were then carried out using a contrasts matrix [R-routine: *contrasts(Species)*] to determine whether there were differences in response to the disturbance treatment between crop progenitors and wild species.

Results

Seed mass and germination

Seed mass at the start of the experiment showed significant variation among species ($F_{8,63} = 95.8$, $P<.001$), ranging from 3.18±0.07 mg (mean±s.e.) in *E. bonaepartis* to 28.9±1 mg in *H. spontaneum* (Fig. 1). Seed mass was significantly larger in the crop progenitors than in the other wild species ($t_1 = 16.12$, $P<.001$, Fig. 1). However, seeds of the crop progenitor *T. boeoticum* weighed 21±0.6 mg, only 10% larger than those of the two largest-seeded wild species, *A. crassa* and *A. tauschii*, which weighed 19±0.8 mg and 19±0.4 mg respectively. Seed mass differed between accessions (Fig. 1), but this variation did not obscure the strong overall difference between crop progenitors and wild species.

Seed mass was negatively correlated with the time to germination, with larger seeded species germinating at a faster rate ($F_{1,7} = 5.6$, $P<.05$, Fig. 2A). Germination was approximately two days slower in the smallest seeded species (*E. bonaepartis*) compared to the largest seeded species (*H. spontaneum*). The larger seed mass in the crop progenitors meant that this group had a faster rate of germination when compared to the other wild species.

Larger seed mass was positively correlated with seedling dry mass in experiment 1 ($F_{1,7} = 120.2$, $P<.001$, Fig. 2B) and dry mass at seven days after germination in experiment 2 ($F_{2,6} = 75.8$, $P<.001$, Fig. 2B). This seed mass effect also carried through to the end of experiment 2, as shown by the positive correlations between dry mass and leaf area at 27 days (Fig. S2).

Seed mass and RGR

RGR was calculated using classical and size-standardised methods, and these models showed contrasting results. Using the classical method there was no relationship between seed mass and RGR (Fig. S3), whereas a weak, positive relationship arose using the size-standardized method ($F_{2,6} = 6.2$, $P<.05$, Fig. 3). This positive relationship was largely driven by the high value of sRGR for *H. spontaneum*, and larger seed mass did not always translate to greater sRGR. In particular, *T. boeoticum* had a much lower sRGR than wild grain species (*A. crassa* and *A. tauschii*) with similar seed masses (Figs. 1 and 3). In comparison, the two other crop progenitors (*T. dicoccoides* and *H. spontaneum*), which had the largest seed masses, also had the highest sRGRs. The sRGR value of 0.31 for *H. spontaneum* was more than double that of the wild species with the lowest sRGR (*E. distans*) and 34% greater than the wild species with the highest sRGR (*A. crassa*). *T. dicoccoides* similarly showed growth advantages of 73% and 10% respectively over these two wild species.

Components of RGR

The best predictor of sRGR was sNAR ($F_{2,6} = 20.1$, $P<.01$, Fig. 4A), followed by sSLA ($F_{1,7} = 6.8$, $P<.05$, Fig 4B) whilst no relationship was found with sLMR (Fig. 4C). Supporting these findings, calculation of the contributions of sLMR, sSLA and sNAR to the variance in sRGR showed that sNAR made the largest contribution (82.5%), whilst sSLA (32.8%) made a small contribution, and sLMR a smaller, negative contribution (−15.3%). Normalization of the variance and covariance for each of these components of sRGR yields an "importance" value [30]. Importance values again showed sNAR to be the main driver of interspecific variation in sRGR (48.3%). However, sLMR was of equal importance to sSLA (26.9 *vs* 24.8% respectively). As expected from these results, sNAR correlates strongly with seed mass, whilst the relationship is weaker for the sLMR and sSLA components (see Fig. S4).

Resilience to disturbance

The defoliation treatment significantly reduced survival ($F_{1,134} = 76.4$, $P<.001$, Fig. 5A), with the magnitude of the response varying among species ($F_{8,135} = 9.5$, $P<.001$, Fig. 5A). The majority of the species defoliated showed 70–100% survival but, in *E. bonaepartis*, only 25% of the plants survived. Comparisons

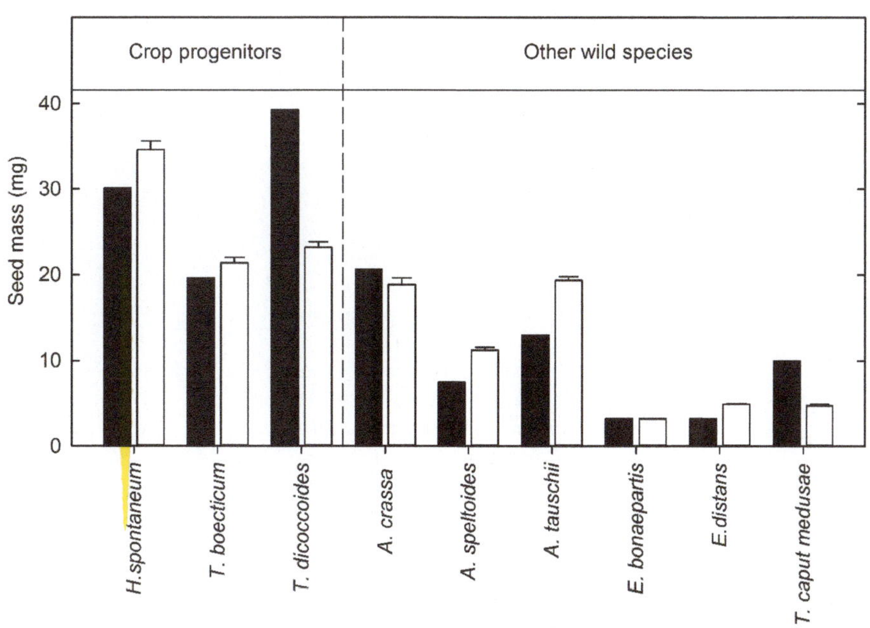

Figure 1. Initial seed mass in the three crop progenitors and six wild species. The black bars show the mean seed mass of accessions used in experiment 1 and the white bars show those used in experiment 2 (+SE). Standard errors are not shown for experiment 1 because seeds were not weighed individually.

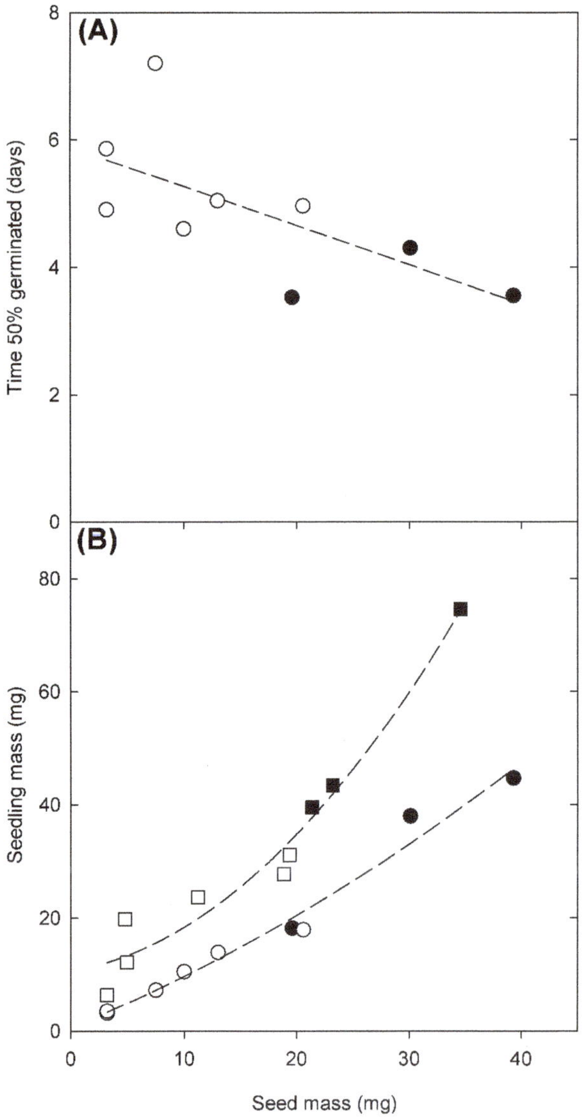

Figure 2. Relationship between seed germination, seedling mass and seed mass. Regression slopes for the relationship between (a) time to 50% of seeds germinated and seed mass ($F_{1,7}$ =5.55, P<.05, R^2=0.44); and (b) seedling mass and seed mass, [experiment 1: circles ($F_{1,7}$=120.156, P<.001, R^2=0.9756), experiment 2: squares ($F_{2,6}$=75.78, P<.001, R^2 0.9619)] for the three crop progenitors (closed symbols) and six wild species (open symbols).

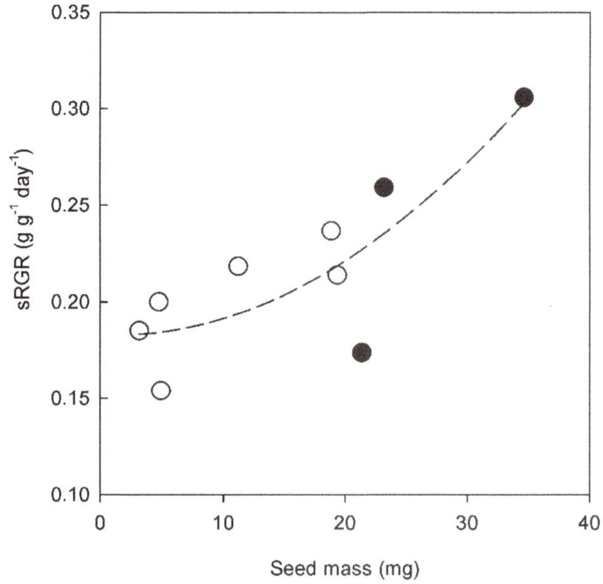

Figure 3. Relationship between size standardised RGR (sRGR) and seed mass. Regression slope for the relationship between sRGR and seed mass ($F_{2,6}$=6.186, P<.05, R^2=0.6734) for the three crop progenitors (closed symbols) and six wild species (open symbols).

of survival between the crop progenitors and wild species showed no significant difference in the response to defoliation.

Defoliation significantly reduced the final height of plants ($F_{1,121}$ = 103.4, P<.001, Fig. 5B), and the response varied between species from a 10% to 48% loss of height compared to the control treatment, leading to a significant interaction between species and treatment ($F_{8,113}$=6.5, P<.001, Fig. 5B). Comparisons of plant height between the crop progenitors and wild species showed that, under the control treatment, the crop progenitors were significantly taller than the wild ones (t_1=10.1, P<.001, Fig. 5B). However, there was no significant difference between the two types in the response of plant height to defoliation.

The defoliation treatment significantly decreased the number of tillers produced by each species (χ^2_1(N = 144) =287.2, P<.001, Fig. 5C). This varied between species from 19% in *T. boeoticum* to 81% in *A. tauschii*, leading to a significant species by treatment interaction (χ^2_8(N = 144) =134.3, P<.001, Fig. 5C). The wild species had a significantly greater number of tillers than the crop progenitors under the control treatment (z = 13.3, P<.001, Fig. 5C). The two types also showed a differential response to the defoliation treatment, with the decline in tiller number being greater in the wild species than crop progenitors (z = 7.1, P<.001, Fig. 5C). The full defoliation treatment led to, on average, a 62% decrease in the number of tillers in the wild species but only a 26% decline in the crop progenitors.

The period of time plants took to flower was extended significantly when a defoliation treatment was applied, with flowering taking 25–49% longer in the defoliated plants compared to the controls (χ^2_1(N = 144) =60.9, P<.001, data not shown). Once flowering was complete and seed set had begun, comparison of seed number between the control and defoliation treatment showed that the decrease in tiller number due to defoliation caused a large decline in seed number (χ^2_1(N = 144) =211.3, P<.001, Fig. 5D) and therefore potential yield (χ^2_1(N = 144) = 119.4, P< .001, Fig. 5E). The proportion of tillers that set seed was not affected by defoliation (data not shown), so the decrease in seed number and potential yield was due to the reduction in tillers alone. The reduction in seed number due to defoliation varied between species (χ^2_8(N = 144) = 118.4, P<.001, Fig. 5D) and was greater in the wild species compared to the crop progenitors (z = 6.6, P<.001, Fig 5D). The decrease in potential yield due to defoliation varied between species from 25% in *T. boeoticum* to 85% in *A. tauschii*, leading to a significant interaction between species and treatment (χ^2_8(N = 144) = 16.1, P<.05, Fig. 5E). Comparison between the two types showed that, in the unperturbed control treatment, the crop progenitors were capable of producing a significantly larger yield than the wild species (z = 6.2, P<.001, Fig 5E). The two groups also showed a differential response to the defoliation treatment, with a significantly greater decline in yield in the wild species than crop progenitors (z = 2.8, P<.01, Fig 5E). The full defoliation treatment led to, on average, a 61% decrease

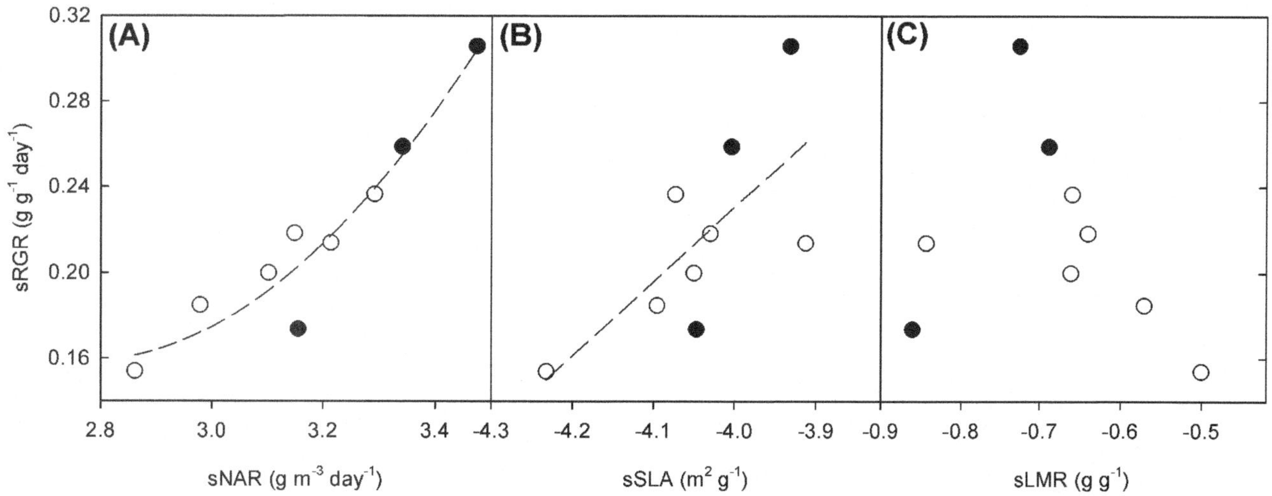

Figure 4. Relationship between sNAR, sSLA and sLMR and sRGR. Regression slopes for the relationships between (a) sNAR and sRGR ($F_{2,6}$ = 31.98, *P*<.001, R^2 = 0.914); (b) sSLA and sRGR ($F_{1,7}$ = 6.781, *P*<.05, R^2 = 0.492); and (c) sLMR and sRGR for the three crop progenitors (closed symbols) and six wild species (open symbols).

in potential yield in the wild species, but only a 31% decline in the crop progenitors.

Discussion

Results from the three experiments showed that crop progenitors have larger seed and seedling mass, earlier germination, and greater biomass and leaf area during early stages of vegetative growth than the other wild species known to have been gathered by pre-agricultural communities. Additionally, the greater seed mass was weakly but positively correlated with a higher sRGR in two of the three crop progenitors tested. The crop progenitors were more resilient in their response to defoliation and were taller at maturity, with a higher potential grain yield than the other wild species.

We hypothesize that this suite of functional traits observed in crop progenitors would have been advantageous in an anthropogenic environment. Initially, gathering would have occurred in the natural habitat of these wild plants and involved collection across a wide range of plant taxa. The gathering of wild plant seeds likely led to their dispersal in and around early human settlements. Seeds dispersed in this way would have been exposed to new selective pressures. Collected species that survived and flourished in such environments would have been those best able to survive disturbance and take advantage of greater levels of soil fertility. These species would also respond well to early attempts at cultivation. Enhanced competitive ability over smaller seedlings [36] and greater resilience to defoliation are of particular importance in this context [49,50].

This suggestion is superficially similar to the so-called 'dump heap hypothesis' that crop plants originated from weeds associated with human refuse heaps and disturbed habitats surrounding pre-agricultural settlements (e.g. [51–53]; and subsequently discussed by, for example: Harlan [54,55], Blumler and Byrne [56] and Abbo *et al.* [57]). It differs from it, however, in that we are not suggesting that crop progenitors were from naturally weedy habitats. Rather our hypothesis rests upon the more subtle differences between collected crop progenitors and other wild species that were also collected but never domesticated, and their relative competitive ability when introduced to a new environment. Our experiments focus specifically on the comparison between these collected species of Southwest Asia.

We argue that by out-competing other species less well adapted to these human-managed environments cereal crop progenitors became more abundant around human settlements and under early cultivation, and therefore more likely to be collected and harvested. This may have driven species selection in the early stages of plant domestication making human populations more dependent on a narrower range of species [18,58], which then became the most likely candidates for, and successful products of, cultivation. Later, as wild progenitors were taken into cultivation, these same functional traits would have continued to offer competitive advantages, *within* cultivated species in deliberately sown, possibly weeded and/or manured cultivation plots.

Larger size may confer a competitive advantage in crop progenitors

In this study, the crop progenitors had a larger seed mass than wild species that were not domesticated. Greater seed mass and seedling size enhance competitive ability [36] through a number of mechanisms. Under conditions of drought, burial or competition, survival is greater due to the longer initial hypocotyl or radicle, termed the 'seedling size effect' [37,59], and we found a strong correlation between seed size and seedling size in experiment 1. If larger seedlings have lower RGR, as shown in previous studies [28], this is hypothesised to give a metabolic effect [37] where a species with a slower growth rate has a slower respiration rate and consumes metabolic resources more slowly. However, in this study and others [30,31], RGR was size-dependent, and the relationship between size-corrected growth rate and seed mass was positive. Larger seeds have an additional benefit of having extra metabolic resources that may serve to better support carbon deficits, the 'reserve effect', where a greater amount of resources is left uncommitted at a given time after germination [37,59]. This is of particular significance when defoliation occurs at a young age [49].

This study shows that larger seed size is weakly but positively associated with a higher RGR at a given plant size. A high sRGR offers larger seeds an advantage by allowing more rapid colonization of an environment. The crop progenitors *H.*

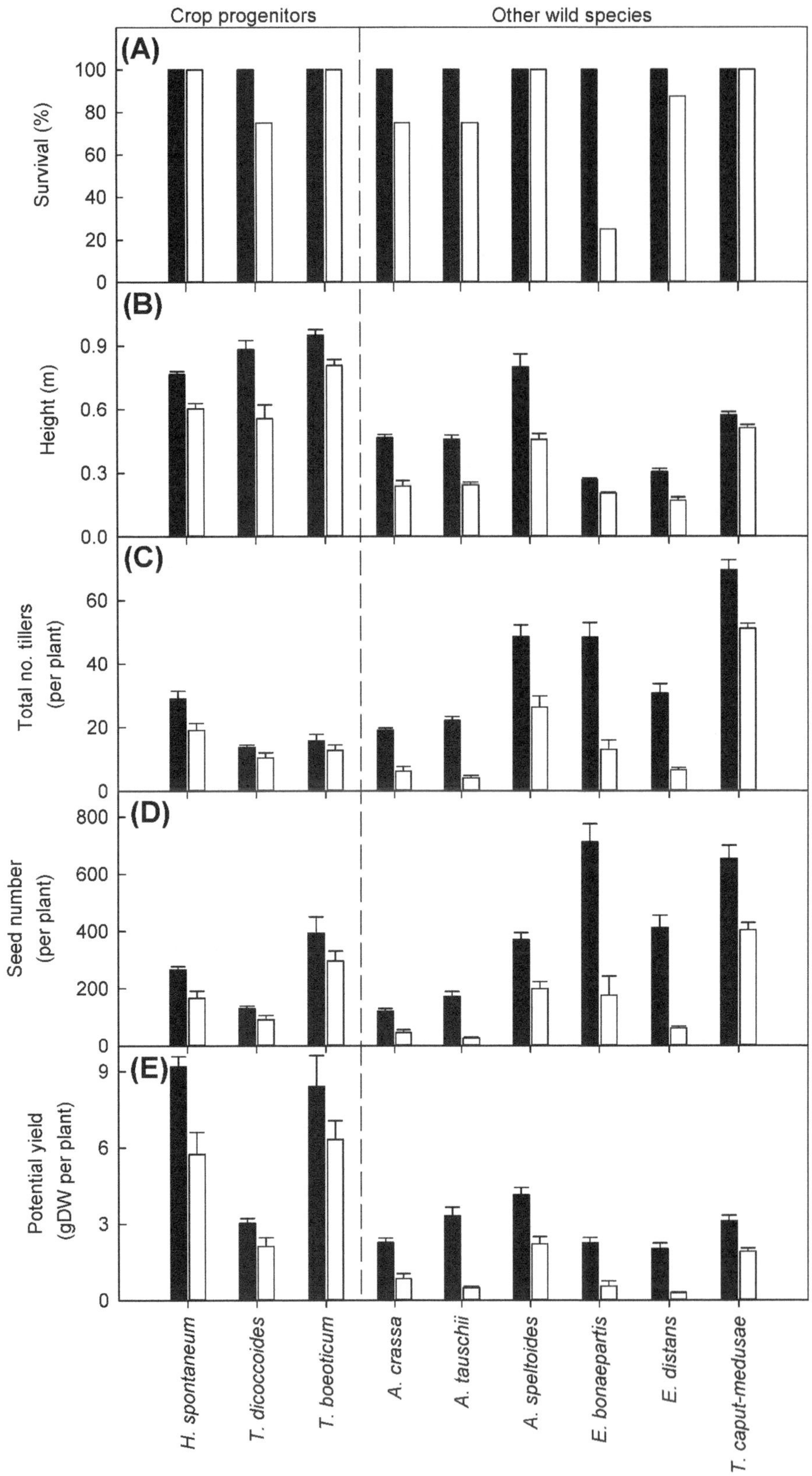

Figure 5. Impact of a defoliation treatment on plant survival, size and yield. Impact of defoliation treatment on (a) survival (%), (b) plant height, (c) number of tillers, (d) number of seeds, and (e) potential yield in crop progenitors and wild species. The defoliation treatment is shown by the white bar and the control treatment (no defoliation) is shown by the black. Data are means + SE of 8 replicates.

spontaneum and T. dicoccoides had the largest seed sizes in this study and the corresponding highest sRGR, the relationship being stronger in *H. spontaneum*. However, the third crop progenitor, *T. boeoticum*, had a similar seed mass to the larger seeded *Aegilops* spp., but a much lower sRGR than expected from its seed size, which weakens the relationship. A useful future addition to this study would be to extend the work to the other primary domesticates including pulses such as pea, lentil, chickpea and bitter vetch to test whether the same relationships exist between seed size and functional traits related to competition and disturbance.

sNAR was the main driver of interspecific variation in sRGR, in contrast to other size-corrected studies [30,42] where sSLA was most important. This dependence on sNAR may relate to high growth irradiance. In a meta-analysis by Shipley *et al.* [47], NAR was the most important driver of RGR in experiments conducted at a DQI of above 25 mol m^{-2} day^{-1}; in our experiment the DQI was 43.6 mol m^{-2} day^{-1}. Below 15 mol m^{-2} day^{-1} (low irradiance) SLA dominates, and between 15 and 25 mol m^{-2} day^{-1} SLA and NAR are of equal importance [47]. Furthermore, other studies have shown NAR to have increased importance at high irradiance [41,60,61]. In fact, there is a strong trade-off between NAR and SLA with changing light intensity [60]. The fast growing species in this study are likely to have a high photosynthetic capacity, which is realised under high irradiance. Photosynthetic rate and NAR are closely correlated, with photosynthesis in different species having different light saturation points, driving species differences in NAR at high light [60,61].

Although large seeds may gain an advantage in seedling competition, small seeds have the advantage of being produced in much greater numbers for a given reproductive effort [27]. For example, under the control treatment in this study, the greatest number of seeds in the crop progenitors was 393 in *T. boeoticum*, whilst in the wild species the maximum was 711 in *E. bonaepartis*. Therefore, although large seeded species may be more competitive, they are often recruitment-limited [36]. However, as humans increasingly become the agents of dispersal for crop species, this limitation diminishes in importance, thereby removing an important cost from the evolutionary trade-off acting on seed size.

The advantages of large seed/seedling size mainly operate during the early stages of plant establishment and growth. However, we found that plant traits in established plants are also correlated with seed mass, albeit less strongly. The larger seed size and higher sRGR translate to mature plants with a taller stature and greater yield. Plant height, measured in the third experiment, differed between the two groups; crop progenitors were significantly taller overall, varying from 0.77 m to 0.95 m, whilst the wild species ranged from 0.27 to 0.80 m in height. As well as offering greater apparency in a landscape for human gatherers, this height advantage confers competitive ability through prior access to light, and may be particularly advantageous in conditions of high fertility where competitive plants are better equipped to take advantage of the higher nutrient availability [62,63]. The ability of wild progenitors to exploit high fertility locations is expected to have been particularly advantageous in and around the human settlements, with nutrient enriched soils, to which they were transported by human gathering from the wild.

Fuller *et al.* [64] list three competing hypotheses to explain the observed increase in grain size during domestication (other than conscious human selection of larger grains within an existing population): (1) that larger grain size is a plastic response to the favourable soil conditions of cultivation, resulting in more fully developed grains [8]; (2) that larger grain varieties were adopted from elsewhere [8]; and (3) that larger size was an adaptive response to disturbance and/or deeper burial during cultivation [65]. Furthermore, genetics is likely to have played a significant role in the increase in seed size, with genetic control of grain size demonstrated not only in wheat [66], but also in other cereals e.g. rice [67,68] and maize [69]. Our experiments provide empirical evidence for an association between grain size and other potentially adaptive plant characteristics and, in particular, indicate that increased grain size could reflect an adaptive response to both improved soil conditions and disturbance.

Resilience to defoliation meant that yields were less impacted in crop progenitors

In this study, large seed mass was associated with substantial increases in potential yield; under the control treatment, the average yield per plant was 2.9 times greater in crop progenitors than wild species, which in itself may have been an important selection criterion for early farmers. Response to the defoliation treatment also differed significantly between the crop progenitors and wild species, with potential yield reduced more in the wild species. Although large seedlings are known to survive defoliation better, with higher subsequent growth rates [49], the plants in this study were beyond the seedling stage before the defoliation treatment was applied, so the response to defoliation cannot be attributed directly to seed mass.

Yield after defoliation was controlled by the number of tillers that re-sprouted; the number of tillers was significantly reduced by the defoliation treatment and this effect was much stronger in the wild species. Re-sprouting after the loss of practically all above-ground biomass requires surviving buds or other meristematic tissue, as well as reserves of carbohydrate and nutrients that can fund expansion of the first leaves of the new sprout [70,71]. Re-sprouting is also related strongly to growth form, with grasses being strong re-sprouters [72]. This explains why the level of survival was high in our experiment. However, considerable between-species variation in the response to defoliation suggests that grasses may employ diverse strategies [73]. The smaller-seeded wild species used in this study could have a reduced ability to re-sprout after defoliation (resilience) because they recruit more resources into producing a larger soil seed bank [74,75]. Alternatively, a smaller reduction in the tillering of crop progenitors could be an adaptation to grazing [76].

Conclusions

Comparison of cereal crop progenitors with other wild grass species exploited by pre-agricultural societies in the Fertile Crescent of Southwest Asia revealed significant differences in functional traits related to competition and disturbance. Supporting our hypothesis (i), the crop progenitors had larger seeds and seedlings than the other wild species, and this was associated with faster germination. The larger seed size was also correlated with a greater size-corrected RGR for two of the three crop progenitors (ii). The height of the crop progenitors and the seed yield were greater at maturity, in agreement with hypothesis (iii) and, although the survival and ability to produce seeds did not differ between the crop progenitors and wild species [hypothesis (iv)], the crop progenitors were more resilient and seed yields were less impacted by defoliation. Hypotheses (i) and (ii) were linked to seed mass, whilst hypotheses (iii) and (iv) were more weakly associated. For the crop progenitors, this combination of traits related to competition and disturbance confers the ability to effectively exploit sites with high levels of fertility and disturbance, potentially allowing them to thrive around early pre-agricultural settlements while other collected species did not. The greater height and potential yield of these species would also have offered a highly apparent and more abundant food resource for gatherers. These same traits would have pre-adapted crop progenitors to the cultivated and managed field environment. Based on this evidence, we argue that the interaction of plant functional traits and ecological processes had the potential to exert a strong influence on the narrowing of the food resource base during the transition to agriculture in the Fertile Crescent. It would be interesting to make similar comparisons of the competitive ability of the progenitors of pulse crops and small-seeded grass crops in anthropogenic conditions, compared with other collected species in their areas of origin,

Supporting Information

Figure S1 Growth functions fitted to plots of ln-total plant mass against time. Linear fits to ln-transformed dry weight per plant, in crop progenitors (a–c) and wild species (d–i). The crop progenitors were: (a) *H. spontaneum*, (b) *T. boeoticum*, (c) *T. dicoccoides*. The wild species were: (d) *A. crassa*, (e) *A. speltoides*, (f) *A. tauschii*, (g) *E. bonaepartis*, (h) *E. distans* and (f) *T. caput-medusae*. Data from experiment 2.

Figure S2 Relationship between dry weight, leaf area and seed mass. Regression slopes for the relationship between (a) dry weight and seed mass (F = 19.001, d.f = 2,6, p = 0.0025, $R^2 = 0.8636$); and (b) leaf area and seed mass (F = 14.32, d.f = 2,6, p = 0.005, $R^2 = 0.827$) for the three crop progenitors (closed circle) and six wild species (open circle). Data from experiment 2.

Figure S3 Relationship between classical RGR and seed mass. Relationship between classical RGR and seed mass for the three crop progenitors (closed circle) and six wild species (open circle). Data from experiment 2.

Figure S4 Relationship between sNAR, sSLA and sLMR and seed mass. Regression slopes for the relationship between (a) sNAR and seed mass [(F = 12.988, d.f = 2,6, p = 0.007, $R^2 = 0.812$)]; (b) sSLA and seed mass; and (c) sLMR and seed mass for the three crop progenitors (closed circle) and six wild species (open circle). Data from experiment 2.

Table S1 Details of species used in the experiments. Table details whether species were a crop progenitor or wild species, the germplasm holding from where seed was obtained and the accession number in the collection. Germplasm holdings are: Leibniz Institute of Plant Genetics and Crop Plant Research (IPK) in Gatersleben, Germany; the National Small Grains Collection (NSGC) of the United States Department of Agriculture (USDA) at the University of Idaho R & E Center, Aberdeen, Idaho; and the Western Regional Plant Introduction Station (WRPIS) of the USDA, Pullman, Washington.

Acknowledgments

We thank the Institute of Plant Genetics and Crop Plant Research (IPK) in Gatersleben, the National Small Grains Collection (NSGC) of the United States Department of Agriculture (USDA), Idaho; and the Western Regional Plant Introduction Station (WRPIS) of the USDA, Pullman, Washington for providing seed.

Author Contributions

Conceived and designed the experiments: JC CPO SW GJ MC. Performed the experiments: JC SW. Analyzed the data: JC MR. Wrote the paper: JC CPO GJ MR MC.

References

1. van Zeist W, Bakker-Heeres JAH (1986) Archaeobotanical studies in the Levant. 3: Late-Palaeolithic Mureybit. Palaeohistoria 26: 171–199.
2. Kislev ME, Nadel D, Carmi I (1992) Epipalaeolithic (19,000 BP) cereal and fruit diet at Ohalo II, Sea of Galilee, Israel. Review of Palaeobotany and Palynology 73: 161–166.
3. Hillman G, Hedges R, Moore A, Colledge S, Pettitt P (2001) New evidence of Lateglacial cereal cultivation at Abu Hureyra on the Euphrates. Holocene 11: 383–393.
4. Weiss E, Kislev ME, Simchoni O, Nadel D (2004) Small-grained wild grasses as staple food at the 23 000-year-old site of Ohalo II, Israel. Economic Botany 58: S125–S134.
5. Weiss E, Wetterstrom W, Nadel D, Bar-Yosef O (2004) The broad spectrum revisited: Evidence from plant remains. Proceedings of the National Academy of Sciences of the United States of America 101: 9551–9555.
6. Savard M, Nesbitt M, Jones MK (2006) The role of wild grasses in subsistence and sedentism: new evidence from the northern Fertile Crescent. World Archaeology 38: 179–196.
7. Fairbairn A, Martinoli D, Butler A, Hillman G (2007) Wild plant seed storage at Neolithic Çatalhöyük East, Turkey. Vegetation History and Archaeobotany 16: 467–479.
8. Willcox G, Fornite S, Herveux L (2008) Early Holocene cultivation before domestication in northern Syria. Vegetation History and Archaeobotany 17: 313–325.
9. Zeder MA (2009) The Neolithic macro-(r)evolution: Macroevolutionary theory and the study of culture change. Journal of Archaeological research 17: 1–63.
10. Nesbitt M (2002) When and were did domesticated cereals first occur in southwest Asia? In: Cappers R, Bottema S, editors. The Dawn of Farming in the Near East. Berlin: Ex Oriente. pp. 113–132.
11. Tanno K, Willcox G (2006) How fast was wild wheat domesticated? Science 311: 1886.
12. Asouti E, Fuller DQ (2013) A contextual approach to the emergence of agriculture in southwest Asia reconstructing early Neolithic plant-food production. Current Anthropology 54: 299–345.
13. Fuller DQ (2007) Contrasting patterns in crop domestication and domestication rates: Recent archaeobotanical insights from the Old World. Annals of Botany 100: 903–924.
14. Hammer K (1984) Das Domestikationssyndrom. Kulturpflanze 32: 11–34.
15. Cohen MN (2009) Introduction: Rethinking the origins of agriculture. Current Anthropology 50: 591–595.
16. Hayden B (2009) The proof is in the pudding: Feasting and the origins of domestication. Current Anthropology 50: 597–601.
17. Purugganan MD, Fuller DQ (2010) Archaeological data reveal slow rates of evolution during plant domestication. Evolution 65: 171–183.
18. Rindos D (1984) The origins of agriculture: An evolutionary perspective. New York: Academic Press. 344 p.
19. Zeder MA (2006) Central questions in the domestication of plants and animals. Evolutionary Anthropology 15: 105–117.
20. Darwin CR (1875) The variation of animals and plants under domestication. London: John Murray.
21. Willcox G (2005) The distribution, natural habitats and availability of wild cereals in relation to their domestication in the Near East: multiple events, multiple centres. Vegetation History and Archaeobotany 14: 534–541.
22. Colledge S (1998) Identifying pre-domestication cultivation using multivariate analysis. In: Damania AB, Valkoun J, Willcox G, Qualset CO, editors. The origins of agriculture and crop domestication. Aleppo: ICARDA. pp. 121–131.
23. Fuller DQ, Willcox G, Allaby RG (2011) Cultivation and domestication had multiple origins: arguments against the core area hypothesis for the origins of agriculture in the Near East. World Archaeology 43: 628–652.
24. Blackman VH (1919) The compound interest law and plant growth. Annals of Botany 33: 353–360.
25. Grime JP, Hunt R (1975) Relative growth-rate: Its range and adaptive significance in a local flora. Journal of Ecology 63: 393–422.
26. Tilman D (1998) Plant strategies and the dynamics and structure of plant communities. Princeton, NJ: Princeton University Press.
27. Westoby M (1998) A leaf-height-seed (LHS) plant ecology strategy scheme. Plant and Soil 199: 213–227.

28. Maranon T, Grubb PJ (1993) Physiological basis and ecological significance of the seed size and relative growth rate relationship in mediterranean annuals. Functional Ecology 7: 591–599.
29. Poorter L, Rose S (2005) Light-dependent changes in the relationship between seed mass and seedling traits: a meta-analysis for rain forest tree species. Oecologia 142: 378–387.
30. Rees M, Osborne CP, Woodward FI, Hulme SP, Turnbull LA, et al. (2010) Partitioning the components of relative growth rate: How important is plant size variation? American Naturalist 176: E152–E161.
31. Turnbull LA, Paul-Victor C, Schmid B, Purves DW (2008) Growth rates, seed size, and physiology: Do small-seeded species really grow faster? Ecology 89: 1352–1363.
32. Rose KE, Atkinson RL, Turnbull LA, Rees M (2009) The costs and benefits of fast living. Ecology Letters 12: 1379–1384.
33. Metcalf CJE, Rees M, Alexander JM, Rose K (2006) Growth–survival trade-offs and allometries in rosette-forming perennials. Functional Ecology 20: 217–225.
34. Turnbull LA, Philipson CD, Purves DW, Atkinson RL, Cunniff J, et al. (2012) Plant growth rates and seed size: a re-evaluation. Ecology 93: 1283–1289.
35. Gómez JM (2004) Bigger is not always better: conflicting selective pressures on seed size in *Quercus ilex*. Evolution 58: 71–80.
36. Turnbull LA, Rees M, Crawley MJ (1999) Seed mass and the competition/colonization trade-off: a sowing experiment. Journal of Ecology 87: 899–912.
37. Westoby M, Leishman M, Lord J (1996) Comparative ecology of seed size and dispersal. Philosophical Transactions of the Royal Society of London Series B-Biological Sciences 351: 1309–1317.
38. Leibniz-Institut Für Pflanzengenetik und Kulturpflanzenforschung (IPK) (2013) Genetic Integrity - IPK Gatersleben [Online]. Available: http://www.ipk-gatersleben.de/en/dept-genebank/resources-genetics-andreproduction/genetic-integrity/ [Accessed May 5th 2013].
39. Savard M, Nesbitt M, Gale R (2003) Archaeobotanical evidence for Neolithic diet and subsistence at M'lefaat (Iraq). Paléorient 29: 93–106.
40. Willcox G (1999) Agrarian change and the beginnings of cultivation in the Near East: evidence from wild progenitors, experimental cultivation and archaeobotanical data. In: Gosden C, Hather J, editors. The prehistory of food: Appetites for change. London: Routledge. pp. 478–500.
41. Ryser P, Wahl S (2001) Interspecific variation in RGR and the underlying traits among 24 grass species grown in full daylight. Plant Biology 3: 426–436.
42. Poorter H, Remkes C (1990) Leaf area ratio and net assimilation rate of 24 wild species differing in relative growth rate. Oecologia 83: 553–559.
43. Poorter H (1989) Plant growth analysis: towards a synthesis of the classical and the functional approach. Physiologia Plantarum 75: 237–244.
44. Hunt R, Cornelissen JHC (1997) Components of relative growth rate and their interrelations in 59 temperate plant species. New Phytologist 135: 395–417.
45. Hendry GAF, Grime JP (1993) Methods in comparative plant ecology: A laboratory manual. London, UK.: Chapman & Hall. 9–28 p.
46. Hewitt EJ (1966) Sand and water culture methods used in the study of plant nutrition. Buckingham, UK.: Farnham Royal: Commonwealth Agricultural Bureaux. 547 p.
47. Shipley B (2006) Net assimilation rate, specific leaf area and leaf mass ratio: which is most closely correlated with relative growth rate? A meta-analysis. Functional Ecology 20: 565–574.
48. Taylor SH, Hulme SP, Rees M, Ripley BS, Woodward IF, et al. (2010) Ecophysiological traits in C3 and C4 grasses: a phylogenetically controlled screening experiment. New Phytologist 185: 780–791.
49. Armstrong DP, Westoby M (1993) Seedlings from large seeds tolerate defoliation better: A test using phylogenetically independent contrasts. Ecology 74: 1092–1100.
50. Harms KE, Dalling JW (1997) Damage and herbivory tolerance through resprouting as an advantage of large seed size in tropical trees and lianas. Journal of Tropical Ecology 13: 617–621.
51. Engelbrecht T (1916) Über die Entstehung einiger feldmässig angebauten Kulturpflanze. Geographische Zeitscgrift 22: 328–334.
52. Anderson E (1952) Man as a maker of new plants and plant communities. In: Thomas WL, editor. Man's role in changing the face of the earth. Chicago: University of Chicago. pp. 763–777.
53. Hawkes JG (1969) The ecological background of plant domestication. In: Ucko PJ, Dimbleby GW, editors. The Domestication and Exploitation of Plants and Animals. London: Gerald Duckworth and Co. pp. 17–29.
54. Harlan JR (1965) The possible role of weed races in the evolution of cultivated plants. Euphytica 14: 173–&.
55. Harlan JR (1992) Crops and Man. Madison, Wisconsin.: American Society of Agronomy and Crop Science Society of America.
56. Blumler MA, Byrne R (1991) The ecological genetics of domestication and the origins of agriculture. Current Anthropology 32: 23–35.
57. Abbo S, Gopher A, Rubin B, Lev-Yadun S (2005) On the origin of Near Eastern founder crops and the 'dump-heap hypothesis'. Genetic Resources and Crop Evolution 52: 491–495.
58. Rindos D (1980) Symbiosis, instability, and the origins and spread of agriculture: a new model. Current Anthropology 21: 751–772.
59. Kidson R, Westoby M (2000) Seed mass and seedling dimensions in relation to seedling establishment. Oecologia 125: 11–17.
60. Shipley B (2002) Trade-offs between net assimilation rate and specific leaf area in determining relative growth rate: relationship with daily irradiance. Functional Ecology 16: 682–689.
61. Poorter L (1999) Growth responses of 15 rain-forest tree species to a light gradient: the relative importance of morphological and physiological traits. Functional Ecology 13: 396–410.
62. Hautier Y, Niklaus PA, Hector A (2009) Competition for light causes plant biodiversity loss after eutrophication. Science 324: 636–638.
63. Vojtech E, Turnbull LA, Hector A (2007) Differences in light interception in grass monocultures predict short-term competitive outcomes under productive conditions. PLoS ONE 2: e499.
64. Fuller DQ, Allaby RG, Stevens C (2010) Domestication as innovation: the entanglement of techniques, technology and chance in the domestication of cereal crops. World Archaeology 42: 13–28.
65. Harlan JR, De Wet JMJ, Price EG (1973) Comparative evolution of cereals. Evolution 27: 311–325.
66. Gegas VC, Nazari A, Griffiths S, Simmonds J, Fish L, et al. (2010) A genetic framework for grain size and shape variation in wheat. Plant Cell 22: 1046–1056.
67. Kovach MJ, Sweeney MT, McCouch SR (2007) New insights into the history of rice domestication. Trends in Genetics 23: 578–587.
68. Fan CH, Xing YZ, Mao HL, Lu TT, Han B, et al. (2006) GS3, a major QTL for grain length and weight and minor QTL for grain width and thickness in rice, encodes a putative transmembrane protein. Theoretical and Applied Genetics 112: 1164–1171.
69. Li Q, Yang X, Bai G, Warburton ML, Mahuku G, et al. (2010) Cloning and characterization of a putative GS3 ortholog involved in maize kernel development. Theoretical and Applied Genetics 120: 753–763.
70. Vesk PA, Westoby M (2004) Sprouting by plants: the effects of modular organization. Functional Ecology 18: 939–945.
71. Vesk PA, Westoby M (2004) Funding the bud bank: a review of the costs of buds. Oikos 106: 200–208.
72. Vesk PA, Warton DI, Westoby M (2004) Sprouting by semi-arid plants: testing a dichotomy and predictive traits. Oikos 107: 72–89.
73. Vesk PA (2006) Plant size and resprouting ability: trading tolerance and avoidance of damage? Journal of Ecology 94: 1027–1034.
74. Thompson K, Band SR, Hodgson JG (1993) Seed size and shape predict persistence in soil. Functional Ecology 7: 236–241.
75. Thompson K, Jalili A, Hodgson JG, Hamzeh'ee B, Asri Y, et al. (2001) Seed size, shape and persistence in the soil in an Iranian flora. Seed Science Research 11: 345–355.
76. Noy-Meir I, Briske DD (2002) Response of wild wheat populations to grazing in Mediterranean grasslands: the relative influence of defoliation, competition, mulch and genotype. Journal of Applied Ecology 39: 259–278.

8

Mercury in the Soil of Two Contrasting Watersheds in the Eastern United States

Douglas A. Burns[1]*, Laurel G. Woodruff[2], Paul M. Bradley[3], William F. Cannon[4]

1 United States Geological Survey, Troy, New York, United States of America, **2** United States Geological Survey, Mounds View, Minnesota, United States of America, **3** United States Geological Survey, Columbia, South Carolina, United States of America, **4** United States Geological Survey, Reston, Virginia, United States of America

Abstract

Soil represents the largest store of mercury (Hg) in terrestrial ecosystems, and further study of the factors associated with soil Hg storage is needed to address concerns about the magnitude and persistence of global environmental Hg bioaccumulation. To address this need, we compared total Hg and methyl Hg concentrations and stores in the soil of different landscapes in two watersheds in different geographic settings with similar and relatively high methyl Hg concentrations in surface waters and biota, Fishing Brook, Adirondack Mountains, New York, and McTier Creek, Coastal Plain, South Carolina. Median total Hg concentrations and stores in organic and mineral soil samples were three-fold greater at Fishing Brook than at McTier Creek. Similarly, median methyl Hg concentrations were about two-fold greater in Fishing Brook soil than in McTier Creek soil, but this difference was significant only for mineral soil samples, and methyl Hg stores were not significantly different among these watersheds. In contrast, the methyl Hg/total Hg ratio was significantly greater at McTier Creek suggesting greater climate-driven methylation efficiency in the Coastal Plain soil than that of the Adirondack Mountains. The Adirondack soil had eight-fold greater soil organic matter than that of the Coastal Plain, consistent with greater total Hg stores in the northern soil, but soil organic matter – total Hg relations differed among the sites. A strong linear relation was evident at McTier Creek ($r^2 = 0.68$; $p < 0.001$), but a linear relation at Fishing Brook was weak ($r^2 = 0.13$; $p < 0.001$) and highly variable across the soil organic matter content range, suggesting excess Hg binding capacity in the Adirondack soil. These results suggest greater total Hg turnover time in Adirondack soil than that of the Coastal Plain, and that future declines in stream water Hg concentrations driven by declines in atmospheric Hg deposition will be more gradual and prolonged in the Adirondacks.

Editor: Stefan Bertilsson, Uppsala University, Sweden

Funding: Funding for this work was provided by the United States Environmental Protection Agency and the National Water Quality Assessment Program of the United States Geological Survey. The funders has no role in study design, data collection and analysis, decision to publish, or preparation of the manuscript.

Competing Interests: The authors have declared that no competing interests exist.

* E-mail: daburns@usgs.gov

Introduction

Soil is the largest reservoir of mercury (Hg) storage in global terrestrial ecosystems, and transfers to and from soil are pivotal in Hg cycling among vegetation, the atmosphere, groundwater, surface water, and the oceans [1]. A recent study estimates the global soil pool size (year 2000) at 240,000 Mg Hg, and indicates that Hg storage has increased by about 20% since 1840 largely due to atmospheric Hg deposition primarily derived from anthropogenic emissions sources such as coal burning, cement manufacturing, and other industrial and mining activities [2]. Soil Hg concentrations vary widely from 10 ng/g to 1000 ng/g in rural and remote areas, and from 100 ng/g to >10,000 ng/g in urban, industrial, and mineralized/mined lands [3,4]. This Hg is largely bound to soil organic matter (SOM), and Hg concentrations are typically strongly related to measures of SOM or soil organic carbon (SOC) [5–7]. Under oxidized conditions, Hg (II) is strongly bound to organic thiol groups, which are typically present in excess of available Hg in most soil [8,9]. Under anoxic conditions, meta-cinnabar and other sulfide minerals may also be important Hg forms [10], though the presence and abundance of these mineral forms in soil without a metal sulfide ore genesis is not clear [11]. Additionally, Hg may be adsorbed either specifically or non-specifically to SOM and to mineral surfaces such as iron and aluminum oxy-hydroxides [12].

Given the strong affinity of Hg for SOM, soil Hg concentrations are typically greatest where SOM is concentrated, in shallow O- or A-horizons, and these concentrations typically decrease with depth [7]. The O-horizon commonly has Hg concentrations an order of magnitude greater than that of the mineral-dominated B-horizon, though Hg/SOM is typically greater in the B horizon [7,13,14]. Similarly, organic-rich soil such as peat typically has greater Hg concentrations than those of soil dominated by inorganic material [6]. Soil in areas with high geogenic Hg concentrations can show the reverse of the typical depth pattern, with greater Hg concentrations in the deep soil than near the surface [15]. Despite the greater Hg concentrations commonly measured in surface horizons, deeper mineral horizons generally store more Hg because of the much greater mass of mineral-dominated subsoil in most landscapes [4,6,16]. Total soil storage of Hg can vary from a few mg/m^2 to >50 mg/m^2, and these values vary, similarly to those of Hg concentrations, as a function of source loads of atmospheric Hg deposition, the amount of SOM in soil, and geogenic concentrations in parent material [3,6]. Disturbance factors related to fire history, agriculture, forest harvesting, mining,

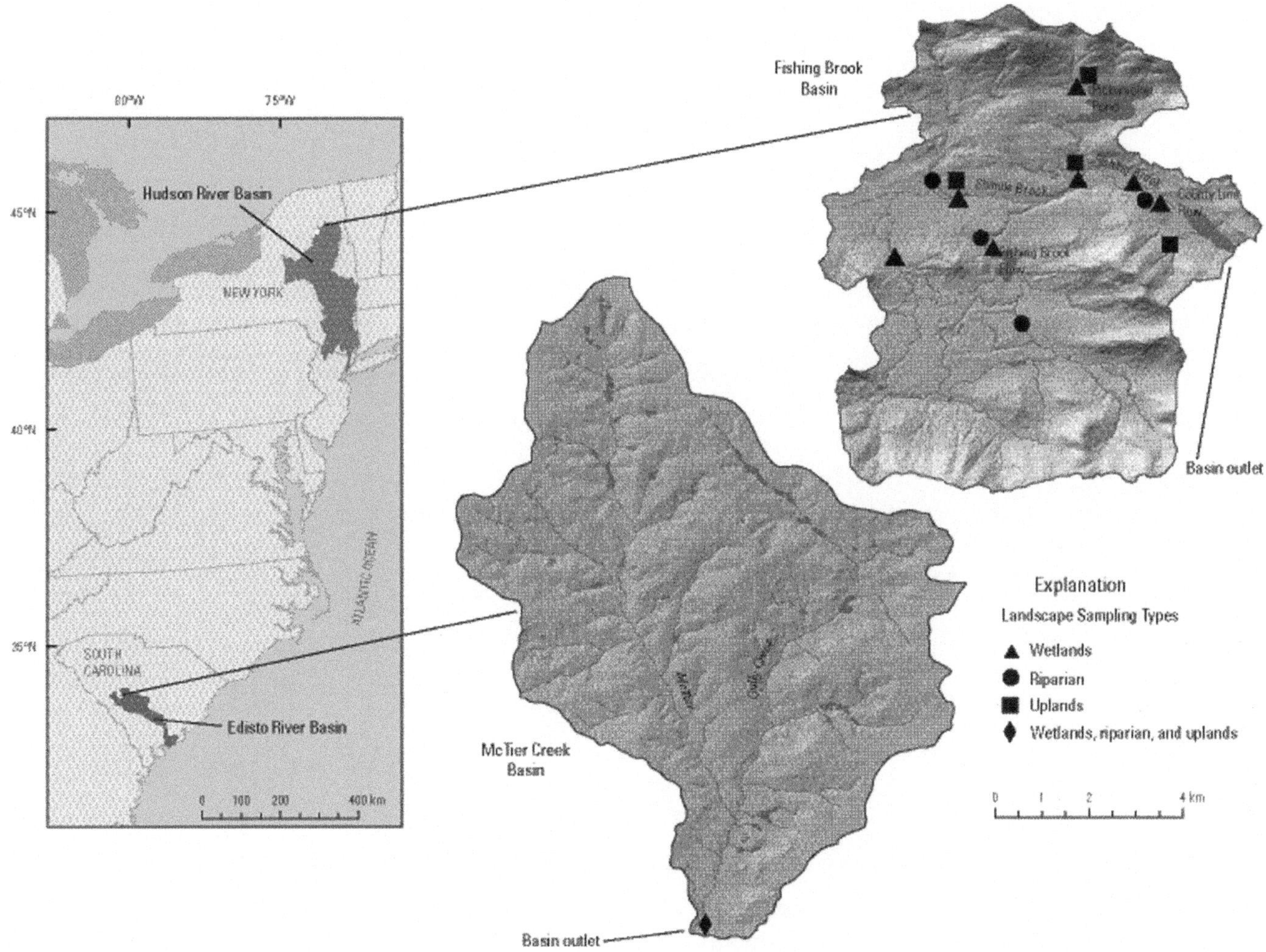

Figure 1. Study site maps. Locations of McTier Creek watershed within the Edisto River basin, and Fishing Brook within the Hudson River basin. Symbols indicate approximate locations where soil samples were collected in different landscape types as described in text. Note that only one symbol appears in the McTier Creek map because the 24 sampling sites were located too close together to be resolved in a map of the whole watershed.

and other human land uses also affect Hg concentrations and storage in soil [17,18].

Several studies have described strong relations between soil Hg concentrations and SOM/SOC, consistent with the strong affinity of Hg for binding with thiol groups [5,7,13,19]. The slopes and intercepts of these Hg–SOM relations vary greatly, however, as a function of regional location, horizon and/or depth, landscape type, and vegetation type [6,16], a finding that is expected given the heterogeneity of SOM and the excess of thiol-like Hg binding sites generally available [9]. Some study results are consistent with variation of Hg/SOM as a function of the rate of atmospheric Hg deposition [5,15], although variation in geogenic Hg may be a contributing factor as well [20].

Although measurements of total Hg (THg) in soil provide important data necessary to better understand the Hg cycle, data on methyl Hg (MeHg) concentrations in soil are more pivotal to understanding bioaccumulation in terrestrial and aquatic food webs. The ratio of MeHg/THg varies widely in soil from $<1\%$ to $>20\%$ [4,6,8]. Wetland soil generally has greater MeHg concentrations and MeHg/THg than soil in steeper upland parts of landscapes, and more complex patterns are often superimposed on this general gradient. One study in a boreal forest catchment in Sweden found greater MeHg concentrations and MeHg/THg in soil at stream bank and near-stream locations than 20 m from the stream [8]. Another more recent investigation showed the highest MeHg concentrations in soil pore water where upland soil borders peaty wetland soil suggesting that the distal riparian area may have greater soil MeHg concentrations than the near-stream riparian area [21]. The MeHg content of soil is important because of the potential for mobilization into soil solution followed by surface and/or subsurface runoff and transport to local surface waters where bioaccumulation in aquatic food webs may then occur.

Despite many published studies on Hg in soil [3–7], little is known about controls on the relation of Hg concentrations and stores in soil to those of adjacent surface waters. Elevated Hg levels are often measured in surface waters with extensive wetlands where high THg and MeHg concentrations and storage in soil have been demonstrated, such as lake watersheds in the US Upper Midwest [22] and the Adirondacks of New York [16].

In this study, we compare Hg concentrations and storage in the soil of two watersheds, located in "hot spot" regions where Hg in predator fish and/or aquatic-feeding birds consistently exceed

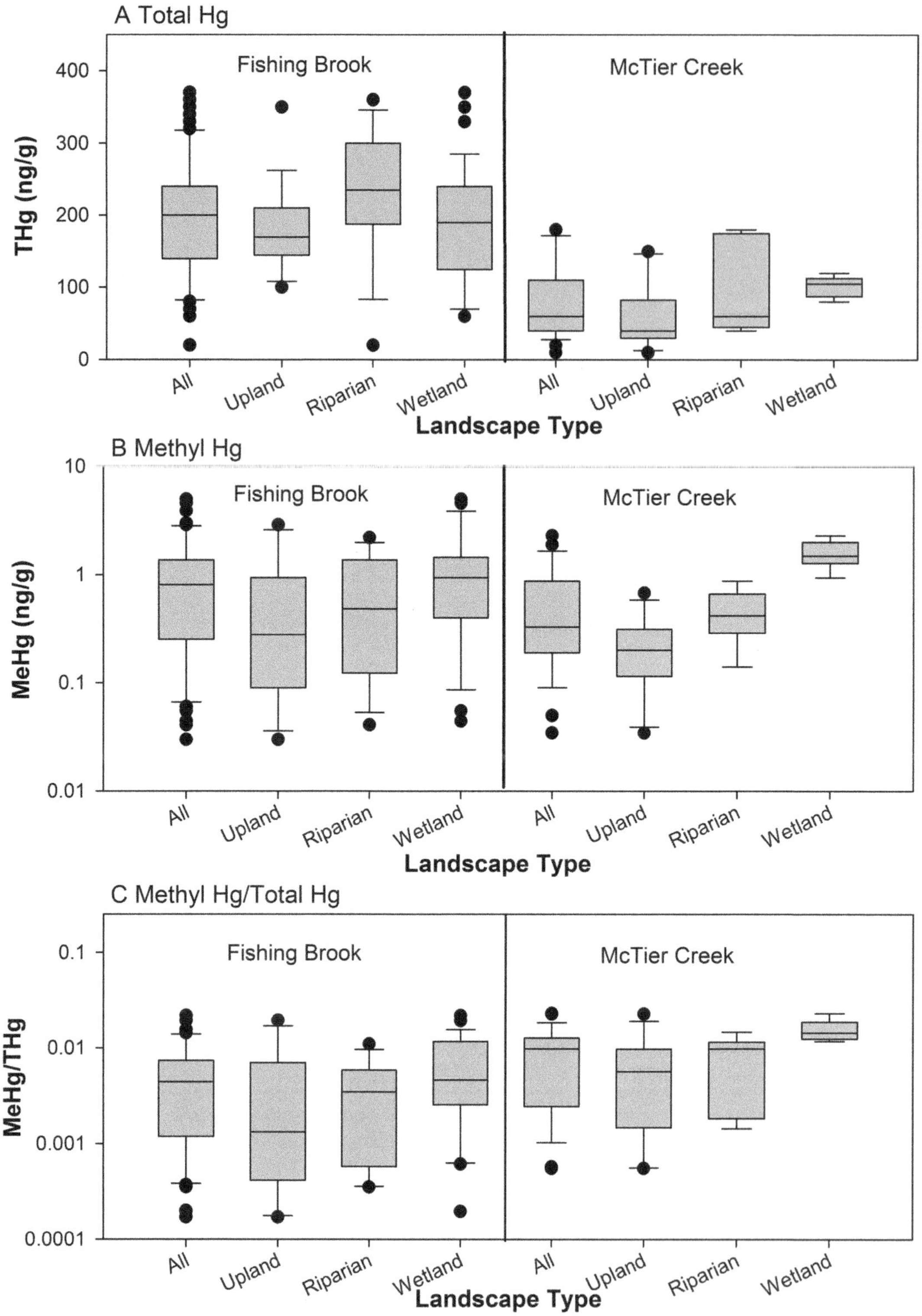

Figure 2. Mercury values in soil Oa/A horizon samples. Box plots of soil Hg concentrations measured in the Fishing Brook, NY and McTier Creek, SC watersheds for samples collected from Oa/A horizons. A Total Hg, B Methyl Hg, and C Methyl Hg/Total Hg.

thresholds for human health warnings and for adverse effects on wildlife [23,24], the Adirondack Mountains of New York, and the Coastal Plain of South Carolina [25,26]. Soil THg and MeHg concentrations and storage are compared across three different landscape types in each watershed, uplands, riparian areas, and wetlands, because hydrologic/transport dynamics among these landscapes are important in mobilization of Hg to local streams and subsequent bioaccumulation in the aquatic ecosystem [25–27]. Our objectives are to describe how Hg concentrations and pool sizes in soil differ between these two watersheds and among

Table 1. Median values of THg, MeHg, and MeHg/THg in Oa/A horizon soil samples from Fishing Brook, NY and McTier Creek, SC.

Constituent	Fishing Brook				McTier Creek			
	All	Upland	Riparian	Wetland	All	Upland	Riparian	Wetland
THg(ng/g)								
n	71	17	16	38	27	12	9	6
median	**200**	**170**	**235**	**190**	60	40	60	105
MeHg (ng/g)								
n	52	11	12	29	27	12	9	6
median	0.81	0.28	0.48	0.94	0.33	0.20	0.42	1.50
MeHg/THg (%)								
n	52	11	12	29	27	12	9	6
median	0.44	0.13	0.35	0.46	**0.98**	0.57	0.98	**1.40**

Note – bold font indicates median value was significantly greater ($p<0.05$) in site-to-site Mann-Whitney pairwise comparison tests done for all samples collected and for each landscape type.

different landscape types, and to discuss the implications of any differences that can be readily identified.

Study Sites

Soil samples were collected in two geographic locations: (1) Fishing Brook watershed in the Adirondack Mountains of New York, and (2) McTier Creek watershed in the Coastal Plain of South Carolina. A brief description of these study sites follows with an emphasis on soil; however, the reader is referred to [28] for more detailed descriptions of these study areas.

Fishing Brook (FB) is a 65.6 km^2 watershed in the westernmost headwaters of the Hudson River basin. The watershed is a heterogeneous landscape characterized by forested uplands, riparian floodplains, and scattered open water bodies. The watershed is dominantly forested (86.3%) by Adirondack Northern Hardwood Forest [29]. Sugar maple (*Acer saccharum*), yellow birch (*Betula alleghaniensis*) and American beech (*Fagus grandifolia*) are the dominant tree species below 980 m elevation, and balsam fir (*Abies balsamea*) and red spruce (*Picea rubens*) are the dominant species above this elevation [29]. Additionally, balsam fir and red spruce are dominant in wet soil at the base of hillslopes and in riparian areas adjacent to streams and ponds. The watershed consists of 8.2% wetland area (National Wetlands Inventory (NWI) approach, [30]), and scrub-shrub is the dominant wetland type, with speckled alder (*Alnus incana*) a dominant species [31]. Open water is 2.7% of watershed area, with numerous small ponds and lakes. Only 0.7% of the watershed is considered developed, as low density residential land use, some dirt roads, and one highway (Rte. 28N).

Upland soils in the FB watershed are mainly Spodosols developed in coarse, bouldery glacial till derived from the Precambrian metamorphic bedrock of the region [32]. The most common soil types found in the watershed include the Skerry, Becket, Adirondack, and Tunbridge Soil Series, generally well-drained, bouldery sandy loams classified as Haplorthods [33]. The most common soil types in riparian wetland areas are broadly classified as Entisols or Histosols and include the Wonsqueak, Rumney, and Bucksport Soil Series, mixtures of loamy sand or silt, muck, and peat with a water table that ranges from just below land surface during the growing season, to near or above the land surface during the dormant season [33]. These riparian soil types are generally developed in fluvial or glaciofluvial sediments, and thus are not as rich in organic matter as the deep peat riparian wetlands in the western Adirondacks described in [34].

McTier Creek (MC) is a 79.4 km^2 watershed (defined by the USGS stream gage near New Holland, SC) tributary to the Edisto River, and located in the Inner Coastal Plain. The watershed consists of 49.6% forested land cover, of which loblolly pine (*Pinus taeda*) is most common. About 6.4% of the watershed is wetland area (NWI approach, [35]), and bottomland hardwoods such as black tupelo (*Nyssa sylvatica*) are dominant in floodplain areas where these wetlands are generally found. Additionally, 1.2% of the watershed consists of open water, 14.9% is agricultural land, and 4.7% is developed [28]. The remaining land cover is mainly herbaceous upland.

Soil at MC is developed in sandy fluviomarine deposits and is generally classified as an Ultisol on ridges and hillslopes and as an Entisol or Inceptisol in riparian floodplains [36]. The Troup, Vaucluse, and Ailey Series are those most common on ridges and hillslopes, generally an excessively drained sandy soil with clay-rich kaolinitic deposits found at depth in isolated areas [36]. Floodplains are dominated by the poorly drained Bibb Series loamy sand and Johnston Series mucky loam, the latter of which sometimes has standing water at the surface signifying wetlands [36].

A comparison of the landscape and soil of these two watersheds reveals some similarities and some sharp differences. Both watersheds are dominated by upland forests, although the FB landscape is generally steeper (median slope = 12.5%) than that of MC (median slope = 5.4%), and MC also includes almost 20% developed land and agriculture, whereas FB has <1% developed land. Narrow floodplains dominated by sandy and loamy deposits as well as some mucky or peaty soil are prevalent in both watersheds, and riparian wetlands are commonly found in these floodplains. Riparian area defined as all land within an elevation of 0.65 m of the stream network [37] is 14.1% of the MC watershed area and 18.4% of the FB watershed area. When all wetland area located within this riparian delineation is excluded from these calculations, these riparian area values decrease to 9.3% of the MC watershed area and 12.3% of the FB watershed area. The two watersheds also have similar proportions of wetland area with a slightly greater value in the FB drainage (8.2%) than in the MC drainage (6.4%) according to NWI methods [35]. This difference narrows further (FB = 9.3%, MC = 8.2%) according to

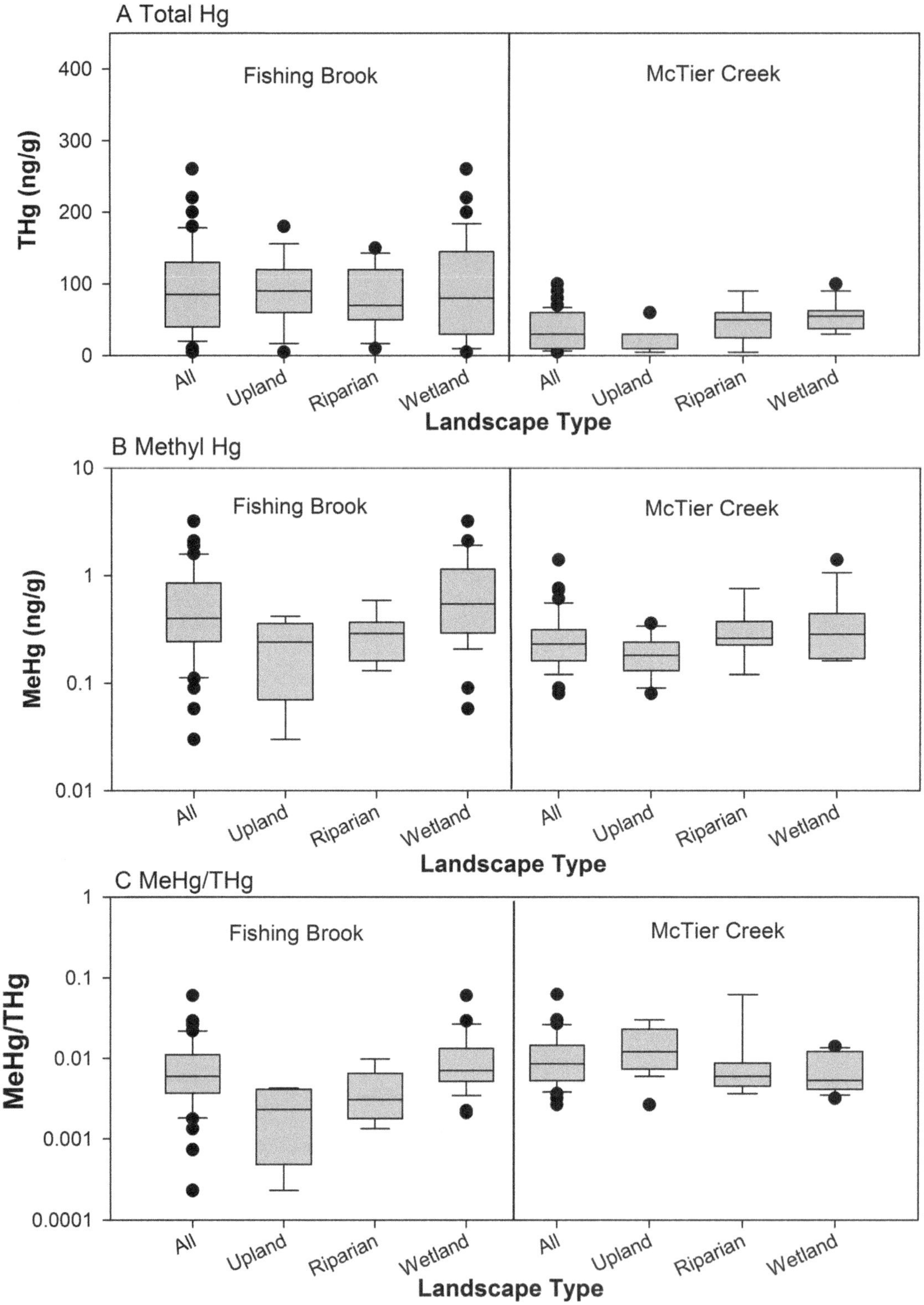

Figure 3. Mercury values in soil B horizon samples. Box plots of soil Hg concentrations measured in the Fishing Brook, NY and McTier Creek, SC watersheds for samples collected from B horizons. A Total Hg, B Methyl Hg, and C Methyl Hg/Total Hg.

the wetland classification approach of the National Land Cover 2001 Dataset [38], but we apply NWI data in the current paper.

Perhaps the greatest difference between these two watersheds is their respective climates. FB experiences a cold, humid continental climate with mean annual air temperature of 4.9°C and mean annual precipitation of 1074 mm (Newcomb, NY, 1981–2010, [39]). Mean minimum daily air temperature is <0°C for >6 months of the year, much of the winter precipitation falls as snow, and spring snowmelt is a major feature of the annual hydrograph [40]. In contrast, MC experiences a subtropical moist climate with

Table 2. Median values of THg, MeHg, and MeHg/THg in B horizon soil samples from Fishing Brook, NY and McTier Creek, SC.

Constituent	Fishing Brook				McTier Creek			
	All	Upland	Riparian	Wetland	All	Upland	Riparian	Wetland
THg(ng/g)								
n	70	17	16	37	42	19	9	14
median	**85**	**90**	70	80	30	10	50	55
MeHg (ng/g)								
n	40	5	7	28	42	19	9	14
median	**0.40**	0.24	0.29	**0.54**	0.23	0.18	0.26	0.28
MeHg/THg (%)								
n	40	5	7	28	42	19	9	14
median	0.60	0.23	0.31	0.71	0.86	**1.20**	0.60	0.53

Note – bold font indicates median value was significantly greater ($p<0.05$) in site-to-site Mann-Whitney pairwise comparison tests done for all samples collected and for each landscape type.

mean annual air temperature of 17.9°C and mean annual precipitation of 1276 mm, of which little falls as snow (Aiken, SC, 1981–2010, [41]). Stream runoff also differs greatly among these two regions. For example, the Edisto River near Givhans, SC of which MC is tributary has mean annual runoff of 273 mm (1981–2010, [42]), whereas the Hudson River at North Creek, NY of which FB is a tributary has mean annual runoff of 735 mm over the same period [42], a difference that is largely attributed to the role of snowmelt in New York, and to higher annual rates of evapotranspiration in South Carolina than in New York.

Materials and Methods

Field

Soil sampling focused on the key landscape types in each watershed. At FB, samples were collected from upland hillslope areas in northern hardwood forest, along riparian areas in spruce-fir forest, and in wetlands [43]. A total of 163 soil samples were collected at 70 distinct locations during Sept. 2008 (Fig. 1). Sampling sites were on private land owned by Finch Pruyn Company (now Finch Company), and verbal permission was provided for collecting environmental samples including a key to access locked gates needed to reach many of the sites. At MC, samples were collected along two transects from upland hilltops and hillslopes, riparian floodplains, and wetlands [43]. A total of 81 samples were collected from 24 distinct locations in Nov. 2008 (Fig. 1). Sampling sites were on private land owned by the family of the late Senator Strom Thurmond who provided verbal permission for access and the collection of environmental samples. For the purposes of this analysis, soil data are distinguished for uplands, riparian floodplains, and wetlands in each watershed.

Samples were collected with a shovel and hand auger and placed into sediment vials immediately upon collection. Nitrile gloves were used when handling soil to avoid contamination. At least two depths/horizons were sampled at each location, either an Oa (FB) or A (MC) horizon, and a B horizon. The B horizon represents a depth integrated sample that reflects the complete horizon to a depth of 40 cm; sampling did not attempt to make fine scale distinctions within the B horizon. At a few of the sites, samples were collected deeper in the profile, either the B/C or C horizon; E-horizon samples were also collected at both Fishing Brook and McTier. In wetland areas without well-developed soil horizons, samples were collected by depth interval. In such soil, the top 10 cm was typically equated to an Oa or A horizon for the purposes of this data analysis, and a second interval below this depth, typically 10 cm to about 20 to 40 cm, was equated to a B horizon for the purposes of this data analysis.

Laboratory

Oa horizon samples were air dried, milled, and split. A portion was set aside for Hg analysis, and the remainder was heated in an oven at 500°C for 13 hr to determine loss-on-ignition. For mineral soil horizons, samples were air dried and sieved to <2 mm. Total carbon (C) in mineral-dominated soil samples was determined by infrared detection of CO_2 in an automated C analyzer [44], carbonate C was determined by coulometric titration [45], and organic C was determined by difference. Organic C data were multiplied by 1.9 to convert these values to percent organic matter for the purposes of data analysis [46].

Mercury was analyzed by cold vapor atomic absorption spectrometry according to a method described in Smith et al. [47]. This method is a modification of U.S. Environmental Protection Agency Method 7471B [48], and is considered an analysis of total Hg. The method detection limit is 10 ng/g. A lab-homogenized sub-sample of soil that was frozen as soon as possible after collection was shipped to the laboratory on dry ice, and held frozen until analysis of MeHg by cold vapor atomic fluorescence spectrometry. Details of the MeHg analysis method, which has a 0.08 ng/g detection limit, are described in [49].

Laboratory quality assurance and quality control procedures for the analytical methods used in this study are described in [43] and [47], except for those of MeHg, which are described in [49]. Sixteen duplicate samples were collected in the FB watershed, and 7 were collected in the MC watershed. Precision for FB duplicates samples as represented by the mean standard deviation and mean relative standard deviation (in parentheses) of these duplicate analyses was 4.9 ng/g (8.4%) for total Hg, and 7.4% (22.0%) for percent organic matter. Precision for MC duplicates was 10 ng/g (21.1%) for total Hg, and 0.38% (12.4%) for percent organic matter. Duplicate analyses were not available for MeHg, but analytical precision data for multiple analyses of a reference solid sample indicate a relative standard deviation of 8.6% for the method used in this study [49].

Soil data used in this paper are available in [43], and can be freely downloaded.

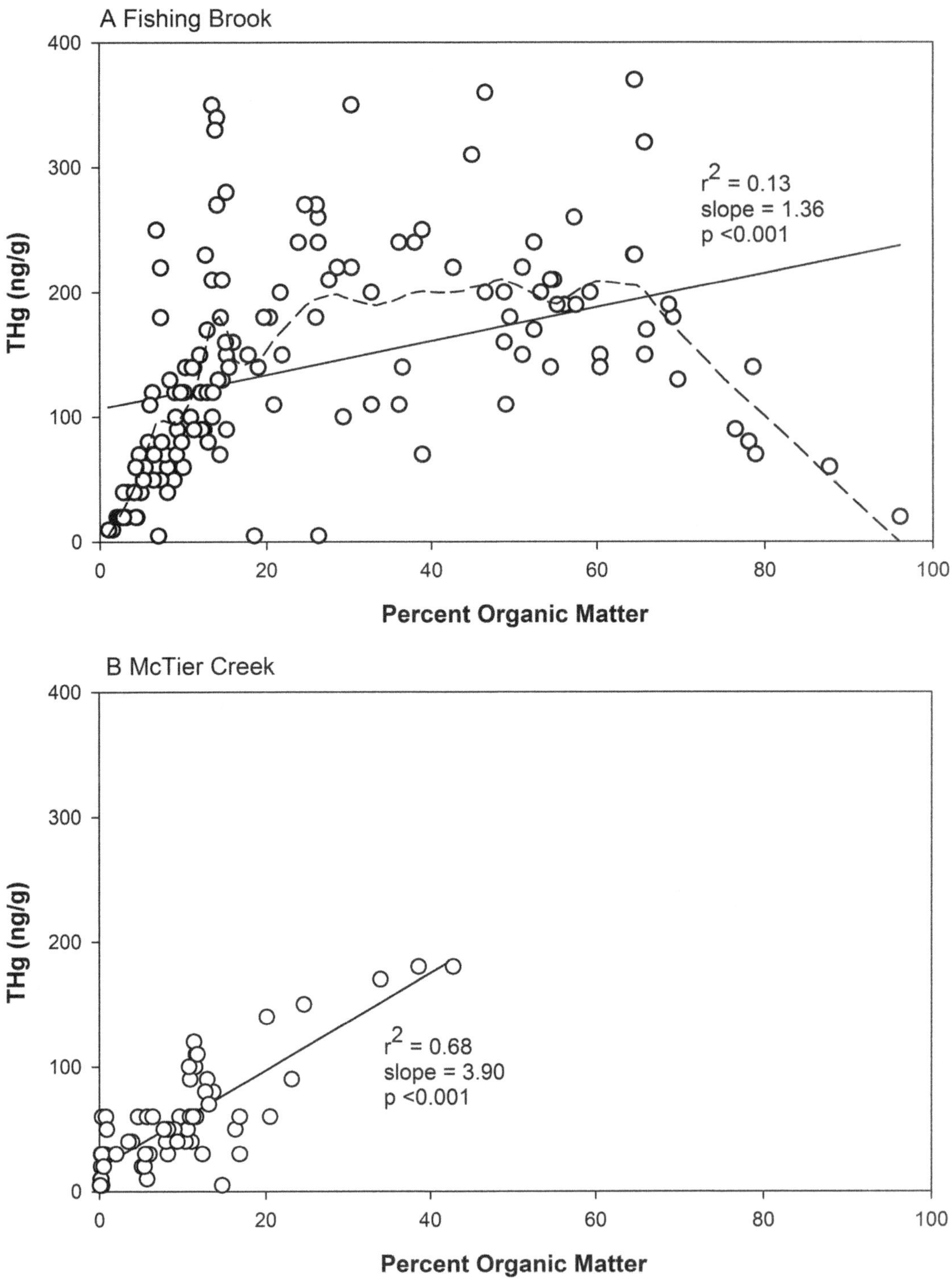

Figure 4. Total Hg and percent organic matter for all soil samples collected in each watershed. Solid line in each panel indicates a significant linear regression relation with r^2, slope, and p value. A Fishing Brook. Dashed line is a Loess smooth obtained by a first order polynomial fitted to a 20% moving window of the data. B McTier Creek.

Storage Estimates

THg and MeHg stores in soil were estimated at each study site. Estimates of soil element stores have high uncertainty due to a variety of factors including the presence of rocks, variations in the depth and thickness of soil horizons, high spatial heterogeneity of chemical and physical variables, and difficulties in accurately measuring bulk density, among others [50]. These sources of uncertainty should be carefully considered when comparing estimates of soil element stores among various studies in which assumptions can vary widely. In this investigation, we did not attempt to quantify soil rock content, a major source of uncertainty in element analyses [51], particularly in formerly glaciated terrain overlain by coarse till such as the FB watershed (rocks were not evident in the MC watershed soil). We developed estimates of THg

Table 3. Spearman rank correlations of THg and percent organic matter and MeHg and percent organic matter in Oa/A horizon and B horizon soil at Fishing Brook, NY and McTier Creek, SC.

Site/Landscape Position	Oa/A Horizon				B Horizon			
	THg		MeHg		THg		MeHg	
	n	r_s	n	r_s	n	r_s	n	r_s
Fishing Brook								
All	71	**−0.315**	52	−0.117	70	**0.759**	40	**0.413**
Upland	17	−0.129	11	0.036	17	**0.780**	5	−0.600
Riparian	16	0.093	12	−0.14	16	**0.793**	7	−0.071
Wetland	38	**−0.402**	29	−0.205	37	**0.734**	28	**0.516**
McTier Creek								
All	27	**0.654**	27	−0.125	42	**0.750**	42	**0.584**
Upland	12	**0.864**	12	−0.070	19	**0.694**	19	0.312
Riparian	9	**0.906**	9	−0.460	9	0.252	9	0.429
Wetland	6	−0.232	6	−0.377	14	**0.648**	14	0.460

Note – bold font indicates that THg – SOM or MeHg – SOM relation was significant.

and MeHg for the uppermost 40 cm of soil at each site, excluding fresh and partially decomposed litter (considered Oi and Oe horizons in some soil types), which was not sampled. These storage estimates were developed for the uplands, riparian area, and wetlands in each watershed using soil samples collected in each of these landscape types.

In Fishing Brook, the Oa horizon was assumed to be the uppermost 5 cm of soil in upland areas, the E horizon the next 5 cm, and the B horizon was assumed to include a depth interval of 10 to 40 cm, broadly consistent with observations in these Spodosols [33] and representing the mean of depth interval measurements made at the time of sampling. In the Entisol and Histosol soils that were dominant in riparian and wetland areas at FB and that frequently had poorly developed horizons, Hg data from Oa samples were applied to the depth interval of 0 to 10 cm, and B-horizon Hg data were applied to a depth interval of 10 to 40 cm.

In MC, the A horizon Hg data were applied to a depth interval of 0 to 4 cm in upland areas, 0 to 11 cm in riparian areas, and 0 to 20 cm in wetland areas based on the mean of depth measurements in each of these landscapes at the time of sampling. The B horizon Hg data were applied to the depth interval from the bottom of the A to 40 cm. The South Carolina soil survey typically includes an E horizon below the A in the soil types that dominate in uplands, and a C or more extensive A horizon below our A horizon classification in the soil types that dominate in riparian and wetland areas [36]. However, we recognize that some samples that were included in the B horizon category may actually be A, E, or C horizons samples based on this classification.

Bulk density was measured at 15 sites in A horizon soil at MC and at 38 sites in B horizon soil at FB. These measurements were made by sampling a known volume of soil and weighing this sample after drying. These measurements did not cover the full range of organic matter content measured in soil at these sites. In the case of MC, low organic matter soil (<5%) was not represented, whereas at FB, high organic matter soil (>24%) was not represented. Available bulk density measurements were averaged within 5 categories each representing a 5% range of soil organic matter (SOM) from 0% to 25% at FB, and at MC were averaged and placed into 4 categories each representing a 10% range up to 40% SOM (first category was 5% to 10%). These categorical data were best fit by a highly significant ($p<0.001$) power law regression of the form: bulk density = a * percent organic matter (−exp b). These equations were used to calculate bulk density values in the range of SOM >5% at MC and SOM< 24% at FB. This approach provided good estimates of bulk density as demonstrated by mean deviations from measured values of + 0.016 g/cm^3 at FB and −0.004 g/cm^3 at MC with no evident bias (organic matter vs. residuals linear regression, slope not significantly different than 0, $p>0.10$) across the OM range. For the SOM ranges of 0–5% at McTier Creek and 24–100% at Fishing Brook, a pedo-transfer equation was applied [52]. This approach is based on a simple mixing model between the bulk density values of 100% organic matter and that of 100% mineral soil and has proven useful in many previous investigations [53,54]. The value used in this equation for 100% SOM at MC was 0.10 g/cm^3 derived from the power law regression, and the value for 100% mineral soil was 1.70 g/cm^3 based on sandy Coastal Plain B-horizon soil with little organic matter [55,56]. The value used in this equation for 100% SOM at FB was 0.11 g/cm^3 based on data from Spodosols in the northeastern US [57] and the value used for 100% mineral soil was 1.99 based on the power law regression equation as it approaches 0% SOM (at 1% SOM).

Statistical Analyses

Seven of 163 soil samples collected at Fishing Brook had THg concentrations less than the 0.1 ng/g laboratory detection limit, and 6 of 81 soil samples collected at McTier Creek also had THg concentrations less than this detection limit. Therefore, non-parametric statistical analyses appropriate for censored data were used for any calculations that included THg data. Although there were no censored MeHg values, non-parametric statistical analyses were applied to these data as well because data were not normally distributed. The Mann-Whitney Rank-Sum test [58] was applied for pairwise comparisons, and the Kruskal-Wallis test for multiple comparisons [59]. Maximum likelihood regression was used instead of least squares linear regression, and a likelihood r^2 value calculated for each model [60]. Finally, relations among variables were determined by Spearman rank correlation analysis. All statistical analyses were performed with SigmaStat, part of the

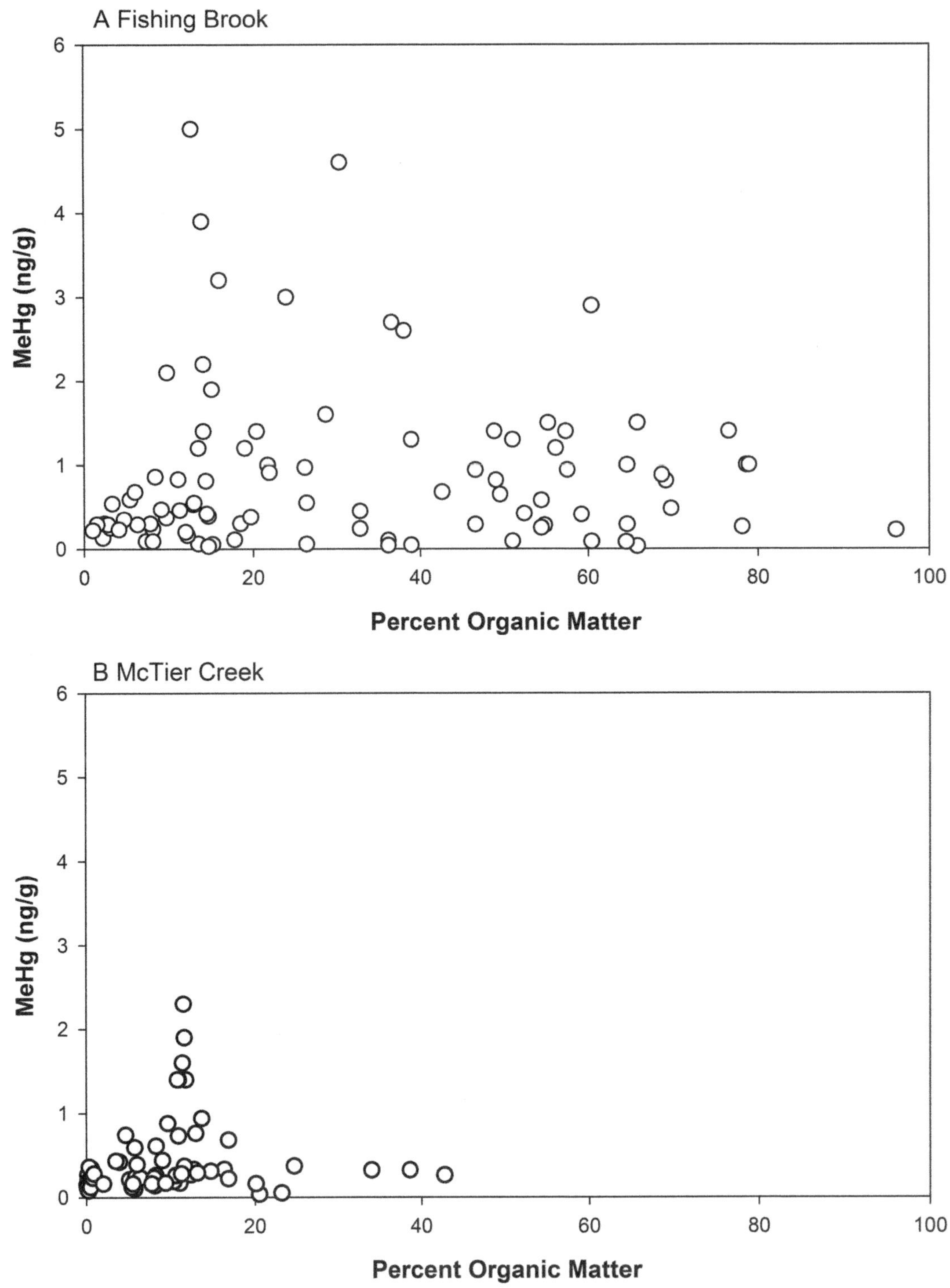

Figure 5. Methyl Hg and percent organic matter for all soil samples collected in each watershed. Absence of a line in panel indicates that a significant linear regression relation was not evident. A Fishing Brook, B McTier Creek.

SigmaPlot software (v. 12, use of product names in this paper is for identification purposes only and does not imply endorsement by the U.S. Government), except for maximum likelihood regression, which was performed with SAS.

Results

THg and MeHg concentrations and the THg/MeHg ratio for the two watersheds are compared first in Oa/A horizon samples and second in B horizon samples. These soil concentrations and ratios are also compared across landscape type within each watershed. The relations of Hg species concentrations to percent

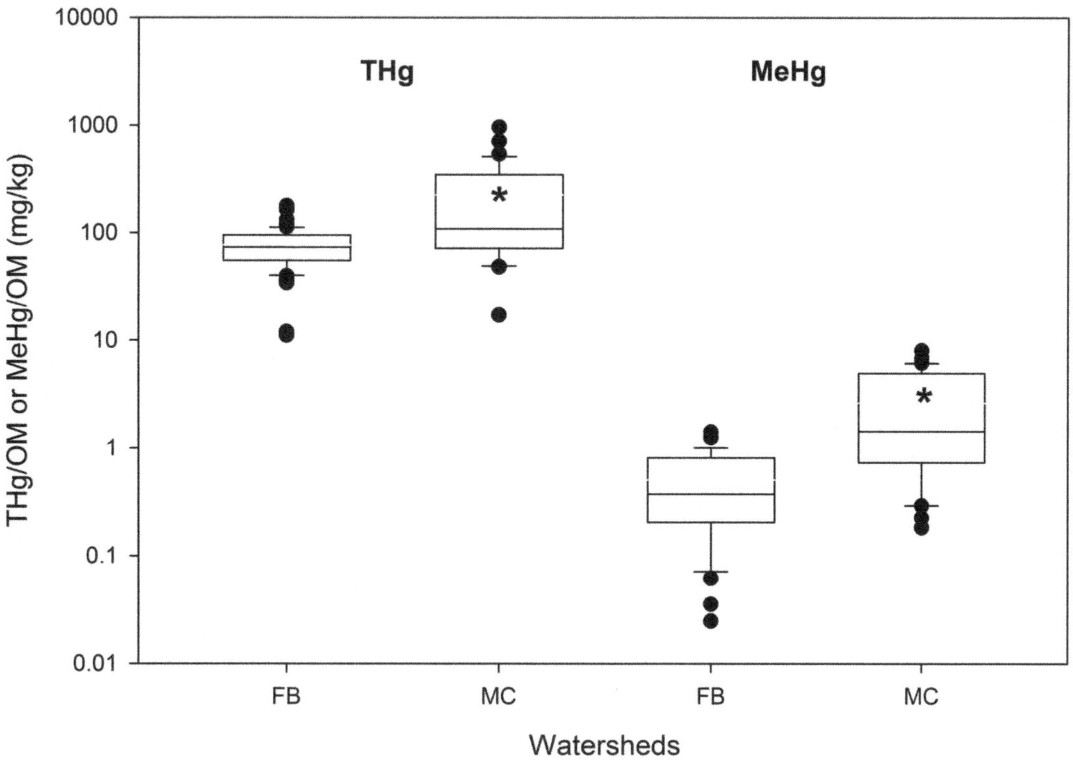

Figure 6. THg/OM and MeHg/OM based on estimates of storage in soil at each watershed. An asterisk in the box indicates the watershed with significantly greater values for each of the quantities shown.

SOM are then presented. Finally, estimated storage of THg and MeHg in soil is presented and compared among the two watersheds.

Comparison of Soil Hg among Watersheds – Oa and A Horizon Soil

THg concentrations were significantly greater ($p<0.05$) in Oa/A horizon soil samples from FB than those from MC (Fig. 2A, Table 1). FB soil also had significantly greater THg concentrations than that of MC across all three landscape types. The median THg concentration of 200 ng/g at FB was more than three-fold greater than the median value of 60 ng/g at MC, and an approximate two-fold to four-fold difference was measured across the three landscape types. The differences in THg among soil from these two watersheds were so great, that the highest concentrations measured at MC were less than the median concentrations at FB

Table 4. Median estimated areally-normalized storage of total Hg, methyl Hg, and organic matter to a depth of 40 cm in the Fishing Brook, NY and McTier Creek, SC watersheds.

Site/Landscape Position	THg (μg/m^2)	MeHg (μg/m^2)	Organic Matter (kg/m^2)
Fishing Brook			
All[1]	**2570**	8.4	**38.8**
Upland	**2540**	7.3	**39.8**
Riparian	**3020**	11.1	**33.5**
Wetland	**2220**	15.1	**37.2**
McTier Creek			
All[2]	690	10.3	4.7
Upland	650	11.0	2.6
Riparian	760	5.1	13.2
Wetland	980	11.1	12.7

Note – bold font indicates median value was significantly greater ($p<0.05$) in site-to-site Mann-Whitney pairwise comparison tests done for all samples collected and for each landscape type.
[1]Represents 96.6% of watershed area excluding land cover not sampled in study (open water+developed land = 3.4%).
[2]Represents 79.3% of watershed area excluding land cover not sampled in study (open water+developed land+agricultural land = 20.7%).

for all samples as well as for those collected in each landscape type (Fig. 2A).

The median MeHg concentration of 0.81 ng/g for all FB soil samples was more than two-fold greater than the median value of 0.33 ng/g for all MC samples, but the difference between these watersheds was not significant ($p = 0.133$; Table 1). The median value for upland and riparian samples was also slightly greater at FB than MC, but these differences were similarly not significant. In contrast, the median MeHg concentration of wetland soil samples collected at MC of 1.5 ng/g was greater than the value of 0.94 ng/g in FB wetland samples, a difference of marginal statistical significance ($p = 0.088$). There were fewer significant site differences for MeHg than for THg because of greater variation in the MeHg data (Fig. 2b); relative standard deviation values were 40.0% and 63.8% for THg at FB and MC, respectively, and increased to 109.4% and 99.5% for MeHg at these watersheds.

The MeHg/THg ratio showed the opposite pattern of the individual constituents with values two- to four-fold greater at MC than at FB (Table 1). This ratio was significantly greater for all samples and for those from the wetlands, but differences were not significant for riparian ($p = 0.110$) and upland ($p = 0.148$) soil. Though MeHg soil concentrations in the Coastal Plain watershed were generally less than those in the Adirondack watershed, MeHg/THg values were generally greater in the south.

Comparison of Soil Hg between Watersheds – B Horizon Soil

Soil Hg concentrations were generally less in B horizon soil than those in Oa/A horizon soil at both watersheds (Table 2, Figs. 2 and 3). The inter-watershed differences in Hg concentrations in the deeper B horizon soil were broadly consistent with the patterns in the shallower Oa/A horizon soil with some minor differences. THg concentrations were significantly greater in FB B horizon soil than those of MC for all samples and for those from uplands and riparian areas, but not for wetland samples. Median MeHg concentrations in B horizon samples were generally greater at FB than at MC as in Oa/A horizon samples. Whereas these inter-watershed differences were not statistically significant in Oa/A horizon soil, the differences were highly significant for all B horizon samples ($p < 0.001$) and significant for wetland samples ($p = 0.023$), but not significant for those from uplands and riparian floodplains. Finally, MeHg/THg was also generally greater in B horizon soil at MC than at FB consistent with the findings for the shallow soil horizons. These differences were highly significant for upland soil ($p = 0.001$), of marginal significance for all soil samples ($p = 0.052$), but not significant for riparian and wetland samples.

Comparison of Soil Hg across Landscape Types

Within the FB watershed, THg concentrations in soil were similar across all landscape types for both Oa/A and B horizon samples (Figs. 2 and 3). MeHg and MeHg/THg values were also similar and statistically indistinguishable across all landscape types for Oa/A horizon soil samples. However, in B horizon samples, MeHg concentrations and MeHg/THg were significantly different across the three landscape types. Pairwise comparisons among the landscape types showed that wetlands>riparian areas = uplands, a consistent finding for both MeHg concentrations and MeHg/THg.

MC showed more differences in soil Hg concentrations as a function of landscape type than did FB (Figs. 2 and 3). All Hg measures were significantly different among the three landscape types in both Oa/A and B horizon samples. In pairwise comparisons, wetland values were nearly always significantly greater than those of the uplands, and the riparian and upland landscape types were most often statistically indistinguishable, despite median THg and MeHg values that were greater in riparian samples. The exception to these generalizations was MeHg/THg in the B horizon samples for which the upland ratio was significantly greater than the wetland ratio, whereas the other pairwise comparisons showed no significant differences.

Relations of Hg to Percent Soil Organic Matter

Some strong differences are evident in the relations of THg to percent SOM among these two watersheds and as a function of soil depth and landscape type (Fig. 4; Table 3). A highly significant ($p < 0.001$) positive maximum likelihood regression relation was evident between THg and SOM for all soil samples collected at FB, but this relation explains only 13% of the variation in these data. At MC, an equally highly significant relation ($p < 0.001$) explains 68% of the variation in these data (Fig. 4). Despite the significance of the regression relation at Fishing Brook, a LOESS smooth better fits the data pattern and confirms a generally non-linear THg-SOM relation that is highly variable throughout the range of SOM data; a generally linear increase in THg to an SOM value of about 15% ($r^2 = 0.46$, $p < 0.001$), little change and much scatter in the range of 15% to about 60%, and a linear decrease in THg at SOM values >60% (Fig. 4A). In contrast, THg values at MC show much scatter at SOM values <20%, but generally increase with SOM at values >20%, though there are few samples with high SOM values (Fig. 4B). These data highlight the generally greater SOM in FB soil where values >90% were measured, whereas the greatest SOM values in soil samples at MC were about 40%.

The inverse relation at FB of THg to SOM at high SOM values (as per LOESS smooth) results largely from the Oa/A horizon samples, and is only evident ($p < 0.05$) among the wetland soil samples (Table 3). In contrast, a strong positive relation of THg and SOM is evident in the B horizon soil samples at Fishing Brook, and is consistent throughout all three landscape types. At MC, THg and SOM were significantly positively related in all landscape types except for the Oa/A horizon wetland samples, and in all landscape types except for the riparian in B horizon samples (Table 3).

Overall, far fewer significant MeHg-SOM relations were evident in these data than were evident in the THg-SOM data. For all soil samples collected at each site, the relation between MeHg and SOM was not significant (Fig. 5). Among soil depths, there were no significant MeHg-SOM relations for any landscape type at either watershed (Table 3). In B horizon soil samples by contrast, MeHg was positively related to SOM among all samples for both watersheds. However, only the wetland soil at FB showed a significant MeHg-SOM relation, whereas the relation at MC was of marginal ($p = 0.094$) significance.

Hg Storage in Soil of the Two Watersheds

Stores of THg in FB were two- to four-fold greater ($p < 0.05$) than those in MC across all landscapes types (Table 4). In contrast, there were no significant differences in MeHg stores between these watersheds among any of the landscape types, although MeHg stores in MC were marginally greater ($p = 0.058$) than those in FB. Stores of SOM were about eight-fold greater overall in FB than in MC, and these differences were highly significant ($p < 0.001$) across all landscape types. These strong differences in SOM storage among the two watersheds suggest that THg and MeHg are more efficiently stored as a function of available SOM in MC than in FB. This suggestion is confirmed by noting that median values of THg/SOM were about three-fold greater in MC than in FB, and median values of MeHg/SOM were nearly four-fold greater in MC than in FB ($p<0.001$; Fig. 6).

The differences in soil stores of THg, MeHg, and SOM among landscape types within each watershed were less distinct than those identified between the watersheds (Table 4), but some differences were noted (values not shown). THg stores were statistically indistinguishable among landscape types in both watersheds. Similarly, there was no difference in MeHg soil stores among landscape types in MC. In contrast, a statistically significant difference in MeHg stores was identified in FB, with values decreasing in the order wetlands = riparian>uplands, according to pairwise comparisons. For SOM stores, no differences among landscape types were noted at FB, but these values differed across landscapes in MC, decreasing in the order uplands>wetlands = - riparian.

Discussion

Comparisons with Data from Other Studies

Hg concentrations and estimates of soil storage in these two watersheds are broadly consistent with values previously reported for the US and globally for rural sites where atmospheric deposition is the dominant input to the landscape [3,6,61,62]. THg concentrations in soil from rural areas reported previously generally range from 50 to 300 ng/g, with greater values generally found in soil with high SOM levels, particularly in shallower soil horizons where SOM is concentrated [6,7,15], consistent with the results reported here. Fewer soil data have been reported for the southeastern US in locations similar to the MC watershed; however, two studies report THg concentrations in rural southeastern US sandy soil of about 10 ng/g [7] and 30 ng/g [63], broadly consistent with the generally lower median Hg values at MC (A = 60 ng/g, B = 30 ng/g) than those of FB (Oa = 200 ng/g, B = 85 ng/g).

Few studies have previously reported MeHg data in soil; however, some have reported MeHg/THg, with values as high as 5% [14], but most <1% [4,6]. Here, we reported values of about 0.5% at FB and close to 1% at MC, consistent with most studies reviewed by Grigal [6] and Obrist [4]. Our data do not show a consistent pattern in MeHg/THg with soil depth across these two watersheds and the three landscape types sampled, but there is a pronounced tendency for the greatest values of MeHg and MeHg/THg to occur in riparian and wetland soil within each watershed. This pattern is consistent with those reported previously in these two watersheds and in other settings in which wetlands and riparian areas have the highest MeHg concentrations in soil, soil water, and shallow groundwater, and are the dominant source of MeHg to nearby surface waters [14,25,37].

The estimates of THg storage in FB soil of about 2500 $\mu g/m^2$ (to 40 cm depth) are within the range of most values reported previously in forested landscapes in the northeastern US [64], and lower than those reported in peat soil and peatlands [3,6,13,65]. The FB THg soil stores are less than those in a nearby watershed in the western Adirondacks of New York [16], but these latter estimates included the Oi and Oe horizons that were not sampled in FB, and also extended to a depth of 1 m, whereas THg stores were estimated only to a depth of 40 cm at FB. The estimated THg soil stores in MC of about 700 $\mu g/m^2$ are generally less than those reported in most previous studies, and at the low end of estimates of mean soil THg storage over large regions in the US and Canada [6,66]. We are not aware of any previously published estimates of soil THg stores from uncontaminated sites in the US Coastal Plain other than that of Obrist [4] for a site near Gainesville, FL. Soil THg concentrations are about 200 ng/g in the Everglades [67], where the soil is rich in peat and not readily comparable to the sandy soil of the South Carolina Coastal Plain.

Comparisons between these Two Watersheds

The study watersheds have been previously identified as Hg "hot spots", and show little inter-watershed difference in THg and MeHg concentrations in surface waters, aquatic invertebrates, and fish [25,26]. For example, median filtered THg and filtered MeHg concentrations were similarly about 2 ng/L and 0.1 ng/L, respectively at MC and FB during 2007–09 [68]. Although soil is the likely dominant immediate source of THg and MeHg to local streams and biota in these two watersheds (Hg in bottom sediment and direct atmospheric deposition are viewed as minor sources), there are large differences in THg concentrations and stores between these two locations, with about three-fold greater values in FB than in MC. As previously noted, however, there are large uncertainties in estimating chemical element storage in watersheds including accounting for rock content, which was not estimated in this study. Despite these uncertainties, we believe that the three-fold difference is greater than the uncertainties that arise in estimating Hg stores.

In contrast, differences in soil MeHg between these two watersheds show a more complex pattern. MeHg concentrations at FB were generally greater overall and in most landscape types than those at MC (though differences were not always statistically significant), but there was little difference in MeHg stores between the watersheds. Furthermore, MeHg/THg was generally greater in the MC watershed than in the FB watershed. These results are broadly consistent with previous studies that have shown increases in concentrations and stores of THg [4] and SOM [69] with increasing latitude in the US, but are inconsistent with previous observations that similarly showed soil MeHg concentrations and stores increasing with latitude [4].

The size of the soil THg pools relative to estimates of annual atmospheric deposition of Hg differs sharply at these two sites. The current soil THg pool is equivalent to about 400 years of stored Hg relative to the current annual atmospheric Hg deposition rate at Fishing Brook of 6.3 $\mu g/m^2/yr$, whereas the THg pool at McTier Creek is equivalent to about 70 years of stored Hg relative to current annual Hg deposition of 9.9 $\mu g/m^2/yr$ [70], a more than 5-fold difference among these sites. In contrast, the same estimates for the soil MeHg pool suggests similar relative storage in these two watersheds (FB = 267 yrs, MC = 208 yrs, assuming MeHg deposition is 0.5% of THg deposition, [71]), though MeHg deposition rates are highly uncertain.

These findings suggest a conceptual model that includes the role of soil in the Hg cycle at these two locations. A much larger proportion of the atmospheric THg deposition that has historically fallen at the Adirondack site remains stored in watershed soil, consistent with the colder climate and larger SOM pool relative to the Coastal Plain. Thus, the soil in the FB watershed represents a large, persistent pool of THg in contrast to the soil of the MC watershed, which turns over more rapidly. Lower rates of annual evapotranspiration and higher annual streamflow in concert with the larger soil Hg pools in the Adirondack watershed imply the potential for greater annual stream THg yields at the more northern location, which is consistent with modeled estimates at these two sites during 2007–09 [70]. In the MC watershed, the rates of other loss processes such as gaseous emissions resulting from respiration of soil organic matter and volatilization of Hg^0 [2] are necessarily much greater than those in the Adirondack watershed to account for the much smaller THg pool in the Coastal Plain soil. Furthermore, the weak relation of THg to SOM in the Adirondack watershed soil with high organic matter content suggests that SOM composition is highly heterogeneous, consistent with the diverse specific ultra-violet absorbance values of DOC and relatively weak relation of DOC concentrations and filtered THg concentrations measured in Fishing Brook [72]. These results also suggest the likelihood of excess Hg binding capacity in this northern soil [8], whereas the retention of THg in the southern soil appears to be more strongly dependent on available SOM, which may be limiting retention.

Greater net methylation efficiency is suggested at the southern Coastal Plain site because the pool sizes are similar among the two sites, the annual stream fluxes are greater at FB than at MC [70], and MeHg/THg is generally greater at MC. However, the role of demethylation rates, a potentially important factor not measured in this study, may also play an important role in mediating MeHg/THg ratios in soil [73]. Nonetheless, a greater proportion of the soil THg pool is likely converted to MeHg on an annual basis at the southern site, necessary to provide the similar MeHg concentrations in streams and aquatic biota observed among these two watersheds [25,26]. We hypothesize that warmer soil temperatures in the southern Coastal Plain than in the Adirondacks is the primary driver of the apparent greater net methylation efficiency at MC. Other factors, however, such as differences in the vegetation and soil between these sites may also play a role.

A potentially important caveat to these results and the conclusions drawn is that soils were sampled only once in each watershed. Little is known from the literature about the extent to which soil concentrations and stores vary seasonally. Our observations of THg and MeHg in springs, streams, and piezometers indicate that aqueous concentrations can vary two- to three-fold throughout the year in these two watersheds [25], and a similar magnitude of variation is possible in soils, though the large size of the stores would be expected to dampen this variation relative to that in solution.

Differences Across Landscape Types

Previously published findings from these two watersheds and nearby indicate that wetlands and riparian areas are major sources of THg and MeHg to local streams, and flow paths that transport Hg species from these landscape settings to streams are the dominant control on the spatial and temporal variations in concentrations and loads of these Hg species [25,27,37,74]. When soil concentrations and stores of THg and MeHg are compared among landscape types, including low Hg-yielding uplands, the differences are not as striking as those observed in surface waters. There were no significant landscape differences in THg concentrations or stores in FB soil, but MeHg values were greater in wetland soil, particularly when compared with those of upland soil. MC had significantly greater THg and MeHg concentrations in wetlands than those in riparian areas and uplands, but the stores of these Hg species were not significantly different among the three landscape types. The data from the current study are consistent with a prominent role for factors other than soil THg and MeHg concentrations and stores as principal controls on landscape variation in Hg cycling rates to streams. These factors include soil drainage characteristics, proximity to the stream channel, and the high water table and fluctuating redox processes (anoxic to oxic) in the relatively flatter wetland and riparian terrain. In particular, differences in the hydrological and biogeochemical processes that affect the concentrations and relative fluxes of dissolved organic matter (DOM) as well as the relative aromatic vs. aliphatic character of this DOM appear to be the principal controls on landscape variation in THg and MeHg concentrations and loads in streams in these two geographic settings [25,27,72].

Conclusions

The results of this study show that two watersheds in different geomorphic and climatic regions with similar THg and MeHg concentrations in surface water and aquatic biota have greatly different concentrations and stores of THg in soil. THg concentrations and stores (to a depth of 40 cm) were about three-fold greater in the Adirondack watershed than those of the Coastal Plain watershed. Despite somewhat higher MeHg concentrations in the Adirondack soil, MeHg stores did not differ significantly among these two locations, although MeHg/THg was generally greater at the Coastal Plain site. These results are consistent with greater annual rates of net methylation and greater methylation efficiency in the soil of the warmer southern site. The greater soil stores of THg in the Adirondack watershed are consistent with an eight-fold greater store of SOM than that of the Coastal Plain watershed, though the relation of THg to SOM was weak in the Adirondacks, suggesting heterogeneous SOM and an excess of Hg-binding thiol groups relative to the net Hg supply. Furthermore, these results indicate that a greater proportion of historical atmospheric Hg deposition is stored in the Adirondack soil and that THg has about a five-fold greater soil turnover time than in the southern Coastal Plain soil. Comparison with known rates of atmospheric Hg deposition and stream export implies that gaseous Hg emissions from the Coastal Plain soil are necessarily much greater than those of Adirondack soil. These results raise questions about the ultimate fate of the Hg stored in the soil of these two sites. Past work has indicated residence times of several decades to several hundred years for atmospherically-deposited Hg, consistent with the results reported here. The overall greater store of THg in the Adirondacks suggests that declines in THg concentrations in stream water of this region will be more gradual and prolonged as future Hg deposition declines and comparatively sharper and less prolonged in Coastal Plain streams.

Author Contributions

Conceived and designed the experiments: DAB LGW PMB WFC. Performed the experiments: DAB LGW PMB WFC. Analyzed the data: DAB LGW PMB WFC. Contributed reagents/materials/analysis tools: DAB LGW PMB WFC. Wrote the paper: DAB LGW PMB WFC.

References

1. Selin NE (2010) Global biogeochemical cycling of mercury: a review. Ann Rev Environ Resour 34: 43–63.
2. Smith-Downey NV, Sunderland EM, Jacob DJ (2010) Anthropogenic impacts on global storage and emissions of mercury from terrestrial soils: insights from a new global model. J Geophys Res Biogeosci 115, G03008, doi:10.1029/2009JG001124.
3. Tipping E, Posakitt JM, Lawlor AJ, Wadsworth RA, Norris DA, et al. (2011) Mercury in United Kingdom topsoils; concentrations, pools, and Critical Limit exceedances. Environ Pollut 159: 3721–3729.
4. Obrist D (2012) Mercury distribution across 14 U.S. forests. Part II: patterns of methyl mercury concentrations and areal mass of total and methyl mercury. Environ Sci Technol 46: 5921–5930.
5. Alriksson A (2001) Regional variability of Cd, Hg, Pb and C concentrations in different horizons of Swedish forest soils. Water Air Soil Pollut Focus 1: 325–341.
6. Grigal DF (2003) Mercury sequestration in forest and soils: a review. J Environ Qual 32: 393–405.
7. Obrist D, Johnson DW, Lindberg SE, Luo Y, Hararuk O, et al. (2011) Mercury distribution across 14 U.S. forests. Part I: spatial patterns of concentrations in biomass, litter, and soils. Environ Sci Technol 45: 3974–3981.

8. Skyllberg U, Qian J, Frech W, Xia K, Bleam WF (2003) Distribution of mercury, methyl mercury and organic sulphur species in soil, soil solution and stream of a boreal forest catchment. Biogeochemistry 64: 53–76.
9. Ravichandran M (2004) Interactions between mercury and dissolved organic matter—a review. Chemosphere 55: 319–331.
10. Wang W, Driscoll CT (1995) Patterns of total mercury concentrations in Onondaga Lake, New York. Environ Sci Technol 29: 2261–2266.
11. Gerbig CA, Ryan JN, Aiken GR (2012) The effects of dissolved organic matter on mercury biogeochemistry. In: Liu G, Cai Y, O'Driscoll NO editors. Environmental Chemistry and Toxicology of Mercury. New York: Wiley. 259–292.
12. Gabriel MC, Williamson DG (2004) Principal biogeochemical factors affecting the speciation and transport of mercury through the terrestrial environment. Environ Geochem Health 26: 421–434.
13. Nater EA, Grigal DF (1992) Regional trends in mercury across the Great Lakes states, north central USA. Nature 358: 139–141.
14. Schwesig D, Ilgen G, Matzner E (1999) Mercury and methylmercury in upland and wetland acid forest soils of a watershed in NE-Bavaria. Water Air Soil Pollut 113: 141–154.
15. Woodruff LG, Cannon WF, Eberl DD, Smith DB, Kilburn JE, et al. (2009) Continental-scale patterns in soil geochemistry and mineralogy: results from two transects across the United States and Canada. Appl Geochem 24: 1369–1381.
16. Demers JD, Driscoll CT, Fahey TJ, Yavitt JB (2007) Mercury cycling in soil and litter in different forest types in the Adirondack region, New York, USA. Ecol Appl 17: 1341–1351.
17. Chen X, Xia X, Wu S, Wang F, Guo X (2010) Mercury in urban soils with various types of land use in Beijing, China. Environ Pollut 158: 48–54.
18. Woodruff LG, Cannon WF (2010) Immediate and long-term fire effects on total mercury in forests soils of northeastern Minnesota. Environ Sci Technol 44: 5371–5376.
19. Fu X, Feng X, Zhu W, Rothenberg S, Yao H, et al. (2010) Elevated atmospheric deposition and dynamics of mercury in a remote upland forest of southwestern China. Environ Pollut 158: 2324–2333.
20. Åkerblom S, Meili M, Bringmark L, Johansson K, Berggren D, et al. (2008) Partitioning of Hg between solid and dissolved organic matter in the humus layer of boreal forests. Water Air Soil Pollut 189: 239–252.
21. Mitchell CPJ, Branfireun BA, Kolka RK (2008) Spatial characteristics of net methylmercury production hot spots in peatlands. Environ Sci Technol 42: 1010–1016.
22. Wiener JG, Knights BC, Sandheinrich MB, Jeremiason JD, Brigham ME, et al. (2006) Mercury in soils, lakes, and fish in Voyageurs National Park (Minnesota): importance of atmospheric deposition and ecosystem factors. Environ Sci Technol 40: 6261–6268.
23. Evers DC, Han Y-J, Driscoll CT, Kamman NC, Goodale MW, et al. (2007) Biological mercury hotspots in the northeastern United States and southeastern Canada. BioScience 57: 29–43.
24. Scudder BC, Chasar LC, Wentz DA, Bauch NJ, Brigham ME, et al. (2009) Mercury in fish, bed sediment, and water from streams across the United States, 1998–2005. Reston, VA: US Geol Surv Sci-Invest Rpt 2009–5109. 86 p.
25. Bradley PM, Burns DA, Riva-Murray K, Brigham ME, Button DT, et al. (2011) Spatial and seasonal variability of dissolved methylmercury in two stream basins in the eastern United States. Environ Sci Technol 48: 2048–2055.
26. Riva-Murray K, Chasar LC, Bradley PM, Burns DA, Brigham ME, et al. (2011) Spatial patterns of mercury in macroinvertebrates and fishes from streams of two contrasting forested landscapes in the eastern United States. Ecotoxicology 9: 1–13.
27. Schelker J, Burns DA, Weiler M, Laudon H (2011) Hydrological mobilization of mercury and dissolved organic carbon in a snow-dominated forested watershed: conceptualization and modeling. J Geophys Res Biogeosci 116, G01002, doi:10.1029/2010JG001330.
28. Scudder Eikenberry BC, Riva-Murray K, Smith MJ, Bradley PM, Button DT, et al. (2012) Environmental settings of streams sampled for mercury in New York and South Carolina, 2005–09. Reston, VA: US Geol Surv Open-File Rpt 2011–1318. 48 p.
29. McGee GG, Mitchell MJ, Leopold DJ, Raynal DJ, Mbila M (2007) Relationships among forest age, composition and elemental dynamics of Adirondack northern hardwood forests. J Torrey Botan Soc 134: 253–268.
30. LaPoint SD, Curren RP, Halasz SS, Barge JW, Spada DM, et al. (2004) Wetlands effects database and GIS for the Adirondack Park. Ray Brook, NY: NY State Adirondack Park Agency. 26 p.
31. Kiernan BD, Hurd TD, Raynal DJ (2003) Abundance of *Alnus incana* ssp. *rugosa* in Adirondack Mountain shrub wetlands and its influence on inorganic nitrogen. Environ Pollut 123: 347–354.
32. Ashwal LD (1982) Mineralogy of mafic and Fe-Ti oxide-rich differentiates of the Marcy anorthosite massif, Adirondacks, New York. Am Mineral 67: 14–27.
33. Silverman MH, Krawiecki KJ (2006) Soil Survey of Hamilton County, New York. Washington DC: US Dept Agric, Nat Resourc Conserv Serv. 635 p.
34. Demers JD, Yavitt JB, Driscoll CT, Montesdeoca MR (2013) Legacy mercury and stoichiometry with C, N, and S in soil, pore water, and stream water across the upland-wetland interface: the influence of hydrogeologic setting, J Geophys Res Biogeosci 118: 825–841.
35. Dahl TE, Dick J, Swords J, Wilen BO (2009) Data collection requirements and procedures for mapping wetland, deepwater and related habitats of the United States. Madison, WI: Div Habitat and Resource Conserv, National Stds and Support Team. 85 p.
36. Rogers VA (1985) Soil Survey of Aiken County area, South Carolina. Washington DC: US Dept Agric, Soil Conserv Serv. 135 p.
37. Burns DA, Riva-Murray K, Bradley PM, Aiken GR, Brigham ME (2012) Landscape controls on total and methyl Hg in the upper Hudson River basin, New York, USA. J Geophys Res Biogeosci 117, G01034, doi:10.1029/2011JG001812.
38. Homer C, Hunag C, Yang L, Wylie B, Coan M (2004) Development of a 2001 National Land Cover Database for the United States. Photogram Engineer Remote Sensing 70: 829–840.
39. NOWData – NOAA Online Weather Data. Available: http://nowdata.rcc-acis.org/BTV/pubACIS-results. Accessed 2014 Jan 20.
40. Park J-H, Mitchell MJ, Driscoll CT (2005) Winter-time climatic control on dissolved organic carbon export and surface water chemistry in an Adirondack forested watershed. Environ Sci Technol 39: 6993–6998.
41. NOWData – NOAA Online Weather Data. Available: http://nowdata.rcc-acis.org/CAE/pubACIS-results. Accessed 2014 Jan 20.
42. USGS Water Data for the Nation. Available: http://waterdata.usgs.gov/nwis/sw. Accessed 2014 Jan 20.
43. Woodruff LG, Cannon WF, Knightes CD, Chapelle FH, Bradley PM, et al. (2010) Total mercury, methylmercury, and selected elements in soils of the Fishing Brook watershed, Hamilton County, New York, and the McTier Creek Watershed, Aiken County, South Carolina, 2008. Reston, VA: US Geol Surv Data Ser Rpt 516. 18 p. Available: http://pubs.usgs.gov/ds/516. Accessed 2014 Jan 20.
44. Brown ZA, Curry KJ (2002) Total carbon by combustion. In: Taggart JE Jr editor. Analytical methods for chemical analysis of geologic and other materials. Denver, CO: US Geol. Surv Open-File Rpt 02–223. 1–4 (Ch R).
45. Brown ZA, Papp C, Brandt E, Aruscavage P (2002) Carbonate carbon by coulometric titration. In: Taggart JE Jr editor. Analytical methods for chemical analysis of geologic and other materials. Denver, CO: US Geol. Surv Open-File Rpt 02–223. 1–6 (Ch S).
46. Broadbent FE (1953) The soil organic fraction. Adv Agron 5: 153–183.
47. Smith DB, Woodruff LG, O'Leary RM, Cannon WF, Garrett RG, et al. (2009) Pilot studies for the North American Soil Geochemical Landscapes Project – site selection, sampling protocols, analytical methods, and quality control protocols. Appl Geochem 24: 1357–1368.
48. US EPA Method 7471B: Mercury in Solid or Semisolid Waste (Manual Cold Vapor Technique). Available: http://www.epa.gov/osw/hazard/testmethods/sw846/pdfs/7471b.pdf. Accessed 2104 Jan 20.
49. DeWild JF, Olund SD, Olson ML, Tate MT (2004) Methods for the preparation and analysis of solids and suspended solids for methylmercury, Reston, VA: U.S. Geol Surv Tech and Meth 5 A-7. 21 p.
50. DeSaules A (2012) Measurement instability and temporal bias in chemical soil monitoring: sources and control measures. Environ Monit Assess 184: 1487–1502.
51. Harrison RB, Adams AB, Licata C, Flaming B, Wagoner GL, et al. (2003) Quantifying deep-soil and coarse-soil fractions: avoiding sampling bias. Soil Sci Soc Am J 67: 1602–1606.
52. Adams WA (1973) The effect of organic matter on the bulk and true densities of some uncultivated podzolic soils. J Soil Sci 24: 10–17.
53. Post WM, Kwan KC (2000) Soil carbon sequestration and land-use change: processes and potential. Global Change Biol 6: 317–28.
54. De Vos B, Van Meirvenne M, Quataert P, Deckers J, Muys B (2005) Predictive quality of pedotransfer functions for estimating bulk density of forest soils. Soil Sci Soc Am J 69: 500–510.
55. Markewich HW, Pavich MJ, Buell GR (1990) Contrasting soils and landscapes of the Piedmont and Coastal Plain, eastern Unites States. Geomorphology 3: 417–447.
56. Duffera M, White JG, Weisz R (2007) Spatial variability of southeastern U.S. Coastal Plain soil physical properties: implications for site-specific management. Geoderma 137: 327–339.
57. Federer CA, Trucotte DE, Smith CT (1993) The organic fraction – bulk density relationship and the expression of nutrient content in forest soils. Can J Forest Res 23: 1026–1032.
58. Mann HB, Whitney DR (1947) On a test of whether one of two random variables is stochastically larger than the other. Ann Math Stat 18: 50–60.
59. Kruskal WH, Wallis WA (1952) Use of ranks in one-criterion variance analysis. J Am Stat Assoc 47: 583–621.
60. Helsel DR (2005) Nondetects and data analysis. Hoboken, NJ: Wiley-Interscience. 250 p.
61. Rundgren S, Ruhling Å, Schlütter K, Tyler G (1992) Mercury in soil–distribution, speciation, and biological effects. Copenhagen: Nordic Council of Ministers. 89 p.
62. Burt R, Wilson MA, Mays MD, Lee CW (2003) Major and trace elements of selected pedons in the USA. J Environ Qual 32: 2109–2121.
63. Aelion CM, Davis HT, McDermott S, Lawson AB (2009) Soil metal concentrations and toxicity: Associations with distances to industrial facilities and implications for human health. Sci Total Environ 407: 2216–2223.
64. Richardson JB, Friedland AJ, Engerbretson TR, Kaste JM, Jackson BP (2013) Spatial and vertical distribution of mercury in upland forest soils across the northeastern United States. Environ Pollut 182: 127–134.

65. Lindqvist O, Johansson K, Aastrup M, Andersson A, Bringmark L, et al. (1994) Mercury in the Swedish environment–recent research on causes, consequences and corrective measures. Water Air Soil Pollut 55: 1–253.
66. Richardson GM, Mitchell IA, Mah-Apulson M, Hackbarth T, Garrett RG (2003) Natural emissions of mercury to the atmosphere in Canada. Environ Rev 11: 17–36.
67. Arfstrom C, MacFarlane AW, Jones RD (2000) Distributions of mercury and phosphorus in Everglades soils from Water Conservation Area 3A, Florida, U.S.A. Water Air Soil Pollut 121: 133–159.
68. Bradley PM, Journey CA, Brigham ME, Burns DA, Button DT, et al. (2013) Intra- and inter-basin mercury comparisons: Importance of basin scale and time-weighted methylmercury estimates. Environ Pollut 172: 42–52.
69. Guo Y, Gong P, Amundson R, Yu Q (2006) Analysis of factors controlling soil carbon in the conterminous United States. Soil Sci Soc Am J 70: 601–612.
70. Journey CA, Burns DA, Riva-Murray K, Brigham ME, Button DT, et al. (2012) Fluvial transport of mercury, organic carbon, suspended sediment, and selected major ions in contrasting stream basins in South Carolina and New York: October 2004– September 2009. Columbia, SC: U.S. Geol Surv Sci-Invest Rpt 2012–5173. 128 p.
71. Bushey JT, Nallana AG, Montesdeoca MR Driscoll CT (2008) Mercury dynamics of a northern hardwood canopy. Atmos Environ 42: 6905–6914.
72. Burns DA, Aiken GR, Bradley PM, Journey CA, Schelker J (2013) Specific ultra-violet absorbance as an indicator of mercury sources in an Adirondack river basin. Biogeochemistry 113: 451–466.
73. Tjerngren I, Karlsson T, Björn E, Skyllberg U (2012) Potential Hg methylation and MeHg demethylation rates related to the nutrient status of different boreal wetlands. Biogeochemistry 108: 335–350.
74. Shanley JB, Schuster PF, Reddy MM, Roth DE, Taylor HE, et al. (2002) Mercury on the move during snowmelt in Vermont, Eos Trans Am Geophys Union 83: 45, 47–48.

Single-Kernel Ionomic Profiles Are Highly Heritable Indicators of Genetic and Environmental Influences on Elemental Accumulation in Maize Grain (*Zea mays*)

Ivan R. Baxter[1]*, Gregory Ziegler[1], Brett Lahner[2], Michael V. Mickelbart[2], Rachel Foley[2], John Danku[3], Paul Armstrong[4], David E. Salt[3], Owen A. Hoekenga[5]*

1 United States Department of Agriculture, Agricultural Research Service, Plant Genetics Research Unit, Donald Danforth Plant Science Center, St. Louis, Missouri, United States of America, **2** Purdue University, Department of Horticulture and Landscape Architecture, West Lafayette, Indiana, United States of America, **3** University of Aberdeen, Institute of Biological and Environmental Science, Aberdeen, United Kingdom, **4** United States Department of Agriculture, Agricultural Research Service, Engineering and Wind Erosion Research Unit, Manhattan, Kansas, United States of America, **5** United States Department of Agriculture, Agricultural Research Service, RW Holley Center for Agriculture and Health, Ithaca, New York, United States of America

Abstract

The ionome, or elemental profile, of a maize kernel can be viewed in at least two distinct ways. First, the collection of elements within the kernel are food and feed for people and animals. Second, the ionome of the kernel represents a developmental end point that can summarize the life history of a plant, combining genetic programs and environmental interactions. We assert that single-kernel-based phenotyping of the ionome is an effective method of analysis, as it represents a reasonable compromise between precision, efficiency, and power. Here, we evaluate potential pitfalls of this sampling strategy using several field-grown maize sample sets. We demonstrate that there is enough genetically determined diversity in accumulation of many of the elements assayed to overcome potential artifacts. Further, we demonstrate that environmental signals are detectable through their influence on the kernel ionome. We conclude that using single kernels as the sampling unit is a valid approach for understanding genetic and environmental effects on the maize kernel ionome.

Editor: Xianlong Zhang, National Key Laboratory of Crop Genetic Improvement, China

Funding: This research was supported in part by the U.S. National Science Foundation, Plant Genome Research Program (grant #IOS 0701119 awarded to David Salt and grant #IOS 1126950 awarded to Ivan Baxter and Owen Hoekenga) and United States Department of Agriculture-Agricultural Research Service Intramural funds to Ivan Baxter. The funders had no role in study design, data collection and analysis, decision to publish, or preparation of the manuscript.

Competing Interests: The authors have read the journal's policy and have the following conflicts: Co-author Dr. Baxter is a PLOS ONE Editorial Board member. This does not alter the authors' adherence to all the PLOS ONE policies on sharing data and materials.

* E-mail: Ivan.Baxter@ars.usda.gov (IB); Owen.Hoekenga@ars.usda.gov (OH)

Introduction

Increasing agricultural sustainability requires improvements in nutrient use efficiency while decreasing fertilizer inputs. These requirements exist as the majority of arable soils have limitations associated with them [1]. Only 16% of crop lands are "without constraint," and most of these constraints are related to elements found in inadequate or excessive amounts [1]. The range of soil elemental concentrations optimal for productive growth of crops is much smaller than that of the wild plants they may have displaced. This is likely due to human selection of crop plants for yield under optimal agricultural conditions and not for adaptive mineral nutrient efficiency on poor soil. Non-optimal concentrations of many elements limit the productivity of crops or necessitate significant inputs to maintain productivity. Major elemental limitations include excessive Na [2]; insufficient N [3], P [4], and K [5]; acid soil syndrome, which causes Al, Mn, and Fe toxicity and Mo, Ca, and P deficiency [6]; and Fe deficiency in alkaline soils [7]. Due to low soil fertility and the effects of poverty (e.g., inability to buy fertilizer), crop yields in most of Africa are less than one-fifth of U.S. yields [8]. In order to meet future food needs, we will need to increase yields while increasing the sustainability of agricultural systems in both developed and developing countries. In order to develop crops that can grow in diverse soils with less fertilizer, we require a deeper understanding of the genes that allow plants to adapt to different soil environments [9].

The elemental composition of a cell, tissue, or organism is referred to as the ionome [10]. The ionome can be profiled using high-throughput, high-accuracy analytical chemistry such as inductively coupled plasma-mass spectrometry (ICP-MS), which can measure the concentrations of 20 elements over 5 logs in ~2 minutes per sample. To apply this systems biology phenotyping platform most efficiently, the best tissue for estimating the ionome of a crop plant must be used [9,11,12]. We assert that mature seeds are the ideal tissue when resources are limited, as mature seeds represent a well-defined developmental end point that summarizes the life history and genetic composition of a particular individual. Seeds are also highly stable and are easy to store, transport, and handle. Furthermore, seeds are feedstocks for people, animals, and industrial processes such that the seed ionome alone is of high value and represents an excellent proxy for a whole plant.

In an ideal world, a survey of tissues could be used to track the ionome through developmental time. However, genetic and environmental determinants make this approach difficult to implement on large populations of field-grown plants, as diverse varieties may progress through development at different rates that may be more or less influenced by daily weather or other environmental factors. Compromises are required to ensure the success of a particular research program, especially one that aims to identify genes that are effective over a range of environments rather than emphasizing a single one. We propose that the analysis of single seeds is the most efficient use of resources to characterize a highly relevant ionome for field-grown crops. Intact seeds are at reduced risk for contamination or preparative artifacts due to sample grinding and are an easily automated sample unit, and the overall reduced cost of preparation and analysis make this the best compromise of efficiency, relevance, and precision. This scheme is not without obvious potential problems, however, not the least of which is heterogeneity between seeds produced by the same plant or related plants within an experimental plot that may shape the estimation of the ionome through single-seed-based observation.

In the present study, we test the premise of confounding heterogeneity to better understand the sources of variance that contribute to the seed ionome using maize kernels. We assert that single-seed-based analysis is a reasonable strategy for phenotypic analysis, especially when resources are limited and considerations of the number of test environments are balanced against accuracy within any single environment. The ionomic profiling workflow described in this study for maize kernels takes advantage of automation for sample handling, weighing, and liquid dispensing to reduce operator time, effort, and overall cost. This optimization allows 576 kernels to be analyzed from start to finish in 3 days. We demonstrate the utility of our workflow as an effective means of collecting ionomic data relevant to increasing agricultural sustainability.

Results

Experimental Design and Analysis

The ionomics pipeline starts with arraying the single kernels in 48 well plates that are then loaded onto the custom-built weighing robot. Each kernel is weighed and deposited into a glass digestion tube. The samples are digested down to the elemental components using heat and acid. The digested material is analyzed by ICP-MS to quantitatively measure the concentrations of 20 elements with high precision. Both internal and external standards are used to correct for instrument drift during and between experiments.

The 26 parents of the maize nested association mapping (NAM) panel (hereafter, the NAM Founders) [13] were grown in a randomized complete block design, in 2010 (six blocks) and 2011 (five blocks), at the Purdue Agronomy Center for Research Education. Kernels from the middle of 1–2 cobs per plot were removed from the cob by hand and analyzed using the ionomics pipeline. For each plot (i.e., each genotype within a block), four to six seeds were analyzed. Outliers within each plot were removed using a conservative cutoff of 15 median absolute deviations from the plot median (derived from [14]) on an element by element basis. To be even more conservative, we used the mean of all the samples in each plot and omitted the one popcorn and two sweet corn entries in the NAM Founders in order to prevent artifacts from low seed weights from biasing the analysis. To analyze the sources of variance, a simple linear model with line and year effects was used to demonstrate that all 20 elements had a significant effect of genotype ($p<0.01$ with a Bonferroni correction; Table 1). Most elements also had significant effects of year as well as line x year interactions. There were four potentially problematic elements characterized by low heritability and or low correlation between seasons: B, Na, Al, and Se. This result was not unexpected, as all are in low abundance in the kernel and have several analytical interferences (see discussion). Narrow sense heritability estimates within a single season were quite high, with 17 and 19 elements having heritabilities >0.5 in each year respectively. Heritability estimates decreased when both years were considered in a single model, likely due to the significant line x year interactions. Twelve elements had a two-year heritability estimate above 0.5. For 16 elements there was a statistically significant ($p<0.01$) correlation between the two years (see examples in Figure 1).

In these ionomic studies, we wished to emphasize the potential to detect QTLs in plants grown across a larger number of different environments. This requires a compromise between population size, number of field locations, and sampling depth for any one accession. To estimate if a single kernel per cob would be a feasible approach to analyzing large populations across multiple environments, we created 100 datasets by randomly sampling one of the four to six analyzed samples from each plot, a process known as jackknifing. Since the outlier removal that we performed before the first analysis was based on the distribution of samples in a given plot, we used the original (without correction for outliers) dataset for this simulation. The distribution of jackknifed heritabilities at the 5% value over all data (i.e., 95% of the single-seed dataset heritabilities were greater than these values) was lower than those calculated for means, but 7 elements had heritability >0.45 and 11 were >0.35 (Table 1). We found this a reasonable compromise, in order to detect potential QTLs from a larger number of environments at different locations rather than focusing all of our attention on a single location.

Potential Confounding effects

To evaluate potential confounding effects that would limit our ability to detect genetic, environment, or gene by environment effects, we examined several potential sources of variance.

Outcrossing. Open pollinated ears are likely to have some kernels fertilized by pollen from nearby rows. If the paternal genotype has a significant effect on the kernel ionome, this genetic contamination may confound the genetic signal from the maternal plant. While we did not specifically test for this effect in these experiments, the random block design of the NAM parent experiment should have randomized the genotypes of the contaminating pollen and therefore contributed to the unexplained variance detected in that experiment. The heritabilities observed in that experiment (Table 1) suggest that this potential contamination will not prevent the detection of genetic effects, at least at the plot spacing used in these experiments (76 cm).

Cob location. Given the differences in seed filling along the ear, elemental accumulation gradients could exist from the base to the tip of the ear. To determine the prevalence and magnitude of these gradients, we analyzed eight kernels from the base, middle, and tip of three independent ears from plants grown at four different agricultural research stations in Missouri, Texas, and Iowa. ANOVA indicated a significant effect ($p<0.01$ with Bonferroni correction) of position within the cob for seven elements (Table 2 and Figure 2). However, the location of the experimental fields accounted for much more of the variance in all elements except Na, S, Ca, and Zn. Additionally, the dynamic range of elemental accumulation (defined by the ratio of maximum accumulation for a cob location to minimum accumulation for a cob location) was well below the dynamic range of the observations from the two years of NAM Founders in Indiana. For

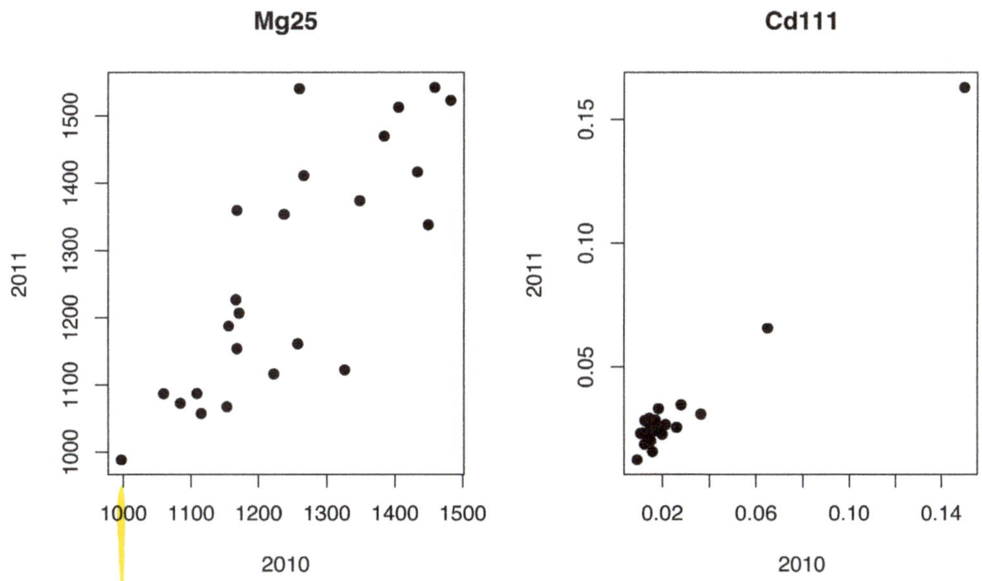

Figure 1. Correlation between 23 NAM parents across two years. The plot means of all 5 or 6 replicate plots for each NAM parent line (excluding Il14H, HP301 and P39) were averaged for each year.

Table 1. Analysis of the NAM founders ionomic profiles from 2010 and 2011.

	Geno	Year	GxY	h2 2010	h2 2011	h2 both	Corr	JK 5% p val	JK 5% h2
Weight	1.3E-31	2.6E-04	1.2E-05	0.64	0.67	0.53	0.65	9.9E-19	0.50
B	3.3E-23	6.6E-17	2.5E-09	0.63	0.56	0.37	NS	6.1E-06	0.30
Na	1.5E-14	9.4E-75	NS	0.28	0.52	0.12	0.76	3.3E-03	0.11
Mg	1.2E-62	NS	5.3E-11	0.81	0.84	0.73	0.80	1.6E-33	0.61
Al	4.2E-09	2.2E-57	4.2E-04	0.3	0.43	0.12	NS	1.5E-02	0.10
P	2.0E-49	6.7E-69	2.7E-07	0.81	0.71	0.38	0.77	1.5E-25	0.39
S	3.7E-40	1.3E-07	8.1E-13	0.81	0.61	0.55	0.55	2.2E-25	0.52
K	4.4E-27	NS	3.8E-11	0.65	0.65	0.46	NS	3.9E-14	0.44
Ca	2.6E-30	7.2E-06	8.5E-08	0.6	0.69	0.5	0.59	9.3E-15	0.34
Mn	2.0E-44	1.8E-22	8.7E-11	0.78	0.71	0.54	0.64	5.6E-33	0.60
Fe	4.0E-51	NS	NS	0.74	0.8	0.7	0.85	1.3E-31	0.21
Co	3.6E-22	NS	NS	0.5	0.54	0.5	0.95	4.1E-17	0.36
Ni	1.8E-24	1.2E-21	2.1E-04	0.6	0.58	0.39	0.74	4.1E-22	0.38
Cu	7.6E-71	6.9E-12	2.5E-06	0.84	0.86	0.77	0.88	3.4E-49	0.66
Zn	3.4E-61	5.2E-23	9.8E-08	0.81	0.83	0.67	0.81	1.0E-46	0.63
As	5.3E-30	NS	4.2E-05	0.66	0.62	0.52	0.64	3.4E-16	0.52
Se	2.7E-08	3.6E-04	NS	0.47	0.28	0.25	NS	3.7E-03	0.27
Rb	2.6E-23	1.4E-36	NS	0.54	0.64	0.32	0.77	1.1E-19	0.30
Sr	4.7E-21	1.9E-31	2.1E-04	0.53	0.57	0.31	0.62	1.2E-08	0.23
Mo	2.5E-29	NS	NS	0.6	0.63	0.55	0.86	4.6E-25	0.51
Cd	5.0E-63	7.5E-05	NS	0.75	0.9	0.79	0.98	4.1E-37	0.32

Geno: p value for the genotype term in an ANOVA of a linear model with genotype and year and their interaction. Year: p value for the year term. GxY: p value for the genotype by year interaction. The significance cutoff for the first three columns was set at $p<0.0005$ to account for the multiple testing correction. NS: not significant. h2: the narrow sense heritability for each year and combined. Corr: the correlation between the line average from 2010 and 2011. The last two rows are the results from the jackknife analysis of 100 datasets with a single seed per plot. JK 5% p val: the 5% most significant p value for the genotype term in the two year model (i.e., 95% of the time the p value was more significant). JK 5% h2: the 5% highest heritability across two years.

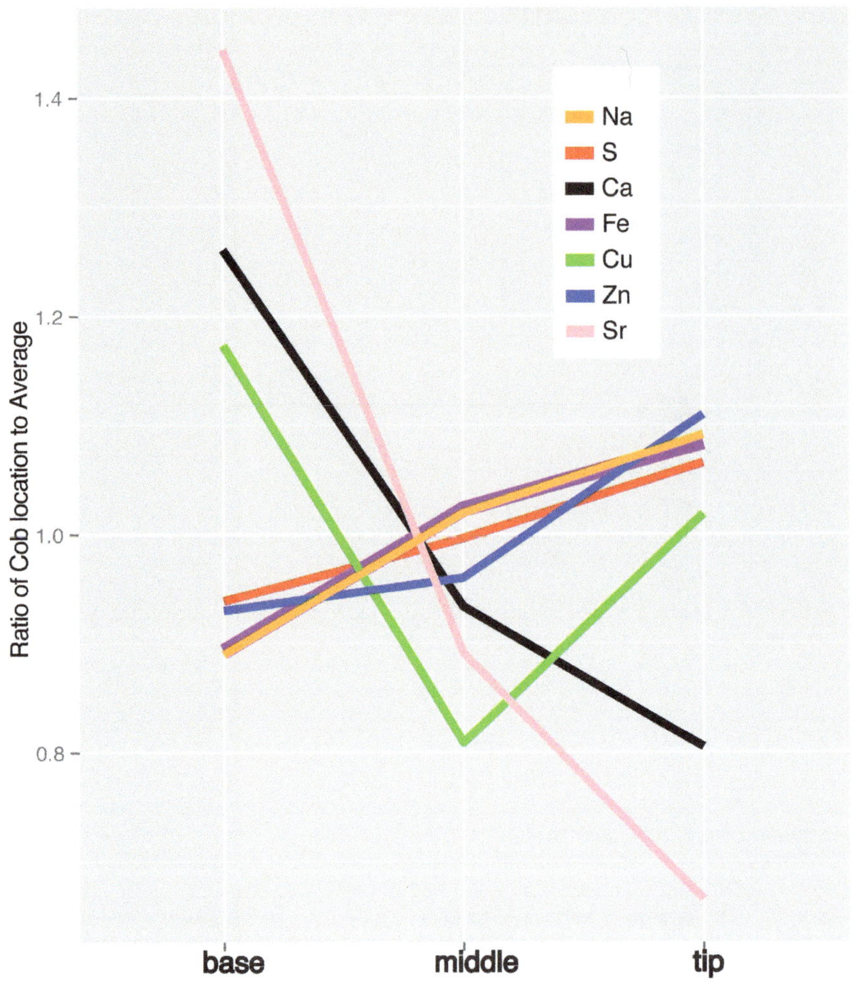

Figure 2. Gradients in elemental accumulation along the cob. For each ear, the samples for each section were averaged and then normalized to the mean of the ear. The values for all ears were then averaged. Only elements with significant effects of cob location were included in the model.

example, while the highest and lowest cob locations differed by an average of 49% for Cu, the highest Cu accumulating line was over 500% higher than the lowest line in both 2010 and 2011. These data suggest that the variation due to cob location will not prevent identification of genetic environmental effects for most elements.

Seed composition. Elements are not distributed uniformly throughout the seeds of grain crops: some elements are concentrated in the embryo (e.g., Fe and Zn), while others (Ca and S) have substantial accumulation in the maternal tissues of the pericarp and endosperm [15–18]. Large differences in amounts of these compartments, such as differences in the organic composition of the seed, could change the total elemental composition of the kernel. The *sugary* locus is a key step in starch biosynthesis, with mutations at *sugary* creating the sweet corn many people eat as a fresh vegetable. Furthermore, mutant kernels also have higher protein levels than wild-type siblings [19]. As *sugary* kernels mature and dehydrate, they shrink, losing more stored carbon than Su+ seeds, and can potentially lose water-soluble ions [20]. To determine the effect of *sugary* on the elemental composition, we identified seven RI lines from the dent x sweet, B73 x IL14H NAM subpopulation where *sugary* was still segregating. We separated the seeds from single plots based on the kernel morphology (collapsed *su/su* seeds from plump *Su/+* seeds) and analyzed their elemental content (Table 3 and Figure 3). For nine of the elements, there was a significant effect of the *sugary* locus, and for all but Fe and As, a significant effect of line could also be detected. We also analyzed a single seed per entry from this population. With this limited sampling, we were able to identify 31 QTLs (Table 4), including 8 strong QTLs with LOD scores >5 (highest 95% permutation threshold for any element was 2.8). We identified QTLs for P, K, Ni, and Cu that localized with *sugary* on Chr 4. The presence of *sugary* did not, however, prevent us from identifying an additional 3, 2, and 1 QTLs for Cu, Ni, and P, respectively, demonstrating that the gross change in kernel composition did not obscure all other genetic signals.

Discussion

A defined developmental stage that is stable at room temperature, the mature kernel has many inherent advantages as a sample tissue. Single kernels are discrete, relatively small samples that can be easily manipulated by technicians or robots. Single-kernel-based analyses also avoid mechanical disruption (e.g., grinding or milling) that are labor intensive and/or introduce the possibility of contamination from the tools used. With these experiments, we have demonstrated that single-seed measurements of elemental profiles can be used to study the genetic and environmental effects on seed elemental accumulation.

Table 2. Analysis of the seed position on the cob on B73 ionomic profiles.

	Dyn Range Cob	Dyn Range 2010	Dyn Range 2011	Farm p val	Field p val	Cob p val	Farm SSperc	Field SSperc	Cob SSperc
B	1.40	1.81	1.69	NS	NS	NS	0.05	0.06	0.31
Na	1.32	2.59	4.00	NS	NS	4.3E-04	0.19	0.04	0.33
Mg	1.08	1.51	1.59	3.2E-11	NS	NS	0.83	0.00	0.01
Al	1.50	2.00	3.00	2.0E-04	NS	NS	0.40	0.00	0.20
P	1.08	1.52	1.43	7.3E-14	NS	NS	0.85	0.01	0.05
S	1.15	1.64	1.56	6.9E-05	NS	3.6E-05	0.36	0.00	0.33
K	1.08	1.80	1.77	5.8E-06	NS	NS	0.54	0.05	0.07
Ca	1.60	3.50	4.07	2.1E-08	NS	1.4E-08	0.44	0.01	0.40
Mn	1.22	2.70	3.63	3.9E-05	NS	NS	0.46	0.02	0.15
Fe	1.24	2.11	2.11	2.6E-10	NS	7.9E-07	0.61	0.01	0.24
Co	1.55	7.84	11.02	NS	NS	NS	0.31	0.04	0.06
Ni	1.26	7.33	9.72	9.7E-16	NS	NS	0.88	0.03	0.02
Cu	1.49	5.78	5.45	2.2E-09	NS	5.7E-07	0.55	0.02	0.28
Zn	1.24	1.98	2.62	NS	NS	3.6E-04	0.18	0.02	0.35
As	1.58	2.65	3.71	8.2E-12	NS	NS	0.84	0.01	0.00
Rb	1.16	2.41	2.77	6.3E-18	NS	NS	0.92	0.01	0.02
Sr	2.30	3.71	3.94	5.8E-11	NS	4.9E-07	0.64	0.00	0.23
Mo	1.21	6.50	3.64	1.5E-21	NS	NS	0.97	0.00	0.00
Cd	1.73	16.30	13.01	NS	NS	NS	0.43	0.00	0.01

Dyn Range Cob: the average dynamic range of values (max accumulation of a position on a single cob/min accumulation on the same cob) from different positions on the cob. Dyn Range 2010 and 2011: the dynamic range in genotype values within that year (max accumulating genotype/min accumulating genotype). Farm p val: p value for the farm term in an ANOVA of a linear model with farm, field location, and location on cob. Field p val: p value for the field location term. Cob p val: p value for the location on cob term. The significance cutoff was set at $p<0.0005$ to account for the multiple testing correction. NS: not significant. Farm, Field and Cob SSperc are the percentage of the variance (sum of squares) accounted for by the various terms. Selenium was not measured in this experiment.

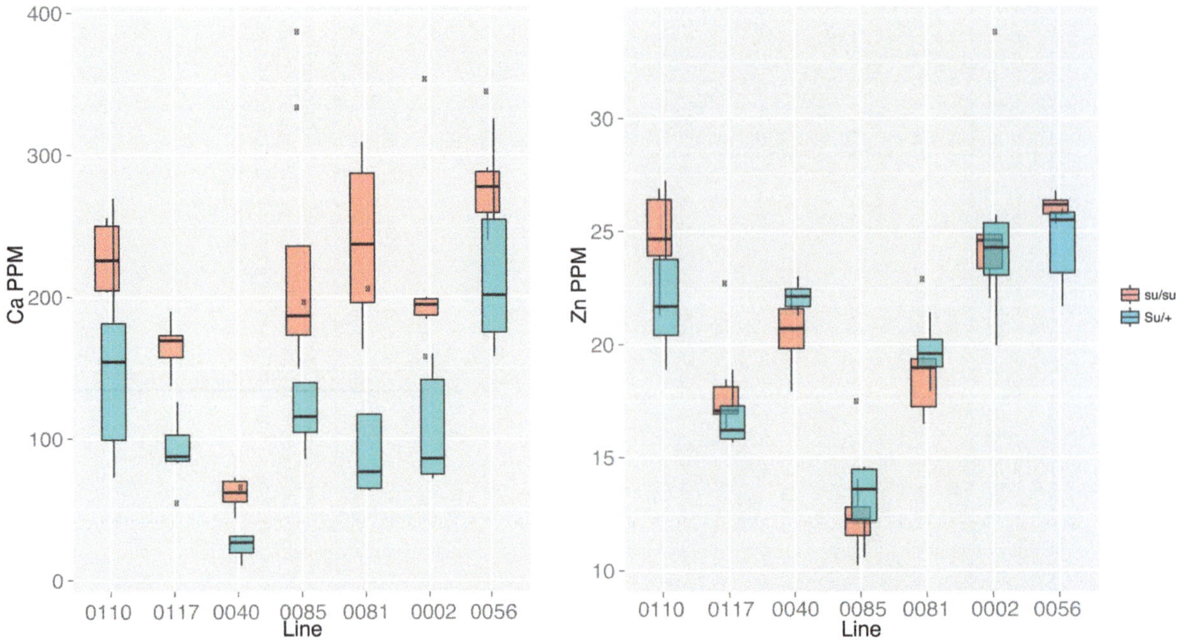

Figure 3. Alphabox plots of the effect of *su* on Ca and Zn accumulation. Five number summaries (median, 1st and 3rd quartiles, and 1.5 interquartile range) are shown of each seed type. Salmon denotes the values for the *su/su* (collapsed) seed, while teal denotes the values for *Su/+* (plump) alleles. Outliers beyond 1.5 IQR are shown as black dots. Line names are the entry numbers in the Panzea database for the B73 x IL14H (Z011) population.

Table 3. Analysis of the sugary status on ionomic profiles.

	Line	Su
weights	2.5E-15	4.02E-09
B	2.2E-05	NS
Na	1.3E-05	NS
Mg	NS	NS
Al	4.7E-05	NS
P	2.9E-12	3.9E-05
S	6.7E-17	1.3E-10
K	2.6E-13	1.2E-16
Ca	8.7E-14	2.3E-09
Mn	2.1E-27	8.6E-08
Fe	NS	1.2E-08
Co	NS	NS
Ni	1.6E-10	NS
Cu	3.3E-23	NS
Zn	2.9E-23	NS
As	NS	4.4E-06
Se	1.9E-06	5.8E-06
Rb	2.7E-32	5.4E-11
Mo	1.9E-27	NS
Cd	1.6E-24	NS

Line: p value for the genotype term in an ANOVA of a linear model with line and *sugary* genotype. Su: p value for the sugary term. The significance cutoff was set at $p<0.0005$ to account for the multiple testing correction. NS: not significant.

Developing the ionomics pipeline around single-kernel sampling has enabled the processing of 1728 samples per week at the USDA-ARS/Danforth Center profiling facility. Multiple kernels can be profiled with less cost and effort than would be required to pool, disrupt and homogenize, subsample, digest, and run as a single composite sample. This provides several advantages from an analytical and statistical perspective. With multiple measurements from a given packet, outliers resulting from the analytical process or machine error can be discarded, as we did here. The measurements can be averaged to a single value or included as nested factors in a statistical model.

We investigated three potential confounding effects: outcrossing, relative position of the seed within the ear, and the composition of the organic components of the seeds. While we were able to detect statistically significant effects for the latter two factors for subsets of elements, neither one of them was large enough to obscure the signal from genetic or environmental factors.

Since we did not explicitly test for outcrossing, instead conducting an experiment, where it could occur, that tested whether the ionomic traits were still heritable, we cannot estimate directly the magnitude of the effects in this study. However, the high heritabilities observed in these experiments suggest that the effects are small. Of the range of processes that can contribute to the seed ionome, root uptake, partitioning, shoot leaf transport, leaf partitioning, leaf remobilization, leaf seed transport, and seed loading are all exclusively maternal, while embryo uptake has contributions from both maternal and paternal alleles. When viewed from this perspective, it seems reasonable that paternal effects are a minor contribution to the ionome. However, we cannot exclude the possibility that there are loci that will have a strong paternal effect.

The high heritability observed in jackknifed datasets of a single seed per plot from the NAM Founders, without the benefit of outlier removal, demonstrates that the genetic variation for many

Table 4. QTLs identified in the B73 x IL14H population.

El	Chm	Pos (cM)	LOD	MI Start	MI Stop	R^2	SumR^2	Add.Eff	PT	Co	Nearest marker
B	3	123.5	3.3	121.9	129.8	0.08	0.13	−58.93	2.5	7	PZB01457.1
B	5	13.3	2.5	13.3	13.3	0.05	0.13	−51.74	2.5	7	PHM13122.43
Cd	2	84.2	17.4	63.0	102.8	0.38	0.38	−0.02	2.6	1	PZA00495.5
Co	1	193.1	2.9	190.0	193.1	0.06	0.15	0.00	2.4	10	PZA00235.9
Co	5	23.2	3.2	13.3	29.1	0.10	0.15	0.00	2.4	10	PZA01925.1
Cu	2	94.5	4.3	88.6	100.0	0.09	0.24	−0.22	2.8	10	PZA01735.1
Cu	4	56.6	3.1	56.0	59.9	0.06	0.24	−0.20	2.8	10	PZA00218.1
Cu	8	80.6	4.5	74.1	86.8	0.10	0.24	0.23	2.8	10	PZA03698.1
K	4	54.9	5.0	49.7	59.9	0.15	0.15	−478.96	2.7	7	PZA01751.2
Mn	1	34.3	4.1	33.9	35.7	0.06	0.57	−0.91	2.8	10	PZA01030.1
Mn	1	89.1	14.4	73.4	106.0	0.30	0.57	−1.45	2.8	10	PZA02135.2
Mn	3	91.8	5.1	83.6	97.9	0.10	0.57	0.91	2.8	10	PZA03735.1
Mn	5	7.2	3.4	7.2	10.3	0.05	0.57	0.61	2.8	10	PZA01438.1
Mn	5	40.8	3.6	34.7	49.2	0.06	0.57	−0.66	2.8	10	PZA01284.6
Mo	1	53.9	3.0	50.6	53.9	0.05	0.33	−0.01	2.8	10	PZA02686.1
Mo	1	147.4	14.6	129.8	155.8	0.29	0.33	−0.03	2.8	10	PZA02269.3.4
Na	5	13.3	3.0	10.3	21.2	0.06	0.14	−25.24	2.5	10	PHM13122.43
Na	5	43.7	3.7	36.8	49.2	0.08	0.14	29.13	2.5	10	PZA01327.1
Ni	1	149.9	8.7	136.2	163.7	0.15	0.35	0.06	2.7	10	PZA02269.3.4
Ni	2	91.7	3.4	70.5	100.0	0.06	0.35	−0.04	2.7	10	PZA01735.1
Ni	4	56.6	3.3	56.0	59.9	0.06	0.35	0.04	2.7	10	PZA00218.1
Ni	5	71.2	3.6	60.7	84.6	0.08	0.35	0.04	2.7	10	PZA03536.1
P	4	53.9	4.0	52.4	56.6	0.08	0.13	−191.73	2.7	10	PZA01751.2
P	6	33.4	3.0	33.4	35.1	0.05	0.13	−127.55	2.7	10	PZA03461.1
Rb	3	119.9	2.6	118.0	121.9	0.05	0.25	−0.23	2.4	10	PZA02516.1
Rb	4	127.1	2.6	127.1	127.1	0.05	0.25	0.24	2.4	10	PZA03322.5.3
Rb	5	101.2	3.0	101.2	104.5	0.05	0.25	−0.27	2.4	10	PZA02411.3
Rb	8	29.6	5.4	17.5	41.2	0.10	0.25	0.34	2.4	10	PZA03178.1
S	1	161.7	4.1	151.9	173.5	0.09	0.09	−117.13	2.5	9	PZA02698.3
Zn	5	99.3	5.5	91.7	109.5	0.12	0.21	−2.51	2.3	7	PZA02411.3
Zn	7	132.4	3.8	123.1	132.4	0.09	0.21	2.12	2.3	7	PZA01744.1

El: Element. Chm: Chromosome. Pos: Position of QTL in cM. LOD: LOD score. MI Start and Stop: the mapping intervals defining where the LOD trace crossed the 95% permutation threshold. R^2: the fraction of variance accounted for by the QTL, Sum R^2: The cumulative fraction of variance accounted for by all of the QTLs for the element. Add.eff: The additive effect of the QTL in PPM. PT: the 95% permutation threshold from 1000 permutations. CO: number of cofactors used in the model. Nearest Marker: The nearest marker.

elemental traits in maize is large enough that most confounding factors will not interfere with analysis. However, if the labor and resources are available, controlled pollination experiments and harvesting only the center of the cob will reduce the variability from pollination and cob location. Analysis of the organic composition of the kernels through NIR spectroscopy or other methods could allow for differences in protein, oil, and starch content to be included as cofactors in statistical models.

The appropriate number of seeds for analysis will depend on the experimental design and expected frequencies of functionally distinct alleles. For quantitative genetic experiments such as recombinant inbred populations with several hundred lines (or thousands. in the case of the NAM), where a given allele may occur over a hundred times, a single seed per line may suffice to identify significant QTLs. From a technical standpoint, more observations will decrease the associated error but be biologically unnecessary to find common alleles, and may reduce the number of lines examined if you have fixed capacity for analysis. In those cases where rare alleles have large phenotypic effects, such as in an association mapping population or when a new mutation is being tested, more seeds will be required to increase the confidence of detection [21]. While we were able to identify significant QTLs with a single seed in the B73 x IL14H population, our standard practice when analyzing a single RIL population is to profile two to four seeds per plot. When analyzing extremely large populations such as the Maize NAM, which has >6000 plots in the standard experimental design, profiling one or two seeds from most lines allows for the analysis of the population in more locations [13,22]. This compromise in technical accuracy of the ionomic phenotype versus increasing the number of environments is a highly reasonable one to make, given that resources are almost always limiting.

There are, of course several drawbacks and limitations to the approach we have taken. By opting for high throughput while analyzing 20 elements simultaneously, we do lose some sensitivity compared to lower throughput efforts focused on fewer elements. Examples of this include using standard glass (sodium borosilicate) tubes for chemical digestion, which increases the background levels of Na and B; not using a dynamic reaction cell on the ICP-MS, which leads to isobaric interferences for Se, As, and Al; and the use of a single sample dilution factor to measure elements with concentrations ranging from several hundred parts per million to low parts per billion.

This speed/number of elements/accuracy trade-off, combined with the low abundance of some of the elements in maize kernels, leads to several elements being at or below the level of detection in many or most of the samples analyzed. We have considered omitting these samples from the analysis pipeline entirely, but currently include them for occasions where a given mutant line or diverse accession accumulates sufficient qualities to exceed the detection threshold, such as for two of the NAM parents for Cd (Figure 1).

The single-seed analysis approach has the potential to create two major artifacts. One is that differences in weights can cause perceived differences in accumulation in elements at or below detection limits when the signal from the instrument is normalized to weight. The second is that some elements may be restricted to a specific subcompartment of the seed, such as the embryo, so alterations in the other compartments may not affect the amount of the element, even though they would affect the weight. However, single-seed analysis, with the weight recorded and included in every dataset, allows for the dataset to be analyzed from an amount per seed perspective, which would alleviate these issues (although it may create artifacts in itself). We have opted to present the data as a concentration in the seed because, for most elements, there appears to be a correlation with weight and total elemental concentration. But depending on the particular question, a hybrid approach may be more appropriate. To ensure that these approaches are possible for other scientists, our data, including kernel weight, are all available from www.ionomicshub.org through the data exchange portal.

Conclusions

Here we have shown that single-seed-based ionomics is a robust and information-rich approach to identifying genetic and environmental determinants of the elemental profile of maize kernels. The limitations and potential artifacts are not large enough to prevent detection of statistically significant effects and represent a reasonable compromise to increase the scope and efficiency of investigation over absolute technical accuracy.

Materials and Methods

Biological Samples

The 25 NAM Founders were received from the U.S. Department of Agriculture Germplasm Resource Information Network (GRIN) at Ames, IA and grown at the Purdue Agronomy Center for Research and Education (West Lafayette, IN). The Founders of the maize Nested Association Mapping Panel were grown in a randomized block design, in 2010 (six blocks) and 2011 (five blocks). The soil type is a Udollic Ochraqualf silt loam. Plots were irrigated in late July of 2011 because of high temperatures and low rainfall. The fertilizer regimen consisted of 336 kg ha^{-1} potash (0-0-60) in the fall of both years and 123 kg ha^{-1} in 2010 and 2011 of anhydrous ammonia (81-0-0) with N-Serve® (DOW Agrosciences, Indianapolis, IN) prior to planting each year. To suppress weeds, the plots were treated at planting with 5.9 L ha^{-1} Harness Xtra® (Monsanto Company, St. Louis, MO), 0.84 kg ha^{-1} Princep® (Syngenta Crop Protection, Greensboro, NC), 0.57 kg ha^{-1} atrazine (Syngenta Crop Protection, Greensboro, NC), 1.75 L ha^{-1} Roundup® (Monsanto Company, St. Louis, MO), cultivated in mid-June, and subsequently hand-weeded through the growing season. Plots consisted of a single 3-m row with 76 cm between rows. Average stand density was 11 plants per row in 2010 and 15 plants per row in 2011. Lines were open pollinated and the seed harvested after maturity. Seed from the middle of the ear of one or two cobs per plot was removed from the cob by hand and analyzed using the ionomics pipeline.

For the heterogeneity within/between cobs experiment, the B73 accessions were received as gifts from Dr. Sherry Flint-Garcia (USDA ARS/Missouri University, Columbia, MO), Dr. Seth Murray (Texas A&M, College Station, TX), and Dr. Michael Muszynski (Iowa State, Ames, IA). Intact, hand-pollinated B73 ears were sampled from their 2009 nursery fields. Lines from the B73 x IL14H population were grown at the Illinois Crop Sciences Research and Education Center farm in 2007 using standard agronomic practices, with a single plot per line. Hand-pollinated ears were used to generate grain samples for this analysis. The soil properties of each site are listed in Table 5.

Table 5. Soil and Yield properties of growth locations.

Site Name	Site Location	Primary Soil Type	Soil Classification	Average Estimated Maize Yield Mt ha^{-1}
ISU Agronomy and Ag Engineering Research Farm	Boone IA	Clarion loam	Typic Hapludoll	11.5
OD Butler Jr Animal Science Complex and University Farm	Burleson TX	Boonville fine sandy loam	Chromic Vertic Albaqualf	2.9
Bradford Research and Extension Center	Columbia MO	Mexico silt loam	Vertic Epiaqualf	11.0
Purdue Agronomy Center for Research Education	West Lafayette IN	Toronto-Millbrook complex	Udollic Epiaqualf	8.6
Musgrave Research Farm	Poplar Ridge NY	Lima silt loam	Oxyaquic Hapludalf	7.5
Crop Sciences Research and Education Center	Urbana IL	Drummer silty loam	Typic Endoaquoll	11.0

Determination of Elemental Concentration by ICP-MS Analysis

The analysis methods used are almost identical to those described for soybeans in Ziegler et al. [23]. As the precise methods of the pipeline are essential and integral to this manuscript, we have included identical or near-identical descriptions in this section. Identical passages are denoted by quotation marks.

Sample preparation and digestion. The seeds for the B73 x IL14H and B73 cob location experiments were weighed by hand. For the "Seeds were sorted into 48-well tissue culture plates, one seed per well. A weight for each individual seed was determined using a custom built weighing robot. The weighing robot holds six 48-well plates and maneuvers each well of the plates over a hole which opens onto a 3-place balance. After recording the weight, each seed was deposited using pressurized air into a 16×110 mm borosilicate glass test tube for digestion. The weighing robot can automatically weigh 288 seeds in approximately 1.5 hours with little user intervention."

"Seeds were digested in 2.5 mL concentrated nitric acid (AR Select Grade, VWR) with internal standard added (20 ppb In, BDH Aristar Plus). Seeds were soaked at room temperature overnight, then heated to 105°C for two hours. After cooling, the samples were diluted to 10 mL using ultrapure 18.2 MΩ water (UPW) from a Milli-Q system (Millipore). Samples were stirred with a custom-built stirring rod assembly, which uses plastic stirring rods to stir 60 test tubes at a time. Between uses, the stirring rod assembly was soaked in a 10% HNO_3 solution. A second dilution of 0.9 mL of the 1st dilution and 4.1 mL UPW was prepared in a second set of test tubes. After stirring, 1.2 mL of the second dilution was loaded into 96 well autosampler trays."

ICP-MS Analysis. For the NAM parent experiments, elemental "concentrations of B, Na, Mg, Al, P, S, K, Ca, Mn, Fe, Co, Ni, Cu, Zn, As, Se, Rb, Mo, and Cd were measured using an Elan 6000 DRC-e mass spectrometer (Perkin-Elmer SCIEX) connected to a PFA microflow nebulizer (Elemental Scientific) and Apex HF desolvator (Elemental Scientific). Samples were introduced using a SC-FAST sample introduction system and SC4-DX autosampler (Elemental Scientific) that holds six 96-well trays (576 samples). " The other experiments were run without the FAST sample introduction system and with a standard Apex desolvator (Elemental Scientific).

"All elements were measured with DRC collision mode off. Before each run, the lens voltage and nebulizer gas flow rate of the ICP-MS were optimized for maximum Indium signal intensity (>25,000 counts per second) while also maintaining low CeO+/Ce+ (<0.008) and Ba++/Ba+ (<0.1) ratios. This ensures a strong signal while also reducing the interferences caused by polyatomic and double-charged species. Before each run a calibration curve was obtained by analyzing six dilutions of a multi-element stock solution made from a mixture of single-element stock standards (Ultra Scientific)." In addition for the NAM parent experiment, "to correct for machine drift both during a single run and between runs, a control solution was run every tenth sample. The control solution is a bulk mixture of the remaining sample from the second dilution. Using bulked samples ensured that our controls were perfectly matrix matched and contained the same elemental concentrations as our samples, so that any drift due to the sample matrix would be reflected in drift in our controls. The same control mixture was used for every ICP-MS run in the project so that run-to-run variation could be corrected. A run of 576 samples took approximately 33 hours with no user intervention. The time required for cleaning of the instrument and sample tubes as well as the digestions and transfers necessary to set up the run limit the throughput to three 576 sample runs per week."

Drift Correction and Analytical Outlier Removal. "Because our internal standard (IS) is added to our digesting acid, we are able to correct for losses due to differential sample evaporation, human error during the dilution process, and any sample introduction variability. So, if the final observed IS concentration is lower or higher than the starting IS concentration, all analyte concentrations are corrected equally for the percent difference between the observed IS concentration and the known starting IS concentration. IS correction is handled automatically by the PerkinElmer Elan 6000 software. Additionally, drift was corrected using the values of the controls run every tenth sample using a method similar to Haugen [24]. In short, drift was corrected by calculating the percentage of concentration change between two controls. This percentage change was assumed to have occurred linearly during the sequence of ten samples run between the two controls. So, for instance, the first sample run after the first control was corrected for $1/10^{th}$ of the drift seen between the two controls. Finally, because responses from the machine may be different between runs, we also corrected for drift between runs. This was performed by calculating a correction factor from the control concentrations in this run and a reference control used for all the runs. After drift correction, samples were corrected for the dilution factor and normalized to the seed weights."

"While biological outliers are of great interest to our analysis, analytical outliers (e.g. from contamination, spurious isobaric and polyatomic interferences, or poor sample uptake) introduce noise and could lead to a higher rate of incorrectly chosen potential mutants. Outlier removal was implemented using the algorithm described in Davies and Gather [14]. To remove outliers while ensuring that we weren't removing biologically significant data points we removed data points on a per element basis from seeds whose reported elemental concentration was greater than a conservatively set 15 median absolute deviations from the median concentration of that element for that line."

Computational Analysis. All calculations were performed using a combination of custom Perl and R scripts. The following R packages were used in the data analysis: reshape [25], ggplot. Standard Perl and R functions were used for drift correction, statistical analysis and creation of the figures. Scripts used in this analysis are available from http://www.ionomicshub.org, in the data exchange portal.

QTL analysis. To create the marker map, we pulled from the core set of 1100 NAM markers all the markers that were segregating in the B73 x IL14H population and removed highly correlated (r^2>0.91) markers. We then computed a map using the Emap function in QTL Cartographer and verified co-linearity with the full map using all 1100 markers. We performed composite interval mapping (CIM) using QTL Cartographer version 1.17f [26], with CIM [27,28] model 6, a walk speed of 2 cM, a window of 5 cM, using the forward and backward regression method. To determine threshold values, the permutation method was used [29] with 1000 permutations per element per population.

Acknowledgments

We gratefully acknowledge Sherry Flint-Garcia, Seth Murray, and Michael Muszynski for providing seed and Jennifer Hard for help with manuscript preparation.

Author Contributions

Conceived and designed the experiments: IB GZ BL MVM RF JD PA DES OAH. Performed the experiments: GZ BL RF JD. Analyzed the data: IB GZ. Contributed reagents/materials/analysis tools: GZ MVM RF PA. Wrote the paper: IB GZ DES OAH.

References

1. Wood S, Sebastian K, Scherr S (2000) Soil Resource Condition. Washington, DC: International Food Policy Research Institute and the World Resources Institute. 45–54 p.
2. Bray E, Bailey-Serres J, Weretilnyk E (2000) Responses to Abiotic Stresses. In: BB W Gruissem RJ, editor. Responses to Abiotic Stresses. Rockville, MD: American Society of Plant Physiologists. pp. 1160.
3. Vance CP (2001) Symbiotic nitrogen fixation and phosphorus acquisition. Plant nutrition in a world of declining renewable resources. Plant Physiol 127: 390–397.
4. Vance CP, Uhde-Stone C, Allan DL (2003) Phosphorus acquisition and use: critical adaptations by plants for securing a nonrenewable resource. New Phytologist 157: 423–447.
5. Graham RD (2008) Micronutrient deficiences in crops and their global significance. In: Alloway BJ, editor. Micronutrient deficiencies in global crop production: Soringer Science + Business Media, B.V. pp. 2–21.
6. Kochian LV, Hoekenga OA, Pineros MA (2004) How Do Crop Plants Tolerate Acid Soils? Mechanisms of Aluminum Tolerance and Phosphorous Efficiency. Annu Rev Plant Physiol Plant Mol Biol 55: 459–493.
7. Mori S (1999) Iron acquisition by plants. Curr Opin Plant Biol 2: 250–253.
8. FAO (2008) FAOSTAT database. Rome, Italy: Food and Agriculture Organization. pp. Agricultural production statistics database.
9. Baxter I, Dilkes BP (2012) Elemental profiles reflect plant adaptations to the environment. Science 336: 1661–1663.
10. Salt D, Baxter I, Lahner B (2008) Ionomics and the study of the plant ionome. Ann Rev Plant Bio 59: 709–733.
11. BAXTER I (2009) Ionomics: studying the social network of mineral nutrients. CURRENT OPINION IN PLANT BIOLOGY 12: 381–386.
12. Baxter I (2010) Ionomics: The functional genomics of elements. Briefings in Functional Genomics 9: 149–156.
13. Yu J, Holland JB, McMullen MD, Buckler ES (2008) Genetic design and statistical power of nested association mapping in maize. Genetics 178: 539–551.
14. Davies L, Gather U (1993) The identification of multiple outliers. Journal of the American Statistical Association 88: 782–792.
15. Lombi E, Scheckel KG, Pallon J, Carey AM, Zhu YG, et al. (2009) Speciation and distribution of arsenic and localization of nutrients in rice grains. The New phytologist 184: 193–201.
16. Lombi E, Smith E, Hansen TH, Paterson D, de Jonge MD, et al. (2011) Megapixel imaging of (micro)nutrients in mature barley grains. Journal of experimental botany 62: 273–282.
17. Pongrac P, Vogel-Mikus K, Jeromel L, Vavpetic P, Pelicon P, et al. (2013) Spatially resolved distributions of the mineral elements in the grain of tartary buckwheat (Fagopyrum tataricum). Food Research International 54: 125–131.
18. Pongrac P, Vogel-Mikus K, Regvar M, Vavpetic P, Pelicon P, et al. (2011) Improved lateral discrimination in screening the elemental composition of buckwheat grain by micro-PIXE. J Agric Food Chem 59: 1275–1280.
19. Goldman IL, Tracy WF (1994) Kernel protein concentration in *sugary-1* and *shrunken-2* sweet corn. HortScience 29: 209–210.
20. Tracy WF (1997) History, genetics and breeding of supersweet (*shrunken2*) sweet corn. In: Janick J, editor. Plant Breeding Reviews. John Wiley & Sons. pp. 189–236.
21. Atwell S, Huang Y, Vilhjálmsson B, Willems G, Horton M, et al. (2010) TitleGenome-wide association study of 107 phenotypes in a common set of Arabidopsis thaliana inbred lines. Nature 465: 627–631.
22. Buckler ES, Holland JB, Bradbury PJ, Acharya CB, Brown PJ, et al. (2009) The genetic architecture of maize flowering time. Science 325: 714–718.
23. Ziegler G, Terauchi A, Becker A, Armstrong P, Hudson K, et al. (2013) Ionomic Screening of Field-Grown Soybean Identifies Mutants with Altered Seed Elemental Composition. The Plant Genome 6: -.
24. Haugen J-E, Tomic O, Kvaal K (2000) A calibration method for handling the temporal drift of solid state gas-sensors. Analytica Chimica Acta 407: 23–39.
25. Wickham H (2007) Reshaping Data with the reshape Package. Journal of Statistical Software 21: 1–20.
26. Basten C WB, Zeng ZB (2000) Cartographer version 1.17f. http://statgen.ncsu.edu/qtlcart/.
27. Jansen RC (1994) High resolution of quantitative traits into multiple loci via interval mapping. Genetics Genetics 136: 1447–1455.
28. Zeng ZB (1994) Precision mapping of quantitative trait loci. Genetics 136: 1457–1468.
29. Churchhill GA, Doerge RW (1994) Empirical threshold values for quantitative trait mapping. Genetics 138: 963–971.

Land-Use History and Contemporary Management Inform an Ecological Reference Model for Longleaf Pine Woodland Understory Plant Communities

Lars A. Brudvig[1]*, John L. Orrock[2], Ellen I. Damschen[2], Cathy D. Collins[3], Philip G. Hahn[2], W. Brett Mattingly[2¤], Joseph W. Veldman[2], Joan L. Walker[4]

1 Department of Plant Biology, Michigan State University, East Lansing, Michigan, United States of America, **2** Department of Zoology, University of Wisconsin, Madison, Wisconsin, United States of America, **3** Department of Biology, Colby College, Waterville, Maine, United States of America, **4** U.S. Forest Service Southern Research Station, Clemson, South Carolina, United States of America

Abstract

Ecological restoration is frequently guided by reference conditions describing a successfully restored ecosystem; however, the causes and magnitude of ecosystem degradation vary, making simple knowledge of reference conditions insufficient for prioritizing and guiding restoration. Ecological reference models provide further guidance by quantifying reference conditions, as well as conditions at degraded states that deviate from reference conditions. Many reference models remain qualitative, however, limiting their utility. We quantified and evaluated a reference model for southeastern U.S. longleaf pine woodland understory plant communities. We used regression trees to classify 232 longleaf pine woodland sites at three locations along the Atlantic coastal plain based on relationships between understory plant community composition, soils (which broadly structure these communities), and factors associated with understory degradation, including fire frequency, agricultural history, and tree basal area. To understand the spatial generality of this model, we classified all sites together and for each of three study locations separately. Both the regional and location-specific models produced quantifiable degradation gradients–i.e., progressive deviation from conditions at 38 reference sites, based on understory species composition, diversity and total cover, litter depth, and other attributes. Regionally, fire suppression was the most important degrading factor, followed by agricultural history, but at individual locations, agricultural history or tree basal area was most important. At one location, the influence of a degrading factor depended on soil attributes. We suggest that our regional model can help prioritize longleaf pine woodland restoration across our study region; however, due to substantial landscape-to-landscape variation, local management decisions should take into account additional factors (e.g., soil attributes). Our study demonstrates the utility of quantifying degraded states and provides a series of hypotheses for future experimental restoration work. More broadly, our work provides a framework for developing and evaluating reference models that incorporate multiple, interactive anthropogenic drivers of ecosystem degradation.

Editor: Ben Bond-Lamberty, DOE Pacific Northwest National Laboratory, United States of America

Funding: This study was funded by the Strategic Environmental Research and Development Program (Project RC-1695) and by funds provided to the Department of Agriculture, Forest Service, Savannah River, under Interagency Agreement DE-AI09-00SR22188 with the Department of Energy, Aiken, SC. The funders had no role in study design, data collection and analysis, decision to publish, or preparation of the manuscript.

Competing Interests: The authors have declared that no competing interests exist.

* E-mail: brudvig@msu.edu

¤ Current address: Department of Biology, Eastern Connecticut State University, Willimantic, Connecticut, United States of America

Introduction

Ecological restoration efforts guided by a target range of reference conditions [1] often fail to achieve these targets [2]. In part, this may be because knowledge of reference conditions, while useful, is by itself insufficient for guiding or prioritizing restoration due to variation in degraded states (unrestored conditions that deviate from reference conditions). Human-modified landscapes support a range of degraded states, resulting from land-use legacies and variation in contemporary management [3,4]. Thus, an important early step in the restoration of human-modified landscapes is the formalization of ecological reference models, which describe both reference conditions and the spectrum of degraded states that are common for a given ecosystem [3,5]. Reference models have been formulated for many ecosystems [6–8]; however, these models are frequently qualitative and models of ecosystem degradation have rarely been quantitatively developed or evaluated [4].

Data-driven ecological reference models that incorporate both the consequences (e.g., altered species compositions) and causes of degradation (e.g., altered disturbance regimes) can promote a better understanding of degraded landscapes, help to prioritize restoration and management activities, and contribute toward the goal of tailoring restoration strategies to specific degraded states [3–5]. For example, in fire-maintained ecosystems, qualitative reference models simply predict increasing degradation with fire suppression [6–8], but do not detail the nature of this relationship, such as the rate at which degradation increases with fire suppression or whether thresholds exist where degradation increases abruptly with fire suppression. In contrast, a quantitative reference model could describe degradation based on thresholds in fire frequency, allowing for sites to be classified along this axis, and restoration planning and approaches could be tailored accordingly [4].

We suggest that a quantitative ecological reference model should have three features to be both ecologically relevant and useful to land managers. First, it would classify sites based on relevant ecological communities, which in many cases will be plants – the basis for many management and restoration decisions [9,10]. Second, suspected drivers of ecosystem degradation would be incorporated into the classification by linking site conditions to factors associated with degradation [4,5]. Together, these two steps would describe the range of degraded states and quantify how, in terms of degrading factors, they differ from each other and from reference conditions. Third, to make such a model applicable to restoration management, it would use data routinely available to land managers across sites spanning landscapes, the scale at which restoration planning and many management efforts typically operate [4,5,11]. We define landscape as "an area that is spatially heterogeneous in at least one factor of interest" [12] – in our case, degradation of sites across a location for which restoration or management might be coordinated. In this study, we incorporate these three features to develop a quantitative ecological reference model for longleaf pine (*Pinus palustris*) woodlands at three locations (i.e., landscapes) in the southeastern United States.

Fire-maintained longleaf pine woodlands support species-diverse understory plant communities that have been widely degraded by human land uses and are an active target for restoration [8]. The starting point for our work is a previously published qualitative reference model [8], which describes a degradation gradient in longleaf pine understory plant diversity and composition caused by past agricultural land use, altered fire regimes, and silvicultural activities (see also [13]). Agricultural legacies can persist for decades following abandonment, leading to reduced understory diversity and modified community composition on post-agricultural sites [14–16]. Fire suppression leads to increased tree abundance, canopy cover, and an accumulation of leaf litter and duff (i.e., forest floor), each of which may reduce understory diversity and modify understory community composition [17–20]. Historically, lightning and human-ignited surface fires burned longleaf pine woodlands as frequently as every 1–6 years, but today fire suppression is widespread and, where fires do occur, they are implemented through prescribed burning [21]. Overstory trees reduce understory plant diversity through competition with understory plants for light and water [18,22]; tree density is also altered by silvicultural management, including tree planting and harvesting [8,23]. Based on these consequences of human land use, we predict longleaf pine understory communities to be most degraded at sites with a history of agriculture, contemporary fire suppression, and a dense overstory (Figure 1), which are also determinants of degradation in many other ecosystems [24,25]. Guided by this past work, we focus on fire history, agricultural legacies, and overstory density as likely degrading factors for longleaf pine understory communities; however, we know of no efforts to quantify a reference model based on these factors.

The goals of our study were to quantify and then evaluate a reference model for longleaf pine woodlands, based on a previously described qualitative model [8] and associated literature (references above). To achieve these goals, we pursued three specific objectives: 1) classify longleaf pine understory plant communities based on a set of previously identified degrading factors (agricultural history, fire frequency, overstory tree basal area; references above), 2) evaluate the spatial generality of this model by evaluating the roles of degrading factors across sites that vary in soil conditions and by comparing models for three different landscape-scale locations and a region-scale model (three locations combined), and 3) evaluate the resulting regional and location-specific reference models by comparing degraded states to a set of reference sites.

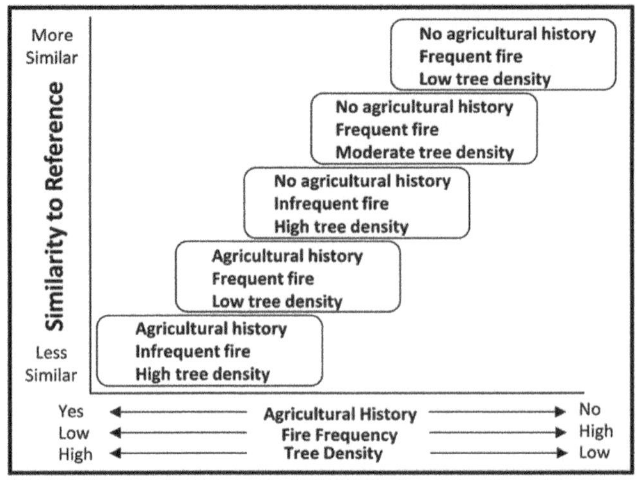

Figure 1. Conceptual model of degradation for understory plant communities in longleaf pine woodlands. Understory degradation (deviation in community composition from reference site conditions) is predicted to increase with occurrence of agricultural history, increasing overstory density, and declining fire frequency. Note: not all combinations of model components are presented in this figure and, while depicted as a linear process in this conceptual diagram, nonlinearities may exist during restoration from degraded states. Model is based on [8].

We pursued our first objective through regression tree analysis and data from 232 longleaf pine woodland sites, which were selected to span a range of biophysical conditions across our three study locations in the southeastern U.S. (Figure 2). This analysis groups sites with similar plant communities based on data-defined levels and combinations of agricultural history, fire frequency, and overstory tree basal area. We also included soil attributes in this classification because of their importance for determining plant community composition in this system [26] and to assist with our second objective. We recognize that additional degrading factors might be identified (e.g., invasive species); however, we selected this set for consideration based on clear linkages with understory degradation (references above), applicability to our study landscapes (e.g., invasive species were in low abundance at all of our study sites), and the likelihood that these data would be easily obtained by land managers, facilitating application of the model to land management and restoration planning. We pursued our second objective in two ways. First, the inclusion of soil variables in our models provides insight into whether degrading factors (e.g., fire frequency) operate generally to determine degraded states, or differently for some types of longleaf pine woodlands compared to others (e.g., those that occur on Entisols vs. Ultisols). Second, by including three locations in our study, we can evaluate the generality of a regional model (spanning all sites) relative to models for three separate landscapes (Table 1). This is important for understanding the degree to which we can generalize our results across a region or to which landscape-to-landscape variation might preclude broad application of the regional model. Finally, to address our third objective, we compared the degraded states resulting from the regional and location-specific models to a set of 38 reference sites based on a suite of biophysical characteristics that are relevant to restoration and land management. In doing so, we evaluated whether each model produces a quantifiable gradient in degradation. Given the geographic distribution of our study locations, our results may be most applicable to Atlantic Coast longleaf pine woodlands, but the degrading factors we study – agricultural legacies, altered fire regimes, and altered overstory tree abundance – are broadly relevant across the longleaf pine ecosystem [8,13].

Methods

Ethics Statement

Prior to conducting field data collection, we obtained approvals from the Fish and Wildlife Branch and Forestry Branch at Fort Stewart, the USDA Forest Service-Savannah River at Savannah River Site (a National Environmental Research Park), and the Endangered Species Branch, Forestry Branch, and the Cultural Resources Program at Fort Bragg. All data collection occurred on publicly owned land. We observed and surveyed, but made no collections of protected species.

Study System

This study was carried out in fire-dependent longleaf pine woodlands on sandy soils of the southeastern United States (Figure 2). Frequently burned longleaf pine woodlands are characterized by a relatively sparse, longleaf pine-dominated overstory and a species-diverse understory of graminoids, forbs, and shrubs [21,26]. Longleaf pine ecosystems, of which woodlands are a component, span from Texas to Virginia (Figure 2), but today less than 3% of historic area remains due to agricultural conversion, urbanization, and fire suppression [21]. As such, longleaf pine ecosystems are of high priority for conservation and restoration [8,21].

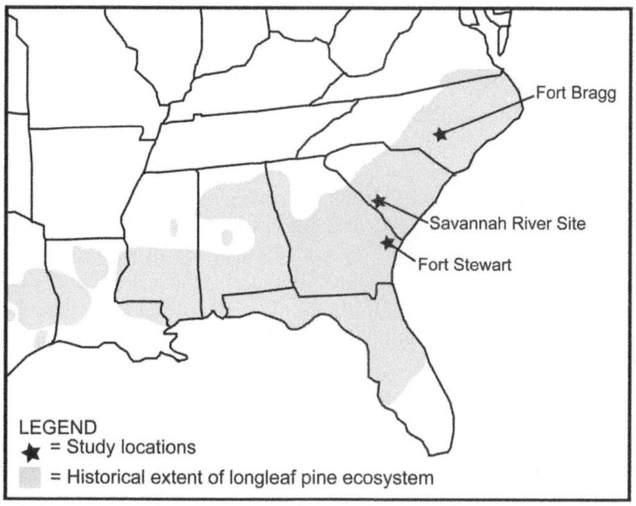

Figure 2. Geographic range of the longleaf pine ecosystem and map of study locations. Study locations (Fort Bragg [NC], 31°11′ N, 79°15′ W; Fort Stewart [GA], 31°56′ N, 81°36′ W; and Savannah River Site [SC], 33°20′ N, 81°40′ W) were in three different physiographic regions (sandhills, southern coastal plain, and Atlantic coastal plain, respectively [26]) allowing for creation and evaluation of an ecological reference model across a range of ecological settings. This model was based on data from 232 sites, which varied in their levels of degradation, with results subsequently compared to data from 38 reference sites.

We sampled longleaf pine woodlands at three locations within the historical range of the longleaf pine ecosystem: Fort Bragg (North Carolina), Fort Stewart (Georgia), and Savannah River Site (South Carolina) (Figure 2). Like much of the longleaf pine region [21], the landscapes at these study locations were historically fragmented by agriculture. Following federal government acquisition (Fort Bragg: 1919, Fort Stewart: 1940, Savannah River Site: 1951) agriculture was abandoned, resulting in contemporary landscapes supporting mosaics of longleaf pine woodlands with and without agricultural land-use histories. Study locations are managed with prescribed fire, but substantial variation in recent fire history exists across the sites, resulting in a range of fire histories from frequently burned to fire suppressed (Table 1).

Site Selection

We selected a set of study sites that characterized the range in variation of degraded conditions (fire history, overstory density, agricultural history) at each location. Sites were each $\geq$1 ha, supported overstory longleaf pines, and lacked firebreaks, drainages, or other features causing abrupt transitions in understory vegetation. Sites varied in overstory tree density, recent fire history (1991–2009), and agricultural land-use history (Table 1); stand age from the 165 of 232 non-reference sites with available data was 62±1.2 years (mean±1SE). We classified each site as having a "forest" or "agricultural" land-use history based on its status in historical aerial photographs (Fort Stewart, Savannah River Site) or maps (Fort Bragg) from the year of federal acquisition. We obtained GIS data from prescribed fire managers at each location to reconstruct fire history (prescribed and wild) between 1991 and 2009 for each site. Sites were considered burned in a given year if they occurred within the boundaries of a fire management unit that was burned in that year. We determined overstory basal area for each site during vegetation sampling.

We quantified reference conditions by sampling a set of reference sites at each location (Fort Bragg n = 15, Fort Stewart n = 14, Savannah River Site n = 9). These sites had been previously identified by regional botanical experts at the Carolina Vegetation Survey (CVS) "to document the composition and status of the natural vegetation of the Carolinas" (http://cvs.bio.unc.edu/). Reference sites had no known history of cultivation, were generally well maintained by prescribed fire, and were located within the boundaries of the respective study locations.

All study sites were located on soils that are primary substrates for longleaf pine communities [26], with three soil orders characterizing the majority of our sites (97%): 212 of our 270 sites (78.5%) were located on Ultisols, 38 sites (14.0%) were on Entisols, 13 sites (4.8%) were on Spodosols, and 7 sites (2.6%) were on Inceptisols.

Data Collection

Between 20 August and 13 November 2009, we surveyed the 270 study sites: 99 at Fort Bragg; 82 at Fort Stewart; 89 at Savannah River Site. We used one randomly located and oriented 20 m×50 m plot at each site. This plot design was a modified version of the CVS protocol [27], which is broadly employed throughout and beyond our region to characterize forest, savanna, and grassland plant communities. We identified and assigned a percent cover value to all understory plant species (herbaceous species and woody species <2.5 cm diameter at 1.4 m height) rooted within or overhanging each of eight 1 m×1 m subplots located within a 20 m×20 m portion of each plot. Taxonomy follows Radford and colleagues [28], except for the genus

Table 1. Attributes across study sites and for each of the three study locations.

Variable	All Sites	Fort Bragg	Fort Stewart	Savannah River Site
Number of sites	232	84	68	80
Canopy cover (%)	51.7±2.8	30.8±1.9	62.7±4.8	64.5±3.5
Total basal area (m^2/ha)	18.8±1.0	19.4±1.5	15.8±1.9	20.7±1.8
Pinus basal area (m^2/ha)	17.0±1.0	17.6±1.7	14.2±1.7	18.8±1.6
Non-Pinus basal area (m^2/ha)	1.8±0.4	1.8±0.8	1.7±0.6	1.8±0.8
Years since fire	3.4±0.8	1.0±0.2	2.3±0.7	6.8±2.1
Number of fires (1991–2009)	4.6±0.3	5.8±0.3	5.0±0.6	3.0±0.5
Soil water holding capacity (%)	39.5±0.8	41.2±1.4	40.8±1.6	36.8±1.0
Soil organic matter	1.8±0.1	2.0±0.2	2.0±0.2	1.5±0.1
Species richness/m^2	5.1±0.4	4.7±0.6	6.7±0.8	4.2±0.6
Species evenness/m^2	0.7±0.02	0.6±0.03	0.7±0.02	0.7±0.02
Vegetation cover (%)	23.2±2.3	13.4±2.0	38.3±4.7	20.5±3.0
Bare ground (%)	8.2±1.5	11.4±2.7	10.2±3.2	3.1±1.2
Down woody debris (%)	5.2±0.7	2.1±0.4	5.2±1.2	8.4±1.5
Litter depth (cm)	2.2±0.3	1.0±0.1	2.7±0.4	3.1±0.4
Duff depth (cm)	0.9±0.2	0.2±0.03	0.6±0.2	1.9±0.3

Values exclude data from reference sites and are mean ±95% confidence interval.

Dichanthelium, which follows Weakley [29], and the genera *Lyonia* and *Persea*, which follow Wunderlin and Hansen [30]. Within each subplot, we estimated the percent cover of green vegetation, bare ground, and down woody debris (logs, sticks, pine cones, and bark). We measured the depth of leaf litter and duff in the center of each subplot. To characterize overstory conditions within each 20 m×50 m plot, we recorded canopy cover with a spherical densiometer held at 1.4 m above six points spaced at 10 m intervals along the plot center line, and identified and measured the diameter of all trees ≥2.5 cm diameter at 1.4 m, within the plot.

To characterize site-level soil conditions, we collected soil cores (2.5 cm diameter by 15 cm deep) at 10 m intervals along the center line of each 20 m×50 m plot. Samples were composited by site and analyzed by Brookside Laboratories, Inc. (New Knoxville, OH) for soil organic matter content (SOM) [31], which is an indicator of soil degradation in post-agricultural forests [3,24]. We also analyzed soil samples for water holding capacity as the proportionate difference between saturated wet weight and oven-dried weight, following the method of [32], as described by [15]. Soil moisture availability correlates with longleaf pine understory diversity and productivity [33].

Statistical Methods

Following data collection, we constructed a site-by-species matrix for subsequent plant community analyses, using the mean cover for each species across the eight subplots at each site. To develop ecological reference models, we used multivariate classification and regression tree analysis to classify non-reference sites based on a combination of plant community composition and environmental data [34]. This analysis creates a dichotomous tree with splits based on environmental data that minimize compositional dissimilarity within groups of sites (i.e., 'classes'). In our case, we used environmental data that corresponded to the degrading factors in Figure 1: agricultural history, tree basal area (total basal area, *Pinus* spp. basal area, non-*Pinus* spp. basal area), and fire frequency (number of burns 1991–2009, time since last fire). We investigated *Pinus* and non-*Pinus* overstory separately because these groups respond differently to fire frequency (*Pinus* is more fire tolerant; [35,36]) and because land managers often remove hardwoods during restoration [8,20]. We also included soil order (Entisol, Inceptisol, Spodosol, Ultisol) and soil water holding capacity in the classification, which are two aspects of soils that broadly structure longleaf pine communities [26]. As we explain above, the inclusion of these soil variables in our models provides an opportunity to evaluate the generality of degrading factors across longleaf pine communities that occur on different substrates. We conducted classifications for all 232 non-reference sites together to accomplish our first objective ('All Sites' analysis) and for each of the three locations separately to address our second objective ('Fort Bragg', 'Fort Stewart', and 'Savannah River Site' analyses). To determine the number of final classes generated by each analysis, we conducted 500 cross-validations of the model and selected the most frequently occurring tree size using the 1-SE rule [37]. For ease of interpretation, we numbered the resulting classes to align with Figure 1, so that Class 1 was most degraded. Classes defined by soil attributes do not align with our conceptual framework in Figure 1 and we do not attempt to attribute degradation to soil conditions; however, we attempt to integrate soil-defined classes into this alignment based on divergence in biophysical factors from reference conditions (see below).

To accomplish our third objective, we evaluated the ecological relevance of the classes resulting from our classification analyses by comparing biophysical attributes of classes to reference sites. We calculated means per site (±95% CI) of understory richness and evenness, canopy cover, basal area (total, *Pinus*, non-*Pinus*), years since last fire, number of fires between 1991 and 2009, percent cover of vegetation, bare ground, and down woody debris, litter and duff depth, SOM, and soil water holding capacity. To visualize plant community composition among classes and to compare these classes to reference sites, we used Canonical Analysis of Principal Coordinates (CAP), with Bray-Curtis similarity as the distance measure following square-root transformation of the raw species abundance data [38]. CAP is a constrained ordination analysis that characterizes multivariate differences among groups (i.e., classes, reference sites) [38]. We note that, while many of the attributes we investigated were unrelated to our classification analyses, some attributes were components of the classifications (e.g., community composition). The purpose of these analyses was to quantitatively compare among classes and reference sites, rather than to formally test research hypotheses. To identify species that distinguished classes and reference sites from one another, we used Indicator Species Analysis (ISA) [39]. ISA results are presented in Tables S1, S2, S3 and S4. We ran separate analyses for each of the four classifications (All Sites, Fort Bragg, Fort Stewart, Savannah River Site). We used PRIMER-E version 6 for CAP analyses and calculation of species richness and Pielou's evenness [40], SAS version 9.2 for calculation of confidence intervals [41], R version 3.0 for multivariate regression trees [42], and PC-ORD version 5.31 for ISA [43].

Results

All Sites Classification

The classification of All Sites from the combined three locations resulted in five splits and six classes (Figure 3a). Sites first split according to fire frequency. Infrequently burned sites (≤4 burns since 1991) were further classified based on soil moisture: Class 1 was characterized by infrequent fire and lower soil moisture (<45.08%), whereas Class 2 was characterized by infrequent fire and higher soil moisture (≥45.08%). Frequently burned sites were further classified based on land-use history, with Class 3 characterized by frequent fire and agricultural history. Among sites with forested history, the model made two splits, with the first based on overstory density and the second based on soil moisture. Class 4 was characterized by frequent fire, forested history, high basal area (≥9.965 m^2/ha), and lower soil moisture (<42.12%), class 5 by frequent fire, forested history, high basal area, and higher soil moisture (≥42.12%), and class 6 by frequent fire, forested history, and low basal area (<9.965 m^2/ha).

Location-specific Classifications

Location-specific classifications illustrated a number of differences from the All Sites classification, in terms of the identity and ordering of degrading factors, as well as the role of soils (Figure 3). The Fort Bragg classification contained a single split and two groups defined by agricultural land-use history. The Fort Stewart classification contained three splits and four groups, which illustrated a prominent influence of soils. The first split was between Entisol/Ultisol sites and Inceptisol/Spodosol sites. Inceptisol/Spodosol sites (Class 1) formed a unique soil-vegetation association in the model, which was not related to degradation. The only degrading factor in this model, non-*Pinus* basal area, distinguished among sites on Entisol and Ultisol soils (low basal area sites were further split by soil moisture). The Savannah River Site classification contained splits based on all three degrading factors, but with differing order of importance relative to the All Sites classification, as basal area was the most important factor, followed by agricultural land-use history.

Degradation Classes in Relation to Reference Conditions

The All Sites classification revealed three main delineations in community composition among classes: sites with forested history (Classes 4–6, reference sites), sites with agricultural history or sites with infrequent fire and high soil moisture (Classes 2, 3), and sites with infrequent fire and low soil moisture (Class 1) (Figure 4a). Species richness declined with degradation, whereas species evenness was generally greater at infrequently burned sites (Figure 4b,c). Among ground cover attributes, vegetation cover was greatest at sites with high soil moisture and at reference sites (Figure 4e), bare ground was greatest at two of the frequently burned classes with forested history, and little difference in down woody debris was apparent (Figure 4e). Forest floor accumulation was related to fire suppression, with greater depths of litter and duff in Classes 1 and 2 (Figure 4f). Soil organic matter was markedly higher on sites with high soil moisture and at reference sites, but not strongly associated with degradation (Figure 4d; Table 2).

Models resulting from the location-specific classifications produced classes that were compositionally distinct, with the exception of Classes 1 and 3 at Fort Stewart (Figures S1, S2, and S3). Reference sites ranged from compositionally unique at Fort Bragg (Figure S1) to comparable to less degraded classes on at least one CAP axis at Fort Stewart and Savannah River Site (Figures S2,S3). Species richness declined with degradation in each location-specific model, whereas many other attributes illustrated location-specific patterns (Figures S1, S2, and S3, Table S5). Metrics used to classify sites (e.g., years since last fire, basal area; Table S5) largely mirrored each classification; however, the ways that these variables changed with degradation varied with the different classification rules at each location. Several variables were associated with degradation at one or more sites, including soil organic matter, vegetation cover, and forest floor depth at Savannah River Site, vegetation cover at Fort Bragg, and forest floor depth at Fort Stewart. Other metrics, such as species evenness and soil variables, illustrated location-specific patterns.

Discussion

An Ecological Reference Model for Longleaf Pine Woodlands

Ecological reference models are commonly employed during restoration, but are frequently qualitative, which limits their utility for prioritizing sites for restoration and guiding restoration efforts on the ground [4]. Using data from 232 sites, we quantified an ecological reference model for longleaf pine woodland understory plant communities spanning a broad geographic region of the Southeastern United States (Figure 3a). Three degrading factors – agricultural legacies, recent fire history, and overstory tree abundance – as well as soil moisture holding capacity, delineated classes of plant communities in this regional model. Location-specific models, however, illustrated substantial variation relative to this regional model and each other (Figure 3). We suggest that our regional model can help prioritize longleaf pine woodlands for restoration across our study region; however, due to substantial landscape-to-landscape variation, location-specific reference models can help guide local management decisions.

Our regional model provides general support for the core components of a qualitative model of longleaf pine degradation (Figure 1); however, by quantifying this conceptual model, we illustrate the relative contributions of degrading factors, as well as threshold values that determine differences between degraded classes. In this regional model, fire frequency was the most important factor (i.e., the first split in the classification), followed by agricultural legacies (Figure 3a). Overstory basal area, while a significant factor in the regional model, was the least important. Based on similarity in community composition to reference sites, Classes 1–3 (infrequently burned and post-agricultural sites) were the most degraded, whereas historically forested sites were less degraded, being both compositionally similar to each other and to reference sites (Figure 4a). Our regional model further quantifies 4

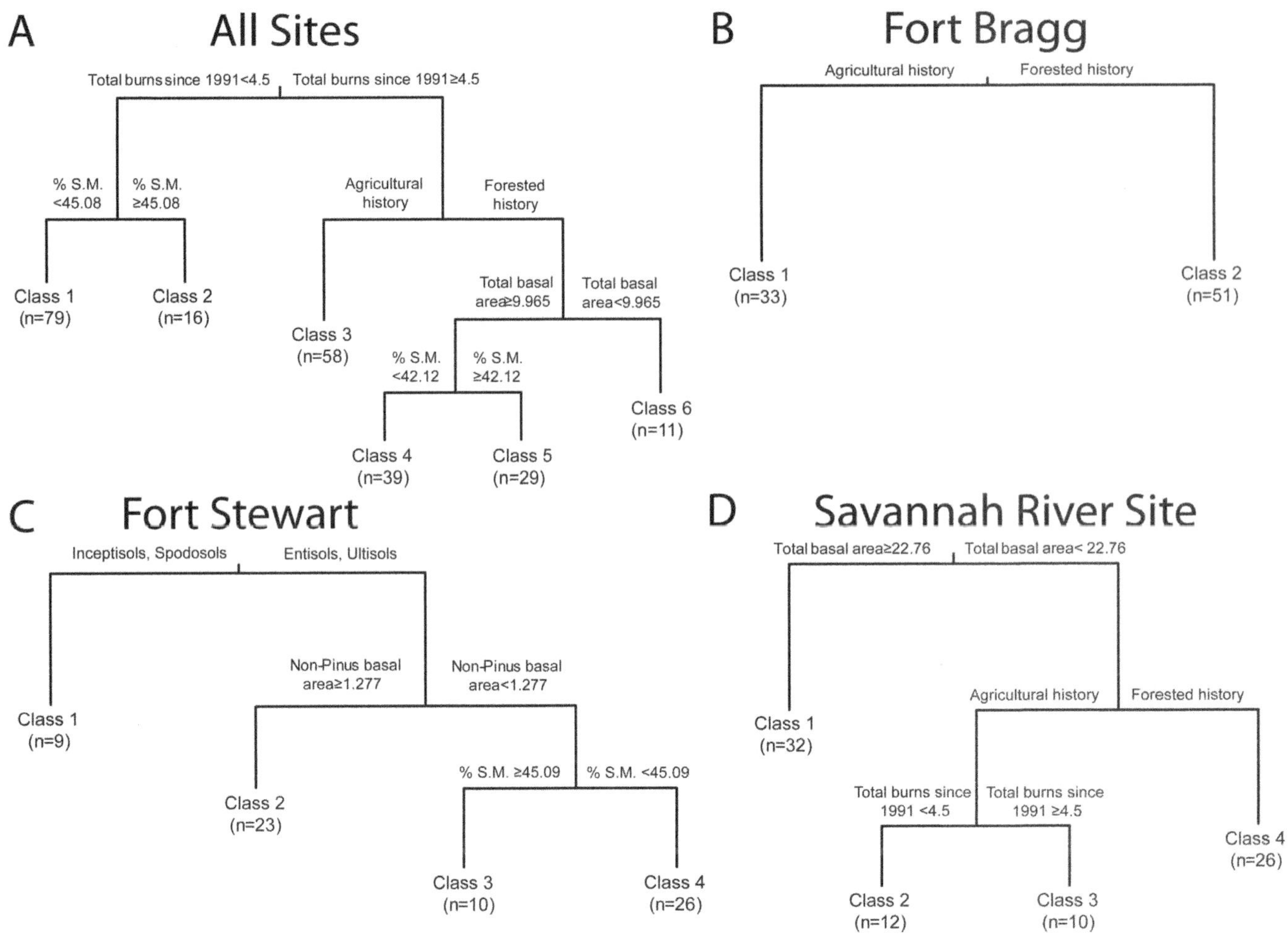

Figure 3. Results of multivariate classification and regression tree analysis. A) All study sites, B-D) separate study locations. In each analysis, sites are classified based on plant community composition and environmental data and classes are ordered to align with the conceptual model in Figure 1 (Class 1 = most degraded), with the exception of Class 1 at Fort Stewart, which represents a soil-vegetation association not related to degradation. Branch length at each split is scaled to the variance explained by the corresponding environmental factor. The number of study sites in each class is presented below each class label.

vs. 5 fires since 1991 and $\sim$10 m^2/ha as breakpoints between degraded classes related to burning and tree basal area, respectively. These results reinforce the importance of frequent prescribed fires for management of longleaf pine woodlands [8,13], because of its positive influence on understory plant communities [15,17]. Our results further highlight the role of agricultural legacies in the degradation of longleaf pine plant communities [14–16]. The pronounced deviation of fire suppressed and post-agricultural plant communities from reference conditions (Figure 4) suggests that sites supporting these degraded conditions may require the greatest efforts to restore.

The resulting classes from our regional and location-specific analyses captured relevant gradients of degradation (Figures 4, S1, S2, and S3). As degradation increased, levels of biotic and abiotic variables became more dissimilar from values at reference sites. This pattern was consistent across the regional and location-specific models for several variables including understory richness, cover, and composition, forest floor accumulation, and tree basal area, whereas other variables including soil organic matter showed a degradation gradient in one or more models. Importantly, a number of these variables were not used in the construction of our ecological reference models (e.g., species diversity metrics, ground cover components, forest floor depth, and soil attributes), yet in many cases these variables corresponded with the degradation gradient.

The explicit consideration of land-use legacies has received increasing attention in conservation and restoration [3,10,24,44,45]. Our results illustrate how multiple drivers of ecosystem degradation, including land-use legacies, can be quantitatively incorporated into ecological reference models. This may be of particular importance when legacies interact with contemporary management to influence population viability and community composition. For example, past agricultural activities can mediate the effects of present-day fire management on plant community diversity, such that frequent burning may lead to increased richness, but only in sites with an agricultural land-use history [15]. Similarly, past land use can modify levels and patterns of soil nutrients, with ensuing effects on plant populations [46]. Future research will be necessary before specific restoration strategies might be tailored to the various degraded states in our regional and location-specific models (see below) and our work shows that consideration of both contemporary and historical causes of degradation will be important during this process.

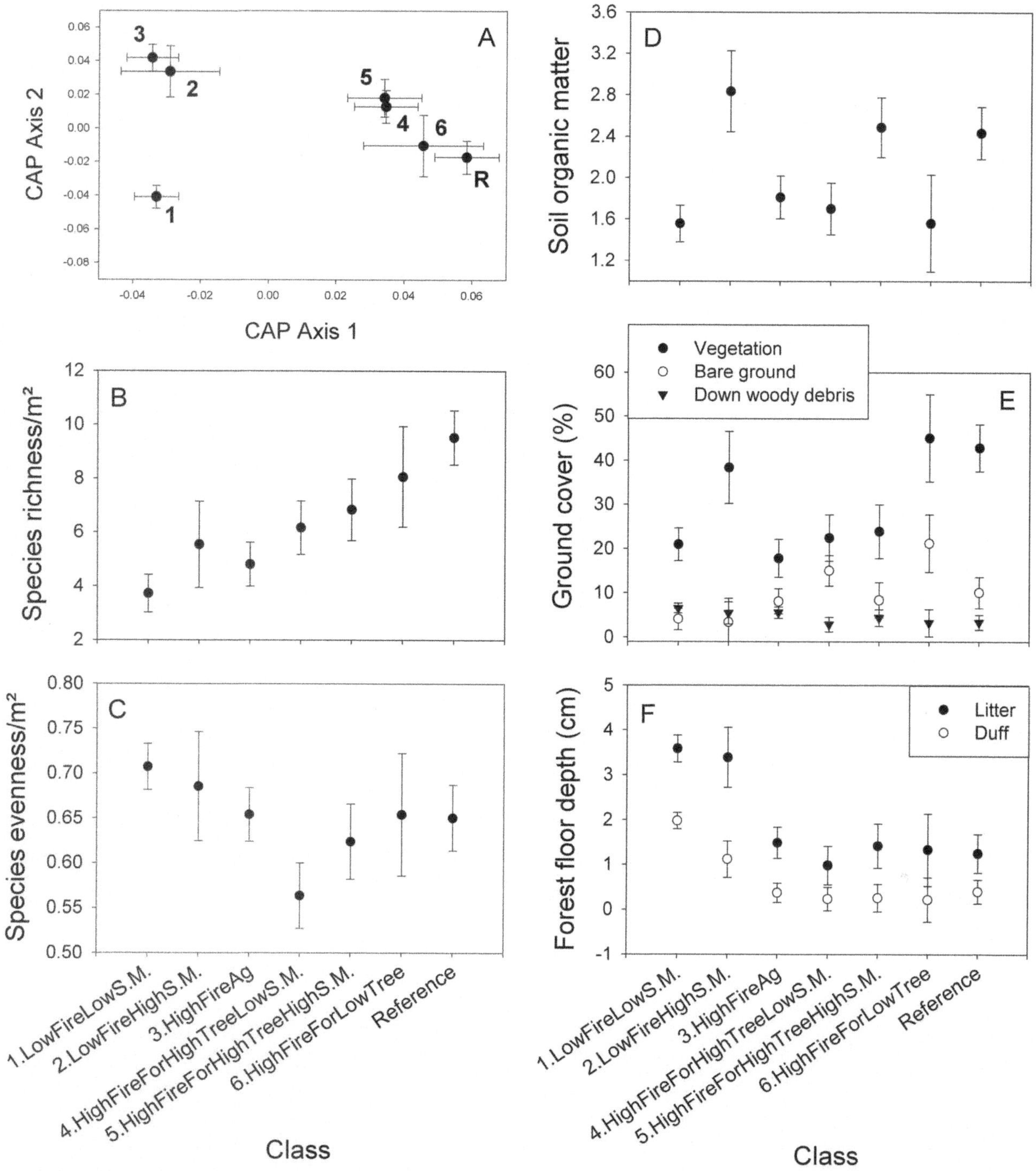

Figure 4. Comparison of Classes (1–6) from the All Sites classification and regression tree analyses to reference conditions. A) Understory community composition, B) understory species richness, C) understory species evenness, D) soil organic matter content, E) ground cover variables, and F) forest floor depth. All values are means ±95% confidence intervals.

Generality of the Ecological Reference Model

Our location-specific models illustrated notable variation, both relative to each other and to the regional All Sites model (Figure 3). Only one location-specific model (Savannah River Site) contained all three degrading factors, but with the reverse order of importance compared to the regional model. Conversely, Fort Bragg supported a simple two class model based only on agricultural land-use history, whereas Fort Stewart's model illustrated a prominent role of soils, in addition to hardwood abundance, for structuring understory plant communities (Figure 3). This variation may reflect differences in underlying environmental factors or differences in land-use and management histories among our three study locations (Table 1). For example, the presence or absence of fire frequency as a variable in location-specific models may be explained in part by variation in prescribed fire management among locations (Table 1). At Fort Bragg, where

Table 2. Attributes of degradation classes (1–6) for the All Sites classification and reference sites.

Class	Canopy cover (%)	Total basal area (m^2/ha)	Pinus basal area (m^2/ha)	Non-Pinus basal area (m^2/ha)	Years since fire	# Fires 1991–2009	Soil water holding capacity (%)
1. Low fire/Low SM[1]	62.4±4.2	20.3±1.6	18.2±1.6	2.0±0.7	7.6±1.2	2.3±0.4	37.0±1.3
2. Low fire/High SM	63.0±9.4	18.3±3.5	16.2±3.5	2.2±1.7	3.1±2.6	2.7±0.8	48.9±3.0
3. Ag/High fire	51.8±4.9	20.5±1.8	18.5±1.9	2.0±0.9	1.0±1.3	5.9±0.4	38.9±1.55
4. Forest/High Fire/High BA[2]/Low SM	34.8±6.0	16.2±2.2	14.8±2.6	1.4±1.1	0.8±1.6	6.4±0.5	37.2±1.9
5. Forest/High Fire/High BA/High SM	45.7±7.0	19.6±2.6	18.2±2.6	1.3±1.2	1.0±1.9	6.4±0.6	46.3±2.2
6. Forest/High fire/Low BA	35.1±11.3	7.3±4.2	7.0±4.3	0.4±2.0	1.5±3.1	5.7±1.0	37.9±3.6
Reference	44.4±6.1	15.3±2.4	14.0±2.3	1.3±1.1	1.0±1.7	6.0±0.5	46.6±2.3

Values are mean ±95% confidence interval.
[1]Soil moisture holding capacity.
[2]Basal area.

fire frequency was not selected as a model component, prescribed fire is highly regimented, resulting in frequent fires and little variation in time since fire among sites (Figure 3). Conversely, substantial variation in fire frequency exists among sites at the Savannah River Site (Table 1), where fire frequency was selected as a model component (Figure 3). Furthermore, our study spanned three physiographic regions of the southeastern U.S. coastal plain [26], capturing variation in important ecological factors such as dominant species (Tables S2, S3, and S4) and soil conditions (Table 1) [26,33]. Longleaf pine communities are broadly structured by soils [26] and this influence of soils was most prominent at Fort Stewart – a location that supports longleaf pine woodlands underlain by a variety of soil orders, including Spodosols and Inceptisols, which formed a unique class in the Fort Stewart model. Fort Stewart also illustrated the lone example of how the influence of a degrading factor depended on soil conditions, as non-*Pinus* basal area was important to plant communities on Entisol and Ultisol soils, but not Inceptisol and Spodosol soils. Importantly, at least one hypothesized degrading factors (Figure 1) was a significant model component at each location, illustrating that these factors structured understory degradation (i.e., departure from reference conditions) at both regional and landscape scales in this study. The differences in reference model details between our three study locations (Figure 3) suggests that, while our regional reference model may provide broad-scale insight into patterns of longleaf pine degradation, locations with available resources should consider collecting data to parameterize location-specific reference models.

Implications for Conservation and Restoration

We suggest that our regional ecological reference model provides a way for managers to broadly infer the degradation status of longleaf pine understory communities in our study region. This model, as well as the location-specific models, employs relatively easily measured data, which serve as proxies for ecological characteristics of interest to land managers (Figure 4, S1, S2, and S3). Agricultural and fire history and, perhaps, overstory basal area, may be available to land managers as GIS data. Thus, our regional and location-specific reference models may be mapped at large spatial scales to assist management and conservation decisions. Longleaf pine understory communities are notable for high levels of species diversity from local (e.g., 1 m×1 m) to regional scales [13,26]. Our findings provide guidance over much of this range in scales, spanning sites, landscapes, and portions of a region (the Atlantic coastal plain). Future work might explore additional reference models to inform small-scale (e.g., within site) restoration decisions.

Our approach to classifying communities based on degraded conditions should be broadly applicable to other ecosystems around the world modified by human land use and altered fire regimes, including fragmented woodlands in Australia [45], fire-suppressed savannas in Brazil [47], and forests in the western United States [6,48], among others. During such application, our framework for developing reference models should be modified to include the relevant suspected drivers of degradation for an ecosystem of interest. For example, the presence of invasive species is an important consideration during longleaf pine woodland restoration [8] and, while invasive species were not abundant at our study sites, their inclusion as a model variable might be important at other longleaf pine sites or in other ecosystems.

The remaining challenge is to determine how to best restore longleaf pine understory communities once patterns of degradation have been assessed. Our regional and location-specific ecological reference models suggest some strategies for restoration, but these should be interpreted with caution. For example, transitioning sites between Classes 4/5 and 6 in our regional model might simply entail mechanical thinning of overstory trees, a strategy that can increase plant diversity in some contexts [49] but has limited effects in others [20,50]. Other transitions might require multiple restoration strategies. For example, restoration of fire regime at post-agricultural sites (e.g., transitioning between Classes 2 and 3 at Savannah River Site) might need to be coupled with introduction of propagules – particularly those of dispersal-limited plant groups, such as passive and ant dispersed species like *Tephrosia virginiana* and *Aristida stricta/beyrichiana* [51] (Tables S1, S2, S3, and S4). Future experimental work will be necessary to evaluate these hypotheses suggested by our models. Further, reinstating processes may lead to unexpected outcomes for some degraded states. For example, reintroducing fire to long-unburned sites may produce novel fire behavior, leading to unexpected outcomes, such as mortality of longleaf pine overstory trees [52]. Moreover, while our conceptual reference model (Figure 1) depicts a simplistic set of linear transitions among degraded states, it remains an open question as to whether a simple linear or a non-linear, such as alternative stable states [53] or state and transition (e.g. [54]), model of restoration will be most appropriate during restoration of longleaf pine woodlands. Finally, more work is needed to understand how to best tailor combinations and sequences of restoration strategies (e.g., seed addition, prescribed fire, overstory thinning) to the variety of degraded conditions illustrated by our reference models for longleaf pine woodlands.

Supporting Information

Figure S1 Comparison of Classes (1–2) from the Fort Bragg classification and regression tree analyses to reference conditions. A) understory community composition, B) understory species richness, C) understory species evenness, D) soil organic matter content, E) ground cover variables, and F) forest floor depth. All values are means ±95% confidence intervals.

Figure S2 Comparison of Classes (1–4) from the Fort Stewart classification and regression tree analyses to reference conditions. A) understory community composition, B) understory species richness, C) understory species evenness, D) soil organic matter content, E) ground cover variables, and F) forest floor depth. All values are means ±95% confidence intervals.

Figure S3 Comparison of Classes (1–4) from the Savannah River Site classification and regression tree analyses to reference conditions. A) understory community composition, B) understory species richness, C) understory species evenness, D) soil organic matter content, E) ground cover variables, and F) forest floor depth. All values are means ±95% confidence intervals.

Table S1 Species with the 10 highest indicator values (from Indicator Species Analysis) for each site class in the All Sites classification. Species identified as indicators of individual site classes ($p<0.05$) are noted by *. P-values are listed for significant indicator species; where not provided p-values were ≥ 0.05. The number of significant indicator species decreased strongly with degradation; Classes 1–6 contained 2, 18, 0, 3, 0, 40 significant indicator species, respectively. Thirteen species were indicative of reference sites.

Table S2 Species with the 10 highest indicator values (from Indicator Species Analysis) for each site class in the Fort Bragg classification. Species identified as indicators of individual site classes are noted by *. Indicator species were primarily present at reference sites (n = 59), with 0 and 4 in Classes 1 and 2, respectively.

Table S3 Species with the 10 highest indicator values (from Indicator Species Analysis) for each site class in the Fort Stewart classification. Species identified as indicators of individual site classes are noted by *. 7, 1, 9, and 10 species were significant indicators of classes 1–4, respectively, and 19 species were indicative of reference sites.

Table S4 Species with the 10 highest indicator values (from Indicator Species Analysis) for each site class in the Savannah River Site classification. Species identified as indicators of individual site classes are noted by *. There were 0, 3, 14, and 4 significant indicators of classes 1–4, respectively, and 26 species indicative of reference sites.

Table S5 Attributes of classes resulting from the location-specific classifications, compared to references sites. Values are mean ±95% confidence interval.

Acknowledgments

For logistical support at field locations, we thank David Heins and the Endangered Species Branch at Fort Bragg; the Fish and Wildlife Branch at Fort Stewart; and John Blake, Chris Hobson, and Ed Olson at SRS. For assistance with data collection, we thank Lauren Bizzari, Cory Christopher, Amanda Powell, and Rick Ranalli.

Author Contributions

Conceived and designed the experiments: EID JLO JLW LAB. Performed the experiments: CDC EID JLO LAB WBM. Analyzed the data: JLO LAB. Contributed reagents/materials/analysis tools: EID JLO LAB. Wrote the paper: CDC EID JLO JLW JWV LAB PGH WBM.

References

1. White PS, Walker JL (1997) Approximating nature's variation: Selecting and using reference information in restoration ecology. Restor Ecol 5: 338–349.
2. Rey Benayas JM, Newton AC, Diaz A, Bullock JM (2009) Enhancement of biodiversity and ecosystem services by ecological restoration: a meta-analysis. Science 325: 1121–1124.
3. Foster D, Swanson F, Aber H, Burke I, Brokaw N, et al. (2003) The importance of land-use legacies to ecology and conservation. Bioscience 53: 77–88.
4. Hobbs RJ (2007) Setting effective and realistic restoration goals: Key directions for research. Restor Ecol 15: 354–357.
5. Hobbs RJ, Harris JA (2001) Restoration ecology: repairing the Earth's ecosystems in the new millennium. Restor Ecol 9: 239–246.
6. Allen CD, Savage M, Falk DA, Suckling KF, Swetnam TW, et al. (2002) Ecological restoration of Southwestern ponderosa pine ecosystems: A broad perspective. Ecol Appl 12: 1418–1433.
7. Asbjornsen H, Brudvig LA, Mabry CM, Evans CW, Karnitz HM (2005) Defining reference information for restoring ecologically rare tallgrass oak savannas in the Midwestern United States. J For 103: 345–350.

8. Walker JL, Silletti AM (2006). Restoring the ground layer of longleaf pine ecosystems. In: Jose S, Jokela EJ, Miller DL, editors. The longleaf pine ecosystem: Ecology, silviculture, and restoration. New York: Springer. 297–325.
9. Bestelmeyer BT, Miller JR, Wiens JA (2003) Applying species diversity theory to landscape management. Ecol Appl 13: 1750–1761.
10. Brudvig LA (2011) The restoration of biodiversity: Where has research been and where does it need to go? Am J Bot 98: 549–558.
11. Palik BJ, Goebel PC, Kirkman LK, West L (2000). Using landscape hierarchies to guide restoration of disturbed ecosystems. Ecol Appl 10: 189–202.
12. Turner MG, Gardner RH, O'Neill RV (2001) Landscape ecology in theory and practice: pattern and process. New York: Springer-Verlang. 401 p.
13. Mitchell RJ, Duncan SL (2009) Range of variability in southern coastal plain forests: Its historical, contemporary, and future role in sustaining biodiversity. Ecol Soc 14(1): 17.
14. Hedman CW, Grace SL, King SE (2000) Vegetation composition and structure of southern coastal plain pine forests: an ecological comparison. For Ecol Manage 134: 233–247.
15. Brudvig LA, Damschen EI (2011) Land-use history, historical connectivity, and land management interact to determine longleaf pine woodland understory richness and composition. Ecography 34: 257–266.
16. Brudvig LA, Grman E, Habeck CW, Orrock JL, Ledvina JA (2013) Strong legacy of agricultural land use on soils and understory plant communities in longleaf pine woodlands. For Ecol Manage 310: 944–955.
17. Brockway DG, Lewis CE (1997) Long-term effects of dormant season prescribed fire on plant community diversity, structure and productivity in a longleaf pine wiregrass ecosystem. For Ecol Manage 96: 167–183.
18. Harrington TB, Edwards MB (1999) Understory vegetation, resource availability, and litterfall responses to pine thinning and woody vegetation control in longleaf pine plantations. Can J For Res 29: 1055–1064.
19. Hiers JK, O'Brien JJ, Will RE, Mitchell RJ (2007) Forest floor depth mediates understory vigor in xeric *Pinus palustris* ecosystems. Ecol Appl 17: 806–814.
20. Kirkman LK, Barnett A, Williams BW, Hiers JK, Pokswinski SM, et al. (2013) A dynamic reference model: a framework for assessing biodiversity restoration goals in a fire-dependent ecosystem. Ecol Appl 23: 1574–1587.
21. Frost C (2006) History and future of the longleaf pine ecosystem. In: Jose S, Jokela EJ, Miller DL, editors. The longleaf pine ecosystem: Ecology, silviculture, and restoration. New York: Springer. 9–42.
22. Harrington TB, Dagley CM, Edwards MB (2003) Above- and belowground competition from longleaf pine plantations limits performance of reintroduced herbaceous species. Forest Sci 49: 681–695.
23. Van Lear DH, Carroll WD, Kapeluck PR, Johnson R (2005) History and restoration of the longleaf pine-grassland ecosystem: Implications for species at risk. For Ecol Manage 211: 150–165.
24. Flinn KM, Vellend M (2005) Recover of forest plant communities in agricultural landscapes. Front Ecol Environ 3: 243–250.
25. Pyke DA, Brooks ML, D'Antonio C (2010) Fire as a restoration tool: A decision framework for predicting the control or enhancement of plants using fire. Restor Ecol 18: 274–284.
26. Peet RK (2006) Classification of longleaf pine woodlands. In: Jose S, Jokela EJ, Miller DL, editors. The longleaf pine ecosystem: Ecology, silviculture, and restoration. New York: Springer. 51–93.
27. Peet RK, Wentworth TR, White PS (1998) A flexible, multipurpose method for recording vegetation composition and structure. Castanea 63: 262–274.
28. Radford AE, Ahles HE, Bell CR (1964) Manual of the vascular flora of the Carolinas. Chapel Hill: University of North Carolina Press. 1183 p.
29. Weakley AS (2008) Flora of the Carolinas, Virginia, and Georgia, and surrounding areas. Chapel Hill: University of North Carolina Press. 924 p.
30. Wunderlin RP, Hansen BF (2003) Guide to the vascular plants of Florida, Second edition. Gainesville: University Press of Florida. 787 p.
31. Schulte EE, Hopkins BG (1996) Estimation of soil organic matter by weight Loss-On-Ignition. In: Magdoff FR, Tabatabai MA, Hanlon Jr EA, editors. Soil organic matter: Analysis and interpretation. Special publication No. 46. Madison: Soil Science Socity of America. 21–32.
32. Salter PJ, Williams JB (1967) The influence of texture on the moisture characteristics of soil. IV. A method of estimating the available-water capacities of profiles in the field. J Soil Sci 18: 174–181.
33. Kirkman LK, Mitchell RJ, Helton RC, Drew MD (2001) Productivity and species richness across and environmental gradient in a fire-dependent ecosystem. Am J Bot 88: 2119–2128.
34. De'ath G (2002) Multivariate regression trees: A new technique for modeling species-environment. Ecology 83: 1105–1117.
35. Glitzenstein JS, Platt WJ, Streng DR (1995) Effects of fire regime and habitat on tree dynamics in north Florida longleaf pine savannas. Ecol Monogr 65: 441–476.
36. Veldman JW, Mattingly WB, Brudvig LA (2013) Understory plant communities and the functional distinction between savanna trees, forest trees, and pines. Ecology 94: 424–434.
37. De'ath G, Fabricius KE (2000) Classification and regression trees: a powerful yet simple technique for ecological data analysis. Ecology 81: 3178–3192.
38. Anderson MJ, Willis TJ (2003) Canonical analysis of principal coordinates: a useful method of constrained ordination for ecology. Ecology 84: 511–525.
39. Dufrene M, Legendre P (1997) Species assemblages and indicator species: the need for a flexible asymmetrical approach. Ecol Monogr 67: 345–366.
40. Clark KR, Gorley RN (2006) PRIMER v6: User Manual/Tutorial. Plymouth: PRIMER-E.
41. SAS Institute (2008) SAS version 9.2. Cary: SAS Institute.
42. R Development Core Team (2005) R: A Language and Environment for Statistical Computing, version 3.0. Vienna: R Foundation for Statistical Computing.
43. McCune B, Mefford MJ (2006) PC-ORD: Multivariate Analysis of Ecological Data, version 5.31. Gleneden Beach: MjM Software.
44. Honnay O, Bossuyt B, Verheyen K, Butaye J, Jacquemyn H, et al. (2002) Ecological perspectives for the restoration of plant communities in European temperate forests. Biodivers Conserv 11: 213–242.
45. Lunt ID, Spooner PG (2005) Using historical ecology to understand patterns of biodiversity in fragmented agricultural landscapes. J Biogeogr 32: 1859–1873.
46. Fraterrigo JM, Turner MG, Pearson SM (2006) Previous land use alters plant allocation and growth in forest herbs. J Ecol 94: 548–557.
47. Roitman I, Felfili JM, Rezende AV (2008) Tree dynamics of a fire-protected cerrado sensu stricto surrounded by forest plantations, over a 13-year period (1991–2004) in Bahia, Brazil. Plant Ecol 197: 255–267.
48. Fulé PZ, Covington WW, Moore MM (1997) Determining reference conditions for ecosystem management of southwestern ponderosa pine forests. Ecol Appl 7: 895–908.
49. Platt WJ, Carr SM, Reilly M, Fahr J (2006) Pine savanna overstorey influences on ground-cover biodiversity. Appl Veg Sci 9: 37–50.
50. Provencher L, Herring BJ, Gordon DR, Rodgers HL, Galley KEM, et al. (2001) Effects of Hardwood Reduction Techniques on Longleaf Pine Sandhill Vegetation in Northwest Florida. Restor Ecol 9: 13–27.
51. Kirkman LK, Coffey KL, Mitchell RJ, Moser EB (2004). Ground cover recovery patterns and life-history traits: implications for restoration obstacles and opportunities in a species-rich savanna. J Ecol 92: 409–421.
52. Varner III JM, Gordon DR, Putz FE, Hiers JK (2005) Restoring fire to long-unburned *Pinus palustris* ecosystems: Novel fire effects and consequences for long-unburned ecosystems. Restor Ecol 13: 536–544.
53. Suding KN, Gross KL, Houseman GR (2004) Alternative stable states and positive feedbacks in restoration ecology. Trends Ecol Evol 19: 46–53.
54. Yates CJ, Hobbs RJ (1997) Woodland restoration in the Western Australian wheatbelt: A conceptual framework using a state and transition model. Restor Ecol 5: 28–35.

11

Cross-Species Extrapolation of Prediction Models for Cadmium Transfer from Soil to Corn Grain

Hua Yang[1,2], Zhaojun Li[1]*, Lu Lu[3], Jian Long[2], Yongchao Liang[1]

1 Institute of Agricultural Resources and Regional Planning, Chinese Academy of Agricultural Sciences, Key Laboratory of Plant Nutrition and Fertilizer, Ministry of Agriculture, Beijing, China, **2** Guizhou Key Laboratory of Mountain Environment, Guizhou Normal University, Guiyang, China, **3** Graduate School of the Chinese Academy of Agricultural Sciences, Beijing, China

Abstract

Cadmium (Cd) is a highly toxic heavy metal for both plants and animals. The presence of Cd in agricultural soils is of great concern regarding its transfer in the soil-plant system. This study investigated the transfer of Cd (exogenous salts) from a wide range of Chinese soils to corn grain (Zhengdan 958). Through multiple stepwise regressions, prediction models were developed, with the combination of Cd bioconcentration factor (BCF) of Zhengdan 958 and soil pH, organic matter (OM) content, and cation exchange capacity (CEC). Moreover, these prediction models from Zhengdan 958 were applied to other non-model corn species through cross-species extrapolation approach. The results showed that the pH of the soil was the most important factor that controlled Cd uptake and lower pH was more favorable for Cd bioaccumulation in corn grain. There was no significant difference among three prediction models in the different Cd levels. When the prediction models were applied to other non-model corn species, the ratio ranges between the predicted BCF values and the measured BCF values were within an interval of 2 folds and close to the solid line of 1:1 relationship. Furthermore, these prediction models also reduced the measured BCF intra-species variability for all non-model corn species. Therefore, the prediction models established in this study can be applied to other non-model corn species and be useful for predicting the Cd bioconcentration in corn grain and assessing the ecological risk of Cd in different soils.

Editor: Xianglin Shi, University of Kentucky, United States of America

Funding: This work was jointly supported by the Special Fund of Public Industry in China (Agriculture, project No. 200903015; website: http://www.most.gov.cn/) and the Natural Science Foundation of Institute of Agricultural Resources and Regional Planning, Chinese Academy of Agricultural Sciences (Project No.2013-9; website:http://www.iarrp.cn). The funders had no role in study design, data collection and analysis, decision to publish, or preparation of the manuscript.

Competing Interests: The authors have declared that no competing interests exist.

* E-mail: lizhaojun@caas.cn

Introduction

Trace metal elements accumulation in crop of agricultural soils is of extensive concern due to its potential risks to human health and detrimental effects on soil ecosystems [1–4]. Cadmium (Cd) is one of the most dangerous trace elements because excessive dietary intake of Cd and its accumulation in human organs over a lifetime can lead to kidney malfunction [5–8].

Numerous studies have demonstrated that a number of factors affect Cd bioavailability in soils, including soil pH, organic matter (OM), cation exchange capacity (CEC), cultivars of crop plants, plant age, and so on [9–13]. Among these soil properties, the pH plays the most important role in determining metal speciation, due to its strong effects on solubility and speciation of metals in the soil and solution [14–15]. Dramatic increases in Cd desorption from soil constituents and its dissolution ability in solution have been observed with the decreasing soil pH [16–17]. The bioavailability of Cd in soil also increases when soil pH decreases [18–23]. Apart from soil pH, OM in soil is another important factor affecting heavy metal availability. It was reported that heavy metal adsorption onto soil constituents declined when organic matter content in soils decreased [24–25], and that the ability of plants to Cd uptake decreased with an increase in soil organic matter [26–27].

The influence of soil factors derived from the soil types on plant uptake of heavy metals and the difficulties in assessing the ecological risk of heavy metals in complex soil types have urged the research on prediction models that can predict trace element transfer to plant [28–30]. These models provide great opportunities to carry out ecological risk assessments and establish soil quality criteria for heavy metals [31–32]. However, the influences of soil factors may result in variability of ecological risk assessments in different soils and of the soil quality criteria for heavy metals [33]. Therefore, when the soil environmental quality standard is established based on the species sensitivity distributions (SSD), which are routinely used in ecological risk assessment procedures [34–35], the toxicology data from the soils with different properties need to be normalized using toxicity prediction models to eliminate the influence of soil factors [36]. The normalization of toxicology data can also improve the accuracy of the sensitivity distribution of species and environmental quality standard values [37]. Ideally, separate species-specific models should be developed for each single species. However, this is unrealistic due to the expenses and efforts involved. Thus, for practical reason, the cross-species extrapolation of biological toxicity model was applied in some cases of pollutants risk assessment. For example, Deleebeeck et al. [38–39] have applied biotic bigand model (BLM) from O. *mykiss* to other fish species and from D. *magna* to other cladocera species. Schlekat et al. [33] applied chronic nickel BLMs

Table 1. Selected soil properties of the 17 soils used in the cadmium phytotoxicity test.

Soil NO.	Location*	pH	OM ($g.kg^{-1}$)	CEC ($cmol. kg^{-1}$)	TN ($g.kg^{-1}$)	TP ($mg.kg^{-1}$)	TK ($g.kg^{-1}$)	Background Cd ($mg.kg^{-1}$)
S1	Hunan	4.90	15.52	10.85	1.14	468.5	15.26	0.18
S2	Chongqing	5.74	17.48	21.34	1.00	547.8	22.61	0.18
S3	Liaoning	5.74	25.84	12.19	1.00	726.6	23.94	0.16
S4	Yunnan	5.92	34.26	11.10	2.01	811.2	4.77	0.26
S5	Jiangxi	6.01	11.69	8.70	0.51	518.0	9.96	0.14
S6	Anhui	6.25	20.04	19.08	0.99	347.1	15.41	0.17
S7	Heilongjiang	6.27	35.69	28.59	1.74	481.1	24.7	0.18
S8	Jilin	6.82	32.85	31.11	1.75	349.2	24.58	0.17
S9	Jiangsu	6.93	47.69	26.20	2.44	689.2	21.03	0.15
S10	Shaanxi	7.90	16.49	22.37	1.36	984.0	24.37	0.24
S11	Hebei	7.98	8.57	8.12	0.68	529.9	24.22	0.19
S12	Henan	8.07	17.79	16.01	1.07	746.1	19.86	0.19
S13	Xinjiang	8.12	19.43	25.25	1.32	780.0	25.49	0.16
S14	Shanxi	8.24	23.17	16.80	1.13	950.7	23.70	0.20
S15	Tianjin	8.29	22.02	24.67	1.42	916.3	24.63	0.19
S16	Shandong	8.65	11.84	13.09	0.93	965.6	21.37	0.18
S17	Neimeng	8.80	16.30	11.61	0.96	376.3	26.40	0.21

*Soil sample locations are listed in the order of increasing pH.

developed for the cladocera such as *Daphnia magna* and *Ceriodaphnia dubia* to predict chronic toxicity of nickel to three other invertebrates including snail (*Lymnaea stagnalis*), insect (*Chironomus tentans*), and rotifer (*Brachionus calyciflorus*). However, little effort has been devoted to establishing models to describe the relationship between Cd uptake of corn grain and soil properties. In addition, the cross-species extrapolation approach has rarely been applied to terrestrial ecosystem. The objectives of this study are: 1) to develop the models for predicting Cd accumulation in corn grain on seventeen soils with greatly different properties, 2) to assess the feasibility of applying these models to other non-model corn species, and 3) to investigate their accuracy in predicting accumulation of Cd in non-model corn species.

Materials and Methods

None of these 17 soil samples were collected from national parks or other protected areas. It is confirmed that no specific permissions were required for the soils sampling activities in the 17 locations in China. It is also can be confirmed that the field studies did not involve endangered or protected species. No tested corn species are under first- or second-class state protection, and they are not listed in the Inventory of Rare and Endangered Plants of China (http://zrbhq.forestry.gov.cn/portal/zrbh/s/3053/content-457748.html), or the Key Protected Inventory of Wild Plants of China (http://zrbhq.forestry.gov.cn/uploadfile/zrbh/2010-10/file/2010-10-14-bb296addeaa047798d6b6c476aaa1da9.doc). These corn species were used for only scientific research as permitted by the Ministry of Agriculture of China.

Table 2. Soil Environmental Quality Standards of China (GB15618–1995) and the content of exogenous Cd ($mg.kg^{-1}$).

pH	<6.5	6.5~7.5	>7.5
Grade Two Standard	0.3	0.3	0.6
Low Cd	0.3	0.3	0.6
High Cd	0.6	0.6	1.0

Soil samples

A set of 17 soils covering a wide range of soil properties was collected from typical locations in China. In each sample location, the soils were collected from the top 20 cm of the soil profile. For analysis of their phys-chemical characteristics, the soils were air dried and sorted to pass a 2-mm sieve. Soil pH was measured in deionized water (soil:solution ratio, 1:5) [40]. Cation exchange capacity (CEC) was determined by the unbuffered silver-thiourea method [41]. OM was measured by dry combustion [42]. The total phosphorus in soils (TP) was measured by colorimetric method [43]. The total nitrogen in soils was determined by the Kjeldahl's method [44]. The background Cd content in the soils was determined by aqua regia (1:3 fresh mixture of concentrated HNO_3 and HCl) digestion [45]. The selected properties of the 17 soil samples were shown in Table 1.

Experimental design

Bioaccumulation factors for Cd by corn with one species from seventeen soils. According to Grade Two Standard for Cd in the Soil Environmental Quality Standards of China (GB15618–1995), the tested levels of Cd in soils were shown in Table 2. To obtain the tested levels, a certain amount of exogenous Cd ($3[CdSO_4].8H_2O$) was added to the 8 kg air-dried soils. All the soils were thoroughly mixed and placed into pots. Then the soils were moistened with deionized water to the 60% of field moisture capacity. Each tested level of Cd was carried out in triplicate. The soils containing Cd at different levels were covered by plastic film with several pores. For Cd aging, all the pots were

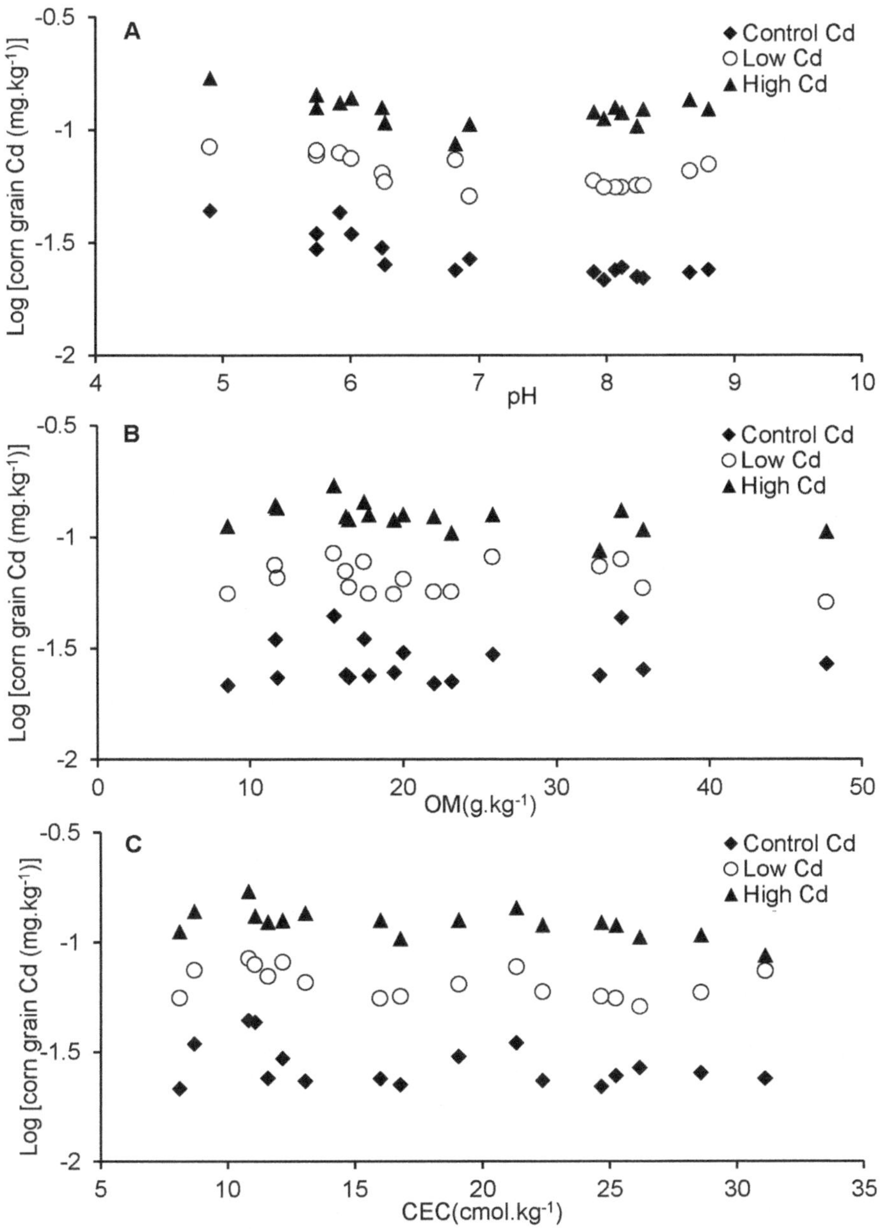

Figure 1. Relationships between Cd in corn grain and (A) soil pH, (B) organic matter (OM) and (C) cation exchange capacity (CEC).

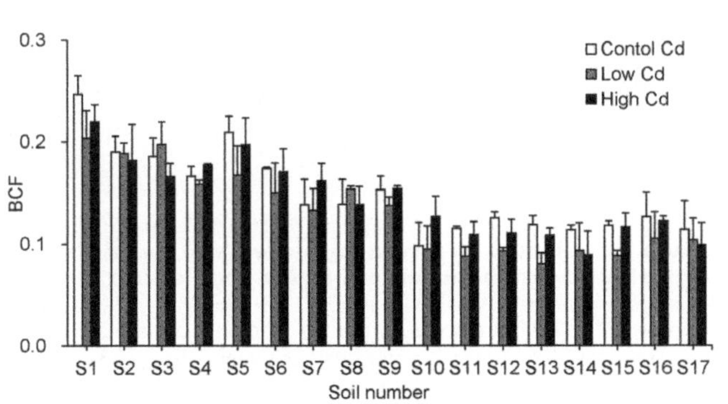

Figure 2. Effects of soil types on the transfer factor based on different Cd treatments in soils.

kept for 3 months at a greenhouse at the temperature of 25±3°C during the daytime and 20±3°C at night with a natural light photoperiod. During the period of aging, water contents of each soil sample were retained by adding deionized water every 3 days.

Table 3. Prediction models for the different Cd levels.

Model No.	Cadmium levels	Prediction models	R^2	*P*
Model 1	Control Cd	Log[BCF] = −0.081pH-0.254	0.728	<0.001
Model 2	Low Cd	Log[BCF] = −0.104pH-0.170	0.811	<0.001
Model 3	High Cd	Log[BCF] = −0.079pH-0.280	0.713	<0.001

Table 4. Intrinsic sensitivity (k) for non-model species fitted by models from cultivar Zhengdan 958.

Prediction Model	Intrinsic sensitivities (k)					
	Jingketian 183	Chuandan 30	Liaodan 565	Tunyu 88	Zhongdan 808	Nongda 84
Model 1	−0.284	−0.284	−0.339	−0.418	−0.324	−0.317
Model 2	−0.128	−0.123	−0.175	−0.254	−0.164	−0.158
Model 3	−0.302	−0.299	−0.353	−0.432	−0.338	−0.331

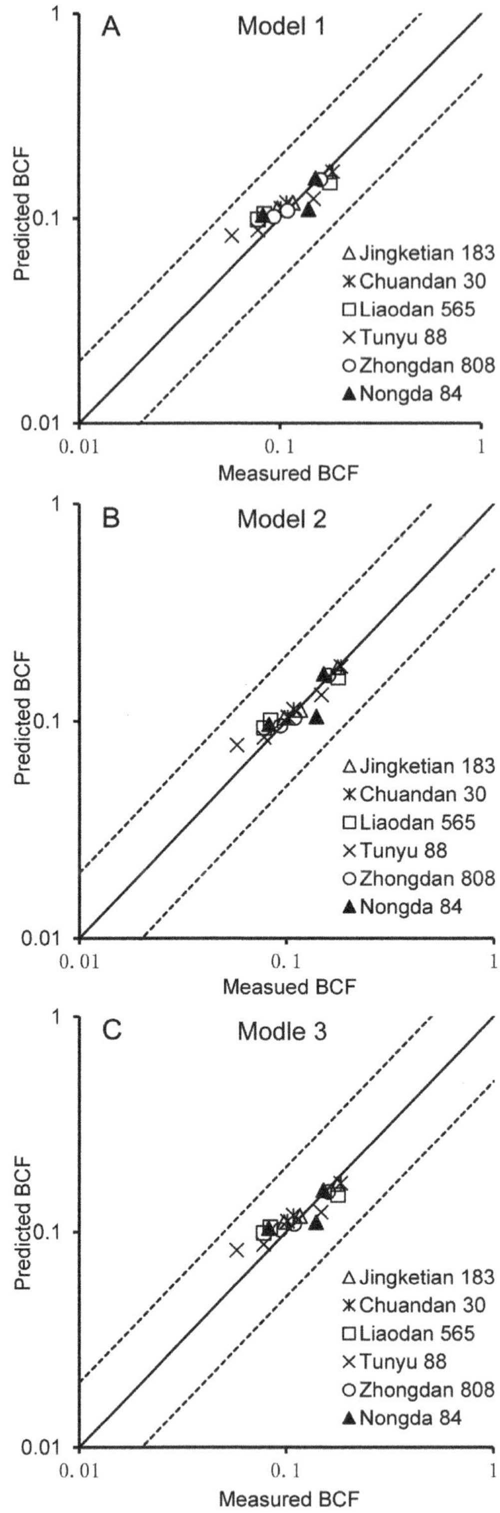

Figure 3. Relativity between measured and predicted BCF values for Cd in non-model corn grains. The predicted BCF values in Figure 3 were estimated by (A) model 1, (B) model 2, and (C) model 3 in Table 3. The solid line represents a 1:1 relationship; the dashed lines indicate a 2-fold prediction interval between the predicted and measured values.

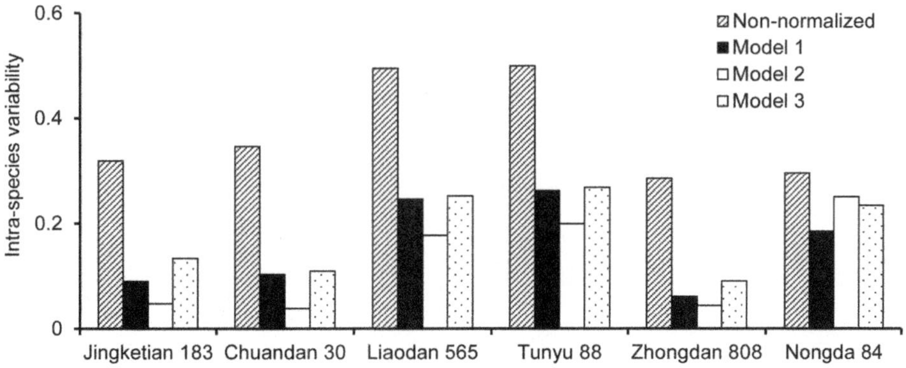

Figure 4. Intra-species variability of Cd BCF The data from all non-model corn species were normalized with the models listed in Table 3.

After 3 months' aging, uniformed seeds of corn (*Zea may* L. cv. Zhengdan 958) were sown in each soil containing Cd.

Bioaccumulation factors for Cd of eight corn species from three soils. Three soils with typical properties sampled from Jiangxi, Shaanxi, and Shanxi provinces were selected for this experiment. The tested levels of Cd in soils and exogenous Cd addition were the same as those in the experiment of bioaccumulation factors for Cd by corn from different soils. Triplicate was conducted on each tested level of Cd. After three months' aging of soils, the seeds of corn (cv. Jingketian 183, Chuandan 30, Liaodan 565, Tunyu 88, Zhongdan 808 and Nongda 84) were surface sterilized with 1% (v/v) NaOCl, rinsed, and soaked in distilled water for 24 h at 33°C in the dark. Then these seeds were sown in each soil containing Cd. All these pots were placed in the greenhouse at the temperature of 25±3°C during the daytime and 20±3°C at night with a natural light photoperiod.

Experimental soil sampling and analysis

After three months' aging, soils in the pots of the above two experiments were sampled, air-dried, passed through a <0.20 mm sieve, and then thoroughly mixed. 0.5 g of the soil sample was weighed and added to digestion tubes containing 9 ml HNO_3 and 3 ml HF. The corn grains were harvested through being oven-dried at 105°C for 30 minutes and kept at 70°C until the weight of the grains was stable. The samples (0.5 g) were digested in 6 ml concentrated HNO_3 (70% w/v) and 3 ml H_2O_2 in a CEM Mars X microwave oven (CEM Mars) at a pressure of 3.1 MPa. The Cd concentrations in the digestion solution from soil and corn grain samples were measured by inductively coupled plasma-mass spectrometry (ICP-MS, Agilent 7500a, Agilent Technologies Co. Ltd., USA).

Data analysis

The bioconcentration factor (BCF) was calculated as the ratio of the content of Cd in the corn grain to that in the soil (Equation 1) [46–47].

$$BCF = C_{grain}/C_{soil} \quad (1)$$

Where C_{grain} is the Cd concentration in the corn grain and C_{soil} is Cd concentration in the soil.

SPSS 16.0 for Windows® 107 (SPSS Inc, Chicago, IL, USA) was used for the regression analysis and statistical analysis of the significant differences. Origin 8.0 (OriginLab Co., Northampton, MA, USA) was employed for figure rendering.

Model

The prediction models for Cd transfer were established through multiple stepwise regression of the Cd BCF from Zhengdan 958 in a wide range of soil properties. These models are based on equation 2.

$$Log[BCF] = a \times pH + b \times \log[OM] + c \times \log[CEC] + k \quad (2)$$

Where Log[BCF], log[OM], and log[CEC] were the logarithm base 10 of the BCF values, the content of organic matter ($g.kg^{-1}$) in soils, and the cation exchange capacity ($cmol.kg^{-1}$) of soils, respectively. The slope of soil property parameters such as a, b and c indicate the impacts of soil properties on heavy metal. The intercept *k* is the intrinsic sensitivity that characterizes the species absorbing Cd.

Cross-species extrapolation

In the process of cross-species extrapolation, an interim alternative was to assume that interactions among Cd and pH, OM, CEC were the same among related species. In other words, the models stability constants including a, b, c were assumed the same among related species, and the only difference between related species was assumed to be their intrinsic sensitivity (*k*) [33]. The variation of intrinsic sensitivities within a species among

Table 5. Soil properties and BCF values for Cd in non-model wheat grain [61].

				BCF			
Cultivation soil	pH	OM(%)	CEC(mequiv./100g)	Anomol	Tj-83	Abadgar	Mehran
Soil	7.53	75.2	15.0	0.241	0.332	0.266	0.417
Sewage	7.70	87.6	20.8	0.179	0.311	0.287	0.375

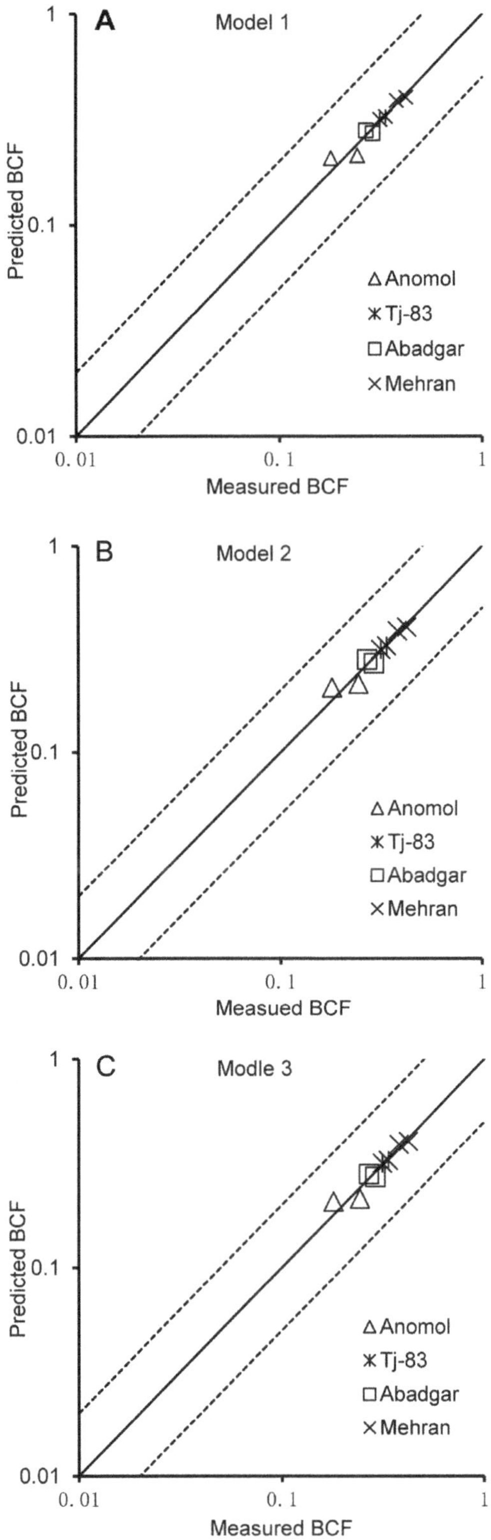

Figure 5. Relativity between measured and predicted BCF values for Cd in non-model wheat grains. The predicted BCF values in Figure 5 were estimated by (A) model 1, (B) model 2, and (C) model 3 in Table 3. The solid line represents a 1:1 relationship; the dashed lines indicate a 2-fold prediction interval between the predicted and measured values.

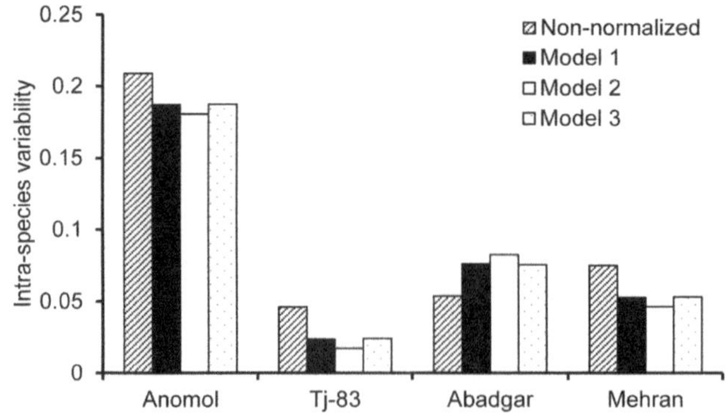

Figure 6. Intra-species variability of BCF for Cd. The data from all non-model wheat species were normalized with the models listed in Table 3.

plants reflects residual variation [36]. According to the minimum squared error between predicted BCF value and the measured BCF value $\sum_{i=1}^{n}(measured\,BCF_i - predicited\,BCF_i)^2$, the intercept (*k*) for different models corresponding to various species were obtained through Excel Solver for linear optimization [32].

The accuracy of the model predictions was evaluated by comparing the measured BCF of the other non-model corn species with the predicted BCF from the model. The predictions for other non-model corn species were calculated with the prediction model developed for Zhengdan 958 [48].

Analysis of the reduction of intra-species variability

The Cd BCF values of non-model corn species were normalized to the specific soil conditions by means of the obtained model for Zhengdan 958 and equation 3.

$$\sqrt{\frac{\sum_{i=1}^{n}(BCFsi-\overline{BCFs})^2}{(n-1)\times(\overline{BCFs})^2}} \qquad (3)$$

Where the BCF of i-th conditions for specific corn specie was normalized to specific soil conditions. *BCFsi* is the BCF of i-th conditions for specific corn species. $\overline{BCFs}$ is the mean of n BCFs. *n* is the number of different conditions for specific corn species.

The BCFs that Cd for different corn species was normalized by the prediction models to a set soil condition should be equal. Therefore, the decrease of intra-species variability indicates that normalization processing eliminated soil properties to a certain extent.

Results

Major factors affecting accumulation of Cd in corn grain in different soils

Simple relationships between log [corn grain Cd] at different Cd levels and soil pH, OM and CEC are shown in Figure 1. At the treatment of low concentration Cd, the Cd content in maize grain decreased with an increase of pH. Similar trends were found in the treatments of control and high concentration Cd. The significantly negative correlations were observed between Cd accumulation in corn grain and soil pH in the treatments with control Cd, low concentration Cd, and high concentration Cd ($P<0.001$). There was no significant correlation between the log [corn grain Cd] and soil OM and CEC. These results indicate that soil pH is the main factor that controls the accumulation of Cd in corn grain.

Effects of different soil types on Cd transfer in soil-corn system

As shown in Figure 2, the BCF values from cultivar Zhengdan 958 with different treatments were significantly different in 17 soils with a wide range of soil properties in China. In general, the variation tendency of BCF values from cultivar Zhengdan 958 was alike over all soil samples from S1 to S17 at different treatments. These BCF values decreased with the increase of soil pH. The BCF values detected in S1 to S9 (pH from 4.90 to 6.93) were significantly higher than those in S10 to S17 (pH from 7.90 to 8.80). In the control soil, the maximum value for BCF (0.24) was achieved in S1 (pH 4.90), and the minimum value for BCF (0.09) in S10. The maximum value was 2.67 folds as high as the minimum value. At the treatment of low Cd, the maximum BCF (0.20) was observed in S1, and the minimum value for BCF (0.08) was found in S13. The maximum value was 2.50 folds as high as the minimum. At the treatment of high Cd, the highest BCF (0.22) values among the 17 soils were once again obtained in S1, and the minimum value for BCF (0.10) was in S17. The maximum was 2.22 folds as high as the minimum. These results indicate that Cd under acidic soil conditions is more highly bioavailable, leading to increased absorption by plants.

Prediction models

The prediction models established by stepwise regression are shown in Table 3. A significantly negative correlation existed between the Log[BCF] and soil pH in the Cd treatments including control, low Cd, and high Cd with the R^2 values ranging from 0.713–0.811 ($P<0.001$). All models displayed the same trend in which Log[BCF] of Cd in corn grain was negatively related to soil pH. There was no significant difference among three prediction models at the different Cd levels.

Cross-species extrapolation

The intercept (k) for different corn species deduced by three different models are shown in Table 4. The intercept (k) means the sensitivity of corn species to Cd accumulation. No significant differences were found to exist either among the k values of all tested non-model corn species deduced by the same model (model 1, 2, or 3) or among those in each of the tested non-model corn species deduced by different models (model 1, 2, and 3).

On the basis of the k values, the BCF values of non-model corn species were predicted by different models developed from cultivar Zhengdan958. The relationship between the predicted BCF values and the measured BCF values of non-model corn species was shown in Figure 3. The ratio between the predicted BCF values and the measured BCF values was within 2 folds interval and close to the solid line of 1:1 relationship, indicating that these models from cultivar Zhengdan 958 can be applied to predict the Cd BCF of non-model corn species.

Reduction of intra-species variability

Figure 4 shows the intra-species variability of Cd BCF of non-model species which were normalized with the models listed in Table 3. The prediction models can reduce the measured BCF intra-species variability for all non-model corn species. The intra-species variability of Cd BCF fitted by model 2 was found much lower than that by model 1 and model 3 except for cultivar Nongda 84. Interestingly, the intra-species variability of Cd BCF for cultivar Nongda 84 normalized with model 1 was much lower than that with model 2 and model 3. No significant differences existed in the intra-species variability of cultivars Chuandan 30, Liaodan 565 and Tunyu 88 which were normalized with model 1 and model 3. Since all models are effective in reducing the uncertainty caused by the soil property differences, they all can be used to establish the soil Cd SSD curves and provide an ecological benchmark in China.

Discussion

The bioavailable concentration of Cd in soil is more important than its total concentration in terms of its uptake and accumulation in plants [49]. In this study, we found that Cd uptake into corn grain increased with an increase of Cd concentration in soil. However, the BCF values tended to be similar when the concentration of Cd in soil was changed responding to the same soil. These findings were in agreement with previous studies [28–30,50–51].

We also found that the soil pH was the major factor influencing the bioavailability of Cd in soil. This may be attributed to Cd uptake by plants from soils through diverse reactions such as absorption, ionic exchange, redox reactions, and precipitation-dissolution etc. [52]. These reactions are mainly affected by soil pH. The pH effects also include its influence on the solution activity of Cd and the distribution of Cd between the soil phase and the solution phase [53–54]. The increased sorption of Cd at high pH values could reduce the solution concentration, which tends to decrease the bioavailable concentration of Cd in soil. In this study, the BCF values observed in acidic soils (S1 to S9) were higher than those in alkaline soils (S10 to S17). These results suggest that Cd is much more bioavailable and leads to increased absorption by plants under acidic conditions [55–57]. No significant differences were observed among three prediction models at the different Cd levels. There were no significant differences among Cd BCF values of corn grain at different Cd levels, either. These results were consistent with the findings of Rezvani et al. [50].

More importantly, we confirmed that the Cd BCF prediction models from cultivar Zhengdan 958 could be used to predict Cd BCF of other non-model corn species according to the soil pH values through cross-species extrapolation approach. Although many prediction models were developed to describe the relationship between concentration of heavy metals in plants and soil properties in terrestrial ecosystems, these models were only suitable for single species [31–32,58–60]. Only a few studies were carried out to develop the prediction models for one species in aquatic ecosystems and subsequently used to normalize ecotoxicity data for other non-model species [33,38,39].

In order to further validate the feasibility of the prediction modes developed in the present study for other non-model plant species, the measured Cd BCF values of wheat grain from soils were selected from the paper of Jamali et al. [61], and shown in Table 5. Models presented in Table 3 were applied to these non-model wheat species. The relativity between measured and predicted BCF values for Cd in non-model wheat grains is shown in Figure 5. The ratios between all of the predicted BCF values and the measured BCF values were also within 2 folds interval and close to 1:1 relationship. In addition, the intra-species variability of Cd BCF of all non-model wheat species was decreased except for Abadgar (Figure 6). This indicates that the models from cultivar Zhendan 958 can be applied not only to non-model corn species but also to non-model wheat species.

Conclusions

The soil pH was the most important factor that can control Cd uptake and lower pH was more favorable for Cd bioaccumulation in corn grain. The BCF of Cd in corn grain can be predicted by the prediction models such as equation 4. There was no significant difference among three prediction models in the different Cd sources. These prediction models could be applied to other non-model corn species,

$$Log[BCF] = a \times pH + k \tag{4}$$

Acknowledgments

We are indebted to Prof. Yibing Ma and Dr. Xiaoqing Wang for their help in the data processing and model derivation.

Author Contributions

Conceived and designed the experiments: ZJL. Performed the experiments: HY LL. Analyzed the data: JL. Contributed reagents/materials/analysis tools: YCL. Wrote the paper: HY ZJL.

References

1. McLaughlin MJ, Singh BR (1999) Cadmium in soil and plants: a global perspective. Kluwer Academic Publishers, Dordrecht, 13–21.
2. Li ZJ, Xu JM, Tang CX, Wu JJ, Muhammad A, et al. (2006) Application of 16S rDNA-PCR amplification and DGGE fingerprinting for detection of shift in microbial community diversity in Cu^{-}, Zn^{-}, and Cd^{-} contaminated paddy soils. Chemosphere 26: 1374–1380.
3. Zhang CB, Huang LN, Luan TG, Jin J, Lan CY (2006) Structure and function of microbial community during the early stages of revegetation near Shaoguan Pb/Zn Smelter, Guangdong, P.R. China. Geoderma 136: 555–565.
4. Chien CC, Kuo YM, Chen CC, Hung CW, Yeh CH, et al. (2008) Microbial diversity of soil bacteria in agricultural field contaminated with heavy metals. Journal of environmental sciences 20: 359–363.
5. Tudoreanu L, Phillips CJC (2004) Empirical models of cadmium accumulation in maize, rye grass and soya bean plants. J Sci Food Agric 84: 845–852.
6. Mushtakova VM, Fomina VA, Rogovin VV (2005) Toxic effect of heavy metals on human blood neutrophils. Biol Bull 32: 276–278.
7. Huang ML, Zhou SL, Sun B, Zhao QG (2008) Heavy metals in wheat grain: Assessment of potential health risk for inhabitants in Kunshan, China. Science of the Total Environment 405: 54–61.
8. Hu JL, Wu FY, Wu SC, Sun XL, Lin XG, et al. (2013) Phytoavailability and phytovariety codetermine the bioaccumulation risk of heavy metal from soils, focusing on Cd-contaminated vegetable farms around the Pearl River Delta, China. Ecotoxicology and Environmental Safety 91: 18–24.
9. Harter RD, Naidu R (2001) Anassessment of environmental and solution parameter impact on trace-metal sorption by soil. Soil Sci Soc Am J 65: 597–612.
10. Jung MC (2008) Heavy metal concentration in soils and factors affecting metal uptake by plants in the vicinity of a Korean Cu-W mine. Sensors 8: 2413–2423.
11. Wang J, Zhang CB, Jin ZX (2009) The distribution and phytoavailability of heavy metal fractions in rhizosphere soils of paulowniu fortune (seem) Hems near a Pb/Zn smelter in Guangdong, PR China. Geodema 148: 299–306.
12. Chapman EEV, Dave G, Murimboh JD (2013) A review of metal (Pb and Zn) sensitive and pH tolerant bioassay organisms for risk screening of metal-contaminated acidic soils. Environmental Pollution 179: 326–342.
13. Barančíková G, Madaras M, Rybár O (2004) Crop contamination by selected trace elements. Journal of Soils and Sediments 4: 37–42.
14. Mühlbachová G, Simon T, Pechová M (2005) The availability of Cd, Pb and Zn and their relationships with soil pH and microbial biomass in soils amended by natural clinoptilolite. Plant, Soil and Environment 51: 26–33.
15. Zhao KL, Liu XM, Xu JM, Selim HM (2010) Heavy metal contaminations in a soil-rice system: identification of spatial dependence in relation to soil properties of paddy fields. Journal of Hazardous Materials 181: 778–787.
16. Sukreeyapongse O, Holm PE, Strobel BW, Magid SJ, Panichsakpatana S, et al. (2002) pH-dependent release of cadmium, copper, and lead from natural and sludge-amended soils. Journal of Environmental Quality 31: 1901–1909.
17. Bang J, Hesterberg D (2004) Dissolution of trace element contaminants from two coastal plain soils as affected by pH. Journal of Environmental Quality 33: 891–901.
18. Badawy SH, Helal MID, Chaudri AM, Lawlor K, McGrath SP (2002) Soil solidphase controls lead activity in soil solution. Journal of Environmental Quality 31: 162–167.
19. Wang AS, Angle JS, Chaney RL, Delorme TA, Reeves RD (2006) Soil pH effects on uptake of Cd and Zn by Thlaspi caerulescens. Plant and Soil 281: 325–337.
20. Du Laing G, Rinklebe J, Vandecasteele B, Meers E, Tack FMG (2009) Heavy metal mobility and availability in estuarine and riverine floodplain soils and sediments: a review. Science of the Total Environment 407: 3972–3985.
21. Kirkham MB (2006) Cadmiumin plantson polluted soils: Effect of soil factors, hyperaccumulation and amendments. Geoderma 137: 19–32.
22. Kim KR, Owens G, Naidu R (2009) Heavy metal distribution, bioaccessibility and phytoavailability in long-term contaminated soils from Lake Macquarie, Australia. Aust J Soil Res 47: 166–176.
23. Zeng F, Ali S, Zhang H, Ouyang Y, Qiu B, et al. (2001) The influence of pH and organic matter content in paddy soil on heavy metal availability and their uptake by rice plants. Environmental pollution 159: 84–91.
24. Hettiarachchi GM, Ryan JA, Chaney RL, La Fleur CM (2003) Sorption and desorption of cadmium by different fractions of biosolids-amended soils. Journal of Environmental Quality 32: 1684–1693.
25. Antoniadis V, Robinson JS, Alloway BJ (2008) Effects of short-term pH fluctuations on cadmium, nickel, lead, and zinc availability to ryegrass in a sewage sludge-amended field. Chemosphere 71: 759–764.
26. Eriksson JE (1988) The effects of clay, organic matter and time on adsorption and plant uptake of cadmium added to the soil. Water Air Soil Pollut 40: 359–373.
27. Pinto AP, Mota AM, De Varennes A, Pinto FC (2004) Influence of organic matter on the uptake of cadmium, zinc, copper and iron by sorghum plants. Science of the total environment.326: 239–247.
28. Brown SL, Chaney RL, Angle JS, Ryan JA (1998) The phytoavailability of cadmium to lettuce in long-term biosolids-amended soils. J Environ Qual 27: 1071–1078.
29. Kim SJ, Chang AC, Page AL, Warneke JE (1988) Relative concentrations of cadmium and zinc in tissue of selected food plants grown on sludge-treated soils. Journal of Environmental Quality 17: 568–573.
30. Ge Y, Murray P, Hendershot WH (2000) Trace metal speciation and bioavailability in urban soils. Environ Pollut 107: 137–144.
31. Li Q, Xu XH, Zuo YB (2009) Effects of biosolid on the accumulation and transfer of trace elements in soil-wheat/maize system. Journal of agro-environment science 28: 2042–2049.
32. Wang XQ (2012) Ecological Thresholds for Copper and Nickel in Chinese Agricultural Soils, Beijing, China Mining University.
33. Schlekat CE, Genderen EV, UGent KDS, Antunes PMC, Rogevich EC, et al. (2010) Cross-species extrapolation of chronic nickel Biotic Ligand Models. Science of the Total Environment 408: 6148–6157.
34. Solomon KR, Baker DB, Richards RP, Dixon KR, Klaine SJ, et al. (1996) Ecological risk assessment of Atrazine in North American surface waters. Environmental Toxicology and Chemistry 15: 31–76.
35. Steen RJCA, Leonards PEG, Brinkman UAT, Barcelo D, Tronczynski J, et al. (1999) Ecological risk assessment of agrochemicals in European estuaries. Environmental Toxicology and Chemistry 18: 1574–1581.
36. Van Sprang PA, Verdonck FAM, Van Assche F, Regoli L, De Schamphelaere KAC (2009) Environmental risk assessment of zinc in European freshwaters: a critical appraisal. Sci Tot Environ 407: 5373–5391.
37. Allard P, Fairbrother A, Hope BK, Hull RN, Johnson MS, et al. (2010) Recommendations for the development and application of wildlife toxicity reference values. Intergrated environmental assessment and management 6: 28–37.
38. Deleebeeck NME, De Schamphelaere KAC, Janssen CR (2007a) A bioavailability model predicting the toxicity of nickel to rainbow trout (*Oncorhynchus mykiss*) and fathead minnow (*Pimephales promelas*) in synthetic and natural waters. Ecotoxicol Environ Saf 67: 1–13.
39. Deleebeeck NME, Muyssen BTA, De Laender F, Janssen CR, De Schamphelaere KAC (2007b) Comparison of nickel toxicity to cladocerans in soft versus hard surface waters. Aquat Toxicol 84: 223–35.
40. Sparks DL, Page AL, Helmke PA, Loeppert RH, Soltanpour PN, et al. (1996) Methods of Soil Anlysis. Part 3. Chemical Methods. Soil Science Society of America, Inc, Madison. WI.
41. Dohrmann R (2006) Cation exchange capacity methodology II: A modified silver–thiourea method. Applied clay science 34: 38–46.
42. Ball DF (1964) Loss-on-ignition as an estimate of organic matter and organic carbon in noncalcareous soils. Journal of Soil Science 15: 84–92.
43. Bray RH, Kurtz LT (1945) Determination of total, organic, and available forms of phosphorus in soils. Soil science 59: 39–46.
44. Bremner JM (1996) Nitrogen total. Methods of soil analysis. Part 3: Chemical methods, American Society of Agronomy-Soil Science Society of America, 1996, 1085–1121.

45. Zarcinas BA, McLaughlin MJ, Smart MK (1996) The effect of acid digestion technique on the performance of nebulisation systems used in inductively coupled plasma spectrometry. Commun Soil Sci Plant Anal 27: 1331–1354.
46. Baker AJM, Reeves RD, Hajar ASM (1994) Heavy metal accumulation and tolerance in British populations of metallophyte Thlaspi caerulescens. New Phytol 127: 61–68.
47. Raskin I, Kumar PBAN, Dushenkov S, Salt DE (1994) Bioconcentration of heavy metals by plants. Curr Opin Biotechnol 5: 285–290.
48. De Schamphelaere KAC, Heijerick D, Janssen C (2006) Cross-phylum comparison of a chronic biotic ligand model to predict chronic toxicity of copper to a freshwater rotifer, Brachionus calyciflorus (Pallas). Ecotoxicol Environ Saf 63: 189–195.
49. Sarwar N, Saifullah, Malhi SS, Zia MH, Naeem A, et al. (2010) Role of mineral nutrition in minimizing cadmium accumulation by plants. J Sci Food Agric 90: 925–937.
50. Rezvani M, Zaefarian F (2011). Bioaccumulation and translocation factors of cadmium and lead in Aeluropus littoralis. Australian Journal Agricutural Engineering 2: 114–119.
51. Mertens J, Luyssaert S, Verheyen K (2005) Use and abuse of trace metal concentrations in plants tissue for biomonitoring and phytoextraction. Environ Pollut 138: 1–4.
52. Smical AI, Hotea V, Oros V, Juhasz J, Pop I (2008) Studies on transfer and bioaccumulation of heavy metals from soil into lettuce. Environmental Engineering and Management Journal 7: 609–615.
53. Christensen TH (1984) Cadmium soil sorption at low concentrations: I. Effect of time, cadmium load, pH, and calcium. Water. Air. Soil Pollut 21: 105–114.
54. Naidu R, Bolan NS, Kookana RS, Tiller KG (1994) Ionic-strength and pH effects on the sorption of cadmium and the surface charge of soils. Euroean Journal of Soil Science 45: 419–429.
55. Hinesly TD, Redborg KE, Pietz RI, Ziegler EL (1984) Cadmium and zinc uptake by corn (Zea mays L.) with repeated applications of sewage sludge. J Agric Food Chem 32: 155–163.
56. Castilho PD, Chardon WJ (1995) Uptake of soil cadmium by three field crops and its prediction by a pH-dependent Freundlich sorption model. Plant Soil 171: 263–266.
57. Adams ML, Zhao FJ, McGrath SP, Nicholsonb FA, Chambersb BJ (2004) Predicting Cadmium Concentrations in Wheat and Barley Grain Using Soil Properties. Journal of Environmental Quality 33: 532–541.
58. Hao XZ, Zhou DM, Li DD (2012) Growth, cadmium and zinc accumulation of ornamental sunflower (*Helianthus annuus* L.) in contaminated soil with different amendments. Pedosphere 22: 631–639.
59. McBride MB (2002) Cadmium uptake by crops estimated from soil total Cd and pH. Soil Sci. 167: 62–67.
60. Wang XP, Shan XQ, Zhang SZ, Wen B (2004) A model for evaluation of the phytoavailability of trace elements to vegetables under the field conditions. Chemosphere. 55: 811–822.
61. Jamali MK, Kazi TG, Arain MB, Afridi HI, Jalbani N, et al. (2009) Heavy metal accumulation in different varieties of wheat (*Triticum aestivum* L.) grown in soil amended with domestic sewage sludge. Journal of Hazardous Materials 164: 1386–1391.

Urban Community Gardeners' Knowledge and Perceptions of Soil Contaminant Risks

Brent F. Kim[1,3], Melissa N. Poulsen[2,4], Jared D. Margulies[1,3,5], Katie L. Dix[6], Anne M. Palmer[1,3,7], Keeve E. Nachman[1,3,8]*

1 Johns Hopkins Center for a Livable Future, Baltimore, Maryland, United States of America, **2** CLF-Lerner Fellow, Johns Hopkins Center for a Livable Future, Baltimore, Maryland, United States of America, **3** Department of Environmental Health Sciences, Johns Hopkins Bloomberg School of Public Health, Baltimore, Maryland, United States of America, **4** Department of International Health, Johns Hopkins Bloomberg School of Public Health, Baltimore, Maryland, United States of America, **5** Department of Geography and Environmental Systems, University of Maryland, Baltimore, Maryland, United States of America, **6** Community Greening Resource Network, Parks & People Foundation, Baltimore, Maryland, United States of America, **7** Department of Health, Behavior and Society, Johns Hopkins Bloomberg School of Public Health, Baltimore, Maryland, United States of America, **8** Department of Health Policy and Management, Johns Hopkins Bloomberg School of Public Health, Baltimore, Maryland, United States of America

Abstract

Although urban community gardening can offer health, social, environmental, and economic benefits, these benefits must be weighed against the potential health risks stemming from exposure to contaminants such as heavy metals and organic chemicals that may be present in urban soils. Individuals who garden at or eat food grown in contaminated urban garden sites may be at risk of exposure to such contaminants. Gardeners may be unaware of these risks and how to manage them. We used a mixed quantitative/qualitative research approach to characterize urban community gardeners' knowledge and perceptions of risks related to soil contaminant exposure. We conducted surveys with 70 gardeners from 15 community gardens in Baltimore, Maryland, and semi-structured interviews with 18 key informants knowledgeable about community gardening and soil contamination in Baltimore. We identified a range of factors, challenges, and needs related to Baltimore community gardeners' perceptions of risk related to soil contamination, including low levels of concern and inconsistent levels of knowledge about heavy metal and organic chemical contaminants, barriers to investigating a garden site's history and conducting soil tests, limited knowledge of best practices for reducing exposure, and a need for clear and concise information on how best to prevent and manage soil contamination. Key informants discussed various strategies for developing and disseminating educational materials to gardeners. For some challenges, such as barriers to conducting site history and soil tests, some informants recommended city-wide interventions that bypass the need for gardener knowledge altogether.

Editor: Sunghun Park, Kansas State University, United States of America

Funding: This work was funded by a Faculty-Community grant from the Johns Hopkins Urban Health Institute (http://www.urbanhealth.jhu.edu/). The Johns Hopkins Center for a Livable Future is supported by a grant from the GRACE Communications Foundation (but did not provide funding specific to this project). The funders had no role in study design, data collection and analysis, decision to publish, or preparation of the manuscript.

Competing Interests: The authors have declared that no competing interests exist.

* E-mail: knachman@jhsph.edu

Background

Urban community gardens—gardens tended by multiple households in an urban neighborhood—may offer a range of benefits. Studies have observed associations between community gardening and health [1–8], social [6,9], and economic benefits [6,7,10], and gardening in general has been associated with cardiovascular [11,12] and mental [13–15] health benefits. Historically, backyard and community gardens have made substantial contributions to the food supply; World War II "Victory Gardens" have been credited with providing an estimated 40% of the U.S. vegetable supply [16]. In urban settings, community gardens—and urban green spaces in general—may confer an additional set of social benefits [17–20] and ecosystem services [21,22]. Urban green spaces also provide educational opportunities for urban residents, for whom parks and gardens may be their primary source of experience, knowledge, and valuation of nature.

Gardening in urban settings may also present health risks, including those stemming from exposure to contaminants such as heavy metals, organic chemicals, and asbestos that may be present in urban soils. Urban soils are often close to pollution sources, such as industrial areas and heavily trafficked roads. As a result, many soil contaminants have been found at higher concentrations with increasing proximity to urban centers [23]. In Baltimore, Maryland, prior soil analyses (Table S1 in File S1) have revealed high concentrations of lead at some sites [24–28], reflecting the city's long history of industrial activity, incinerators, and vehicular traffic, and raising concerns about lead exposure [24]. Table S2 in File S1 summarizes some of the more common urban soil contaminants, their sources, and health effects associated with exposure.

Gardeners can be exposed to contaminants by inadvertently ingesting soil, inhaling soil particles, or via dermal contact. Soil ingestion is of particular concern among children, who may ingest larger quantities of soil than adults (e.g., by putting their hands in their mouths), absorb higher levels of certain contaminants into their bloodstream [29], and are generally more sensitive to their effects. People who consume produce grown in contaminated environments also risk ingesting soil particles on the surfaces of plants [30]. Contaminants in soils including lead [31], cadmium [32], and arsenic [33] may accumulate in the tissues of vegetables grown in contaminated soils, posing another potential route of ingestion.

Urban gardeners may be unaware of how to manage these risks. Harms et al. [34,35] surveyed 121 urban farmers and gardeners from Kansas, Indiana, and Washington, most of whom indicated they do not have sufficient knowledge of how to minimize health risks associated with gardening in contaminated environments and want more information on soil testing and best management practices. Gardeners may also benefit from information on soil remediation, i.e. removing, destroying, detoxifying, immobilizing or containing soil contaminants [36].

The purpose of our study is to characterize urban community gardeners' knowledge of risks associated with contaminated garden soils, their perceptions of these risks, their knowledge of how to assess and reduce these risks, the sources they draw upon for information on soil contamination, and the information and training needs they have related to soil contamination.

Methods

To characterize urban community gardeners' knowledge and perceptions of soil contamination risks, we conducted surveys among urban community gardeners and semi-structured interviews with key informants in the gardening community.

Gardener surveys

We conducted brief verbal surveys in-person or by phone with Baltimore community gardeners. The survey included questions regarding demographics, garden site history, and knowledge, perceptions, and practices related to soil contamination. To be eligible to participate, gardeners had to be at least 18 years of age and have been gardening at their current site for at least 6 months.

We partnered with the Community Greening Resource Network (CGRN) to identify gardens from which to recruit survey participants. CGRN is Baltimore's gardening support network and maintains a registry of community gardens in the Baltimore metropolitan area. We randomly selected 30 gardens from the CGRN registry of 70 food-producing community gardens, contacting leaders at the selected gardens to identify opportunities to survey gardeners. After experiencing difficulty reaching some garden leaders, we included additional gardens – recommended to us by representatives in the gardening community or identified through personal contacts – whose leaders were willing to help us arrange surveys.

As an incentive for participating gardeners, we collected soil samples from represented community gardens, sent the samples for analysis at the U.S. Department of Agriculture (USDA) soils lab, shared the results with garden leaders, and offered guidance in interpreting the results.

Key informant interviews

We conducted semi-structured interviews with 18 purposively selected informants knowledgeable about community gardening and soil contamination in Baltimore City: representatives from City government urban agriculture-related programs (4), federal agency employees (2), a representative of a Baltimore community gardening organization (1), agricultural extension employees (2), Master Gardeners - trained volunteers who advise and educate the public on gardening (2), community garden leaders (4), and urban farmers (3). We distinguish farming from gardening by the intent to produce goods for sale.

Interviews focused on informants' perceptions of community gardeners' concerns about soil contamination, barriers to soil testing, and information needs related to soil contamination. When applicable, informants were also asked relevant questions about their roles and perspectives related to their employment in city, state, or federal agencies.

To identify major themes in the qualitative data, three members of the research team first developed a codebook that was organized by axial codes and sub-codes. Two researchers coded each transcript using Atlas/ti (v7); when discrepancies arose, we included all quotes assigned to a particular code by either researcher. We then generated reports of the text assigned to each code, writing reflective memos and pulling out illustrative quotes for each theme.

Ethical considerations

The Johns Hopkins School of Public Health Institutional Review Board (IRB) approved this study. Study participants provided verbal informed consent prior to participating in surveys and interviews. Oral consent was deemed adequate by the IRB, eliminating the need to record identifying information in study documents. An IRB-approved oral consent script was read by trained investigators to study participants. A dated questionnaire served as a record that the oral consent process had been completed.

Results

Gardener demographics

Seventy gardeners, representing 15 community gardens from a range of socioeconomic census tracts, responded to our survey. Most were female (66%), lived within a quarter-mile of their garden plots (76%), and had been at their current gardens for less than four years (76%). The median age of surveyed gardeners was 45. See Figure 1 for additional gardener demographic information.

Gardening practices and harvest use

Most (73%) surveyed gardeners indicated they avoid using commercial pesticides and fertilizers, relying instead on practices such as composting and mulching to promote soil fertility and suppress pests. Others (19%) reported using one or more commercial fertilizers and/or insecticides, several of which are allowed for use under USDA organic standards. Interview informants indicated that Baltimore's community gardeners generally garden without chemical inputs.

Among surveyed gardeners, 86% grew produce for home consumption, while the other 14% grew produce primarily for soup kitchens and other charitable uses. Among gardeners who grew produce for home consumption, almost half (45%) supplied over 60% of their family's produce intake from their community garden during the growing season.

General knowledge and concerns about contaminants

To assess their knowledge about chemical contaminants, surveyed gardeners were asked to list the soil contaminants they are aware of (Table 1). Most (66%) gardeners mentioned lead, and to a lesser extent, other trace elements (19%) and some types of organic chemicals (36%).

Gardeners were also asked to list any health concerns they have as community gardeners (Table 2); half (51%) cited soil contaminants as among their concerns. When asked to express

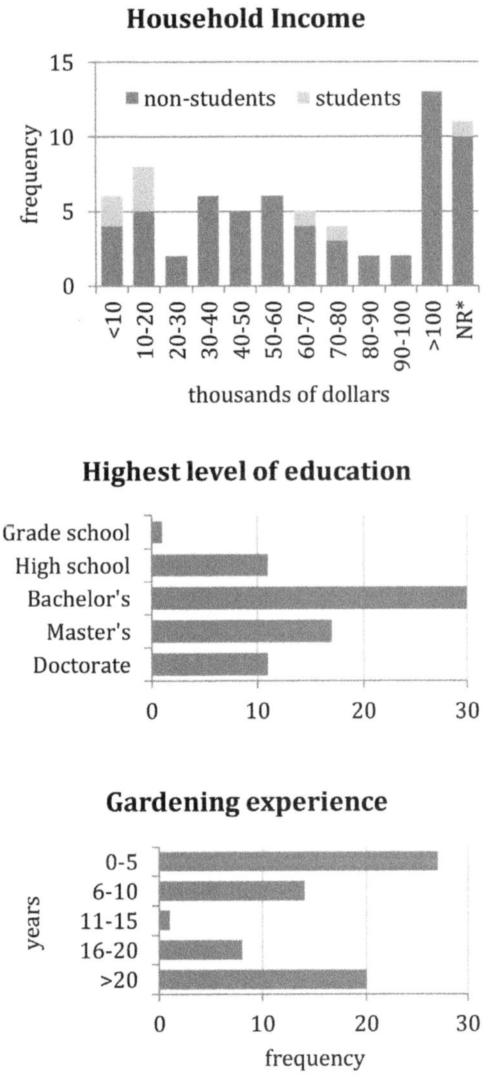

Figure 1. Additional gardener demographic information. * NR = No response.

their overall levels of concern about contaminants in their gardens, the average response was 2.3 on a scale of one to five, with five being the most concerned. There was no apparent association between levels of concern and whether gardeners thought their soil had been tested, or what they thought the test results indicated (Table 3).

When asked to list the ways in which one might come into contact with contaminants, most (70%) surveyed gardeners mentioned ingestion (e.g., "eating crops"). They did not specifically mention incidental ingestion, e.g., accidentally swallowing small amounts of soil while gardening. Other responses included dermal contact with (63%), and inhalation of (39%) contaminants.

Through interviews it became clear that lead is the contaminant of greatest concern among informants and, based on informants' perceptions, also the most common contaminant concern among gardeners. Informants expressed particular concerns about the vulnerability of children to contaminants—and specifically to lead—as compared to adults; among gardens where children may be present, some informants emphasized the heightened importance of testing their soil and making sure children do not ingest it. Informants were also concerned about other contaminants such as trash, drug paraphernalia, and animal feces, as well as potential contaminants in fill dirt, compost, and water. Informants also expressed concerns about chemical inputs, such as pesticides, and indicated that gardeners may view their use as more harmful to

Table 1. Open-ended responses to questions about soil contaminant knowledge: "What soil contaminants are you aware of that urban gardeners should be concerned about in general?" and "As a community gardener, do you have any concerns about hazards to your health?"

Response	%
Heavy metals and other trace elements	71
Non-specific	20
Lead	66
Arsenic	11
Mercury	4
Chromium	4
Cadmium	3
Copper	1
Organic chemicals	36
Petrochemicals (e.g., fuel, oil)	19
Pesticides	13
Persistent organic pollutants	7
Automotive fluids	6
Chemicals (non-specific)	16
Biological hazards	11
Human excreta	7
Animal excreta	6
Building materials (e.g., asbestos, asphalt, roofing tar)	11
Foreign objects (e.g., trash, needles)	21
Other	9
No response	10

Table 2. Open-ended responses to "As a community gardener, do you have any concerns about hazards to your health?"

Response	%
Soil contaminants	51
Non-specific	20
Organic chemicals (e.g., pesticides)	16
Heavy metals and other trace elements	11
Trash	11
Discarded needles	9
Human or animal excreta	4
Crime	6
Animal pests	4
Injury	3
Air quality	3
No concerns	44

Table 3. Gardeners' levels of concern about contaminants in their garden, by perceived soil test status.

"Has your soil been tested for contaminants?"	"Did the results reveal a problem with contamination?"	Number of respondents	Average level of concern (1–5)* among respondents
No	NA	14	2.5
Unsure	NA	16	1.7
Yes	No	20	2.6
Yes	Unsure	9	2.0
Yes	Yes**	11	2.4

NA: Not applicable.
*On a scale of one to five, with five being the most concerned.
**8 of these 11 respondents indicated they discontinued growing food crops in contaminated areas; two indicated the soil was remediated; one was unsure whether corrective action was taken.

health than contaminants directly in soil. Table S3 in File S1 illustrates the range of concerns noted in these discussions.

Most informants did not express a high level of concern about soil contaminants. Frequently, informants indicated that their concerns about soil contamination were alleviated by the use of raised beds or after seeing safe results come back from soil tests. Informants also made repeated comments about soil quality, suggesting issues of soil fertility may be more salient than contaminant concerns.

> *For me, the most important thing is that we have good soil structure here. … So, really my energies have been not around contaminations, but just building healthy soil so that we get the best vegetable production out of here. (Community garden leader)*

In discussing gardeners' levels of awareness or concern about soil contamination, informants' views varied widely. Several had little confidence that gardeners think about soil contamination as an issue or are aware that soil testing can and should be conducted. Informants working with municipal programs related to community gardening noted that few people ask about soil contamination when starting a garden. At the other end of the spectrum, some informants noted broad concern about soil contamination among community gardeners.

> *We talk about air. We talk about water. … Nobody talks about soil, and, essentially, it became very obvious that soil's probably the most contaminated thing that we have in our environment, particularly in urban areas. (MasterGardener)*

> *I think each year, the new gardeners ask do we have to be concerned about soil, and I say we tested it and everything was okay. (Community garden leader)*

> *And I've noticed that most gardens want raised beds, because they think there is a lead issue. (City government representative)*

Informants noted that knowledge and concerns about contaminants vary with different populations. Younger and more educated individuals, for example, were thought to have greater awareness of soil contaminant issues. A few informants noted that soil contamination was of less concern for gardeners and volunteers at urban farms, because they trusted that the appropriate steps had been taken to ensure the safety of the soil. One urban farmer noted that community members were concerned about soil contamination when the farm was first getting started, but once it became "established", these concerns disappeared.

Site history

One of the first steps in determining soil safety is learning how a particular site was used. This was a top concern of one City government representative, who worried that by testing soil without the knowledge of a site's history, gardeners may be unaware of potential contaminants:

> *If you do a test for lead and other heavy metals and you pat yourself on the back and you go on, are you missing the fact that there used to be a gas station on that site and there could be types of contaminants that you don't even know how to test for that could pose a risk?*

Most (73%) surveyed gardeners said they knew the site history of their gardens. Likewise, the community garden leaders and urban farmers we interviewed indicated they knew the past use of their garden and farm sites.

When asked how they learned about their site's history, most community garden leaders and urban farmers we interviewed indicated they spoke to residents in the surrounding area. One urban farmer said they would start by referencing Sanborn Maps, which are available through local libraries and depict historical land uses from 1867 to 1970. Surveyed gardeners reported obtaining information on site histories primarily from other gardeners (42%), from neighbors (26%) or based on their own observations (23%). A small proportion (7%) had obtained information from a government office, such as the Department of Planning.

When asked if site history is important information in determining if a site is suitable for gardening, nearly all (99%) surveyed gardeners agreed. In contrast, one City government representative suggested that most gardeners would not be interested in trying to uncover information on site history, noting, "I think, based on the people I've met and talked to, they just wanna grow something."

Informants also suggested that gardeners may lack the expertise necessary to conduct a site history. One City government representative noted that in an ideal world, assistance would be provided to new gardens to test the soil and "sit down with somebody and go over the site history in a way that's simple and doesn't take too long and is very clear." One federal agency employee indicated that expert guidance is the "crucial part" of a Phase 1 Environmental Site Assessment, in which an auditor reviews historical and other records, visits the site, and interviews previous landowners and other informants:

> *[T]he site history will get you … maybe 85% there.... [T]here's nothing about [a Phase 1 assessment], less judgment, that a person can't do at the library. But that is still missing that crucial part of the equation—the expertise piece.... Because you'll see maybe it was a paint factory. Well, will everybody think about, "Okay, well, how was a paint made at that time"? Well, it was linseed oil and the white lead was actually mixed in, and it was about at 40% lead.*

Another federal agency employee described additional limitations of site histories, noting that "the use and dumping on backyards and gardens is so idiosyncratic that it's impossible to learn what happened even ten years before," and that conducting sporadic site histories on a per-garden basis may not reveal evidence of contaminants from other parts of the city (e.g., from nearby industries that have since shut down). For these reasons, this informant recommended a more comprehensive approach of conducting site histories and soil testing at a city-wide level.

Soil testing

One theme that emerged during informant interviews was the barriers that deter gardeners from testing their soil for contaminants.

Cost was perceived to be prohibitive, particularly in situations in which a gardener wanted to test for a contaminant outside the scope of the usual metals panel, such as asbestos. According to informants, these situations also required additional knowledge about what to test for and how to find a service that offered those tests.

Informants also suggested that gardeners might perceive the steps involved in taking and sending away soil samples to be cumbersome or too complicated. Some informants suggested the need for a local testing service or a government-funded public service for soil testing, although not all agreed that a local testing service would be worth the cost.

> *There should be an immediate way to get testing … because some people will not go through the process of sending something away. (Community garden leader)*

> *I don't really see a huge a demand out there for this kind of information … it takes a whole lot to get the right people together to have a soil-testing lab here, in the city. And then, again, it might not even be used. (City government representative)*

Other barriers mentioned by informants included fear of discovering contamination after having already made investments in the land, and the need to document the exact locations from where soil samples are taken within a garden (since contaminant levels may vary spatially across a garden).

A few informants also noted that once soil tests are conducted, gardeners might have difficulty interpreting the results. Our survey results also hinted at this – in responding to questions about tests for contaminants, several gardeners referenced nutrient levels and soil tilth, suggesting that some gardeners may conflate tests of soil fertility with tests for contaminant levels.

Several informants also perceived a lack of scientific consensus about what levels of contamination are considered safe.

> *[H]ow much lead is too much lead? I have read different numbers. In Canada, the safe level is different than in the United States, and I think that in Europe it's different. (City government representative)*

> *[T]here's a lot of conflicting information from EPA and different universities about what an action level would be for total lead...(Agricultural extension employee)*

Both federal agency employees – experts in contamination – also noted imperfections in the science in determining risk standards.

Reducing exposure to contaminants

Another aim of this study was to explore knowledge and practices related to reducing exposure when working in potentially contaminated gardens.

We asked surveyed gardeners how they thought one should approach working in contaminated environments, and what they would do if their soil was found to be contaminated; these results are summarized in Table 4.

Among interviewed informants, most cited building raised beds and filling them with clean, imported soil as the safest and easiest way to manage potential soil contamination. Among surveyed gardeners, using raised beds was a common practice. The majority (74%) reported growing at least some crops in raised beds, and 50% said they use raised beds exclusively. Some informants, however, alluded to concerns regarding limitations of raised beds. One federal agency employee noted the possibility that the soil used to fill raised beds may be contaminated, particularly if it was taken from a questionable source, and that plant roots may extend down into contaminated soils below the raised bed, potentially allowing contaminants to enter plant tissues. One urban farmer also suggested that people may be exposed to contaminants from soil not covered by the raised bed (e.g., if native soil is kicked up), and that gardeners who use raised beds might underestimate these risks and not test the underlying soil.

We also asked informants and surveyed gardeners about soil remediation. One City government representative suggested that gardeners may not necessarily know how to remediate the soil and that guidance is needed to provide direction. When surveyed gardeners were asked, for example, whether they thought planting sunflowers in contaminated soil would effectively remediate it, 9% incorrectly said yes and 51% were unsure.

Information sources

Another aim of this study was to understand where gardeners obtain information about soil contamination. Surveyed gardeners (see Table 5) and interviewed informants most commonly mentioned gardening support institutions, particularly the agricultural extension office and its Master Gardener program.

One theme that emerged from informant interviews was the need for a central repository where gardeners could access information about soil contamination. Most informants thought this should be offered through an organization that gardeners already associate with gardening information. Compared to the more formal services of the agricultural extension, the community-based CGRN was cited as having "the biggest direct communication with community gardeners in the city" and being "more accessible" than more "bureaucratic" organizations. One informant noted, however, that while it provides a valuable network for gardeners, within CGRN "a lot of misinformation gets shared." The agricultural extension was thought to be the traditional place where gardeners and farmers would think to access soil

Table 4. Open-ended responses to questions about reducing exposure in contaminated environments.

Response		%
Stop growing produce in contaminated areas, and/or stop eating produce grown in contaminated areas		50
Remove surface contaminants		26
	Wash produce	26
	Peel root crops	3
Remediate soil		26
	Install a barrier over contaminated soil	9
	Add soil amendments (e.g., compost or minerals)	9
	Bioremediate, phytoremediate, and/or mycoremediate	9
	Remove contaminated soil	9
	Remediate (non-specific)	4
Grow in raised beds or containers		17
Only grow certain crops (e.g., not root vegetables)		13
Wear gloves		9
Wash hands		6
Apply mulch (e.g., to reduce splashing on crops)		3
Continue using the same methods		3
Seek out more information		29
Don't know		24

"What methods should one use to grow, harvest or handle produce grown in contaminated environments?" and "What would you do if you found out your soil was contaminated?"

contamination information, but informants noted it had "not historically had a real urban presence" and was not prepared to deal with issues common in cities, such as urban soil contamination. One City government representative saw the municipal government as best positioned to gather, hold, and disseminate information on soil contamination:

[G]overnment can play a really important role in doing the due diligence and gathering all of that stuff and then making it publicly available. … I think we have the institutional longevity to hold on to that [information]. … And we need to have this historical data so if that [soil] test has already been done you don't need to go do it again.

Table 5. Open-ended responses to "Where do you get information on gardening practices?"

Response		%
Gardening support organizations		84
	Extension office/Master Gardeners	37
	CGRN	19
	Other	4
Online		53
Other gardeners		44
Books/magazines		29
Friends/family		16

Information needs

Through informant interviews, we aimed to identify urban community gardeners' information needs related to soil contamination. Responses fell under four main topics: site history, soil testing, remediation, and minimizing exposure (see Table 6).

Regarding how best to present this information to gardeners, the overriding theme was that it must be concise, clearly organized, and use simple, straightforward language that speaks to people of varying educational levels. One City government representative noted that the information that is needed already exists, but should be combined "in a document with clear instructions that a layperson would feel comfortable using."

One community gardening organization representative also placed value in communicating the "fluid" nature of determining what levels of contaminants are considered safe, and that the "best way … [to approaching contaminant issues is] not black and white." Furthermore, informants emphasized the need to balance risk reduction messages related to soil contamination with the health, social, and environmental benefits of gardening, as well as "the values in gardening that are beyond measure."

[W]e don't want to create barriers. We want more people to be growing food. (City government representative)

Disseminating information

Another consideration is how best to disseminate information about soil contamination to community gardeners. Informants had a broad range of suggestions, and several noted that a combination of dissemination strategies – including print, online, and face-to-face information – is needed in order to reach all types of gardeners.

Table 6. Community gardeners' information needs related to soil contamination, as reported by key informants.

Site history	How to find information about past uses of a plot of land
	Which contaminants to test for, given specific past land uses
	Geographic areas of the city where there are likely to be high levels of contamination
Soil testing	Importance of obtaining a soil test prior to gardening
	Which contaminants to test for
	Why to test for certain contaminants and not others
	Where to get soil testing done
	How much soil testing costs
	How to correctly take a soil sample for a soil test
Remediation	Best practices for remediating contaminated urban soils
Minimizing exposure	How to reduce exposure risks when gardening
	Contamination risks associated with imported materials such as compost or mulch

Most informants expressed a need for more interactive, face-to-face methods of dissemination, such as individual consultations, workshops, or a citywide gardening conference. The important role of Master Gardeners was noted, for, as stated by one City government representative, they are "already tapped into the lives of the gardeners out there in the world."

[N]o matter how hard we try we may never reach half the people who want to garden. Because they're not going to ask, they're not going to look, they're not going to read, they're not going to do a Web search. … So only by person-to-person communication pushed out from the community master gardeners. (Federal agency employee)

Training needs

While we did not ask informants about their level of knowledge related to soil contamination, many spontaneously acknowledged confusion and a lack of understanding about the issue. One federal agency employee recommended that to help urban community gardeners make decisions about the safety of garden sites, training is needed to provide plant and soil experts with expertise on soil contamination.

[T]he only really education I've had about soil is what we had in our Master Gardener class. And to be honest, that was very cursory, tip of the iceberg, basic soil composition. And I don't really remember … there being a lot of information shared about contamination. (Community garden leader)

Discussion

Through surveys with urban community gardeners and key informant interviews, we characterized urban community gardeners' knowledge and perceptions of soil contaminant risks, including their knowledge of how to assess and reduce these risks, sources of information on soil contamination, and information needs related to soil contamination.

Knowledge and concerns

Our results suggest that concern about soil contaminants among community gardeners in Baltimore is generally low, particularly among established gardens. Informants indicated this is likely because gardeners assume soil contamination has already been addressed through safe soil test results, remediation, or the use of raised beds. Concern may be warranted, however, since prior studies of Baltimore soils suggest that soil contaminant levels vary widely [24–26]—even within the same garden plot [25]—and at some sites lead levels greatly exceed EPA screening levels (Table S1 in File S1). Soil lead levels have also been shown to increase over time due to atmospheric deposition [37]. Informants called for extra precautions where children may be present in gardens, echoing evidence of children's vulnerability to soil contaminants [24,29].

Surveyed gardeners' knowledge and concerns related to soil contaminants largely focused around lead. Their awareness may have been informed by recent state and city interventions in Baltimore aimed at raising awareness of child lead poisoning [38,39]. Gardeners' concerns were in accordance with tests of Baltimore soils (Table S1 in File S1), which identified high levels of lead more often than other trace elements included in analyses. Gardeners demonstrated inconsistent awareness about other potential contaminants.

Among key informants, issues that affect gardeners' ability to cultivate plants often appeared to be more salient than contamination concerns. Additionally, our results suggest that Baltimore's gardeners are more concerned about chemicals added to the gardening environment than what contaminants may already be present in soil. Gardeners' concerns about pesticides reflect the results of prior surveys that found the primary reason consumers purchase organic produce is to reduce exposure to pesticides and other chemicals [40].

Our results identified areas where gardeners' knowledge and concerns may not be concomitant with the potential health risks associated with urban soil contaminants. Efforts to address discrepancies in gardeners' knowledge, however, must be carefully crafted so as to not elevate levels of concern beyond those at which they would discontinue gardening altogether. Informants also made the important point that risk reduction messages must be balanced with the health, social, and environmental benefits of gardening.

Site history

Our results suggest gardeners recognize the importance of knowing a garden site's prior uses. Several informants indicated, however, that gardeners may lack the motivation, information and expertise to determine accurately the prior use of their garden site, or to anticipate the contaminants that may be present as a result. City-wide documentation of site histories may be an effective means to alleviate these responsibilities from gardeners.

Soil testing

Our findings also revealed potential barriers to soil testing, including not knowing how to properly sample soil from a garden, where to send soil samples for testing, and which contaminants to test for under various circumstances, as well as the perception that testing is too expensive, complicated, or cumbersome. Even when soil tests are conducted, gardeners may have difficulty interpreting the results. Given surveyed gardeners' knowledge and concerns were largely centered around lead, they may interpret a negative test result for lead as a "clean bill of health" and neglect to consider the presence of other contaminants. These and other concerns speak to the potential value of providing gardeners with cost-assistance and guidance on testing soil and interpreting results. Such services could, for example, be included as part of a lease to adopt city-owned vacant lots.

Reducing exposure to contaminants

When surveyed gardeners were asked to list practices to reduce exposure in contaminated environments (Table 4), several best management practices [25,41,42] were notably absent from responses, including reducing soil ingestion among children (e.g., by washing their hands, and reducing hand-to-mouth contact), growing produce away from busy streets, railways, and older buildings, and removing shoes to avoid tracking contaminants into the home. Few respondents alluded to the use of mulch (3%) to reduce splashing on crops; or the removal of surface contaminants, e.g., by washing produce (26%) and peeling root crops (3%). Other gaps in practices included the application of soil amendments (9%) to dilute contaminants and/or reduce their mobility or bioavailability (e.g., applying phosphorus to reduce the bioavailability of lead [29]), though such amendments may also *increase* the mobility or bioavailability of certain contaminants [43]. Some gardeners were quick to acknowledge their limited knowledge on how to approach contaminated environments, and indicated they would seek out more information (29%) or take the conservative approach of not growing produce in and/or eating produce from contaminated areas (50%).

Contaminant concerns among gardeners and informants were often alleviated by the use of raised beds, which were viewed as an easy and effective solution to managing soil contamination. As some informants noted, raised beds do not remedy the presence of contaminated soil surrounding the bed, which may be kicked up during gardening activities. Clark and colleagues [37] raise particular concerns regarding children: based on a model specific to lead-contaminated gardens that considered incidental soil ingestion, inhalation of ambient air, and consumption of tap water and garden-grown produce, an estimated 72–91% of children's lead exposure is via incidental soil ingestion. Because raised beds only cover a small percentage of land, they offer limited protection against incidental ingestion via hand-to-mouth behavior among children playing in areas with contaminated soils. Raised bed soils may also become contaminated over time; lead levels measured in urban raised beds in Dorchester, Massachusetts were found to increase by roughly 185 parts per million over a four year period due to wind-transported fine grain soil [37]. Another study explored concerns related to the use of timbers in the construction of raised beds, detecting elevated levels of arsenic in garden plots framed by chromated copper arsenate treated lumber and elevated levels of polycyclic aromatic hydrocarbons in plots framed by railroad ties [44]. In addition to these concerns, as one informants noted, soil used to fill raised beds may be contaminated, and plant roots may extend into contaminated soils below the bed. While raised beds provide some protection against contaminant exposure, they are not a panacea, and recommendations for their use should be tempered with information about their limitations.

Information and training needs

Informants voiced a need for specific information related to the management of soil contamination and indicated that gardeners need a central place where they can access such information. They noted several potential organizations that could serve such a role. Informants suggested collaboration between such organizations and City government to develop and disseminate a single set of information could yield the greatest reach.

Informant interviews suggest the major challenges in providing such information are the complexity and uncertainty surrounding the issue. Thus, there may be a role for two levels of guidance: additional training for gardening experts, to help them better understand the issues around soil contamination and how to effectively communicate risks to community gardeners; and concise, straightforward messaging for gardeners.

Study limitations

Our sample population was small, and the median income bracket among surveyed gardeners ($50,000 – 60,000) and percentage with a bachelor's degree (83%) were high relative to the population of Baltimore City in 2007–2011 [45]. For these reasons, our study population may not be representative of the average Baltimore gardener. Our qualitative findings may also be unique to the Baltimore context; however, given the lack of research on this topic, we believe the results of this study can be used as a starting point to inform educational interventions for reducing soil contamination risks among gardeners in a variety of urban contexts.

Conclusions

Through this study, we identified a range of factors, challenges, and needs related to Baltimore community gardeners' perceptions of risk related to soil contamination, including low levels of concern and inconsistent levels of knowledge about heavy metal and organic chemical contaminants, barriers to investigating a garden site's history and conducting soil tests, limited knowledge of best practices for reducing exposure, and a need for clear and concise information on how best to prevent and manage soil contamination. Key informants discussed various strategies for developing and disseminating educational material to gardeners. For some challenges, such as barriers to conducting site history and soil tests, some informants recommended city-wide interventions that bypass the need for gardener knowledge altogether. In cases where public health messages about the risks from soil contaminants are implemented, informants stressed the importance of crafting messages in ways that do not dissuade gardeners from continuing to garden in urban environments. Given the health, social, environmental, and economic benefits associated with participating in and supporting urban green spaces, it is critical to protect the viability of urban community gardens while also ensuring a safe gardening environment.

Supporting Information

File S1 Supporting Information Tables. Table S1, Levels of selected heavy metals and other trace elements detected in Baltimore soils. Heavy metals panels do not detect for the range of contaminants that may be present in soil (e.g., organic chemicals), and should be paired with information on a site's prior use before proceeding with gardening activities. Reported levels may be capped at minimum and maximum detectable levels. **Table S2, Examples of potential urban soil contaminants, recognized sources, and health effects associated with exposure. Table S3, Examples of contaminant concerns expressed by key informants.**

Acknowledgments

The authors thank the Community Greening Resource Network for their partnership on the project and access to the Baltimore urban gardening community, and Rufus Chaney of the U.S. Department of Agriculture's Agricultural Research Service for analyzing soil samples from participating gardens.

Author Contributions

Conceived and designed the experiments: JM KD AP KN. Performed the experiments: BK MP JM KD AP KN. Analyzed the data: BK MP AP. Wrote the paper: BK MP JM KD AP KN. Acquired funding for the study: KN.

References

1. Alaimo K, Packnett E, Miles RA, Kruger DJ (2008) Fruit and vegetable intake among urban community gardeners. J Nutr Educ Behav 40: 94–101. doi:10.1016/j.jneb.2006.12.003.
2. Blair D, Giesecke CC, Sherman S (1991) A dietary, social and economic evaluation of the Philadelphia urban gardening project. J Nutr Educ 23: 161–167.
3. McCormack LA, Laska MN, Larson NI, Story M (2010) Review of the nutritional implications of farmers' markets and community gardens: a call for evaluation and research efforts. J Am Diet Assoc 110: 399–408. doi:10.1016/j.jada.2009.11.023.
4. Litt JS, Soobader MJ, Turbin MS, Hale JW, Buchenau M, et al. (2011) The influence of social involvement, neighborhood aesthetics, and community garden participation on fruit and vegetable consumption. Am J Public Health 101: 1466–1473. doi:10.2105/AJPH.2010.300111.
5. Armstrong D (2000) A survey of community gardens in upstate New York: implications for health promotion and community development. Health Place 6: 319–327.
6. Patel IC (1991) Gardening's socioeconomic impacts. J Ext 29.
7. Wakefield S, Yeudall F, Taron C, Reynolds J, Skinner A (2007) Growing urban health: community gardening in South-East Toronto. Health Promot Int 22: 92–101. doi:10.1093/heapro/dam001.
8. Twiss J, Dickinson J, Duma S, Kleinman T, Paulsen H, et al. (2003) Community gardens: lessons learned from California Healthy Cities and Communities. Am J Public Health 93: 1435–1438.
9. Kuo FE, Sullivan WC, Coley RL (1998) Fertile ground for community: inner-city neighborhood common spaces. Am J Community Psychol 26: 823–851.
10. Voicu I, Been V (2006) The effect of community gardens on neighboring property values. Real Estate Econ 36: 241–283.
11. Caspersen CJ, Bloemberg M, Saris WHM, Merritt RK, Kromhout D (1991) The prevalence of selected physical activities and their relation with coronary heart disease risk factors in elderly men: The Zutphen Study, 1985. Am J Epidemiol 133: 1078–1092
12. Magnus K, Matroos A, Strackee J (1979) Walking, cycling, or gardening, with or without seasonal interruption, in relation to acute coronary events. Am J Epidemiol 110: 724–733.
13. Kaplan R (1973) Some psychological benefits of gardening. Environ Behav 5: 145–162. doi:10.1177/001391657300500202.
14. Fabrigoule C, Letenneur L, Dartigues JF, Zarrouk M, Commenges D, et al. (1995) Social and leisure activities and risk of dementia: a prospective longitudinal study. J Am Geriatr Soc 43: 485–490.
15. Simons LA, Simons J, McCallum J, Friedlander Y (2006) Lifestyle factors and risk of dementia: Dubbo Study of the elderly. Med J Aust 184: 68–70.
16. Brown KH, Jameton AL (2000) Public health implications of urban agriculture. J Public Health Policy 21: 20–39.
17. Kuo FE (2001) Coping with poverty: impacts of environment and attention in the inner city. Environ Behav 33: 5–34. doi:10.1177/00139160121972846.
18. Kuo FE, Taylor AF (2004) A potential natural treatment for attention-deficit/hyperactivity disorder: evidence from a national study. Am J Public Health 94: 1580–1587.
19. Wells NM (2000) At home with nature: effects of "greenness" on children's cognitive functioning. Environ Behav 32: 775–795.
20. Kuo FE, Sullivan WC (2001) Environment and crime in the inner city: does vegetation reduce crime? Environ Behav 33: 343–367. doi:10.1177/0013916501333002.
21. Bolund P, Hunhammar S (1999) Ecosystem services in urban areas. Ecol Econ 29: 293–301. doi:10.1016/S0921-8009(99)00013-0.
22. Andersson E, Barthel S, Ahrné K (2007) Measuring social-ecological dynamics behind the generation of ecosystem services. Ecol Appl A Publ Ecol Soc Am 17: 1267–1278.
23. Meuser H (2010) Causes of soil contamination in the urban environment. In: Alloway BJ, Trevors JT, editors. Contaminated urban soils. Springer. pp. 29–94.
24. Mielke HW, Anderson JC, Berry KJ, Mielke PW, Chaney RL, et al. (1983) Lead concentrations in inner-city soils as a factor in the child lead problem. Am J Public Health 73: 1366–1369.
25. Chaney R, Sterrett SB, Mielke HW (1984) The potential for heavy metal exposure from urban gardens and soils. In: Preer JR, editor. Proc. symp. heavy metals in urban gardens. USDA ARS. pp. 37–84.
26. Pouyat R V, Yesilonis ID, Russell-Anelli J, Neerchal NK (2007) Soil chemical and physical properties that differentiate urban land-use and cover types. Soil Sci Soc Am J 71: 1010. doi:10.2136/sssaj2006.0164.
27. Schwarz K, Pickett STA, Lathrop RG, Weathers KC, Pouyat R V, et al. (2012) The effects of the urban built environment on the spatial distribution of lead in residential soils. Environ Pollut 163: 32–39. doi:10.1016/j.envpol.2011.12.003.
28. Yesilonis ID, Pouyat RV, Neerchal NK (2008) Spatial distribution of metals in soils in Baltimore, Maryland: role of native parent material, proximity to major roads, housing age and screening guidelines. Environ Pollut 156: 723–731. doi:10.1016/j.envpol.2008.06.010.
29. Hettiarachchi GM, Pierzynski GM (2004) Soil lead bioavailability and in situ remediation of lead-contaminated soils: a review. Environ Prog 23: 78–93. doi:10.1002/ep.10004.
30. Morgan R (2013) Soil, heavy metals, and human health. In: Brevik EC, Burgess LC, editors. Soils and human health. CRC Press. pp. 59–82.
31. Finster ME, Gray KA, Binns HJ (2004) Lead levels of edibles grown in contaminated residential soils: a field survey. Sci Total Environ 320: 245–257. doi:10.1016/j.scitotenv.2003.08.009.
32. Smolders E (2001) Cadmium uptake by plants. Int J Occup Med Environ Health 14: 177–183.
33. Ramirez-Andreotta MD, Brusseau ML, Artiola JF, Maier RM (2013) A greenhouse and field-based study to determine the accumulation of arsenic in common homegrown vegetables grown in mining-affected soils. Sci Total Environ 443: 299–306. doi:10.1016/j.scitotenv.2012.10.095.
34. Harms AMR (2011) Determining and meeting the educational needs of students and urban gardeners and farmers on urban soil quality and contamination topics.Master's Thesis, Kansas State University.
35. Harms AMR, Presley DR, Hettiarachchi GM, Thien SJ (2013) Assessing the educational needs of urban gardeners and farmers on the subject of soil contamination. J Ext 51.
36. Meuser H (2012) Soil remediation and rehabilitation: treatment of contaminated and disturbed land. Alloway BJ, Trevors JT, editors. Spinger. p. 128.
37. Clark HF, Hausladen DM, Brabander DJ (2008) Urban gardens: lead exposure, recontamination mechanisms, and implications for remediation design. Environ Res 107: 312–319. doi:10.1016/j.envres.2008.03.003.
38. Maryland Department of the Environment (2000) State, city unveil new lead awareness campaign. MD Environ 4.
39. City of Baltimore Health Department (2010) Data watch: Maryland Department of the Environment releases 2009 lead poisoning data. Available: http://baltimorehealth.org/press/2010-08-27-lead-report-pr.pdf. Accessed 01 January 2014.
40. Hughner S, McDonagh P, Prothero A, Shultz CJ, Stanton J (2007) Who are organic food consumers? A compilation and review of why people purchase organic food. J Consum Behav 6: 94–110. doi:10.1002/cb.
41. U.S. Environmental Protection Agency (2011) Brownfields and urban agriculture: interim guidelines for safe gardening practices. Available: http://www.epa.gov/brownfields/urbanag/pdf/bf_urban_ag.pdf. Accessed 2014 Jan 1..
42. Kessler R (2013) Urban gardening: managing the risks of contaminated soil. Environ Health Perspect 121. A327–A333.
43. Scheckel KG, Diamond GL, Burgess MF, Klotzbach JM, Maddaloni M, et al. (2013) Amending soils with phosphate as a means to mitigate soil lead hazard: a critical review of the state of the science. J Toxicol Env Heal B Crit Rev 16: 337–380. doi:10.1080/10937404.2013.825216.
44. Heiger-Bernays W, Fraser A, Burns V, Diskin K, Pierotti D, et al. (2009) Characterization and low-cost remediation of soils contaminated by timbers in community gardens. Int J Soil, Sediment Water 2: 1–19.
45. U.S. Census Bureau (2013) Baltimore City, Maryland. state & county quickfacts. Available: http://quickfacts.census.gov/qfd/states/24/24510.html. Accessed 2013 June 28.

Microbial Community Diversities and Taxa Abundances in Soils along a Seven-Year Gradient of Potato Monoculture Using High Throughput Pyrosequencing Approach

Xing Liu[1,2], Junlian Zhang[3,4], Tianyu Gu[5], Wenming Zhang[1,2], Qirong Shen[6], Shixue Yin[5]*, Huizhen Qiu[1,2]*

1 College of Resources and Environmental Sciences, Gansu Agricultural University, Lanzhou, Gansu Province, China, **2** Gansu Provincial Key Laboratory of Aridland Crop Science, Gansu Agricultural University, Lanzhou, Gansu Province, China, **3** College of Agronomy, Gansu Agricultural University, Lanzhou, Gansu Province, China, **4** Gansu Key Laboratory of Crop Genetic and Germplasm Enhancement, Gansu Agricultural University, Lanzhou, China, **5** College of Environmental Science and Engineering, Yangzhou University, Yangzhou, Jaingsu Province, China, **6** Jiangsu Provincial Key Lab of Organic Solid Waste Utilization, Nanjing Agricultural University, Nanjing, China

Abstract

Background: Previous studies have focused on linking soil community structure, diversity, or specific taxa to disturbances. Relatively little attention has been directed to crop monoculture soils, particularly potato monoculture. Information about microbial community changes over time between monoculture and non-monoculture treatments is lacking. Furthermore, few studies have examined microbial communities in potato monoculture soils using a high throughput pyrosequencing approach.

Methodology/Principal Findings: Soils along a seven-year gradient of potato monoculture were collected and microbial communities were characterized using high throughput pyrosequencing approach. Principal findings are as follows. First, diversity ($H_{Shannon}$) and richness (S_{Chao1}) indices of bacterial community, but not of fungal community, were linearly decreased over time and corresponded to a decline of soil sustainability represented by yield decline and disease incidence increase. Second, *Fusarium*, the only soilborne pathogen-associated fungal genus substantially detected, was linearly increased over time in abundance and was closely associated with yield decline. Third, *Fusarium* abundance was negatively correlated with soil organic matter (OM) and total nitrogen (TN) but positively with electrical conductivity (EC). Fourth, *Fusarium* was correlated in abundances with 6 bacterial taxa over time.

Conclusions: Soil bacterial and fungal communities exhibited differential responses to the potato monoculture. The overall soil bacterial communities were shaped by potato monoculture. *Fusarium* was the only soilborne pathogen-associated genus associated with disease incidence increase and yield decline. The changes of soil OM, TN and EC were responsible for *Fusarium* enrichment, in addition to selections by the monoculture crop. *Acidobacteria* and *Nitrospirae* were linearly decreased over time in abundance, corresponding to the decrease of OM, suggesting their similar ecophysiologial trait. Correlations between abundance of *Fusarium* with several other bacterial taxa suggested their similar behaviors in responses to potato monoculture and/or soil variables, providing insights into the ecological behaviors of these taxa in the environment.

Editor: Kevin McCluskey, University of Missouri, United States of America

Funding: This work was supported by the Agricultural Ministry of China (201103004), the Natural Science Foundation of China (31360500) and the National Nature Science Foundation of China (41071177). The funders had no role in study design, data collection and analysis, decision to publish, or preparation of the manuscript.

Competing Interests: The authors have declared that no competing interests exist.

* E-mail: hzqiu@gsau.edu.cn (HQ); sxyin@yzu.edu.cn (SY)

Introduction

Soil microbial community is extremely important in maintaining soil quality, functions, sustainability and global nutrient cycling. Majority of past studies focused on linking the microbial community structure, diversity, or specific taxa to soil variables/disturbances, such as soil pH [1,2], soil types [3], soil texture [4], soil layers [5], plant species [6], nitrogen level [7], temperature [8], water regime [9], tillage [10,11], amendments [12], biocontrol agents [13], land uses [14], and contaminations [15]. However, relatively limited attention has been directed to the links between crop monoculture and soil microbial communities.

Crop monoculture is when the same crop is grown year after year on the same soil, without interruption crops that could potentially induce different crop-specific microbial communities [16]. Crop monoculture can be viewed as a simple and an ideal model system to study the relationship between agricultural practices and soil microbial communities. When crop monoculture is practiced, the microbial community is continuously exposed to the roots of the same crop that selects and enriches certain groups of microorganisms including yield-debilitating populations (i. e. soilborne pathogens) of that crop [17]. This consequently shapes the soil microbial community that can often be easily observed. Several studies have determined differences of soil microbial communities between rotation/mixed culture and monoculture of cotton [18], maize [19], wheat (and maize) [20], rice [21], soybean [22], oilseed rape [23] and potato [24]. With a few notable exceptions [24], these studies typically used an one time-point soil sampling strategy. Soil samples were collected at a given time-point after a certain time span, generally years, over which the monoculture was practiced. The microbial communities were then compared between monoculture and non-monoculture treatments. This sampling strategy is useful for revealing differences in communities between treatments, but it does not allow measurements of community changes over time (i. e. the number of years a crop monoculture has been maintained). This is important information to understand the assembly processes of microbial communities under continuous selection of crop monoculture practice [25,26,27,28].

An alternative strategy is to sample soils along a year-gradient of monoculture. This strategy allows observations of changes in patterns of different microbial taxa over the years that may occur in a predictable manner within the experimental time span. In addition, the change pattern of one taxon may be correlated with another in abundance, allowing us to perceive the behaviors of these taxa in response to potato monoculture. In spite of these benefits, studies using year-gradient samples are limited in literature. Some early reports followed this strategy (see review [29]), but community changes were based on culture-dependent methods only, which have been proven to have strong biases against yet-to-cultured majorities [30]. The relationship between crop monoculture practices and the soil microbial community is thus not sufficiently understood. Relevant knowledge on potato monoculture agroecosystems is particularly lacking.

In a previous study [31], we examined the fungal community in potato monoculture soils. Further re-examination of our original data revealed that the diversity (H_{Shannon}) and richness (S_{Chao1}) indices of the fungal community were nearly un-changed over the years of monoculture (time). Furthermore, *Fusarium* was the only soilborne pathogen-associated fungal genus substantially detected using a 454 pyrosequencing approach. *Fusarium* was significantly enriched over time, and closely associated with disease incidence observed in the field (see the Results section below). These findings encouraged us to further examine whether the diversity and richness indices of the bacterial community were also significantly changed. Specifically we are interested in knowing change patterns of different bacterial taxa with reference to that of *Fusarium.* Based on the above findings, we hypothesized that the overall soil bacterial community were significantly changed under the continuous monoculture pressure, and abundance of some bacterial taxa were correlated with *Fusarium* abundances over time. As soil variables play key roles in microbial community structuring[1,7,14], associations between soil variables and the microbial change patterns were also studied.

Materials and Methods

Ethics Statement

Authors confirm that no specific permissions were required for the locations/activities involved in this study, because the land accessed is publically owned and no protected species were sampled. Authors also confirm that the field studies did not involve endangered or protected species.

Field experiment, soil sampling and variable measurements

Detailed descriptions of the field experiment, soil sampling and variable measurements can be found elsewhere [31]. Only a brief description is given here. The experimental field, located at the Tiaoshan Farm, a warm terrestrial arid area with a mean annual precipitation 186 mm is located in Gansu Province, China. This field contained 21 plots, each being 9.0×6.1 m in size. Prior to potato monoculture, the field was under a corn-potato rotation. Each year, starting from year 2005, 3 plots (replicates) were selected for practicing potato ("Atlantic" cultivar) monoculture, leaving other plots to continue with the corn-potato rotation. The previous crop of selected plots or the potato year of the rotation was always corn. By 2011, all plots were used up. This experimental design provides opportunity to collect soil samples after potato monoculture from 1 to 7 years simultaneously.

Potato was typically seeded in April 25 with a few days variation among years. Tuber pieces were buried on the top of raised paths (~40 cm in height and ~135 cm in bottom width) and plants were spaced 17 cm apart along the row. Two rows were planted on each raised path with 70 cm between the two rows. This resulted in a plant density of 460 plants per plot (equivalent to 84,075 plants ha^{-1}). Blended fertilizer (15-15-15), additionally supplemented with urea and K_2SO_4, was used at the rate of 210 kg N ha^{-1}, with the ratio $N:P_2O_5:K_2O$ being 1.4:1.0:2.0. Nitrogen form in blended fertilizer was $(NH_4)_2SO_4$. Fertilizers were applied before seeding by machine. All other field management activities were performed manually. Once seeded and fertilized, the raised paths were covered manually with a plastic film. The field was irrigated by introducing water into the ditch between two raised paths through pipes. Irrigation was done three times during the growth period, typically at June 1 (seedling stage), July 1 (early flowering stage) and July 20 (tuber enlargement stage) at the rate of 2,700 t ha^{-1} each time. Potato was harvested in later August. For yield evaluation, the tubers were dug out manually. Stolons, underground stems, true roots, scale leaf rhizomes and adhered soils were removed. Tubers smaller than 2 cm in diameter were considered as developing tubers and were excluded from the yield calculations. Disease incidence was calculated by totaling the number of plants with symptoms of any recognizable disease, and converting to percent by dividing by the total number of plants and then multiplying by 100. Fusarium wilt was the major disease, covering ~90% of diseased plants. The disease incidence was recorded three times during the growth period, but only the records at harvest time are presented.

Soil samples were collected in July 20, 2011. The plots with 5 years of monoculture were not sampled. From each plot, 15 spots were randomly sampled from the 0–20 cm soil layer and the soil from these spots were placed into a single aggregate sample and then well mixed. Totally, 18 samples were obtained. Samples were put into sterile plastic bags, placed into an ice box, transported to laboratory and used as soon as possible, or stored in a refrigerator at −80°C if not immediately used. Selected soil variables measured included organic matter (OM, by dichromate oxidation), total nitrogen (TN, by total Kjeldahl N), NH_4^+–N, NO_3^-–N (by 1 M KCl extraction), pH and electrical conductivity (EC) (both at 5:1, water:soil ratio).

DNA extraction, amplification, sequencing and sequence treatment

For each soil sample, bulk DNA was extracted from three 0.5 g aliquots of soils (dry weight basis) with a FastDNA SPIN Kit for Soil (Bio 101, Carlsbad, Calif.) following the manufacturer's instructions. The triplicate DNA samples were pooled together, resulting 18 DNA samples in total, representing 18 field plots. The integrity of the extracted DNA was confirmed by running on 0.8% agarose gel with 0.5 TBE buffer (45 mM Tris-borate, 1 mM EDTA, pH 8.0). The extracted DNA (as well as PCR products) was quantified using TBS-380 Mini-Fluorometer (Turner Biosystems, CA, USA). All DNA samples were diluted to 100 ng μl^{-1} and used as PCR templates.

For the fungal community, detail description on primer pairs, amplification, sequencing and sequence treatment can be found elsewhere [31]. Operational taxonomic units (OTUs) were assigned at the 97% identity level.

For the bacterial community, the primer pair 563F (5′-AYTGGGYDTAAAGVG-3′) and 802R (5′-TACNVGGG-TATCTAATCC-3′) [32] was used for amplifying the V4 hypervariable region of the 16 S rRNA gene. The PCR mixture (20 μl) contained 5×FastPfu buffer 4 μl, 2.5 mM dNTPs 2 μl, each primer(5 μM) 0.4 μl, DNA 0.5 μl, FastPfu polymerase 0.4 μl, and ddH_2O (TransGen Biotech, Beijing, China). DNA samples were amplified using the following program: 95°C 2 min, 30 cycles: 95°C 30 sec, 55°C 30 sec, 72°C 30 sec; 72°C 5 min; 10°C forever. A second round of PCR was performed under the same conditions using Amplicon Fusion Primers 5′-*A*-*x*-563F-3′ and 5′-*B*-802R-3′, where *A* and *B* represent the pyrosequencing adaptors (CCATCTCATCCCTGCGTGTCTCCGACGACT and CCT-ATCCCCTGTGTGCCTTGGCAGTCGACT) and *x* represents an 10 bp-tag for sample identification. For each sample, three independent PCRs were performed. The triplicate products were pooled and purified using AxyPrep PCR Clean-up Kit (Axygen Biosciences, CA, USA). The purified PCR products from different soil samples contained 1.06~3.30 ng μl^{-1} DNA and were pooled in equal quantities for pyrosequencing in one run, which was performed commercially at Shanghai Majorbio Bio-pharm Biotechnology Co., Ltd., on a Roche GS 454 FLX platform (http://www.majorbio.com).

Sequence reads were treated using MOTHUR v. 1.29.2 [33]. The forward primer (563F), tags and adaptor *A* and *B* were removed. Low quality nucleotides were filtered using qwindowaverage 27 and qwindowsize 50. Reads shorter than 120 bp were removed. Filtered sequences were aligned against Silva database (106 v.). Chimera reads were removed using both *chimera.slayer* and *chimera.uchime* scripts. A few reads belonging to *Archaea* were excluded. OTUs were assigned at the 97% identity level. On average, 11,950 original reads per sample were obtained, and after filtering 8,750 high-quality sequences per sample were left, of which 97.5% were 155 bp in length. The OTU counts were normalized into relative abundances (%).

Statistics was done with SPSS (v 19) and curve fitting with Sigmaplot (v. 12.5). Unless stated, *p* values <0.05 were considered significant. Raw sequence data for bacteria and fungi were submitted to DDBJ database under the accession number DRA001116 and DRA000962 respectively.

Results

Overall microbial communities

The overall bacterial communities are summarized in Figure 1. For comparison purposes, the summarized fungal community is also shown. Average relative abundances are listed in a column to the right of the pie chart. As for bacteria, *Acidobacteria*, *α-Proteobacteria* and *Gemmatimonadetes* were the top three abundant phyla with average relative abundances over 10%. As for fungi, "unidentified fungus", *Hypocreales* and *Sordariales* were the top three abundant taxa. All the fungal taxa in Figure 1A, except "unidentified fungus" and some "others", belong to phylum *Ascomycota*, which accounted for 58.2% in average relative abundance. *Basidiomycota* accounted for only 0.6%. The richness (S_{Chao1}) and diversity ($H_{Shannon}$) indices of the fungal community showed nearly flat lines over time, with slightly increasing trends by fitting to a linear model. Note that the *p* value of the fitted line for the fungal diversity index was not significant. In contrast, the richness and diversity indices of the bacterial community showed significant decreasing trends over time.

The change trends in relative abundances of these fungal taxa over 18 samples were shown in Figure S1. By curve fitting, the abundances of *Hypocreales*, *Pezizomycotina*, *Onygenales* and "fungi unclassified" showed significant increasing trends over time whereas *Mortierellales* and "*Ascomycota* incertae sedis" showed decreasing trends. By *p* values, *Hypocreales* was best fitted. *Hypocreales* and *Sordariales* were further broken down into genera and their relative abundances across soil samples were listed in Table S1. Within order *Hypocreales*, *Fusarium* was the top abundant genus with an average relative abundance 8%. All other genera were 1% or below in abundance. Within order *Sordariales*, *Chaetomium* and *Thielavia* were the top abundant genera with average relative abundances over 3%. All other genera in this order (except "unidentified fungus") were 1% or lower in abundance.

Yield trend and *Fusarium* abundance

Potato yield declined substantially after 2 years of monoculture, and was accompanied by an increase in disease incidence (Figure 2). Major diseases found in the field were fungi-associated, such as fusarium dry rots, late bright and black scurf/stem canker [31] that are caused by *Fusarium* spp., *Phytophthora infestans*, and *Rhizoctonia solani*, respectively. Because all these pathogens are soilborne, we further looked at how abundant they were in the soils. Because phytopathogens are often species-, even variety-specific, a pathogen in a strict taxonomic sense must be identified by pure culture, and can not be identified by environmental DNA-based approaches only (as in this present study). Therefore, we grouped the fungal abundances by genus that were detected in the soil samples, and checked them from a list of well-recognized potential soilborne potato pathogens [34,35] by genus. If found among the list, the genera detected were referred to as pathogen-associated genera. Obviously we are not saying all *Fusarium* are pathogens, but viewing the data in this way showed that *Fusarium* was the only pathogen-associated fungal genus substantially detected in our soil samples with a range of relative abundance from 1.9% to 14.1% (Table S2). Other pathogen-associated genera were either below the detection limit (e. g. *Phytophthora* and *Rhizoctonia*) or detected but with too low relative abundances (e. g. *Phoma* and *Pythium*). The relative abundance of *Fusarium* increased with the increase of monoculture time, with the increase of disease incidence, and with the decrease of yield (Figure 2).

Correlation between *Fusarium* abundance and soil variables

Using a linear model to fit soil variables with the year of monoculture, soil EC and pH were significantly increased over time, whereas OM, TN, and NH_4^+-N were decreased (Figure S2). NO_3^--N was also decreased but did not fit the model. Further fitting soil variables with *Fusarium* abundance, it was shown that

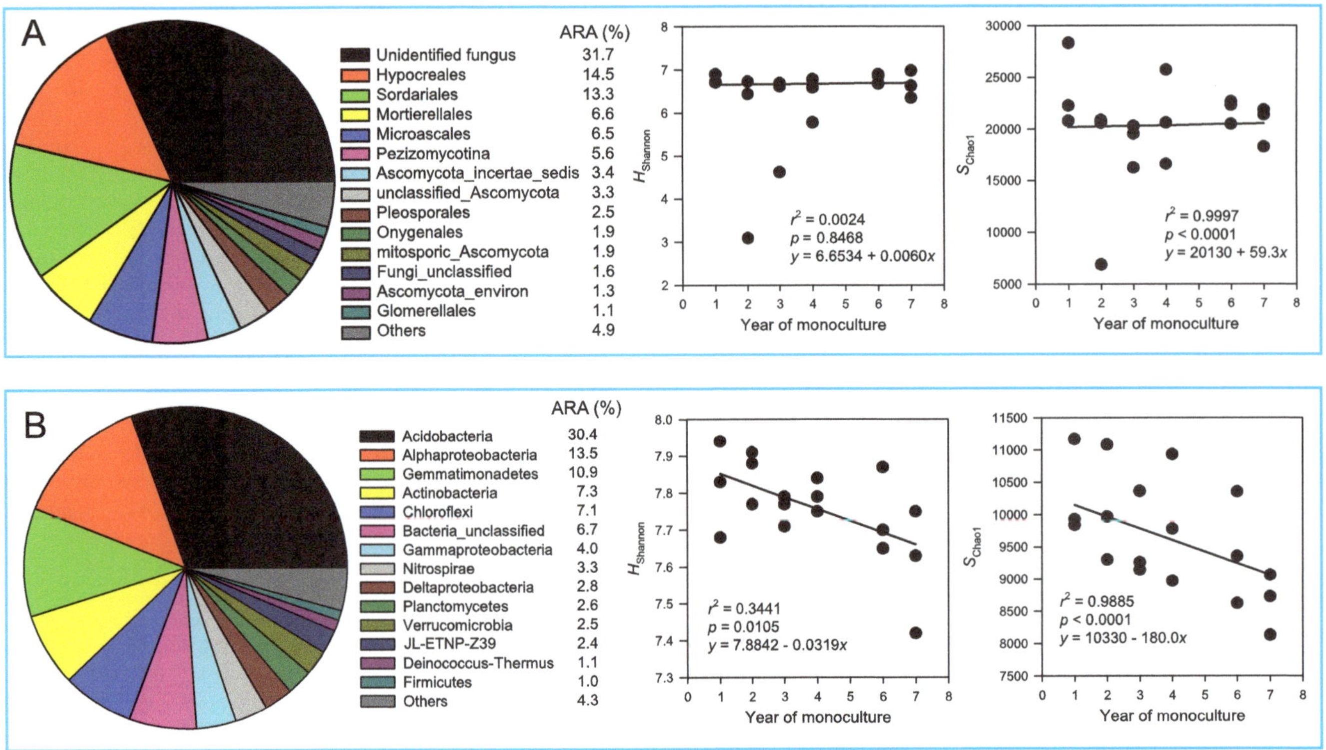

Figure 1. Overall fungal (A) and bacterial (B) communities in the soil samples along a year-gradient of potato monoculture. Pie charts show the average relative abundance over 18 samples. OTUs were assigned at the 97% identity level, based on which the $H_{Shannon}$ and S_{Chao1} indices were calculated, which are shown by graphs. Phylogenetic lineages of bacteria were assigned by phylum/subphylum and of fungi by order or higher levels. ARA stands for average relative abundance. The "unidentified fungus" is a taxon able to be assigned to kingdom Fungi but unable to any lower taxon.

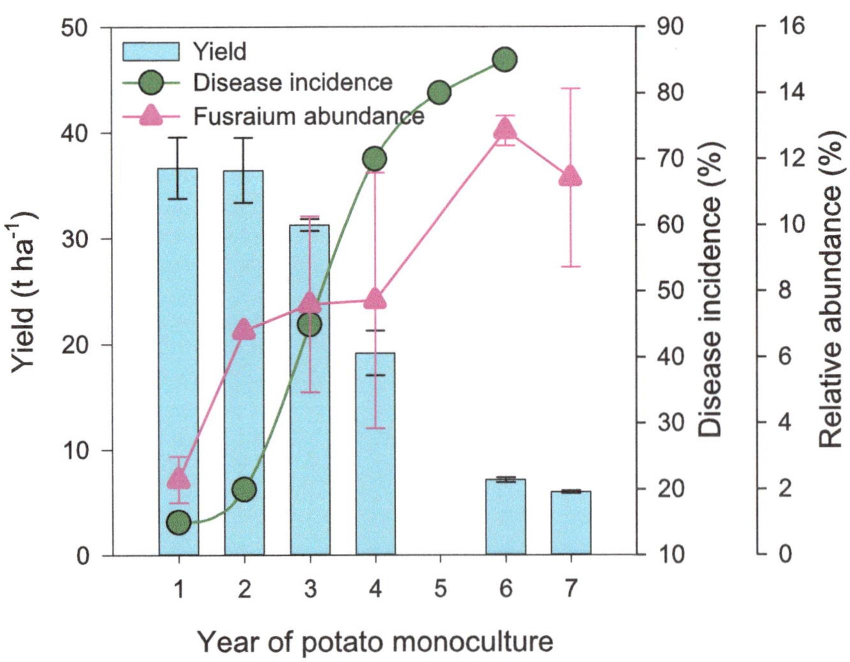

Figure 2. The change of potato yield, disease incidence and *Fusarium* abundance over the years of potato monoculture. Disease incidences shown in this figure were measured at harvest time.

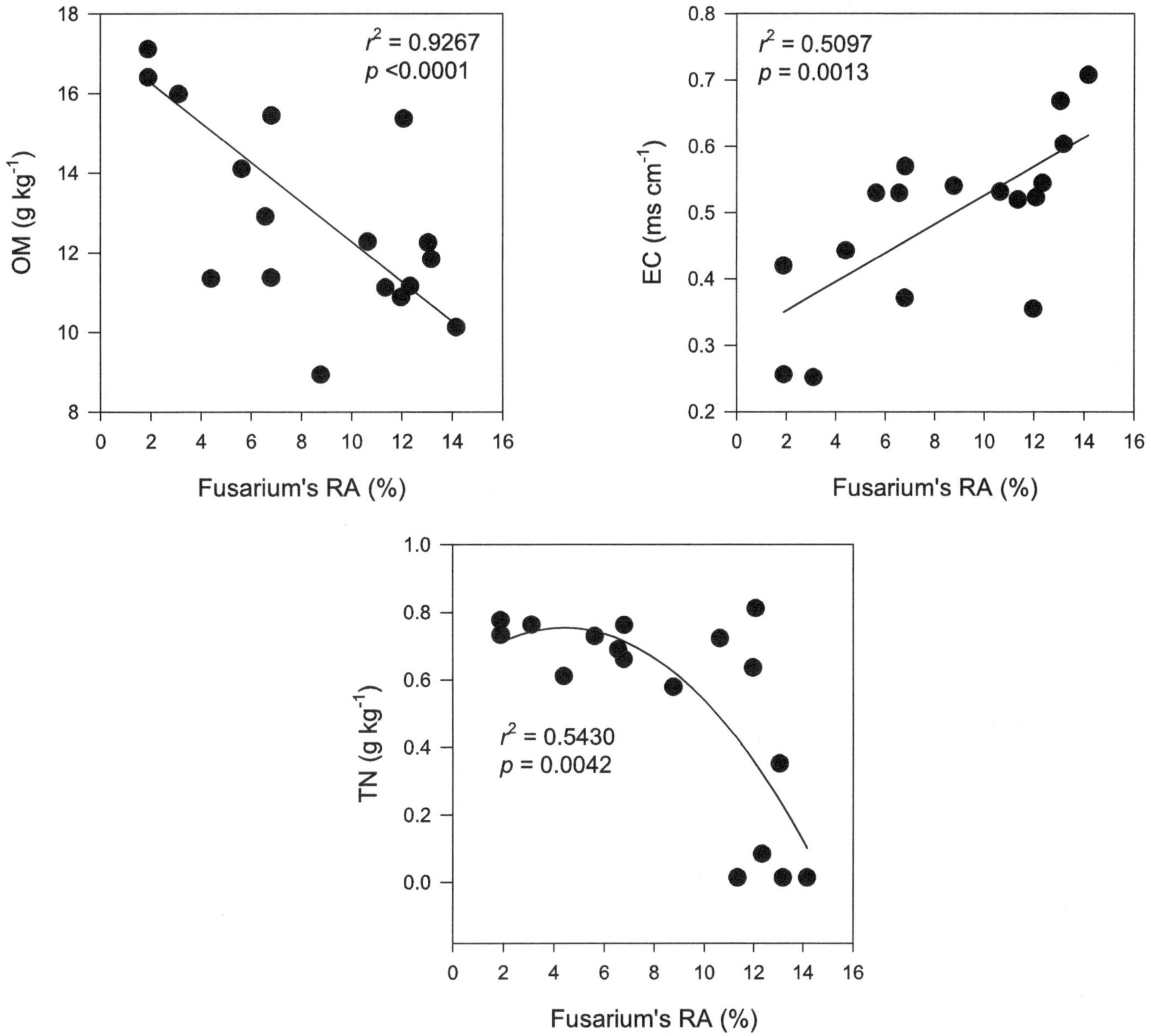

Figure 3. The correlations between *Fusarium* abundances and soil variables.

only OM, TN and EC were significantly correlated with the relative abundance of *Fusarium* (Figure 3). Other soil variables (NH_4^+-N and pH) were not significantly correlated, or not fitted with any of the models integrated in the software (NO_3^--N).

Correlation between abundance of bacterial taxa and the year of monoculture

To determine what bacterial taxa were significantly changed over time (the year of monoculture), correlations between taxa abundances and time were conducted. In total, seven taxa showed significant correlations (Table 1). The sum of average relative abundances of these taxa was 57.6%. *Acidobacteria* and *Nitrospirae* were linearly decreased in abundance (Figure 4), whereas *Gemmatimonadetes*, *Chloroflexi*, *Planctomycetes*, and JL-ETNP-Z39 were linearly increased. β-*Proteobacteria* was also decreased but was best fitted with a non-linear model.

Regression between *Fusarium* and other bacterial taxa in abundances

Regressions between *Fusarium* and the bacterial taxa abundances showed that among the seven taxa that significantly changed over time (Table 1), six showed significant regressions (Figure 5). *Acidobacteria* and *Nitrospirae* showed negative linear regressions whereas β-*Proteobacteria* showed negative non-linear regression. *Gemmatimonadetes*, *Chloroflexi* and JL-ETNP-Z39 showed positive linear regressions.

Discussion

The relative abundance rank profiles of fungal and bacterial phyla reported in this study generally agree with previous pyrosequencing surveys of soil microbial communities. As for bacteria, *Acidobacteria* and α-*Proteobacteria* were the top two abundant phyla across all samples (Figure 1), agreeing with the abundance rank profiles derived from 88 soils collected from North and South America [1], as well as from forest and grassland soils in Germany[2]. Acosta-Martínez et al [36] also reported that *Actinobacteria* and *Bacteriodetes* were the top two phyla in Texas (USA) soils with different managements. The average relative abundance of *Gemmatimonadetes* in our samples was 10.9% (Figure 1), which is much higher than the <3% reported in previous studies [1,2,36]. DeBruyn et al [37] reported that *Gemmatimonadetes* is a drier soil-adapted phylum, which is probably the reason why its abundance was so high in our soils. As for fungi, *Ascomycota* was the top abundant phylum in our samples, accounting for 58.2% in average relative abundance, generally agreeing with the results reported by Buée et al [38] and by McGuire et al [39] for forest soils. However, our results contrasted with those derieved from tallgrass prairie soils [40], in which basidiomycetes (*Basidiomycota*) accounted for 54% relative abundance and ascomycetes (*Ascomycota*) accounted for 41%. In our samples, *Basidiomycota* accounted for 0.6% only. The difference was probably due to different soils in nature. Different primers targeting different genes/regions and different databases used for phylogentical assignment also affected the results.

Using the soil samples along a year-gradient of potato monoculture, several facts about potato monoculture and soil microbial communities were revealed. First, Pearson, Kendall, and Spearman correlation methods were used simultaneously in order to catch all possible significant correlations between bacterial taxa and the years of monoculture. Different methods have different merits. Pearson method measures the degree of linear dependence between two variables, Kendall measures the similarity of the orderings of the variables when ranked by each of the quantities (rank correlation), and Spearman measures how well the relationship between two variables can be described using a monotonic function. Generally speaking, Pearson measures linear relationships, whereas Kendall and Spearman measure nonlinear relationships. Results showed that the majority of the bacterial community (57.6% by abundance) was significantly changed over time (Table 1), including the top abundant taxon *Acidobacteria*. Bacterial diversity and richness indices were significantly and linearly decreased with an increase of monoculture time (year), but the corresponding values for fungal community was nearly unaffected (Figure 1). This suggests differential responses of bacterial and fungal communities to potato monoculture. Using fatty acid methyl ester profiles to characterize soil microbial communities, several other studies [41,42,43,44,45] also reported differential responses of fungal and bacterial communities to potato monoculture. Larkin and Honeycutt [42], for example, reported that continuous potato resulted in lower proportions of polyunsaturated fatty acids, which are most closely associated with fungi, resulting in a lower fungi-to-bacteria ratio than several other rotations. By substrate utilization approach, Larkin et al [44] reported that potato monoculture produced the lowest values in soil microbial activity, richness and diversity among other rotation treatments. Although these studies did not use high throughput sequencing approach as present study did, similar conclusion was drawn, suggesting that the differential responses of fungal and bacterial communities were common in potato monoculture agroecosystems. However, whether differential responses exist in other crop monoculture agroecosystems is not clear. A few studies have reported decreased bacterial diversities in monoculture ecosystems of rice [21], cotton [36], soybean [22] and watermelon [46]. Unfortunately, these studies did not include fungal communities, or they were included but such differential responses were not looked for. Nevertheless, our results point out that the link between microbial diversity and soil quality/sustainability depends on the focus of the microbial community, i.e. bacterial or fungal. In the present case, the soil sustainability decline represented by a potato yield decline and disease incidence increase (Figure 2) corresponded to a decrease in bacterial, but not fungal, diversity.

Second, *Fusarium* was the only soilborne pathogen-associated fungal genus substantially detected by our present high-throughput pyrosequencing protocol (Table S1 and Table S2), and was continuously and significantly enriched (Figure 2). The change pattern of *Fusarium* was closely associated with disease incidence and yield decline, suggesting its reponsibility for yield decline. *Fusarium* enrichment in soils is reportedly related to disease resistance of cultivars used [46,47]. The cultivar "Atlantic" used in the present experiment was reportedly susceptible to *Fusarium roseum* [48]. Root exudates and debris left in soils from previous crops, especially a monoculture potato crop that was sampled in this study, would likely favor *Fusarium* proliferation and enrichment in soils.

Third, *Fusarium* enrichment over time was correlated with soil variables. Among the six selected variables, OM and TN were significantly and negatively correlated with *Fusarium* abundances, and EC was significantly and positively correlated (Figure 3). These three correlated soil variables are all associated with field management. The site of the experimental plots for this study was located in an arid region where the mean annual precipitation is only 186 mm. Cations introduced from fertilizers are not easily leached out of the soils, leading to an increase of EC values (Figure S2). Arid soils are also water limited. Irrigation improves the soil water regime, promotes microbial activity, thus enhancing decomposition of soil OM [49]. Without organic amendments to compensate OM loss (as in the present experiment), soil OM content will be inevitably decreased (Figure S2), instead of being preserved by dryness. TN will be decreased as well because it is mostly associated with soil OM. Therefore, soil management causes change in the soil variables of OM, TN and EC, thus causing *Fusarium* enrichment in soils.

Fourth, among the seven taxa that were significantly changed over time under continued potato monoculture pressure (Table 1), *Acidobacteria* and *Nitrospirae* were linearly decreased over time in abundance (Figure 4). This is probably relevant to their ecophysiologies. As for *Acidobacteria*, it has been shown that the abundance of *Acidobacteria* was negative correlated with soil carbon mineralization rates [50]. Although the carbon mineralization rate was not measured in this study, it was reasonable to believe that microbial activity and carbon mineralization rates in the arid soils

Table 1. Correlation coefficient between abundance of microbial taxa and the year of monoculture[a].

Taxa	Average RA (%)	Pearson	Kendall τ_b	Spearman ρ
Acidobacteria	30.4	−0.550*	−0.424*	−0.561*
Gemmatimonadetes	10.9		0.369*	0.486*
Chloroflexi	7.1	0.524*	0.466**	0.649**
Nitrospirae	3.3	−0.554*	−0.452*	−0.561*
Planctomycetes	2.6	0.618**	0.424*	0.567*
JL-ETNP-Z39	2.4	0.706**	0.522**	0.680**
β-Proteobacteria	0.9	−0.500*	−0.369*	−0.505*

[a]RA is relative abundances. * and ** represent significance level at <0.05 and < 0.01 respectively. The taxa with average relative abundance over 18 samples < 0.5% and with significance level >0.05 were not shown.

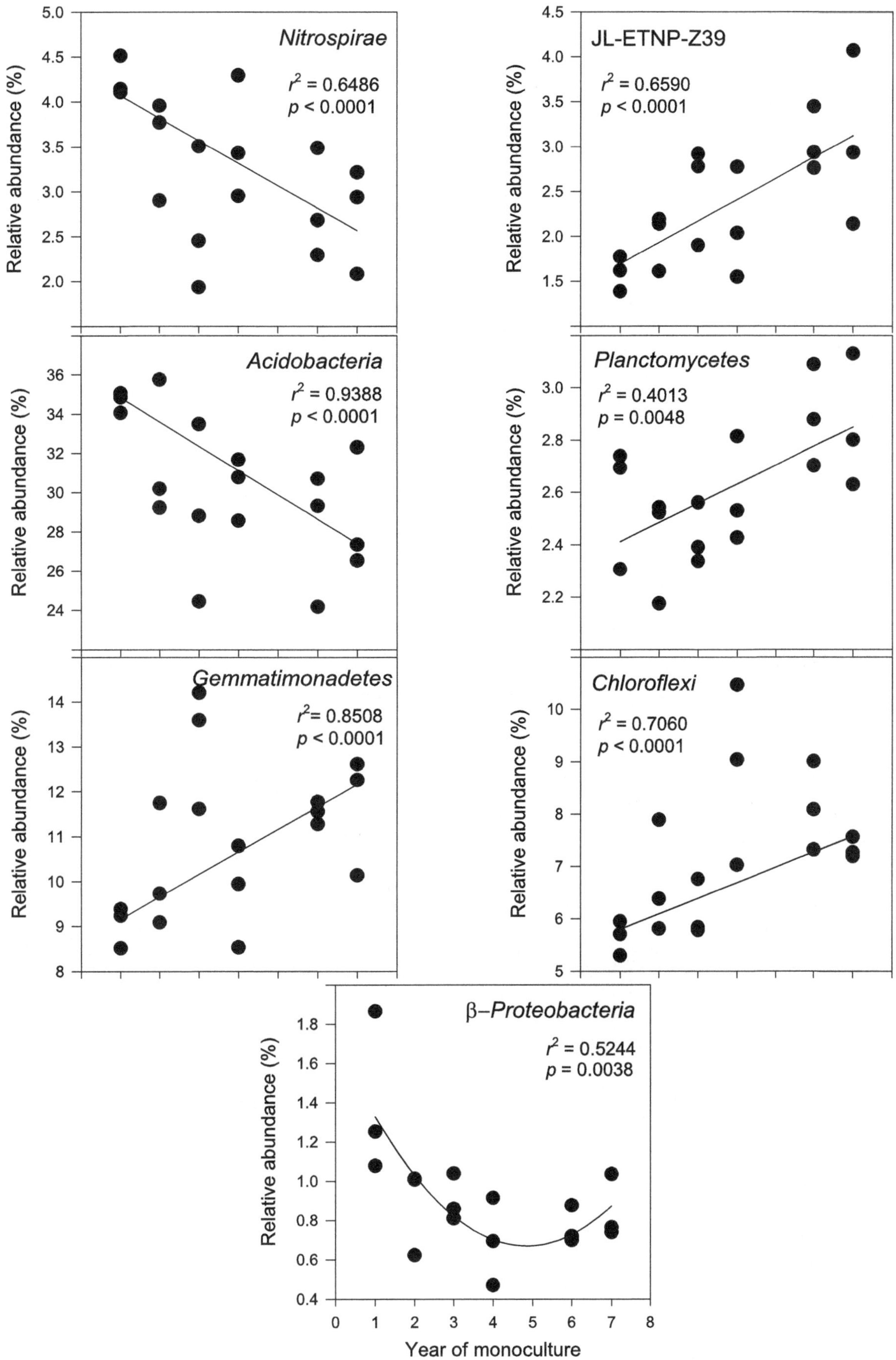

Figure 4. Regressions between bacterial taxa abundances and the year of potato monoculture.

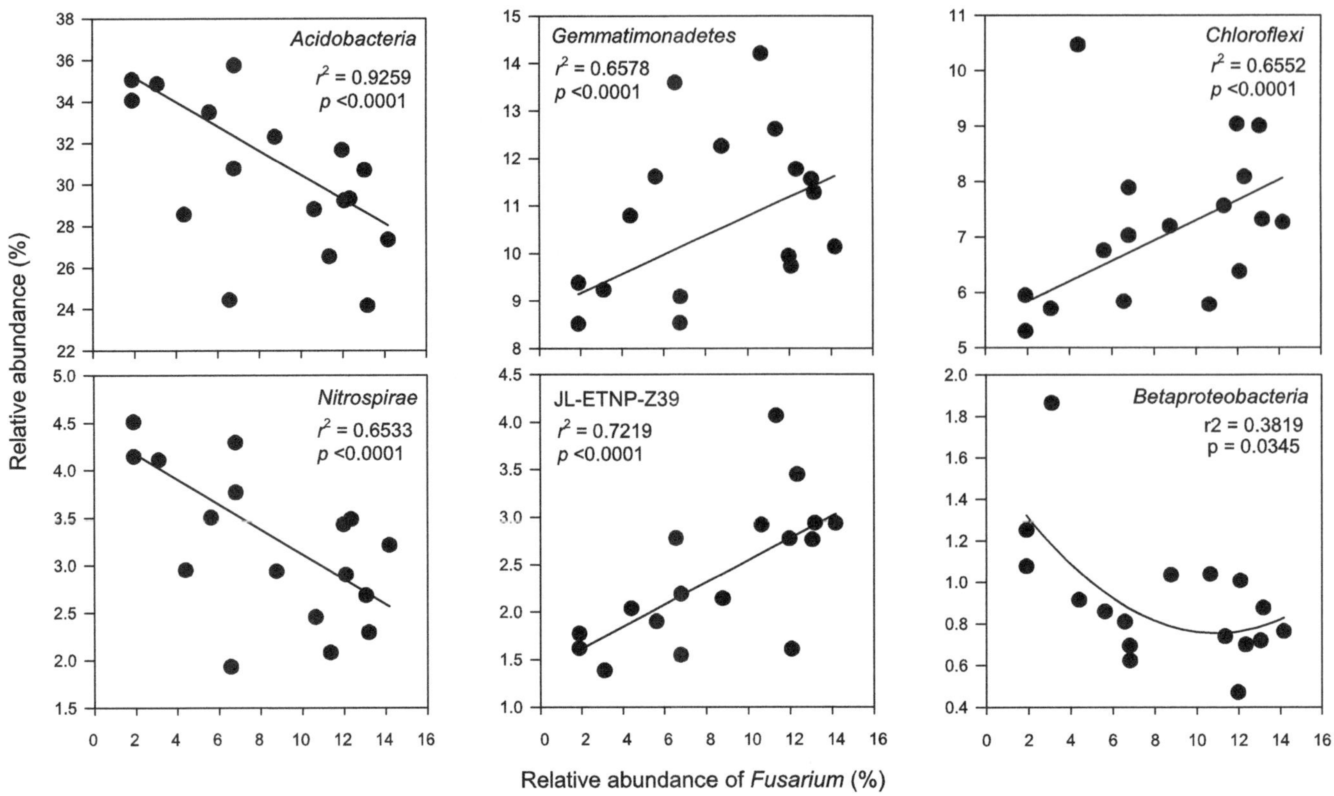

Figure 5. Regressions between relative abundance of *Fusarium* and bacterial taxa. Regressions with *p* values >0.05 are not shown.

were increased due to the improved water regime by irrigation, which consequently led to the decrease of soil OM contents (Figure S2). Therefore our data indirectly supported the ecological classification of *Acidobacteria* as an oligotrophic phylum. As for *Nitrospirae*, its ecophysiology is not clear. However, there have been evidences indicating that it tended to adapt to substrate-limited and well-aerated conditions [51,52], implying that the ecophysiology of *Nitrospirae* was similar to that of *Acidobacteria*. Furthermore, *Nitrospirae* is a phylum most likely involved in nitrite oxidation [51,52], although phylogenetic heterogeneity implies it has diverse functions. Nevertheless, the decrease of *Nitrospirae* abundance suggested negative effects on the process of soil nitrification.

Fifth, six bacterial taxa were significantly correlated with the abundance of *Fusarium* (Figure 5). Positive correlations were found between *Fusarium* and *Gemmatimonadetes*, *Chloroflexi* and JL-ETNP-Z39. Negative correlations occurred between *Fusarium* and *Acidobacteria*, *Nitrospirae* and β-*Proteobacteria*. The sum of average relative abundances of these correlated bacterial taxa was 55% (Figure 5), suggesting that potato monoculture not only enriched *Fusarium* but also shaped the overall soil bacterial communities by enriching and impoverishing certain bacterial taxa. Furthermore, the data also suggested these taxa exhibited similar behaviors in responses to potato monoculture and/or soil variables, providing additional insights into the ecological behaviors of different taxa in the environments.

Conclusion

Fungal and bacterial communities exhibited different responses to potato monoculture. Fungal diversity was nearly unaffected, whereas bacterial diversity was significantly decreased over time (years). Seven bacterial taxa were significantly changed, including the top abundant taxa *Acidobacteria* and *Gemmatimonadetes*. *Fusarium* was the only soilborne pathogen-associated fungal genus substantially detected by our high throughput pyrosequencing approach. *Fusarium* became enriched over the experimental time span, and was closely associated with potato yield decline and disease increases. *Fusarium* abundance was also significantly and negatively correlated with soil OM and TN, but significantly and positively with EC. *Fusarium* was positively correlated with abundances of *Gemmatimonadetes*, *Chloroflexi*, and JL-ETNP-Z39, and negatively correlated with *Acidobacteria*, *Nitrospirae*, and β-*Proteobacteria* abundances. These correlations suggested their similar behaviors in responses to potato monoculture and/or soil variables, providing further insights into the effects of potato monoculture on soil microbial communities.

Supporting Information

Figure S1 The changing trends in abundances of fungal taxa over time. Phylogenetic lineages were assigned by order or higher levels.

Figure S2 Regression between soil variable and the year of monoculture.

Table S1 Relative abundance of genus within the order Hypocreales and Sodariales.

Table S2 The relative abundance of pathogen-associated fungal genus detected in the soil samples.

Acknowledgments

Authors are grateful for Professor Warren A. Dick, School of Natural Resources, The Ohio State University, for his comments and language editing, and Professor Yuhua Shan, College of Environmental Science and Engineering, Yangzhou University, for his professional consultancies on statistical analyses. Many graduate students and staff members who were involved in maintaining the field plots, collecting field data, and collecting soil samples, but are not listed as co-authors are grateful. Authors would like to thank three anonymous reviewers for their comments, which significantly improve the manuscript.

Author Contributions

Conceived and designed the experiments: QRS HZQ. Performed the experiments: XL WMZ JLZ. Analyzed the data: SXY TYG. Contributed reagents/materials/analysis tools: JLZ. Wrote the paper: SXY. Pyrosequencing data and data analyses: SXY.

References

1. Lauber CL, Hamady M, Knight R, Fierer N (2009) Pyrosequencing-based assessment of soil pH as a predictor of soil bacterial community structure at the continental scale. Appl Environ Microbiol 75: 5111–5120.
2. Nacke H, Thurmer A, Wollherr A, Will C, Hodac L, et al. (2011) Pyrosequencing-based assessment of bacterial community structure along different management types in German forest and grassland soils. PLoS One 6: e17000.
3. Girvan MS, Bullimore J, Pretty JN, Osborn AM, Ball AS (2003) Soil type is the primary determinant of the composition of the total and active bacterial communities in arable soils. Appl Environ Microbiol 69: 1800–1809.
4. Sessitsch A, Weilharter A, Gerzabek MH, Kirchmann H, Kandeler E (2001) Microbial population structures in soil particle size fractions of a long-term fertilizer field experiment. Appl Environ Microbiol 67: 4215–4224.
5. Will C, Thurmer A, Wollherr A, Nacke H, Herold N, et al. (2010) Horizon-specific bacterial community composition of German grassland soils, as revealed by pyrosequencing-based analysis of 16S rRNA genes. Appl Environ Microbiol 76: 6751–6759.
6. Garbeva P, Postma J, van Veen JA, van Elsas JD (2006) Effect of above-ground plant species on soil microbial community structure and its impact on suppression of Rhizoctonia solani AG3. Environ Microbiol 8: 233–246.
7. Fierer N, Lauber CL, Ramirez KS, Zaneveld J, Bradford MA, et al. (2012) Comparative metagenomic, phylogenetic and physiological analyses of soil microbial communities across nitrogen gradients. ISME J 6: 1007–1017.
8. Norris TB, Wraith JM, Castenholz RW, McDermott TR (2002) Soil microbial community structure across a thermal gradient following a geothermal heating event. Appl Environ Microbiol 68: 6300–6309.
9. Brockett BFT, Prescott CE, Grayston SJ (2012) Soil moisture is the major factor influencing microbial community structure and enzyme activities across seven biogeoclimatic zones in western Canada. Soil Biol Biochem 44: 9–20.
10. Jansa J, Mozafar A, Kuhn G, Anken T, Ruh R, et al. (2003) Soil tillage affects the community structure of mycorrhizal fungi in maize roots. Ecol Appl 13: 1164–1176.
11. Yin C, Jones KL, Peterson DE, Garrett KA, Hulbert SH, et al. (2010) Members of soil bacterial communities sensitive to tillage and crop rotation. Soil Biol Biochem 42: 2111–2118.
12. Stark C, Condron LM, Stewart A, Di HJ, O'Callaghan M (2007) Influence of organic and mineral amendments on microbial soil properties and processes. Appl Soil Ecol 35: 79–93.
13. Luo J, Ran W, Hu J, Yang X, Xu Y, et al. (2010) Application of Bio-organic fertilizer significantly affected fungal diversity of soils. Soil Sci Soc Am J 74: 2039–2048.
14. Ederson CJ, Marsh TL, Tiedje JM, de S Moreira FM (2009) Changes in land use alter the structure of bacterial communities in Western Amazon soils. ISME J 3: 1004–1011.
15. Lee J, Lee T, Löffler F, Park J (2011) Characterization of microbial community structure and population dynamics of tetrachloroethene-dechlorinating tidal mudflat communities. Biodegradation 22: 687–698.
16. Garbeva P, van Veen JA, van Elsas JD (2004) Microbial diversity in soils: selection of microbial populations by plant and soil type and implications for disease suppressiveness. Annu Rev Phytopathol 42: 243–270.
17. Cook RJ (2006) Toward cropping systems that enhance productivity and sustainability. P Natl Acad Sci USA 103: 18389–18394.
18. Acosta-Martínez V, Dowd SE, Bell CW, Lascano R, Booker JD, et al. (2010) Microbial community composition as affected by dryland cropping systems and tillage in a semiarid sandy soil. Diversity 2: 910–931.
19. Ceja-Navarro J, Rivera F, Patiño-Zúñiga L, Govaerts B, Marsch R, et al. (2010) Molecular characterization of soil bacterial communities in contrasting zero tillage systems. Plant Soil 329: 127–137.
20. Govaerts B, Mezzalama M, Unno Y, Sayre KD, Luna-Guido M, et al. (2007) Influence of tillage, residue management, and crop rotation on soil microbial biomass and catabolic diversity. Appl Soil Ecol 37: 18–30.
21. Xuan D, Guong V, Rosling A, Alström S, Chai B, et al. (2012) Different crop rotation systems as drivers of change in soil bacterial community structure and yield of rice, *Oryza sativa*. Biol Fert Soils 48: 217–225.
22. Li C, Li X, Kong W, Wu Y, Wang J (2010) Effect of monoculture soybean on soil microbial community in the Northeast China. Plant Soil 330: 423–433.
23. Hilton S, Bennett AJ, Keane G, Bending GD, Chandler D, et al. (2013) Impact of shortened crop rotation of oilseed rape on soil and rhizosphere microbial diversity in relation to yield decline. PLoS One 8: e59859.
24. Larkin RP (2003) Characterization of soil microbial communities under different potato cropping systems by microbial population dynamics, substrate utilization, and fatty acid profiles. Soil Biol Biochem 35: 1451–1466.
25. Pereira e Silva MC, Dias AC, van Elsas JD, Salles JF (2012) Spatial and temporal variation of archaeal, bacterial and fungal communities in agricultural soils. PLoS One 7: e51554.
26. Gobet A, Boer SI, Huse SM, van Beusekom JE, Quince C, et al. (2012) Diversity and dynamics of rare and of resident bacterial populations in coastal sands. ISME J 6: 542–553.
27. Pereira e Silva MC, Poly F, Guillaumaud N, van Elsas JD, Falcão Salles J (2012) Fluctuations in ammonia oxidizing communities across agricultural soils are driven by soil structure and pH. Front Microbiol 3: 77.
28. Bruggen AC, Semenov A, Diepeningen A, Vos O, Blok W (2006) Relation between soil health, wave-like fluctuations in microbial populations,and soil-borne plant disease management. Plant disease epidemiology: facing challenges of the 21st Century: Springer Netherlands. pp. 105–122.
29. Shipton PJ (1977) Monoculture and soilborne plant pathogens. Annu Rev Phytopathol 15: 387–407.
30. Amann RI, Ludwig W, Schleifer KH (1995) Phylogenetic identification and in situ detection of individual microbial cells without cultivation. Microbiol Rev 59: 143–169.
31. Lu L, Yin S, Liu X, Zhang W, Gu T, et al. (2013) Fungal networks in yield-invigorating and -debilitating soils induced by prolonged potato monoculture. Soil Biol Biochem 65: 186–194.
32. Cardenas E, Wu WM, Leigh MB, Carley J, Carroll S, et al. (2010) Significant association between sulfate-reducing bacteria and uranium-reducing microbial communities as revealed by a combined massively parallel sequencing-indicator species approach. Appl Environ Microbiol 76: 6778–6786.
33. Schloss PD, Westcott SL, Ryabin T, Hall JR, Hartmann M, et al. (2009) Introducing mothur: open-source, platform-independent, community-supported software for describing and comparing microbial communities. Appl Environ Microbiol 75: 7537–7541.
34. Fiers M, Edel-Hermann V, Chatot C, Le Hingrat Y, Alabouvette C, et al. (2012) Potato soil-borne diseases. A review. Agron Sustain Dev 32: 93–132.
35. Jeger M, Hide G, Van Den Boogert P, Termorshuizen A, Van Baarlen P (1996) Pathology and control of soil-borne fungal pathogens of potato. Potato Res 39: 437–469.
36. Acosta-Martínez V, Dowd S, Sun Y, Allen V (2008) Tag-encoded pyrosequencing analysis of bacterial diversity in a single soil type as affected by management and land use. Soil Biol Biochem 40: 2762–2770.
37. DeBruyn JM, Nixon LT, Fawaz MN, Johnson AM, Radosevich M (2011) Global biogeography and quantitative seasonal dynamics of *Gemmatimonadetes* in soil. Appl Environ Microbiol 77: 6295–6300.
38. Buée M, Reich M, Murat C, Morin E, Nilsson RH, et al. (2009) 454 Pyrosequencing analyses of forest soils reveal an unexpectedly high fungal diversity. New Phytol 184: 449–456.
39. McGuire K, Fierer N, Bateman C, Treseder K, Turner B (2012) Fungal Community Composition in Neotropical Rain Forests: the influence of tree diversity and precipitation. Microbial Ecol 63: 804–812.
40. Jumpponen A, Jones KL, Blair J (2010) Vertical distribution of fungal communities in tallgrass prairie soil. Mycologia 102: 1027–1041.
41. Larkin RP (2008) Relative effects of biological amendments and crop rotations on soil microbial communities and soilborne diseases of potato. Soil Biol Biochem 40: 1341–1351.
42. Larkin RP, Honeycutt CW (2006) Effects of different 3-year cropping systems on soil microbial communities and rhizoctonia diseases of potato. Phytopathol 96: 68–79.
43. Larkin RP, Griffin TS, Honeycutt CW (2010) Rotation and cover crop effects on soilborne potato diseases, tuber yield, and soil microbial communities. Plant Dis 94: 1491–1502.
44. Larkin RP, Honeycutt CW, Griffin TS, Olanya OM, Halloran JM, et al. (2011) Effects of different potato cropping system approaches and water management on soilborne diseases and soil microbial communities. Phytopathol 101: 58–67.
45. Bernard E, Larkin RP, Tavantzis S, Erich MS, Alyokhin A, et al. (2012) Compost, rapeseed rotation, and biocontrol agents significantly impact soil microbial communities in organic and conventional potato production systems. Appl Soil Ecol 52: 29–41.
46. Hopkins DL, Lobinske RJ, Larkin RP (1992) Selection for *Fusarium oxysporum* f.sp. niveum race 2 in monocultures of watermelon cultivars resistant to *Fusarium* wilt. Phytopathol 82: 290–293.

47. Hopkins DL, Elmstrom GW (1984) Fusarium wilt in watermelon cultivars grown in a 4-year monoculture. Plant Dis 68: 129–131.
48. Leach SS, Webb RE (1981) Resistance of selected potato cultivars and clones to *Fusarium* dry rot. Phytopathol 71: 623–629.
49. Arroita M, Causapé J, Comín FA, Díez J, Jimenez JJ, et al. (2013) Irrigation agriculture affects organic matter decomposition in semi-arid terrestrial and aquatic ecosystems. J Hazard Mater 263, Part 1: 139–145.
50. Fierer N, Bradford MA, Jackson RB (2007) Toward an ecological classification of soil bacteria. Ecology 88: 1354–1364.
51. Lücker S, Wagner M, Maixner F, Pelletier E, Koch H, et al. (2010) A *Nitrospira* metagenome illuminates the physiology and evolution of globally important nitrite-oxidizing bacteria. P Natl Acad Sci USA 107: 13479–13484.
52. Schramm A, de Beer D, van den Heuvel JC, Ottengraf S, Amann R (1999) Microscale distribution of populations and activities of *Nitrosospira* and *Nitrospira* spp. along a macroscale gradient in a nitrifying bioreactor: quantification by in situ hybridization and the use of microsensors. Appl Environ Microbiol 65: 3690–3696.

14

Spatial Distribution of Soil Organic Carbon and Total Nitrogen Based on GIS and Geostatistics in a Small Watershed in a Hilly Area of Northern China

Gao Peng[1]*, Wang Bing[2]*, Geng Guangpo[1], Zhang Guangcan[1]

1 Shandong Agricultural University, College of Forestry/Taishan Mountain Forest Ecosystem Research Station/Shandong Provincial Key Laboratory of Soil Erosion and Ecological Restoration, Tai'an, Shandong, P.R. China, 2 Research Institute of Forest Ecology Environmental Protection, Chinese Academy of Forestry, Beijing, P.R. China

Abstract

The spatial variability of soil organic carbon (SOC) and total nitrogen (STN) levels is important in both global carbon-nitrogen cycle and climate change research. There has been little research on the spatial distribution of SOC and STN at the watershed scale based on geographic information systems (GIS) and geostatistics. Ninety-seven soil samples taken at depths of 0–20 cm were collected during October 2010 and 2011 from the Matiyu small watershed (4.2 km^2) of a hilly area in Shandong Province, northern China. The impacts of different land use types, elevation, vegetation coverage and other factors on SOC and STN spatial distributions were examined using GIS and a geostatistical method, regression-kriging. The results show that the concentration variations of SOC and STN in the Matiyu small watershed were moderate variation based on the mean, median, minimum and maximum, and the coefficients of variation (CV). Residual values of SOC and STN had moderate spatial autocorrelations, and the Nugget/Sill were 0.2% and 0.1%, respectively. Distribution maps of regression-kriging revealed that both SOC and STN concentrations in the Matiyu watershed decreased from southeast to northwest. This result was similar to the watershed DEM trend and significantly correlated with land use type, elevation and aspect. SOC and STN predictions with the regression-kriging method were more accurate than those obtained using ordinary kriging. This research indicates that geostatistical characteristics of SOC and STN concentrations in the watershed were closely related to both land-use type and spatial topographic structure and that regression-kriging is suitable for investigating the spatial distributions of SOC and STN in the complex topography of the watershed.

Editor: Ben Bond-Lamberty, DOE Pacific Northwest National Laboratory, United States of America

Funding: This research was funded by the Special Fund for Forestry Scientific Research in the Public Interest (No. 201204101), the National Basic Research and Development Program of China (973 Program, Nos. 2011CB403201 and 2012CB416904), and was subsidized by CFERN&GENE Award Funds on Ecological Paper. Funders had a role in the data collection and decision to publish the manuscript.

Competing Interests: The authors have declared that no competing interests exist.

* E-mail: gaopengy@163.com (GP); wangbing@caf.ac.cn (WB)

Introduction

As important factors in global biogeochemical cycling of the terrestrial ecosystem, soil organic carbon (SOC) and total nitrogen (STN) are important in alleviating global warming, mitigating land degradation, improving food security, and enhancing crop production [1,2,3]. They are also at the heart of the global carbon-nitrogen cycle and climate change research. SOC and STN are carbon and nitrogen sources for plant growth. They also affect soil biodiversity and the structure and physical stability of soil that enables it to resist erosion [4]. SOC and STN have strong spatial heterogeneity, with internal changes in the vertical and horizontal directions and external exchanges with the atmosphere and biosphere [5]. Many factors, such as topography, land-use type, field management and vegetation, can control SOC and STN spatial variability at various scales. Understanding and incorporating such heterogeneity and spatial distribution characteristics can improve the precision of carbon-nitrogen budgets and assist in implementation of effective measures toward vegetation recovery.

There is considerable research into SOC and STN spatial distributions on different scales [6,7,8,9,10], and results show that SOC and STN have a changing continuum with a non-uniform spatial distribution. Several studies have shown variability in topography, vegetation, cultivation, land use and parent material [11,12,13,14].

However, current research mainly relies on the ordinary kriging method, which produces large uncertainty in the prediction of soil spatial distribution because of the impacts of land use, topographic characteristics and other factors [15]. In recent years, some methods have been proposed to solve this problem, such as the regression-kriging method, the geographically weighted regression method (GWR) and the geographically weighted regression kriging method (GWRK) [16,17]. However, among these methods, only the regression-kriging method can incorporate topographic factors, vegetation coverage and other elements and thereby improve the accuracy of spatial prediction [18,19,20].

Some researchers have studied SOC and STN spatial variability in different regions of China, including the black-soil region in the northeast and the Loess Plateau in the northwest [21,22,23]. However, there is generally little quantitative information on this spatial variability on small-watershed scales in the rocky mountain area of the northern part of the country. This is especially true regarding the few spatial variability studies using regression-kriging

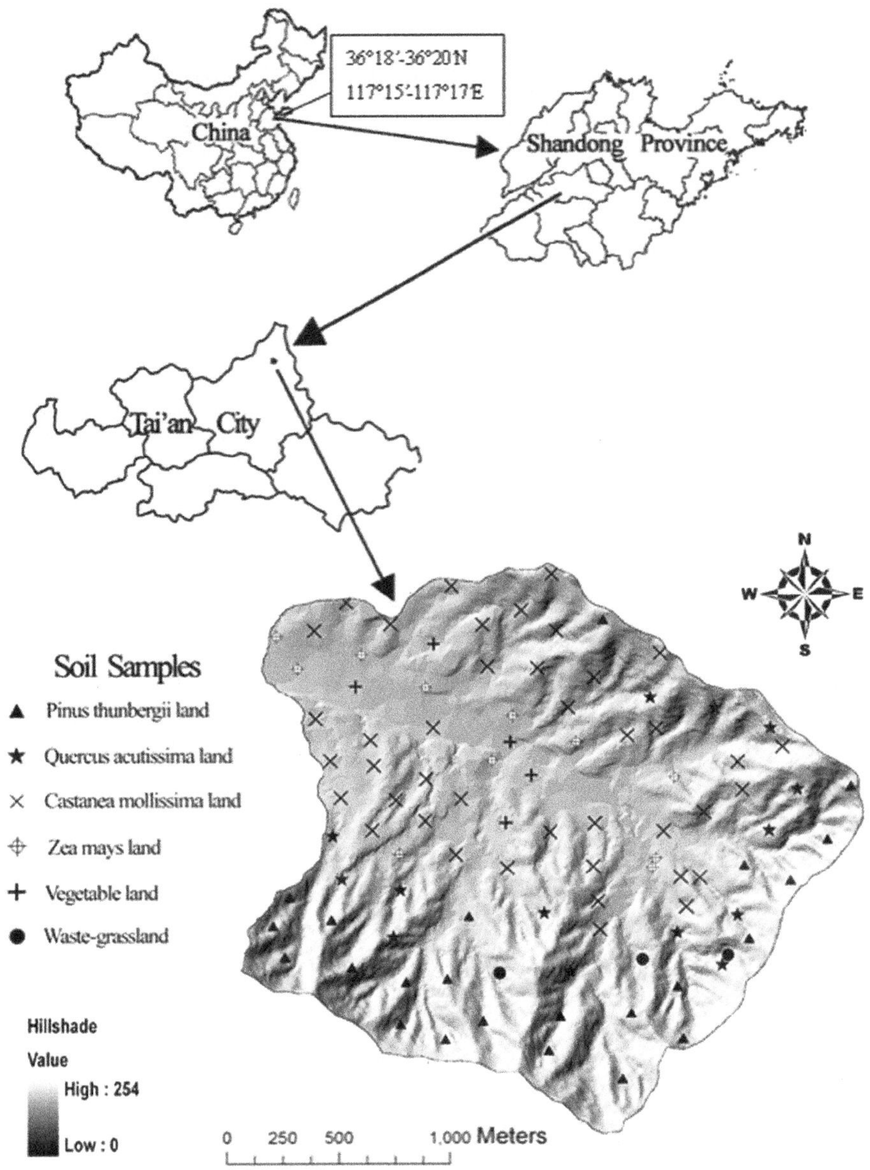

Figure 1. Distribution map of soil sample points (0–20 cm) in Matiyu small watershed.

and geographic information system (GIS) technology, limiting the capability of evaluating the carbon budget and predicting the ecosystem response to environmental and climate change.

In this paper, we selected the Matiyu small watershed, which is typical of the rocky mountain areas of northern China, as a research site. We used regression-kriging and GIS to achieve the following objectives: 1) to reveal and analyze spatial SOC and STN distributions at the scale of a small watershed; 2) to address the impacts of different land-use types, elevation, vegetation coverage, and other factors on those spatial distributions; and 3) to compare SOC and STN prediction accuracy using regression-kriging and ordinary kriging methods.

Materials and Methods

Ethics Statement

The research station for this study is managed by Shandong Agricultural University. The study was approved by the Mountain Tai Forest Ecosystem Research Station of the State Forestry Administration.

Study area condition

The Matiyu watershed is in Tai'an city, Shandong Province. It has an area of 4.2 km^2 and lies between 36°18′–36°20′N and 117°15′–117°17′E (Figure 1), with elevations ranging from 219 m to 595 m. The climate is semiarid, with an average of 765 mm of annual precipitation; 70% of this precipitation falls from June through September. Annual average air temperature is 12.8°C, the accumulated temperature greater than 10°C is 4120°C, and there are 200 frost-free days.

The dominant soil type is cinnamonsoil. The study area includes the following major land-use types: forest (including *Pinus thunbergii*, *Quercus acutissima*, and *Castanea mollissima*), farmland (including *Zea mays* and vegetables), and waste-grassland.

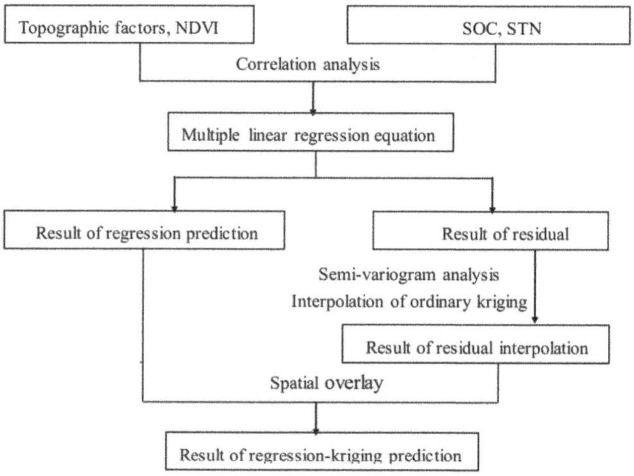

Figure 2. Flowchart of regression-kriging.

Sampling and measurement methods

The soil sampling design was based off of a digital topographic map of the watershed at a scale of 1:10000, which was used to generate a raster digital elevation model (DEM) map with a resolution of 10 m×10 m. A grid square crossed by the study boundary was regarded as a distinct unit if more than half of it lay inside the study area; otherwise, it was merged into a neighboring grid. A Real Time Kinetic – Global Positioning System (RTK-GPS) receiver was used to locate each grid center in the field and to record latitudes, longitudes and elevations of sampling points. The terrain information was then recorded in detail. Within a radius of approximately 20 cm at each site, four separate sub-samples were taken from the top 20-cm of soil using a 5.0-cm diameter soil auger. These samples were then mixed and were considered representative of the soil at the site.

In total, 97 soil samples (0–20 cm) were collected in Matiyu watershed during October 2010 and October 2011 (Figure 1) [24,25]. Soil samples were air-dried for approximately 7 days and then passed through a 0.25-mm sieve. SOC and STN contents were determined in duplicate for each sample, using the potassium dichromate oxidation and micro-Kjeldahl methods [26].

Acquisition of topographic factor attributes and NDVI

Topographic factors were computed using a spatial analysis model and digital topography analysis, and included elevation (H), slope (β), sinα and cosα of aspect, compound topographic index (CTI), stream power index (SPI), and sediment transport index (STI). The normalized difference vegetation index (NDVI) was acquired from a Landsat 5 image from August 30, 2010.

Interpolation of regression-kriging

Figure 2 shows a flowchart of the regression-kriging method. Taking into account the spatial distribution law and geographic influence factors, the method simulates both spatial distribution trends and uncertainty. Equations were established between the assistant variable and target variables, and then ordinary kriging was used to separate the trend item and interpolate the residual. Finally, the spatial overlay was carried out between the trend item of regression prediction and the residual value of ordinary kriging to obtain the predicted value of the target variable [27,28].

Precision evaluation of prediction results

Twenty-six of the 97 soil samples were randomly extracted from the data to test the model's predictive accuracy based on data from the remaining 71 soil samples. Prediction accuracy was evaluated by comparing observed and predicted values of SOC and STN from validation point locations. Mean prediction error (*MPE*) and root mean square prediction error (*RMSE*) were selected as evaluation indexes, and R_I was used to analyze improvement of the prediction accuracy by comparing regression-kriging with ordinary kriging [29,30].

$$MPE = \frac{1}{n}\sum_{j=1}^{n}(Z_{oi} - Z_{pi})$$

$$RMSE = \sqrt{\frac{1}{n}\sum_{j=1}^{n}(Z_{oi} - Z_{pi})^2}$$

$$R_I = \frac{RMSE_R - RMSE_{RK}}{RMSE_R} \times 100\%$$

where *MPE* is the mean prediction error; *RMSE* is the root mean square prediction error; *n* represents sampling points of the test dataset; Z_{oi} and Z_{pi} are observed and predicted values of the sampling points, respectively; and R_I is the improvement of prediction accuracy from comparing regression-kriging with ordinary kriging. If R_I is positive, it means that prediction accuracy of regression-kriging was higher than that of ordinary kriging, and vice-versa for a negative R_I value. $RMSE_R$ is the root mean square prediction error of ordinary kriging, and $RMSE_{RK}$ is that of regression-kriging.

Software platform. SPSS17.0 software was used for classical statistical and regression analyses of SOC and STN and for testing their differences under different land use types via ANOVA.

Table 1. Statistical results of mean concentrations of SOC and STN in Matiyu small watershed[†].

Soil nutrients‡	Mean ($g \cdot kg^{-1}$)	Median ($g \cdot kg^{-1}$)	Min. ($g \cdot kg^{-1}$)	Max. ($g \cdot kg^{-1}$)	Range ($g \cdot kg^{-1}$)	Std. Deviation	CV (%)
SOC	11.91	11.73	6.20	21.67	15.47	2.87	24.10
STN	1.18	1.16	0.54	2.57	2.03	0.28	23.73

"†" Numbers represent the mean of 2 years (2010–2011a). The following is the same.
"‡" SOC = soil organic carbon; STN = soil total nitrogen.

Table 2. Correlation analysis of SOC and STN with the environmental factors.

Soil nutrients‡	Elevation	slope	cosα	sinα	CTI	SPI	STI	NDVI
SOC	0.495^{**}	0.339^{**}	0.298^{*}	−0.052	−0.094	0.030	0.117	0.199^{*}
STN	0.415^{**}	0.249^{*}	0.400^{**}	−0.020	−0.083	0.016	0.042	0.215^{*}

*and ** indicate significance at $P<0.05$ and $P<0.01$, respectively.
"‡" SOC = soil organic carbon; STN = soil total nitrogen; cosα = cosine of aspect; sinα = sine of aspect; CTI = compound topographic index; SPI = stream power index; STI = sediment transport index; NDVI = Normalized difference vegetation index.

ArcGIS 9.2 was used to establish the small-watershed DEM and extract its topographic factors. GS+7.0 was used for semi-variogram analysis of regression prediction and residual values. ArcGIS9.2 was then used to perform the spatial plus operation between the trend item of regression prediction and the residual value of ordinary kriging [31]. Finally, the spatial prediction distribution map of SOC and STN in the Matiyu watershed was produced.

Results

Descriptive statistics of SOC and STN in Matiyu watershed

Mean SOC and STN concentrations in the watershed were 11.91 $g \cdot kg^{-1}$ and 1.18 $g \cdot kg^{-1}$, respectively. Respective coefficients of variation (CVs) were 24.10% and 23.73%, respectively, and both CVs were moderate (Table 1).

Correlations of SOC and STN with environmental factors

Table 2 shows that both SOC and STN had highly significant positive correlations with elevation (correlation coefficients of 0.495 and 0.415, respectively). This indicates that the higher the elevation, the larger the SOC and STN concentrations. SOC and STN also had highly significant positive and significant positive correlations with slope, respectively. This reveals that the greater the slope, the larger the two concentrations. The increasing trend of SOC was much more significant than that of STN. Both concentrations had little correlation with the sine of aspect. However, SOC and STN had highly significant positive and significant positive correlations with the cosine of aspect, respectively. This cosine measures the degree of northward orientation; thus, the greater the northward orientation, the larger the SOC and STN concentrations.

Both SOC and STN had significant positive correlations with NDVI, which indicates that the greater the vegetation cover, the higher the two concentrations. Both SOC and STN had little correlation with CTI, SPI or STI.

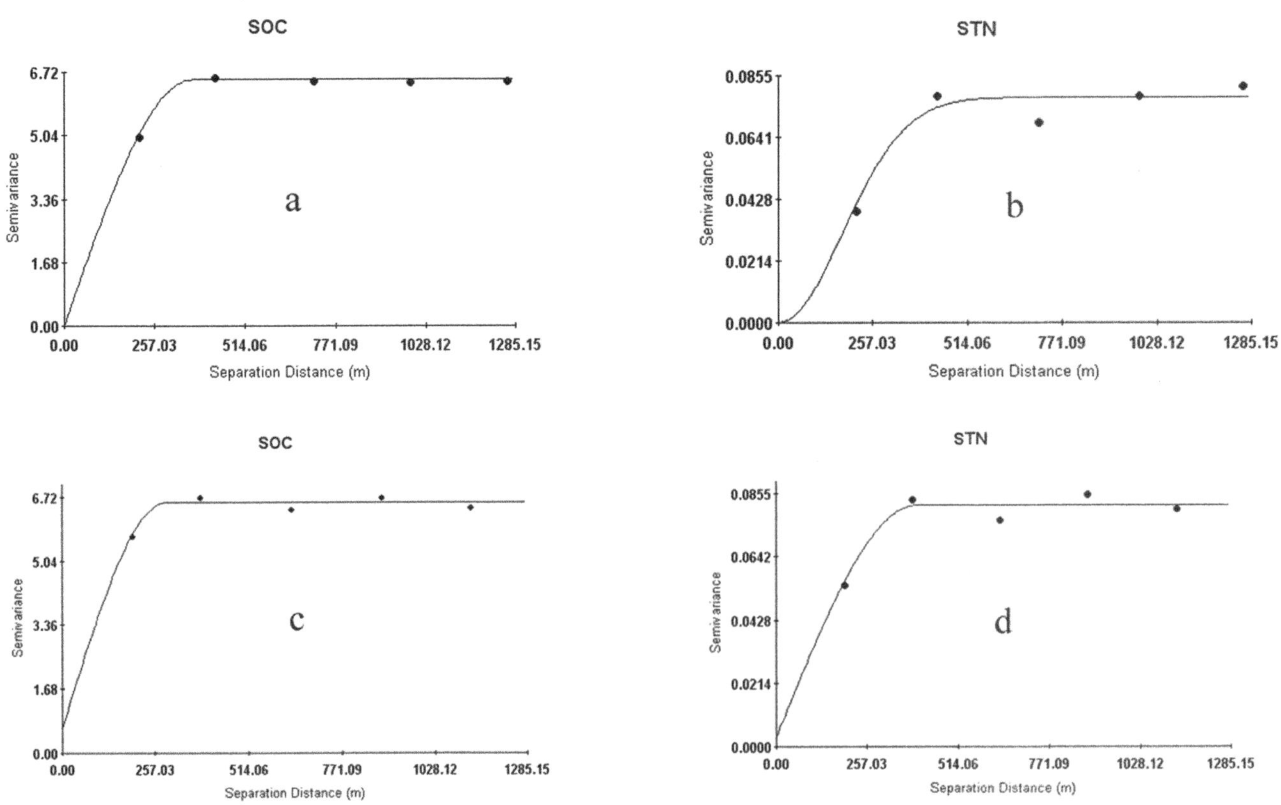

Figure 3. Semi-variogram analysis of SOC and STN for regression-kriging (a, b) and ordinary kriging (c, d).

Table 3. Semi-variogram analysis of SOC and STN and its parameters.

Soil nutrients‡	Data Item	Active lag distance (m)	Lag interval (m)	Model	Nugget C_0	Sill C_0+C	Nugget/Sill C_0/(C_0+C) (%)	Range (m)	R^2
SOC	residual error(RK)	1285.15	280	Spherical	0.0100	6.5230	0.2	365	0.996
STN	residual error(RK)	1285.15	280	Gaussian	0.0001	0.0779	0.1	424	0.915
SOC	raw data(OK)	1285.15	250	Spherical	0.6500	6.5740	9.9	288	0.880
STN	raw data(OK)	1285.15	250	Spherical	0.0031	0.0817	3.8	399	0.908

"‡" SOC = soil organic carbon; STN = soil total nitrogen; RK = regression-kriging; OK = ordinary kriging.

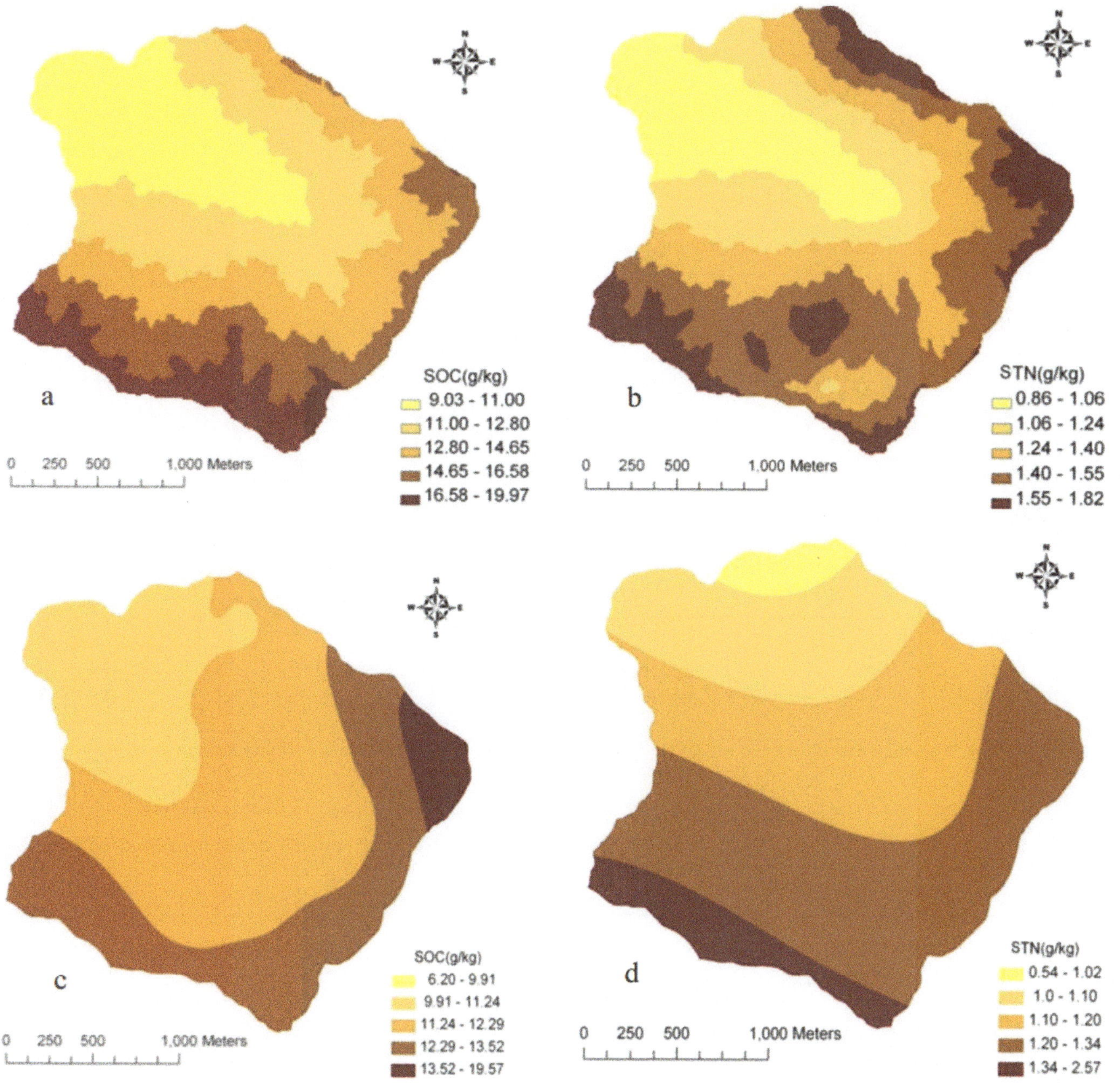

Figure 4. Distribution map of SOC and STN concentrations by regression-kriging (a, b) and ordinary kriging (c, d) in Matiyu small watershed.

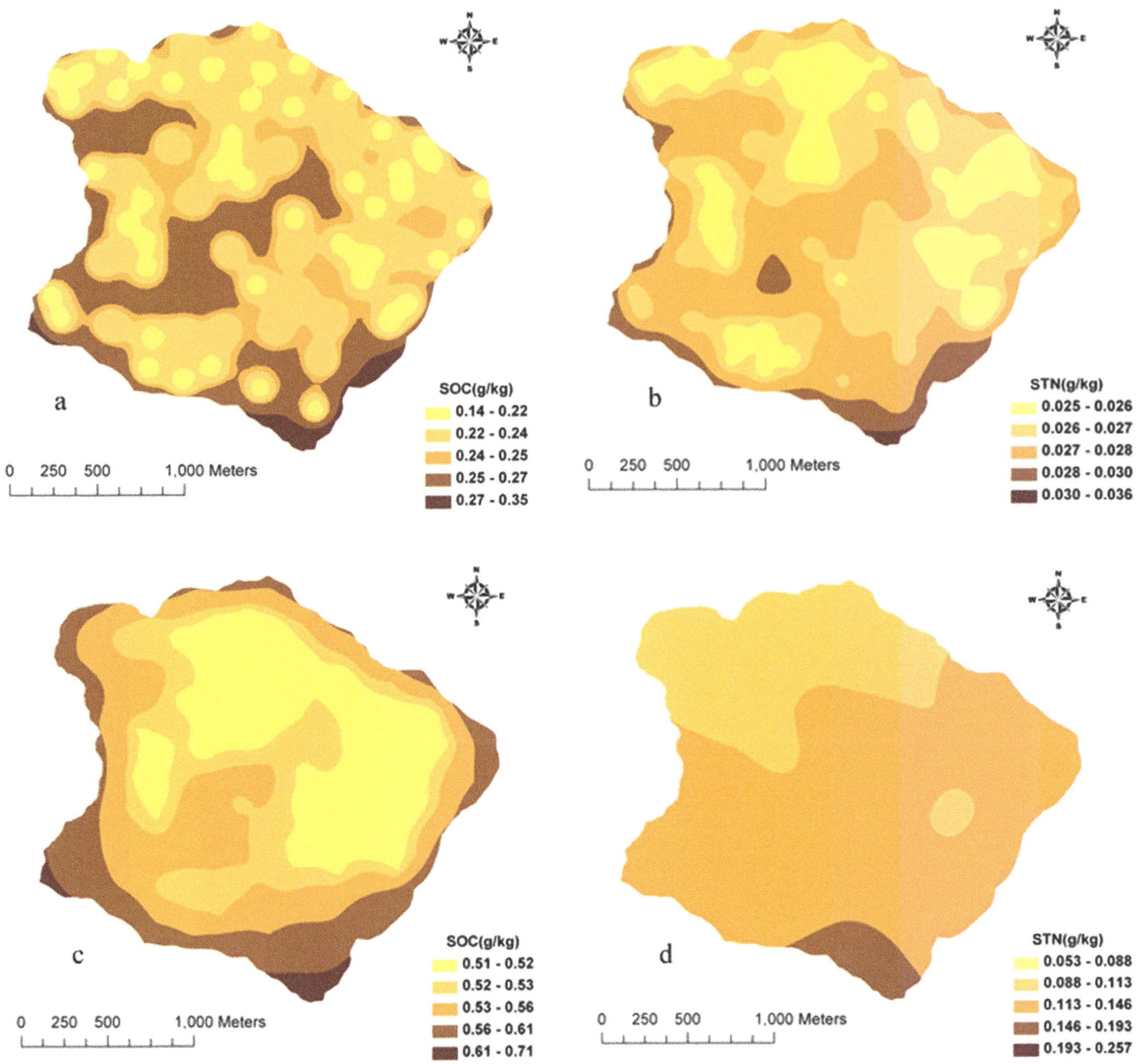

Figure 5. Maps of Prediction Standard Error of SOC and STN by regression-kriging (a, b) and ordinary kriging (c, d) in Matiyu small watershed.

Table 4. Comparison of prediction accuracy between regression-kriging and ordinary kriging.

Soil nutrients‡	Ordinary kriging		Regression-kriging		R_I(%)
	MPE	*RMSE*	*MPE*	*RMSE*	
SOC	−0.5	3.63	0.39	3.02	16.8
STN	−0.06	0.29	−0.01	0.24	17.24

"‡" SOC = soil organic carbon; STN = soil total nitrogen; MPE = mean prediction error; RMSE = root mean square prediction error; RI = improvement of prediction accuracy.

Regression-kriging models and semi-variogram analysis

Using the multiple linear stepwise regression method, all topographic factors and NDVI were used to explain the variability of SOC and STN concentrations. Equation fitting results were as follows:

$r^2 = 0.330$ ($P<0.05$)

$r^2 = 0.328$ ($P<0.01$)

The elevation and cosine of aspect that joined the prediction of SOC and STN were the optimal factors from all selected environmental factors. The determination coefficients of SOC and STN (r^2) were 0.330 and 0.328, and the equation fits were good.

Spatial autocorrelations of SOC and STN from regression-kriging (RK) and ordinary kriging (OK) were both low; however,

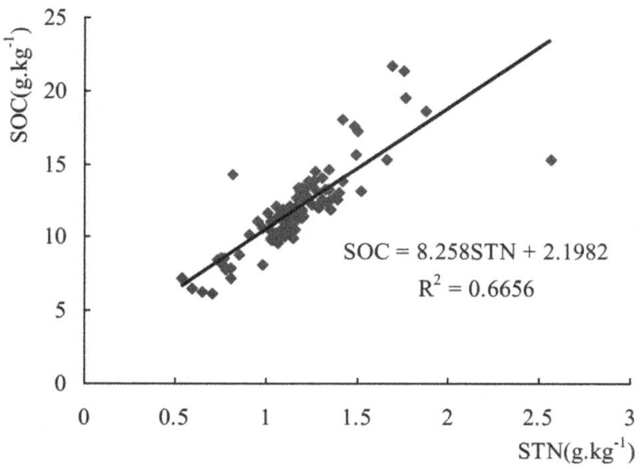

Figure 6. Regression equations of SOC and STN in Matiyu small watershed.

the effect of regression-kriging was better than that of ordinary kriging, and the Nugget/Sill were 0.2% and 0.1%, respectively (Table 3, Figure 3). The semi-variogram test showed that structural factors such as topography, vegetation and climate were the primary causes of SOC and STN spatial variation; random factors had little influence.

Evaluation of prediction accuracy using different prediction methods

In accordance with the flow chart of regression-kriging, linear combinations of environmental factors were considered as the external drift trend term to separate residual error and eliminate non-stationary properties to increase prediction accuracy. Seventy-one samples were randomly selected to conduct ordinary kriging interpolation for regression residual error of SOC and STN in the Matiyu small watershed. Meanwhile, as a control, those 71 samples were used to conduct ordinary kriging interpolation. From the prediction errors and distribution maps of regression-kriging and ordinary kriging (Figure 4, Figure 5), the result showed that the regression-kriging was better than that of ordinary kriging. The regression-kriging predictions were much more detailed concerning the partly variation and topographical relationships and much closer to the observed spatial distribution of SOC and STN. The other 26 samples were used to compare the accuracy of the two prediction methods (Table 4). The regression-kriging predicted values were very close to the measured ones, and its prediction accuracy was superior to that of ordinary kriging. The improvements of prediction accuracy (R_I) of SOC and STN were 16.80% and 17.24%, respectively (Table 4).

SOC and STN spatial distribution characteristics and their correlations in small watershed

The distribution maps of regression-kriging showed that both SOC and STN concentrations decreased from southeast to northwest in the watershed, similar to the DEM results (Figure 4). SOC and STN were strongly correlated, with a correlation coefficient of 0.6656 ($P<0.01$). Regression analysis revealed that SOC and STN had a highly significant linear relationship (Figure 6).

Table 5. Changes of SOC and STN concentrations from different land use types*.

Land use types	Samples	SOC(g·kg^{-1}) ‡	TN(g·kg^{-1})
Pinus thunbergii land	14	14.07±0.57c	1.41±0.07c
Quercus acutissima land	23	13.05±0.93bc	1.20±0.07bc
Castanea mollissima land	41	10.65±0.33ab	1.03±0.03ab
Zea mays land	11	10.70±0.41ab	1.16±0.04abc
Vegetable land	5	13.12±1.18bc	1.37±0.08c
Waste-grassland	3	9.65±1.52a	0.93±0.15a

*means within a column that contain different letters are significantly different at p<0.01;.
means within a column that contain the same letters are not significantly different at p<0.05.
"‡" SOC = soil organic carbon; STN = soil total nitrogen.

Table 6. Changes of SOC and STN concentrations at different elevations*.

Elevation(m)	samples	SOC(g·kg^{-1})‡	STN(g·kg^{-1})
220–290	35	10.70±0.36a	1.11±0.04a
290–370	29	11.32±0.42a	1.08±0.04a
370–500	33	13.71±0.55b	1.32±0.06b

*Means within a column that share different letters are significantly different at p<0.05; the same letters are not significantly different at p<0.05.
"‡" SOC = soil organic carbon; STN = soil total nitrogen.

Table 7. Changes of SOC and STN concentrations from different aspects*.

Aspect	samples	SOC(g·kg^{-1})‡	STN(g·kg 1)
shady slope	36	12.06±0.50a	1.25±0.06b
half-sunny slope	44	12.03±0.45a	1.16±0.03ab
sunny slope	17	11.26±0.53a	1.08±0.05a

*Means within a column that share different letters are significantly different at p<0.05; the same letters are not significantly different at p<0.05.
"‡" SOC = soil organic carbon; STN = soil total nitrogen.

Discussion

Influences on SOC and STN spatial distributions under different land use types

A change of land-use type can generate corresponding changes of microenvironment for every land type and subsequently alter soil nutrients [32,33]. Table 5 reveals that SOC concentrations in varying land use types were ranked as *P. thunbergii* > *Q. acutissima* > vegetable > *Z. mays* > *C. mollissima* > waste-grassland. STN concentrations were ranked as *P. thunbergii* > vegetable > *Q. acutissima* > *Z. mays* > *C. mollissima* > waste-grassland. This suggests that land-use type had a significant impact on spatial SOC and STN patterns.

The variance analysis showed that differences of SOC and STN concentrations for varying land use types were identical ($P<0.01$), whereas differences among vegetable, *Z. mays* and *C. mollissima* lands were not significant ($P>0.05$). We speculate that these results may be attributed to the following four factors: 1) Forest land (such as *P. thunbergii* and *Q. acutissima*) could fix plentiful SOC and STN because of flourishing soil plant roots and a thicker forest litter layer, which were easily resolved and beneficial to SOC and STN accumulation; 2) Litterfall on *C. mollissima* land was often collected by humans, which reduced land cover and increased soil erosion with rainfall, and therefore, SOC and STN concentrations were lower than on other land use types (except waste-grassland); 3) Farmland, especially vegetable land, was close to residential areas and had increased application of chemical fertilizers, so SOC and STN concentrations were higher than in waste-grassland; and 4) Soil and water loss was very serious in the waste-grassland, so its SOC and STN concentrations were the lowest of all land use types.

Influences on SOC and STN spatial distributions at different elevations and aspects

The spatial distributions of soil nutrients were correlated with many environmental factors. Some research indicates that SOC and STN concentrations had significant positive correlations with elevation, cosine of aspect, and other topographic factors [34]. However, this research was mainly on pure forest land or farmland and used the ordinary kriging method. Little work has been conducted for a small watershed with varying elevations and aspects. Table 6 again shows that SOC and STN concentrations had significant positive correlations with elevation, which is consistent with the distribution map of those concentrations (Figure 4).

The variance analysis showed that the effects of elevation on SOC and STN concentrations were identical. For both nutrients, the 370–500 m elevation band was significantly different from 220–290 m and 290–370 m ($P<0.05$), but 220–290 m was not significantly different from 290–370 m ($P>0.05$). We suggest that these results may be due to the following factors: 1) The sample area at high elevation (370–500 m) supported SOC and STN accumulation because it had a low average soil temperature, slow nutrient decomposition and fewer human impacts; and 2) The sample area at low elevation (220–290 m) was not beneficial for accumulation due to rapid nutrient decomposition and greater human influence.

Aspect can also modify the spatial distribution of soil nutrients [35,36]. In the study, the aspects were divided into three classes at 90° intervals from due north; 0°–45° and 315°–360° were shady slopes, 45°–135° and 225°–315° were half-sunny slopes, and 135°–225° were sunny slopes. Table 7 shows that SOC and STN concentrations with different aspects were ranked as follows, shady slopes > half-sunny slopes > sunny slopes. These results are consistent with the distribution map of SOC and STN concentrations (Figure 1, Figure 4).

The variance analysis revealed that the differences of SOC concentrations with aspect were not significant ($P>0.05$). STN concentrations on half-sunny slopes were significantly different from sunny slopes ($P<0.05$), and shady slopes were significantly different from half-sunny and sunny slopes ($P>0.05$). These findings are attributed to the following factors: 1) On sunny slopes, SOC decomposes rapidly and STN concentrations are low, the opposite of conditions on shady and half-sunny slopes; and 2) fewer samples, which caused slight changes among shady, half-sunny and sunny slopes, and little significant difference of SOC and STN. Therefore, to improve the accuracy of SOC and STN spatial distributions in the study area, more attention should be given to the influences of soil temperature and the total number of samples.

Conclusion

1) Compared with the ordinary kriging method, improvement of SOC and STN prediction accuracies by regression-kriging was 16.80% and 17.24%, respectively. This demonstrates that regression-kriging is suitable for studying SOC and STN spatial distributions in a watershed with a complex topography.
2) The semi-variogram test of the regression-kriging models revealed intermediate spatial autocorrelations of both SOC and STN, and the Nugget/Sill were 0.2% and 0.1%, respectively. The test also showed that structural factors such as topography, vegetation and climate were the main factors causing SOC and STN spatial variation in the small Matiyu watershed.
3) The distribution maps of regression-kriging revealed that both SOC and STN concentrations decreased from southeast to northwest in the watershed, similar to the DEM trend.
4) Both land use types and elevation factors had significant influences on the spatial distributions of SOC and STN concentrations in the watershed. Aspect had a smaller but still appreciable effect.

Acknowledgments

The corresponding author sincerely thanks Prof. Chen Fusheng for reviewing the manuscript.

Author Contributions

Conceived and designed the experiments: GP WB. Performed the experiments: GP GG. Analyzed the data: WB GG GP. Contributed reagents/materials/analysis tools: WB GG. Wrote the paper: GP GG ZG.

References

1. Elliott (1986) Aggregate Structure and Carbon, Nitrogen, and Phosphorus in Native and Cultivated Soils. Soil Science Society of America Journal, 50: 627–633.
2. Kucharik CJ, Brye KR, Norman JM, Foley JA, Gower ST, et al. (2001) Measurements and modeling of carbon and nitrogen cycling in agroecosystems of Southern Wisconsin: potential for SOC sequestration during the next 50 years. Ecosystems, 4: 237–258.
3. Lal R (2004) Soil carbon sequestration impacts on global climate change and food security. Science, 304: 1623–1627.
4. Cao LH, Zhao SW (2007) Progress of study on factors of affecting the SOC pool and measures for its regulating and controlling. Journal of Northwest Agriculture and Forestry University (Nat. Sci. Ed.), 35 (3): 177–182.
5. Zhao Z, Li P, Wang NJ (2000) Distribution patterns of root systems of main planting tree species in Weibei Loess Plateau. Chinese Journal of Applied Ecology, 11 (1): 37–39.
6. Wang HJ, Shi XZ, Yu DS, David C, Huang B (2009) Factors determining soil nutrient distribution in a small-scaled watershed in the purple soil region of Sichuan Province, China. Soil & Tillage Research, 105: 300–306.
7. Dou FG, Yu X, Chen LP, Gary M, Guo LG, et al. (2010) Spatial variation of tundra soil organic carbon along the coastline of northern Alaska. Geoderma, 154: 328–335.
8. Gebreyesus BT, Lulseged T, Paul LG (2011) Catchment-scale spatial variability of soil properties and implications on site-specific soil management in northern Ethiopia. Soil & Tillage Research, 117: 124–139.

9. Mohammadi J, Motaghian MH (2011) Spatial Prediction of Soil Aggregate Stability and Aggregate-Associated Organic Carbon Content at the Catchment Scale Using Geostatistical Techniques. Pedosphere, 21(3): 389–399.
10. Fantappiè M, L'Abate G, Costantini EAC (2011) The influence of climate change on the soil organic carbon content in Italy from 1961 to 2008. Geomorphology, 135 (3–4): 343–352.
11. Wang HQ, Hall CAS, Cornell JD, Hall MHP (2002) Spatial dependence and the relationship of soil organic carbon and soil moisture in the Luquillo, Experimental Forest, Puerto Rico. Landscape Ecol., 17: 671–684.
12. Tan ZX, Lal R (2005) Carbon sequestration potential estimates with changes in land use and tillage practice in Ohio, USA. Agr. Ecosyst. Environ., 111: 140–152.
13. Liu DW, Wang ZM, Zhang B, Song KS, Li XY, et al. (2006) Spatial distribution of soil organic carbon and analysis of related factors in croplands of the black soil region, Northeast China. Agr. Ecosyst. Environ., 113: 73–81.
14. Su ZY, Xiong YM, Zhu JY (2006) Soil organic carbon content and distribution in a small landscape of Dongguan, South China. Pedosphere, 16: 10–17.
15. Stacey KF, Lark RM, Whitmore AP, Milne AE (2006) Using a process model and regression kriging to improve predictions of nitrous oxide emissions from soil. Geoderma, 135: 107–117.
16. Harris P, Fotheringham AS, Crespo R, Charlton M (2010) The use of geographically weighted regression for spatial prediction: An evaluation of models using simulated data sets. Mathematical Geosciences, 42: 657–680.
17. Kumar S, Lal R, Liu DS (2012) A geographically weighted regression kriging approach for mapping soil organic carbon stock. Geoderma, 189: 627–634.
18. Hengl T, Heuvelink GBM, Stein A (2004) A generic framework for spatial prediction of soil variables based on regression-kriging. Geoderma, 120: 75–93.
19. Sumfleth K, Duttmann R (2008) Prediction of soil property distribution in paddy soil landscapes using terrain data and satellite information as indicators. Ecological Indicators, 8(5): 485–501.
20. Simbahan GC, Dobermann A, Goovaerts P, Jianli P, Haddix ML (2006). Fine-resolution mapping of soil organic carbon based on multivariate secondary data. Geoderma, 132 (3–4):471–489.
21. Wang YQ, Zhang XC, Zhang JL, Li SJ (2009) Spatial Variability of Soil Organic Carbon in a Watershed on the Loess Plateau. Pedosphere, 19 (4): 486–495.
22. Li Y (2010) Can the spatial prediction of soil organic matter contents at various sampling scales be improved by using regression kriging with auxiliary information? Geoderma, 159, 63–75.
23. Sun W, Minasny B, McBratney A (2012) Analysis and prediction of soil properties using local regression-kriging. Geoderma, 171–172, 16–23.
24. Wang K (2013) Application of Regression Kriging on the Spatial Prediction of Total Soil Nitrogen. Chinese Agricultural Science Bulletin, 29(20): 142–147.
25. Liu S, Ren HY, Shi XZ, Pan JJ (2010) Different methods for prediction of spatial patterns of paddy soil organic carbon density in Changxing County of Zhejiang Province. Journal of Geo-information Science, 12(2): 180–185.
26. Nelson DW, Sommers LE (1982) Total carbon, organic carbon and organic matter. Agronomy Monograph No. 9. ASA and SSSA, Madison, WI., 534–580.
27. Lu WD (2005) SPSS 12.0 for windows statistics analysis. Beijing: Electron Industry Press, 1–135.
28. Zhao YC, Shi XZ, Yu DS (2005) Different methods for prediction of spatial patterns of soil organic carbon density in Hebei Province, China. Acta Pedol. Sin., 42: 379–385.
29. Batjes NH, Sombroek WG (1997) Possibilities for carbon sequestration in tropical and subtropical soils. Global Change Biology, 3 (2): 161–173.
30. Guo LB, Gifford RM (2002) Soil carbon stocks and land use change: a meta analysis. Global Change Biology, 8 (4): 345–360.
31. Fang X, Xue Z, Li B, An S (2012) Soil organic carbon distribution in relation to land use and its storage in a small watershed of the Loess Plateau, China. Catena 88: 6–13.
32. McGrath DA, Smith CK, Gholz HL, Oliveira FA (2001) Effects of land-use change on soil nutrient dynamics in Amazonia. Ecosystems, 4: 625–645.
33. Odeh IOA, McBratney AB, Chittleborough DJ (1995) Further results on prediction of soil properties from terrain attributes: heterotopic cokriging and regression-kriging. Geoderma, 67, 215–225.
34. Hicks RR, Frank PS (1984) Relationship of aspect to soil nutrients, species importance and biomass in a forested watershed in West Virginia. Forest Ecology and Management, 8 (3–4): 281–291.
35. Wang J, Fu B, Qiu Y, Chen LD (2003) Analysis on soil nutrient characteristics for sustainable land use in Danangou catchment of the Loess Plateau, China. CATENA, 54 (1–2): 17–29.
36. Zhang S, Huang Y, Shen C, Ye H, Du Y (2012) Spatial prediction of soil organic matter using terrain indices and categorical variables as auxiliary information. Geoderma, 171–172, 35–43.

15

Effect of Optimal Daily Fertigation on Migration of Water and Salt in Soil, Root Growth and Fruit Yield of Cucumber (*Cucumis sativus* L.) in Solar-Greenhouse

Xinshu Liang, Yinan Gao, Xiaoying Zhang, Yongqiang Tian, Zhenxian Zhang, Lihong Gao*

Beijing Key Laboratory of Growth and Developmental Regulation for Protected Vegetable Crops, Department of Vegetable Science, China Agricultural University, Beijing, P.R. China

Abstract

Inappropriate and excessive irrigation and fertilization have led to the predominant decline of crop yields, and water and fertilizer use efficiency in intensive vegetable production systems in China. For many vegetables, fertigation can be applied daily according to the actual water and nutrient requirement of crops. A greenhouse study was therefore conducted to investigate the effect of daily fertigation on migration of water and salt in soil, and root growth and fruit yield of cucumber. The treatments included conventional interval fertigation, optimal interval fertigation and optimal daily fertigation. Generally, although soil under the treatment optimal interval fertigation received much lower fertilizers than soil under conventional interval fertigation, the treatment optimal interval fertigation did not statistically decrease the economic yield and fruit nutrition quality of cucumber when compare to conventional interval fertigation. In addition, the treatment optimal interval fertigation effectively avoided inorganic nitrogen accumulation in soil and significantly ($P<0.05$) increased the partial factor productivity of applied nitrogen by 88% and 209% in the early-spring and autumn-winter seasons, respectively, when compared to conventional interval fertigation. Although soils under the treatments optimal interval fertigation and optimal daily fertigation received the same amount of fertilizers, the treatment optimal daily fertigation maintained the relatively stable water, electrical conductivity and mineral nitrogen levels in surface soils, promoted fine root (<1.5 mm diameter) growth of cucumber, and eventually increased cucumber economic yield by 6.2% and 8.3% and partial factor productivity of applied nitrogen by 55% and 75% in the early-spring and autumn-winter seasons, respectively, when compared to the treatment optimal interval fertigation. These results suggested that optimal daily fertigation is a beneficial practice for improving crop yield and the water and fertilizers use efficiency in solar greenhouse.

Editor: Jin-Song Zhang, Institute of Genetics and Developmental Biology, Chinese Academy of Sciences, China

Funding: This work was supported by the earmarked fund for Modern Agro-industry Technology Research System in China (CARS-25) and the Special Fund for Research in the Public Interest provided by the Chinese Ministry of Water Resources (201001061). The funders had no role in study design, data collection and analysis, decision to publish, or preparation of the manuscript.

Competing Interests: The authors have declared that no competing interests exist.

* E-mail: gaolh@cau.edu.cn

Introduction

Protected vegetable production systems have been rapidly developed in recent decades in China. Solar-greenhouse, an unheated plastic greenhouse using solar light energy to make sure that vegetables grow normally, plays more and more important roles in China's vegetable production and supplication during the winter [1]. However, excessive irrigation and fertilization are commonly practiced in the solar-greenhouse production systems. For instance, some investigations have revealed that irrigation water rate is 1000 mm each year and fertilizer N apparent recovery efficiency can be less than 10% using conventional management practices [2,3]. Consequently, redundant water and fertilizers affect environmental protection by nutrient accumulation and soil salinization [4–6].

It has become the focus of agricultural field that the efficient water and fertilizer management methods are applied in vegetable production in greenhouse. For example, Cabello et al. [7] demonstrated that under moderate deficit irrigation (90% evapotranspiration) condition, reducing the inputs of nitrogen fertilizer did not reduce yield of melon, but increased water and nitrogen fertilizer use efficiencies. Similarly, Mahajan and Singh [8] found that fertigation through reducing rates of water and nitrogen fertilizer promoted fruit yield and quality, and water and fertilizer use efficiencies in greenhouse tomato cultivation system. In addition, both subsurface drip irrigation and alternate furrow irrigation can significantly improve the root growth, and thus enhance the yield and water and fertilizer use efficiencies [9–12].

However, conventional interval fertigation is still common in intensive vegetable production systems in China. In general, the period between successive irrigation and fertilization events is 7 to 10 days, or even longer. Consequently, the nutrient concentration in the root-zone soil may be in excess of plant requirement for growth on the first day after fertigation, and then decreases gradually to reach deficit levels before the day preceding the next fertigation event, and eventually inhibit crop growth and development [13]. Therefore, there is a need for ensuring water and nutrient in root-zone soil stable to reduce adverse effects caused by their fluctuation on crop growth under intensive vegetable management practices.

Drip fertigation, a technique used to manage soil water and nutrition supplies according to the actual water and nutrient requirement of the plants, can be applied to vegetable production to maintain stable water and nutrition contents in the root zone of crops. It applies frequent and small amounts of soluble fertilizers along with water by exerting the soil buffer characteristic and by reducing the time interval between successive irrigations [14,15]. This technique is still rarely implemented in soil cultivation systems in developing countries, despite of the fact that it has been applied for many years in developed countries.

The efficiency of a fertigation treatment can be evaluated with a combination of the availability and distribution of soil water and nutrients, the growth and development of crop root, and the formation of crop yield. These parameters are relatively simple, rapid, easy to obtain and cost effective. Among these parameters, particular attention should be paid to plant root since the morphology and spatial configuration of root can significantly affect soil water and nutrient transformation, mobilization and use efficiency by plant and crop yield [16–18]. However, due to the complexity of root growth environment and the limitations of root research methods, little information is available regarding the vegetable root growth under daily fertigation based on crop requirement under greenhouse soil cultivation conditions.

Cucumber is one of the major greenhouse vegetables in China. However, inappropriate irrigation and fertilizing practices have caused soil nutrient imbalance that negatively affects cucumber growth and the reduction of soil water and fertilizer use efficiencies. It is important for farmers to manage soil water and nutrition supplies according to the actual water and nutrient requirement of the plants during greenhouse cucumber production seasons. Thus, the objectives of this study were to investigate how the migration of water and salt in soil, and root growth and fruit yield of cucumber are affected by the daily fertigation based upon the actual water and nutrient requirement of the plants.

Materials and Methods

Ethics statement

No specific permissions were required for the described field study.The experiment was carried out in 2011 in a typical solar greenhouse of Fangshan District Agricultural Science Research Institute, Beijing, China (39.7°N; 116.1°E), which is an experimental station of China Agricultural University. The field study did not involve endangered or protected species.

Site description and experimental design

The greenhouse was covered with polyethylene film (ground area 56 m×7 m) without supplementary lighting and heating. Daily average soil temperatures and air temperature in solar-greenhouse were shown in Fig. 1. Both the two temperatures in April gradually increased until each reached the highest value in July and August, and then gradually decreased. Furthermore, Daily average soil temperatures at 10 cm depth from conventional interval fertigation (CK), optimal interval fertigation (OIF) and optimal daily fertigation (ODF) were not significantly different. Field experiments were conducted on a sandy loam soil (36% sand, 48% silt and 16% clay). The topsoil (0–30 cm layer) had a pH (1:2.5 soil/water) value of 7.37, an electrical conductivity (EC) (1:5 soil/water) value of 1.17 dS m^{-1}, a field capacity of 20.8% and a bulk density of 1.36 g cm^{-3}, and contained 15.8 g kg^{-1} organic matter, 0.96 g kg^{-1} total nitrogen (N), 195 mg kg^{-1} inorganic N, 193 mg kg^{-1} available phosphorus (P) and 275 mg kg^{-1} available potassium (K).

The experimental period comprised two growth cycles including the early-spring (ES) (February 1-July 29) and autumn-winter (AW) (August 1-December 25) seasons. The varity of cucumber was Zhongnong No. 26. This varity was a high-quality and largely fruit-bearing cucumber hybrid. The average fruit length and width of this varity were 30 and 3.3 cm, respectively. Due to its high resistance to several diseases (including powdery mildew, downy mildew and gray mold) and tolerance to both low temperature and weak light, this varity was very suitable for cultivation in solar-greenhouse. Cucumber seedlings with two leaves were transplanted by hand, before which, soils were incorporated with organic fertilizer (basal fertilizer) at designed rates (Table 1), and were ploughed and harrowed to a depth of 30 cm. Plant pruning was performed as follows: all lateral branches were removed by hand, however, the axial shoot of cucumber was remained and it climbed upward along a vertical rope. Planting density of cucumber was 5 plants m^{-2}.

The experiment consisted of three treatments:

(1) Conventional interval fertigation (CK): Conventional organic manure (basal fertilizer) and chemical fertilizer were applied at rates based on average fertilization level used by greenhouse cucumber growers in the suburb of Beijing (Table 1). A detailed description of chemical fertilizer topdressing/irrigation rates under the CK was given in Fig. 2.

(2) Optimal interval fertigation (OIF): The organic manure (basal fertilizer) application rate was half the rate conventionally applied per hectare (Table 1). Based on the N requirement for cucumber growth [19–21] and N fertilizer recommendation, the total N rates applied by topdressing were 427.5 and 151.5 kg N ha^{-1} in the ES and AW seasons (Table 1), respectively. These total mineral N (N_{min}) application rates were calculated using the method of soil N balance, where expected yield of solar greenhouse cucumber were 120 and 60 t ha^{-1} in the ES and AW seasons, respectively. The equation [22] was as follows:

$$N_{recommend} = N_{crop} + N_{safty} + N_{loss} - N_{initial} - N_{manure} - N_{mineralization} \quad (1)$$

Where $N_{recommend}$ is Recommended fertilizer N; N_{crop} is crop N uptake; N_{safty} is soil N_{min} safety margin; N_{loss} is N loss; $N_{initial}$ is soil N_{min} in the root zone before transplanting; N_{manure} is N_{min} from the mineralization of organic fertilizer; $N_{mineralization}$ is N_{min} from the mineralization of organic nitrogen in soil.

Chemical fertilizer topdressing/irrigation frequencies and irrigation rates were the same as the CK, except fertilizer topdressing rates. A detailed description of chemical fertilizer topdressing/irrigation rates under the treatment OIF was given in Fig. 2.

(3) Optimal daily fertigation (ODF): The total amounts of organic manure (basal fertilizer) and chemical fertilizer were the same as treatment OIF (Table 1), however, chemical fertilizer was applied automatically daily according to the actual water and nutrient requirement of the plants. A detailed description of chemical fertilizer topdressing/irrigation rates under the treatment ODF was given in Fig. 2.

The total irrigation amount was the same for three treatments in whole growth period of cucumber. Fertigation was used under transparent thin film for all treatment. Irrigations were applied 16

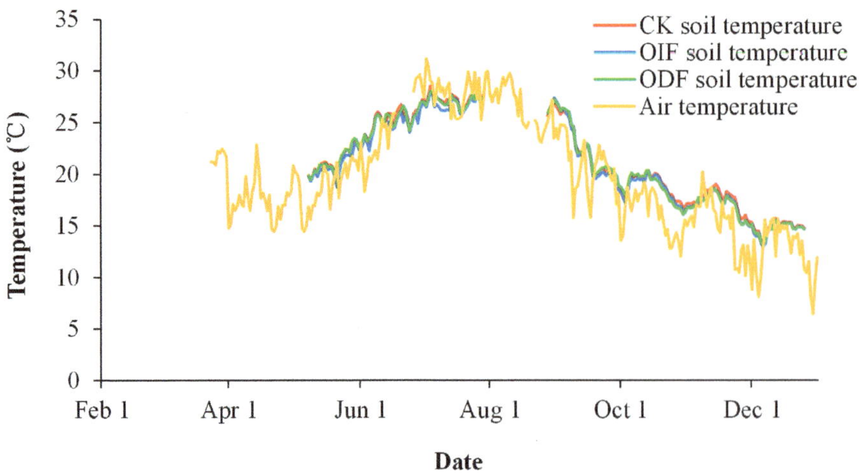

Figure 1. Daily average soil temperatures at 10 cm depth from conventional interval fertigation (CK), optimal interval fertigation (OIF) and optimal daily fertigation (ODF) and daily average air temperature at 150 cm height in cucumber greenhouse cropping system in 2011 at Fangshan, Beijing suburbs. These data were determined by RTH-1010 TPE rensin-shield sensor and RT-12 Thermo Recorder made in Japan. Time interval for data recording was set to 10 minutes.

and 7 times in the ES and AW seasons, respectively. The irrigation rate each time for CK and OIF teatments was 22.5–30 mm according to requirement rule of water for cucumber [23–25].

The chemical fertilizer used in this study was a special water-soluble fertilizer for cucumber called Shengdanshu (a local fertilizer). This fertilizer had three $N:P_2O_5:K_2O$ formulations, i.e., 20:20:20, 19:8:27 and 18:6:34, which were used at the early harvest stage (April 21 – May 12 during ES season, September 25 – October 11 during AW season), the middle harvest stage (May 13 – July 5 during ES season, October 12 – November 30 during AW season) and the late harvest stage (July 6 – July 29 during ES season, December 1 – December 25 during AW season), respectively. Fertilizers were applied 12 and 6 times in the ES and AW seasons, respectively. Except different fertigation methods related to different treatments, the same local management practices were applied in all treatments. The experiment was a randomized block design with four replications and the size of each replicate plot was 3.9 m×4.8 m. Each replicate plot had three cultivation furrows and was separated from the adjacent plots by plastic films buried at a depth of 50 cm.

Soil water content, EC and mineral N content

To evaluate the migration of soil water and salt under different fertigation treatments, soil samples from five cores per subplot were collected five times within a single fertigation cycle during the middle fruit harvest period when daily fruit production was very high in each cropping season. For the ES season, soils were sampled on May 27, May 28, May 31, June 2 and June 4. For the AW season, the corresponding sampling times were October 26, October 27, November 1, November 6 and November 12. Soil samples were taken at 0–15, 15–30 and 30–45 cm depth. Soil samples of each plot at each depth were mixed thoroughly and passed through a 2-mm sieve. Sub-samples of 20 g fresh soil were dried for 12 h at 105°C, and then soil water content was determined at the ratio of water and dry soil weight. Sub-samples of about 300 g fresh soil used to measure EC were air-dried passed through a 1-mm sieve. Soil EC was analyzed from a 1:5 (*w/v*) soil (air-dried) to water ratio using an EC meter and combination glass electrodes (FE30, METTLER TOLEDO, Shanghai, China). Sub-samples of 12 g fresh soil used to estimate mineral N were submerged into 100 ml 0.01 M $CaCl_2$ solution and shaken for 1 hour to extract inorganic N. The extracts were filtered and

Table 1. Amounts of fertilizers used in the treatments CK (conventional interval fertigation), OIF (optimal interval fertigation) and ODF (optimal daily fertigation) in the ES (early-spring) and AW (autumn-winter) seasons.

Cropping season	Treatment	Organic manure[a]	Chemical fertilizer (kg ha^{-1})			Irrigation water
		(t ha^{-1})	N	P_2O_5	K_2O	(mm)
ES	CK	45.0	819.0	690.4	1006.2	349.2
	OIF	22.5	427.5	361.3	524.1	349.2
	ODF	22.5	427.5	361.3	524.1	349.2
AW	CK	60.0	465.0	312.0	593.4	147.2
	OIF	30.0	151.5	123.1	183.1	147.2
	ODF	30.0	151.5	123.1	183.1	147.2

[a]For the ES season, total N, total P and total K contents of the organic manure were 2.09% (N), 2.06% (P_2O_5) and 1.34% (K_2O), respectively. For the AW season, total N, total P and total K contents of the organic manure were 1.30% (N), 0.5% (P_2O_5) and 1.06% (K_2O), respectively.

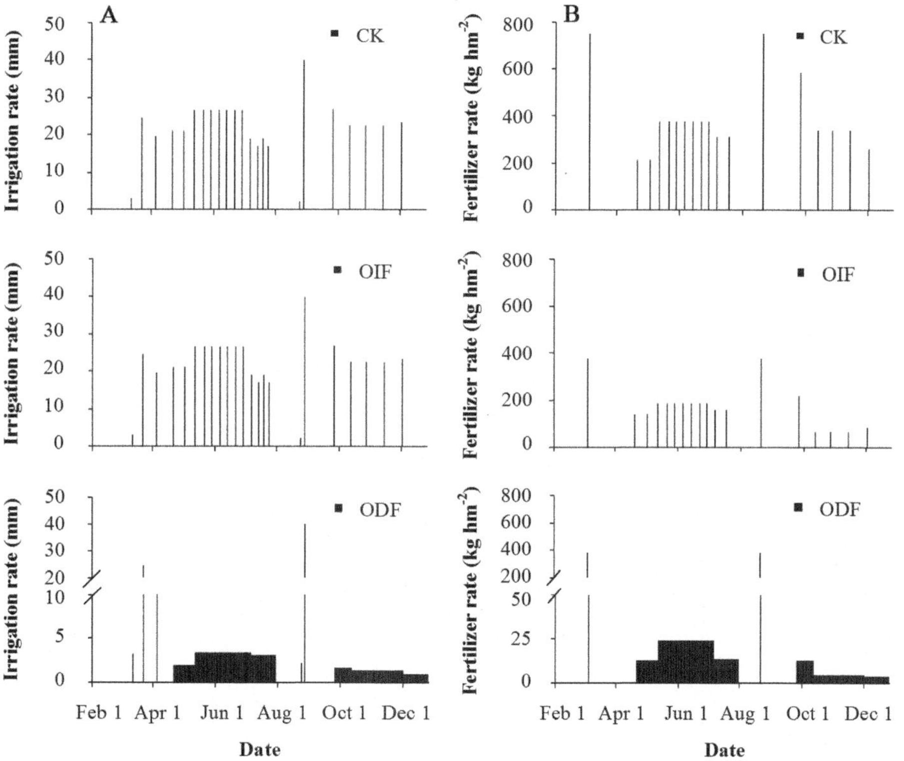

Figure 2. Irrigation scheduling (A) and fertilization scheduling (B) under three different fertigation ways in a solar greenhouse cucumber cultivation system. All fertilizer varity were compound fertilizer. The two $N:P_2O_5:K_2O$ formulations of basal fertilizer were respectively (18:46:0) and (15:15:15) during ES and AW seasons. The three $N:P_2O_5:K_2O$ formulations of topdressing fertilizer were respectively (20:20:20), (19:8:27) and (18:6:34) for early fruit stage, middle fruit stage, late fruit stage during ES and AW seasons.

analyzed using an continuous flowing analyzer (TRAACS2000, USA) to determine NO_3^--N and NH_4^+-N contents [26].

Cucumber root morphologic characters

Once the final harvest was completed, one typical cucumber root in each plot was collected with an Eijelkamp root auger (length = 0.15 m, diameter = 0.08 m) from the 0–15, 15–30 and 30–45 cm soil layers. For each layer, 9 holes were drilled around the cucumber main root in a shape of 3×3 cross square. The roots in the soil were carefully selected and washed to acquire the roots from different soil layers. Root morphology was analyzed by using fresh roots and a root scanner system (EPSON EXPRESSION 4990, Japan). Data were then analyzed with the WinRHIZO root analysis software (LC4800-II LA2400; Saint foy, Canada) to determine the root characteristics, including root length, root surface, root volume and average diameter. The root characteristics could be divided into four classification according to the root diameters aggregated into classes of 0.0–0.5, 0.5–1.0, 1.0–1.5 and ≥1.5 mm. The scanned roots were oven-dried at 65°C until weight constancy and weighed.

Cucumber fruit yield, irrigation water use efficiency and partial factor productivity of applied nitrogen

Economic yield was measured for whole cucumber growth cycles in each plot and translated into economic yield weight per hectare. The ratio of yield to water supply was referred to as irrigation water use efficiency (*IWUE*, kg mm^{-1}):

$$IWUE = Y/W \tag{2}$$

where *Y* and *W* represent the economic yield (kg ha^{-1}) and the amount of water (mm) applied to the cucumber during the growing cycle [7], respectively. The ratio of yield to N supply is referred to as partial factor productivity of applied N (PFP_N, kg kg^{-1}):

$$PFP_N = Y/F \tag{3}$$

where *F* is the amount of fertilizer N (kg) applied to the cucumber during the growing cycle [27].

Vitamin C, soluble sugar and nitrate contents in cucumber fruit

To estimate cucumber fruit quality under different fertigation treatments, three quality parameters which were concerned by local residents were made. Fresh fruit samples of cucumber in each plot were collected when daily fruit production was very high, and their appearance should be similar and marketable. Fresh fruits were washed, chopped and mixed in the lab to assess different quality indexes. Contents of Vitamin C and soluble sugar were determined by the methods of 2, 6-dichloro-indophenol titration [28] and anthrone ethyl acetate colorimetic [29], respectively. Content of nitrate was measured by the method of sulfuric acid-acid [30].

Statistical analysis

SPSS 17.0 was used to analyse data. Treatment means were separated using the least significant difference (LSD) test at $P<0.05$. Principal component analysis (PCA) was done to comprehensively determine the whole effect of different fertigation methods on several root morphologic characters (Root dry weight, length, surface area, average diameter and volume).

Results

Soil water content, EC and mineral N content

Soil water content under the treatments OIF and CK increased rapidly and then decreased gradually within an irrigation cycle (Fig. 3). However, it maintained a relatively stable level under the treatment ODF. No significant difference was found in soil water content between the treatments OIF and CK in all tested soil layers in both ES and AW seasons. The treatment ODF decreased the amplitude (i.e. the difference between maximum and minimum values) of soil water content in all tested soil layers during ES season and that in the 0–15 cm soil layer during AW season in an irrigation cycle, when compared to the treatments OIF and CK (Table 2). In addition, the relative soil water content under the treatment ODF was higher in the 0–15 cm soil layer than in the 15–30 and 30–45 cm soil layers, suggesting that the soil water was concentrated by the treatment ODF in the 0–15 cm soil layer.

The EC values in the 0–15 cm soil layer were higher and more variable than those in the 15–30 and 30–45 cm soil layers, suggesting the salts were concentrated in the 0–15 cm soil layer (Fig. 3). The CK showed the highest and most variable EC values in the 0–15 cm soil layer, resulting in the highest amplitude of soil EC value (Table 2). Generally, in the ES season, the treatments OIF and ODF did not show significant differences in EC values in the 0–15 cm soil layer. However, in the AW season, EC values under the treatment OIF were significantly higher and more variable than those under the treatment ODF in most sampling times. The soil EC values were not significantly different (P>0.05) between any two treatments in the 15–30 and 30–45 soil layers in both ES and AW seasons.

Similar to soil EC values, generally, soil mineral N (N_{min}) contents in the 0–15 cm soil layer were higher and more variable than those in the 15–30 and 30–45 cm soil layers (Fig. 3). In addition, the CK had the higher and more variable soil N_{min} contents in the 0–15 cm soil layer. However, the amplitude of soil N_{min} content was significantly higher under the CK than under the treatments OIF and ODF in the 15–30 and 30–45 soil layers in ES and AW seasons, respectively (Table 2).

Principal component analysis of root growth properties

Root dry weight, length, surface area, average diameter and volume were used to evaluate the plant root growth. Since there were statistical links among each other of these root properties (all P<0.05), we used the principal component analysis (PCA) to transform these correlated root variables (Table 3) into two principal components (i.e. PC_1 and PC_2) to obtain two principoal expressions as follow:

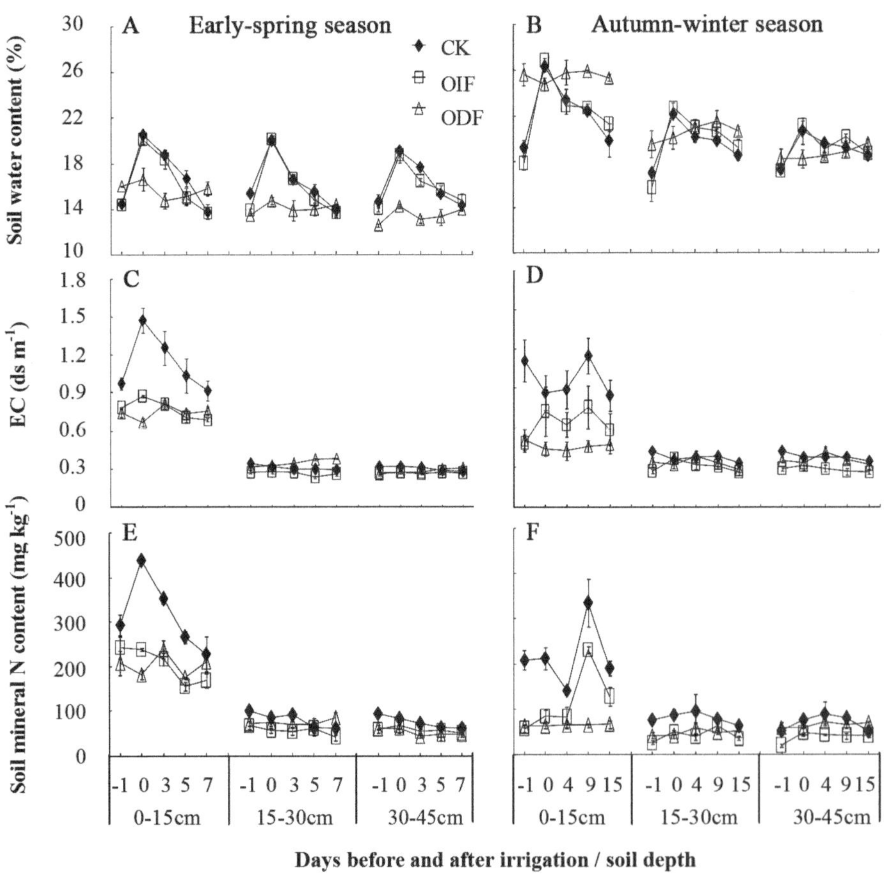

Figure 3. Effects of conventional interval fertigation (CK), optimal interval fertigation (OIF) and optimal daily fertigation (ODF) on changes of soil water content (A and B), EC value (C and D) and mineral N content (E and F) in the early-spring (ES) and autumn-winter (AW) seasons. The numbers on the abscissa represent the days before (negative value), during (zero) and after (positive value) irrigation. Bars represent standard errors.

Table 2. Effects of conventional interval fertigation (CK), optimal interval fertigation (OIF) and optimal daily fertigation (ODF) on the amplitudes (the difference between maximum and minimum values) of soil water content, EC value and mineral N content in the early-spring (ES) and autumn-winter (AW) seasons.

Cropping season	Treatment	Soil water (%)			Soil EC (mS cm^{-1})			Soil mineral N (mg kg^{-1})		
		0–15 cm[a]	15–30 cm	30–45 cm	0–15 cm	15–30 cm	30–45 cm	0–15 cm	15–30 cm	30–45 cm
ES	CK	7.0 a[b]	6.1 a	5.0 a	0.717 a	0.066 a	0.071 a	230.7 a	59 a	43.9 a
	OIF	6.5 a	6.6 a	5.0 a	0.204 b	0.062 a	0.033 a	114.5 b	39.9 b	35.6 a
	ODF	3.4 b	2.2 b	2.1 b	0.154 b	0.087 a	0.077 a	91.6 b	32.6 b	41.8 a
AW	CK	8.4 a	5.2 a	3.8 a	0.682 a	0.108 a	0.094 a	205.2 a	77.5 a	76.1 a
	OIF	9.0 a	7.0 a	4.0 a	0.495 a	0.130 a	0.076 a	177.6 a	38.7 a	36.4 b
	ODF	3.5 b	4.0 a	2.2 a	0.223 b	0.112 a	0.097 a	23.9 b	29.6 a	25.2 b

[a]Soil sampling layers
[b]The same letter in the same data column denotes no significant difference (P≤0.05) by LSD.

$$PC_1 = 0.452X_1 + 0.489X_2 + 0.544X_3 - 0.243X_4 + 0.449X_5 \quad (4)$$

$$PC_2 = 0.418X_1 - 0.357X_2 - 0.076X_3 + 0.705X_4 + 0.441X_5 \quad (5)$$

where X_1, X_2, X_3, X_4 and X_5 represent the root dry weight, length, surface area, average diameter and volume, respectively. The eigenvalue contribution rates of PC_1 and PC_2 were 65.95% and 31.60%, respectively.

Since the greater values of root dry weight, length, surface area and volume and the thinner root diameter, the root system is more powerful for absorption of water and nutrient, the comprehensive principal component (CPC) was received as follow:

$$CPC = 0.6774PC_1 - 0.3155PC_2 = 0.201X_1 + 0.444X_2 + 0.389X_3 - 0.349X_4 + 0.219X_5 \quad (6)$$

The treatment CK showed the lowest PC_1 and CPC values in the ES season, but the highest PC_2 values in the ES and AW seasons and PC_1 and CPC values in the AW season (Table 4). However, the treatment OIF showed the lowest PC_1 and CPC values in the AW season. The treatment ODF showed the lowest PC_2 value in the ES and AW seasons, but the highest PC_1 and CPC values in the ES season.

Table 4. Effects of conventional interval fertigation (CK), optimal interval fertigation (OIF) and optimal daily fertigation (ODF) on the first (PC1) and second (PC2) principal components and comprehensive principal component (CPC) values of root characteristic parameters in the early-spring (ES) and autumn-winter (AW) seasons.

Principal component	ES season			AW season		
	CK	OIF	ODF	CK	OIF	ODF
PC_1	−0.09	1.05	1.61	0.50	−2.62	−0.46
PC_2	1.28	1.26	−0.81	−0.22	−0.58	−0.93
CPC	−0.46	0.30	1.32	0.40	−0.93	−0.01

Root length and distribution in different soil layers

In general, root length of four diameter classes (i.e. 0.0–0.5, 0.5–1.0, 1.0–1.5 and >1.5 mm) decreased with rooting depth under all fertigation treatments (Fig. 4). However, while the fine roots (<0.5 mm diameter) mainly concentrated at 0–15 cm depth, the thicker roots (>0.5 mm diameter) were more evenly distributed

Table 3. Effects of conventional interval fertigation (CK), optimal interval fertigation (OIF) and optimal daily fertigation (ODF) on the root weight, length, surface area, average diameter and volume in the early-spring (ES) and autumn-winter (AW) seasons.

Cropping season	Treatment	X_1	X_2	X_3	X_4	X_5
		Root weight (g)	Root length (cm)	Root surface area (cm^2)	Root average diameter (mm)	Root volume (cm^3)
ES	CK	1.68 a[a]	4913.69 b	653.73 a	0.416 a	8.58 a
	OIF	1.86 a	5641.86 ab	732.78 a	0.407 a	9.31 a
	ODF	1.67 a	7368.76 a	781.54 a	0.370 b	8.47 a
AW	CK	1.51 a	6104.37 a	715.41 a	0.388 a	8.35 a
	OIF	1.09 b	3870.42 b	481.92 b	0.399 a	5.92 b
	ODF	1.31 a	5855.82 ab	659.25 a	0.384 a	7.08 b

[a]The same letter in the same data column denotes no significant difference (P≤0.05) by LSD.

with soil depth. Root length of 0.0–0.5 mm diameter was statistically affected by fertigation treatment within the uppermost 15 cm of the soil depth, however, root length of 0.5–1.5 mm diameter was statistically affected by fertigation treatment within the uppermost 30 cm of the soil depth ($P<0.05$; Fig. 4A,B). Root length of >1.5 mm diameter was only statistically affected by fertigation treatment in the 15–30 cm soil layer in the AW season ($P<0.05$; Fig. 4B). The treatment OIF significantly increased the root length of 0.5–1.0 mm diameter in the 0–15 cm soil layer in the ES season, however, the treatment ODF significantly increased the root length of 0.0–1.5 mm diameter in the 0–15 cm soil layer in the ES season and the root length of 1.0–1.5 mm diameter in the 0–15 cm soil layer in the AW season, when compared to the CK. Generally, the root lengths were not significantly different ($P>0.05$) between the treatment OIF and CK in the 15–30 and 30–45 soil layers in both ES and AW seasons. However, the treatment ODF significantly decreased the root length in the 15–30 and 30–45 soil layers in both ES and AW seasons, when compared to the CK.

Cucumber economic yield, IWUE and PFP_N

In both ES and AW seasons, cucumber economic yield, IWUE, PFP_N from the treatment ODF were significantly (P<0.05) higher than those from the CK (Table 5). No significant (P>0.05) difference was found in both cucumber economic yield and IWUE between the treatments OIF and CK, however, PFP_N from the treatment OIF was significantly (P<0.05) higher than that from the CK in both ES and AW seasons. The treatment ODF increased cucumber economic yield, IWUE and PFP_N by 6.2%, 6.1% and 103% in the ES season, and 8.3%, 8.4% and 232% in the AW season when compared to the CK, respectively. The treatment OIF increased PFP_N by 88% and 209% in the ES and AW seasons when compared to the CK, respectively.

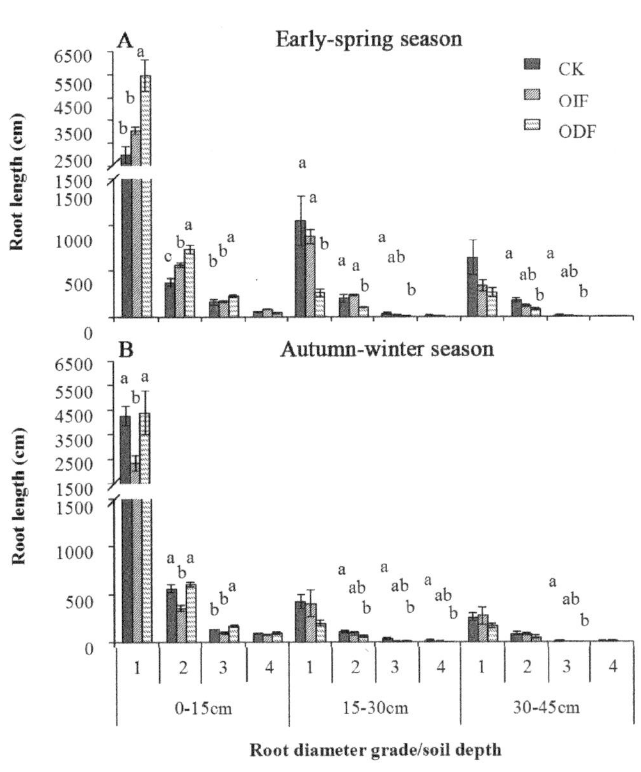

Figure 4. Effects of conventional interval fertigation (CK), optimal interval fertigation (OIF) and optimal daily fertigation (ODF) on the length and distribution of root with different diameter grades in different soil layers in the early-spring (A) and autumn-winter (B) seasons. The numbers 1, 2, 3 and 4 on the abscissa represent four root diameter scales, 0.0–0.5, 0.5–1.0, 1.0–1.5 and ≥1. 5 mm, respectively. Bars represent standard errors. The same letter in the same data column denotes no significant difference (P≤0.05) by LSD.

Vitamin C, soluble sugar and nitrate in cucumber fruit

In general, there was no significant ($P>0.05$) difference in soluble sugar in cucumber fruit between the optimal fertigation treatments (i.e. OIF and ODF) and the CK (Table 6). However, the nitrate in cucumber fruit was significantly ($P<0.05$) decreased by the treatment OIF in the ES and AW seasons, and by the treatment ODF in the AW season, when compared to the CK. In addition, vitamin C in cucumber fruit was significantly ($P<0.05$) increased by the treatment ODF in both ES and AW seasons, when compared to the treatment OIF.

Discussion

Effects of different fertigation treatments on soil water and nutrients

The spatial and temporal distribution of water and nutrients is very heterogeneous in soil. Plant root growth not only can be induced by nutrient supply intensity on the whole, but also can be effected by spatiotemporal variation of water and nutrients [31,32]. Thus, it is important to maintain a relative stable nutrient supply in plant root-zone. Our results clearly showed that increasing the frequency of fertigation could decrease the amplitudes of water and nutrient contents in soils (Table 2). In general, optimal daily fertigation significantly decreased the amplitudes of water content, EC value and mineral N content in the 0–15 cm soil layer in the AW season, when compared to optimal interval fertigation (ODF *vs* OIF; Table 2). Thus, the water and nutrient limitation for plant root growth is probably not likely in soils under optimal daily fertigation. In addition, since the topsoil (0–15 cm) received more water and nutrient than the subsoil (15–45 cm) under fertigation conditions, optimal daily fertigation significantly promoted the root length of <1.5 mm diameter in the 0–15 cm soil layer when compared to optimal interval fertigation (ODF *vs* OIF; Fig. 4). These results are in accordance with previous researches [33–36] and further demonstrate that using the traditional interval fertigation may lead to larger fluctuation of water and nutrients in soil and inhibit plant root growth [37,13]. Furthermore, since soil relative water contents in the 0–15 cm soil layer generally exceeded 25.5% and were higher than soil field capacity (20.8%) under optimal daily fertigation in the AW season (Fig. 3), there is a possibility for our study to further reduce the irrigation rate.

Effects of different fertigation treatments on plant root growth

Plant roots are able to make different responses to different water and nutrient contents of soils. Although excessive fertilizer application can increase crop yield, it can inhibit plant root growth due to the high nutrient concentration in root-zone. Furthermore, unused nutrients will accumulate in the soils and finally enhance the potential threat to the environment. In contrast, root growth can be depressed when soil nutrient is in deficit [38,2,39,40]. In general, optimal fertilizer application can promote plant root growth and improve fertilizer use efficiency. In this study, however, optimal fertilizer application had no promoting effect on both cucumber root growth (Fig. 4B) and economic yield

Table 5. Effects of conventional interval fertigation (CK), optimal interval fertigation (OIF) and optimal daily fertigation (ODF) on cucumber economic yield, irrigation water use efficiency (IWUE) and partial factor productivity of applied nitrogen (PFP_N) in the early-spring (ES) and autumn-winter (AW) seasons.

Cropping season	Treatment	Economic yield (t ha^{-1})	IWUE (kg mm^{-1})	PFP_N (kg kg N^{-1})
ES	CK	132.6b[a]	379. 8b	161. 9c
	OIF	130.2b	372. 8b	304. 5b
	ODF	140.8a	403. 1a	329. 2a
AW	CK	41.2b	279. 8b	88. 6c
	OIF	41.5b	281. 7b	273. 7b
	ODF	44.6a	303. 2a	294. 5a

[a]The same letter in the same data column denotes no significant difference ($P\leq0.05$) by LSD.

(Table 5) in the AW season under same fertilizer application ways (OIF *vs* CK). The explanation is that although the amount of inorganic fertilizer application was sufficient for plant root growth under the treatment OIF during middle fruit harvest period, it was insufficient for plant root growth during the late fruit harvest period, resulting in significant inhibition of root growth and no significant effect on the cucumber economic yield. For instance, contents of mineral nitrogen in soil laye of 0–30 cm of the treatments CK, OIF and ODF during the late fruit harvest period were respectively 93, 35 and 61 mg kg^{-1}. The contents of mineral nitrogen of the treatment OIF was obviousely less than 50 mg kg^{-1}, which is the lower limit of soil N_{min} content for depth 0–30 cm for normal growth and development of cucumber [41], leading to significantly inhibit root growth in the AW seasnon. The results presented here suggested that fertilizer should be applied according to the actual nutrient requirement of the crops at different growing stages.

Effects of different fertigation treatments on cucumber fruit yield

Root growth is closely coordinated with shoot growth [42]. Generally, spatial distribution of roots in soils, together with distribution of water and nutrients in root-zone and water and nutrients requirements for root and shoot growth, determine the water and nutrient use efficiency of crops and the economic yield formation [43]. The data on economic yield in this study showed that although the amount of fertilizer applied to soil was visibly lower under optimal interval fertigation than under conventional interval fertigation (Table 1), optimal interval fertigation still met the water and nutrient requirements of cucumber (Fig. 3) and maintained root growth (Table 3 and Fig. 4), resulting in the higher efficiency of roots as absorbing organs and no significant effect on cucumber economic yield (Table 5). Thus, excessive fertilizer applied to soil under conventional interval fertigation did not make much contribution to increasing crop yield. In contrast, it can inhibit plant root growth due to the high nutrient concentration in root-zone.

Although soils under the treatments OIF and ODF received the same amount of fertilizer (Table 1), the treatment ODF significantly increased economic yield when compared to the treatment OIF (Table 5). This can be explained by the fact that ODF maintained sufficient and relatively stable water and nutrients in soil (Table 2) and promoted proliferation of fine roots which increased root surface area (Table 3). In addition, ODF reduced the spatiotemporal variation of water in root-zone when compare to OIF (Fig. 3). It was also supported by two recent studies: Gao et al. [14] found that the technology of solution daily application based on the requirement of crops had a promoting effect on tomato yield in a plotted system; Yoshida [44] reported that this technology could increase both the yield of tomato fruit and the water and fertilizers use efficiency by plant.

Root function is related to the morphological and physiological characteristics of root. However, this study did not consider the root physiological function, hence, further work is required test the root physiological function to elucidate the mechanism of communication of root and shoot growth.

Table 6. Effects of conventional interval fertigation (CK), optimal interval fertigation (OIF) and optimal daily fertigation (ODF) on the contents of vitamin C, soluble sugar and nitrate in cucumber fruit in the early-spring (ES) and autumn-winter (AW) seasons.

Cropping season	Treatment	Vitamin C (mg 100 g^{-1})	Soluble sugar (mg g^{-1})	Nitrate (ug g^{-1})
ES	CK	6. 94ab[a]	20. 78a	92. 23a
	OIF	6. 71b	19. 81a	76. 97b
	ODF	7. 31a	21. 12a	84. 94ab
AW	CK	5. 51ab	32. 56a	137. 68a
	OIF	5. 33b	32. 31a	78. 65b
	ODF	5. 76a	31. 02a	63. 89b

[a]The same letter in the same data column denotes no significant difference ($P\leq0.05$) by LSD.

Effects of different fertigation treatments on nitrate content in cucumber fruit

In general, vegetables can provide 80% of nitrate absorbed into the human body [45]. However, part of these nitrate can be reduced to nitrite by bacteria, which may induce lower oxygen-carrying capacity of blood and methemoglobinemia. In addition, nitrite can produce a reaction with secondary amines, such as amides and amino acids, possibly causing the formation of nitrosamines, which may induce cancer in the digestive system of human body [46]. Therefore, it is important to control and maintain low nitrate content in edible part of vegetables. In this study, both OIF and ODF treatments significantly decreased nitrite content, compared to CK treatment. The relative high nitrate content in fruit under CK can be explained by the previous research that surplus nitrate can be partly absorbed and stored by the plant than its need, in order to maintain normal growth requirement when nitrate supply is in deficit [47]. The results presented here suggested that reducing the fertilizer amounts, according to the actual nutrient requirement of the crops at different growing stages, is beneficial to fruit quality and human healthy without having a significant effect on cucumber fruit yield.

Conclusions

On the premise of optimal management of irrigation rate for cucumber, optimal interval fertilization based on nutrient target values, including soil mineral nutrient safety margin and crop uptake, and soil nutrients in root-zone had no significant effect on economic yield, but increased partial factor productivity of applied N. Optimal daily fertigation based on the actual requirement of crops maintained the relatively stable water and nutrients in soils, reduced the spatiotemporal variation of water and nutrients in root-zone and promoted cucumber fine root (<1.5 mm diameter) growth. In addition, optimal daily fertigation maintained significantly higher cucumber economic yield, irrigation water use efficiency and partial factor productivity of applied N than optimal interval fertigation.

Acknowledgments

The authors thank you very much to all the staff of Fangshan District Agricultural Science Research Institute.

Author Contributions

Conceived and designed the experiments: XL ZZ LG. Performed the experiments: XL XZ. Analyzed the data: XL YG. Wrote the paper: XL YT ZZ LG.

References

1. Gao LH, Qu M, Ren HZ, Sui XL, Chen QY, et al. (2010) Structure, function, application and ecological benefit of single-slope, energy-efficient solar greenhouse in China. Horttechnology 20: 626–631.
2. He F, Chen Q, Jiang R, Chen X, Zhang F (2007) Yield and nitrogen balance of greenhouse tomato (Lycopersicum esculentum Mill) with conventional and site-specific nitrogen management in Northern China. Nutrient Cycling in Agroecosystems 77: 1–14.
3. Hvistendahl M (2010) China's Push to Add by Subtracting Fertilizer.Sci. 327: 801–801.
4. Shi WM, Yao J, Yan F (2009) Vegetable cultivation under greenhouse conditions leads to rapid accumulation of nutrients, acidification and salinity of soils and groundwater contamination in South-Eastern China. Nutrient Cycling in Agroecosystems 83: 73–84.
5. Thompson R, Martinez-Gaitan C, Gallardo M, Gimenez C, Fernandez M (2007) Identification of irrigation and N management practices that contribute to nitrate leaching loss from an intensive vegetable production system by use of a comprehensive survey. Agricultural Water Management 89: 261–274.
6. Tian Y, Liu J, Zhang X, Gao L (2010) Effects of summer catch crop, residue management, soil temperature and water on the succeeding cucumber rhizosphere nitrogen mineralization in intensive production systems. Nutrient Cycling in Agroecosystems 88: 429–446.
7. Cabello MJ, Castellanos MT, Romojaro F, Martinez-Madrid C, Ribas F (2009) Yield and quality of melon grown under different irrigation and nitrogen rates. Agricultural Water Management 96: 866–874.
8. Mahajan G, Singh KG (2006) Response of Greenhouse tomato to irrigation and fertigation. Agricultural Water Management 84: 202–206.
9. Jokinen K, Sarkka LE, Nakkila J, Tahvonen R (2011) Split root fertigation enhances cucumber yield in both an open and a semi-closed greenhouse. Sci Hortic (Amsterdam) 130: 808–814.
10. Wang Y, Liu F, Andersen MN, Jensen CR (2010) Improved plant nitrogen nutrition contributes to higher water use efficiency in tomatoes under alternate partial root-zone irrigation. Funct Plant Biol 37: 175–182.
11. Zhang L, Gao L, Zhang L, Wang S, Sui X, et al.(2012) Alternate furrow irrigation and nitrogen level effects on migration of water and nitrate-nitrogen in soil and root growth of cucumber in solar-greenhouse. Sci Hortic (Amsterdam) 138: 43–49.
12. Zotarelli L, Scholberg JM, Dukes MD, Munoz-Carpena R, Icerman J (2009) Tomato yield, biomass accumulation, root distribution and irrigation water use efficiency on a sandy soil, as affected by nitrogen rate and irrigation scheduling. Agricultural Water Management 96: 23–34.
13. Matsuo N, Mochizuki T (2009) Assessment of three water-saving cultivations and different growth responses among six rice cultivars. Plant Prod Sci 12: 514–525.
14. Gao YM, Li JS, Cao YE (2006) Study on the fertigation of tomato soil culture by drip irrigation in greenhouse. Acta Agriculturae Boreali-Occidentalis Sinica 15: 121–126. (in Chinese with English abstract)
15. Zhao SM, Li BM (2001) Introduction to present status,development and new techniques of nutri-culture systems in Japan. Transactions of the Chinese Society of Agricultural Engineering 17: 171-173. (in Chinese with English abstract)
16. Lynch JP (1995) Root architecture and plant productivity. Plant Physiol 109: 7–13.
17. Lynch JP (2007) Roots of the second green revolution. Australian Journal of Botany 55: 493–512.
18. Lynch JP, Brown KM (2012) New roots for agriculture: exploiting the root phenome. Philos Trans R Soc Lond B Biol Sci 367: 1598–1604.
19. Pei XB, Zhang FM, Wang L (2002) Effect of light and temperature on uptake and distribution of nitrogen,phosphorus and potassium of solar greenhouse cucumber. Zhongguo Nong Ye Ke Xue 35: 1510–1513. (in Chinese with English abstract)
20. Wang JF, Xing SZ, Chen SY, Yu QY, Li XL (2005) Dry matter accumulation in and NPK uptake by non-polluted cucumber in protected cultivation. Chinese Journal of Soil Science 36: 708–711. (in Chinese with English abstract)
21. Yu SF, Gao XB, Lu LP (2000) The study on nutrient absorption of cucumber in chinese lean-to solar greenhouse. China Vegetables: 10–11. (in Chinese with English abstract)
22. Chen Q, Zhang FS, Li XL (2005) Application of N management strategies in vegatable production. China Vegetables: 57–63. (in Chinese with English abstract)
23. An SW, Wang YQ, Li HL, Wang SJ, Gao LH (2010) Effects of different irrigation quantities on growth, yield and fruit qual ity of tomato in solar greenhouse. Acta Agriculturae Boreali-Occidentalis Sinica: 188–192. (in Chinese with English abstract)
24. Sun LP, Wen YG, Wang SZ, Zhao JW, Gao LH (2010) Effects of irrigation on water distribution of cucumber in solar greenhouse. Acta Agriculturae Boreali-Occidentalis Sinica: 173–178. (in Chinese with English abstract)
25. Wei Y, Sun L, Wang S, Wang Y, Zhang Z, et al. (2010) Effects of different irrigation methods on water distribution and nitrate nitrogen transport of cucumber in greenhouse. Transactions of the Chinese Society of Agricultural Engineering 26: 67–72. (in Chinese with English abstract)
26. Houba VJG, Temminghoff EJM., Gaikhorst GH (2000) Soil analysis procedures using 0.01M calcium chloride as extraction regent. Commun Soil Sci Plant Anal 31: 1299–1396.
27. Zhang FS,Wang JQ, Zhang WF, Cui ZL, Ma WQ, et al. (2008) Nutrient use efficiencies of major cereal crops in China and measures for improvement. Tu Rang Xue Bao 45: 915–924.
28. Li H (2000) Principles and techniques of plant physiological biochemical experiment. Beijing: Higher Education Press. 246–248 p.
29. Li H (2000) Principles and techniques of plant physiological biochemical experiment. Beijing: Higher Education Press. 195–197 p.
30. Cataldo D, Maroon M, Schrader L, Youngs V (1975) Rapid colorimetric determination of nitrate in plant tissue by nitration of salicylic acid 1. Commun Soil Sci Plan Anal 6: 71–80.
31. De Kroon H, Huber H, Stuefer J F, Van Groenendael J M (2005) A modular concept of phenotypic plasticity in plants. New Phytol 166: 73–82.
32. Hodge A (2006) Plastic plants and patchy soils. J Exp Bot 57: 401–411.
33. Rajput TBS, Patel N (2006) Water and nitrate movement in drip-irrigated onion under fertigation and irrigation treatments. Agricultural Water Management 79: 293–311.

34. Silber A, Xu G, Levkovitch I, Soriano S, Bilu A, et al. (2003) High fertigation frequency: the effects on uptake of nutrients, water and plant growth. Plant Soil 253: 467–477.
35. Silber A, Bruner M, Kenig E, Reshef G, Zohar H, et al. (2005) High fertigation frequency and phosphorus level: Effects on summer-grown bell pepper growth and blossom-end rot incidence. Plant Soil 270: 135–146.
36. Xu G, Levkovitch I, Soriano S, Wallach R, Silber A (2004) Integrated effect of irrigation frequency and phosphorus level on lettuce: P uptake, root growth and yield. Plant Soil 263: 297–309.
37. Belder P, Spiertz J, Bouman B, Lu G, Tuong T (2005) Nitrogen economy and water productivity of lowland rice under water-saving irrigation. Field Crop Res 93: 169–185.
38. Boomsma C, Santini J, Tollenaar M, Vyn T (2009) Maize per-plant and canopy-level morpho-physiological responses to the simultaneous stresses of intense crowding and low nitrogen availability. Agron J 101: 1426–1452.
39. Mi GH, Chen FJ, Wu QP, Lai NW, Yuan LX, et al. (2010) Ideotype root architecture for efficient nitrogen acquisition by maize in intensive cropping systems. Sci China Life Sci 53: 1369–1373.
40. Shen J, Li C, Mi G, Li L, Yuan L, et al. (2012) Maximizing root/rhizosphere efficiency to improve crop productivity and nutrient use efficiency in intensive agriculture of China. J Exp Bot.
41. Guo R, Li X, Christie P, Chen Q, Zhang F (2008) Seasonal temperatures have more influence than nitrogen fertilizer rates on cucumber yield and nitrogen uptake in a double cropping system. Environ pollut 151: 443–451.
42. Wang H, Inukai Y, Yamauchi A (2006) Root development and nutrient uptake. CRC Crit Rev Plant Sci 25: 279–301.
43. Garnett T, Conn V, Kaiser B N (2009) Root based approaches to improving nitrogen use efficiency in plants. Plant Cell Environ 32: 1272–1283.
44. Yoshida C, Iwasaki Y, Makino A, Ikeda H (2011) Effects of Irrigation Management on the Growth and Fruit Yield of Tomato under Drip Fertigation. Horticultural Research (Japan) 10: 325–331.
45. Jackson T, Westermann D, Moore D (1966) The effect of chloride and lime on the manganese uptake by bush beans and sweet corn. Soil Science Society of America Journal 30: 70–73.
46. Wang XZ, Cheng BS, Zhang GZ (1991) Nitrate and its influence factors in vegetables. Chinese Bulletin of Botany 8: 34–37. (in Chinese with English abstract).
47. Koch G, Schulze ED, Percival F, Mooney H, Chu C (1988) The nitrogen balance of Raphanus sativus x raphanistrum plants. II. Growth, nitrogen redistribution and photosynthesis under NO^{3-} deprivation. Plant Cell Environ 11: 755–767.

16

Maize Yield Response to Water Supply and Fertilizer Input in a Semi-Arid Environment of Northeast China

Guanghua Yin[1], Jian Gu[1], Fasheng Zhang[1]*, Liang Hao[1,2], Peifei Cong[1,2], Zuoxin Liu[1]

1 Institute of Applied Ecology, Chinese Academy of Sciences, Shenyang, China, **2** University of Chinese Academy of Sciences, Beijing, China

Abstract

Maize grain yield varies highly with water availability as well as with fertilization and relevant agricultural management practices. With a 311-A optimized saturation design, field experiments were conducted between 2006 and 2009 to examine the yield response of spring maize (Zhengdan 958, *Zea mays* L) to irrigation (*I*), nitrogen fertilization (total nitrogen, urea-46% nitrogen,) and phosphorus fertilization (P_2O_5, calcium superphosphate-13% P_2O_5) in a semi-arid area environment of Northeast China. According to our estimated yield function, the results showed that *N* is the dominant factor in determining maize grain yield followed by *I*, while *P* plays a relatively minor role. The strength of interaction effects among *I*, *N* and *P* on maize grain yield follows the sequence *N+I* >*P+I*>*N+P*. Individually, the interaction effects of *N+I* and *N+P* on maize grain yield are positive, whereas that of *P+I* is negative. To achieve maximum grain yield (10506.0 kg·ha^{-1}) for spring maize in the study area, the optimum application rates of *I*, *N* and *P* are 930.4 m^3·ha^{-1}, 304.9 kg·ha^{-1} and 133.2 kg·ha^{-1} respectively that leads to a possible economic profit (*EP*) of 10548.4 CNY·ha^{-1} (CNY, Chinese Yuan). Alternately, to obtain the best *EP* (10827.3 CNY·ha^{-1}), the optimum application rates of *I*, *N* and *P* are 682.4 m^3·ha^{-1}, 241.0 kg·ha^{-1} and 111.7 kg·ha^{-1} respectively that produces a potential grain yield of 10289.5 kg·ha^{-1}.

Editor: Hany A. El-Shemy, Cairo University, Egypt

Funding: This work was funded by the Special Fund for Agro-scientific Research in the Public Interest (201303125-9) and National Major Technology Program (2011BAD16B12, 2012BAD09B02, 2013BAD05B07). The funders had no role in study design, data collection and analysis, decision to publish, or preparation of the manuscript.

Competing Interests: The authors have declared that no competing interests exist.

* E-mail: fasheng.zhang@yahoo.com

Introduction

Maize is the third most important grain crop after rice and wheat grown in China [1,2]. In order to ensure food security for its vast population, the Chinese government and its research institutions have made extensive efforts to improve maize grain production in North China since the 1950s [3–5].

Water scarcity and soil infertility are two critical factors limiting maize grain yield over most regions of North China [6–8]. Although irrigation and fertilization are widely applied to improve maize productivity [9,10], maize production in China has not been able to keep pace with grain demand [11,12]. At the same time, low water use efficiency aggravates water stress in North China [13–15] while excessive inputs of chemical fertilizer result in surplus nitrogen and phosphorus in soils that cause eutrophication of surface water as well as greenhouse gas emissions [16–20]. In modern agriculture, such consequences arise mainly from a limited understanding of how irrigation and fertilization affect maize production and a biased estimation of the yield function for identifying maize yield variation. In this context, there is a need to investigate the combined effect of water supply and fertilizer input on maize productivity in North China.

Many field studies have been conducted since the 1990s to examine main and interaction effects of irrigation and fertilization on maize productivity around the world, including North China [21–26]. The optimum coupling or combination of water supply and fertilizer inputs has been derived to seek maximum maize grain yield or to achieve maximum water and fertilizer use efficiency [27–30]. However, these studies mostly focused on the individual influences of irrigation (*I*), nitrogen application (*N*), phosphorus application (*P*) and/or their binary combination effects on maize productivity. A holistic understanding of the ternary combination effect of *I*, *N* and *P* on maize productivity is still developing. The economic efficiency of growing maize is another important factor influencing maize grain production [31–34]. Farmers will grow more maize if the economic profits of growing maize are higher than for other crops. Profits associated with maize production, however, decrease with improper management practices as well as with increasing energy, material and human labor costs in the context of global climate change [35–39]. The declining profit rate dampens farmers' enthusiasm for growing maize and consequently impacts maize grain production [40,41]. Thus, it is important to improve maize productivity while taking into account the economic evaluation of growing maize.

The relationship between maize grain yield and management practices varies over time and space depending on the maize cultivars, climatic conditions and cropping systems. Knowledge obtained from studies in other regions may not be valid in any specific area of North China. Therefore, the objectives of this study were (1) to construct a yield function to examine the combination effect of *I*, *N* and *P* on maize productivity using field experimental data collected from 2006 to 2009 in a semi-arid environment of Northeast China and (2) to use the estimated yield function for further deriving optimum application rates of *I*, *N* and

P based on the criteria of maximum grain yield and best economic profit.

Materials and Methods

Site and Soil

The field study was conducted from 2006 to 2009 at the field experimental station of Liaoning Key Laboratory of Water-Saving Agriculture in Fuxin County of Northeast China (42°08′14″ N, 121°44′21″ E). This region is a warm temperate zone with a temperate continental monsoon climate. According to the Fuxin Weather Station, the average annual temperature is 7.2°C with an average of 2865.5 hrs of annual sunshine. It is a typical semi-arid area with average annual precipitation of 480 mm, over 60% of which occurs from June to August. The compensation of water resources depends mainly on precipitation of atmosphere. Annual precipitation and precipitation during the maize growing season of Fuxin County are shown in Figure 1.

The main agricultural soil in the region is cinnamon soil which develops through a combination of calcium carbonate leaching, illuviation and humification. It is characterized by a thin humus layer and a medium or thick solum. Its bulk density is 1.51 $g \cdot cm^{-3}$, pH (H_2O) is 7.5−8.5 and the average soil organic matter content is 10.2 $g \cdot kg^{-1}$. The average soil total nitrogen and available phosphorus concentrations are 6.1 $g \cdot kg^{-1}$ and 4.0 $g \cdot kg^{-1}$, respectively.

Experimental Design and Treatments

To reduce cost and size, the experiment in this study was implemented according to a 311-A optimized saturation design [42]. This system consisted of three factors at five levels. There were 11 treatments with 3 replicates each for a total of 33 experimental plots. Each experimental plot was 40 m^2 in size (10 m×4 m). The water supply and fertilizer inputs were standardized for comparability by applying a non-dimensional linear code substitution (Table 1). Rates of *N* and *P* in Table 1 were expressed in format of pure nitrogen and P_2O_5 that were supplemented by urea (46% total nitrogen) and calcium superphosphate (13% P_2O_5). One unit of *I*, *N* and *P* represented 225 $m^3 \cdot ha^{-1}$ of *I*, 112.5 $kg \cdot ha^{-1}$ of total nitrogen and 67.5 $kg \cdot ha^{-1}$ of P_2O_5, which indicated 225 $m^3 \cdot ha^{-1}$ of water supply, 244.6 $kg \cdot ha^{-1}$ of urea and 519.2 $kg \cdot ha^{-1}$ of calcium superphosphate, respectively. One third of the urea used in each treatment was applied at the sowing stage and the remaining amount at the early jointing stage. All calcium superphosphate was applied at the sowing stage in each treatment. Experimental plots were variously irrigated at the jointing stage. All necessary permits were obtained for the described field experiments. The land user and owner approved the field-work activities at each experiment plot. The field employed in this study is not protected in any way, and the study did not involve any endangered or protected species.

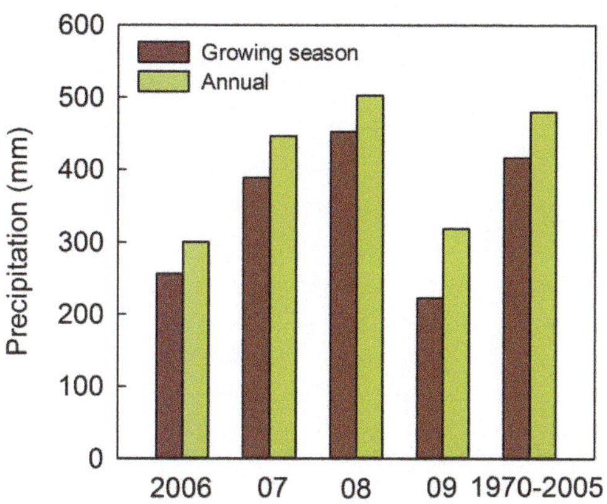

Figure 1. Annual precipitation and precipitation during maize growing season during 2006–2009 in Fuxin County.

Data Collection and Analysis

The maize cultivar in this experiment was Zhengdan 958 (*Zea mays* L). In China, the planting area of Zhengdan 958 was 4.54 million ha in 2009, and the planting area of this variety is still the largest in 2012 [43]. Zhengdan 958 has outstanding yield performance. It can generate relatively stable yield under various environmental conditions and has good disease resistance. This variety can be planted in high density and the ideal planting density is 60,000 to 75,000 plants per ha. In this study, maize plant density was 60,000 plants per ha with 50 cm between rows. The maize was planted in late April and harvested in late September. At maize maturity, the outer two rows in each experimental plot were considered as edge effects and not harvested, while the remaining middle rows were hand-harvested for analysis of maize grain yield. The effective area of each experimental plot was approximately 20 m^2. The average fresh ear weight (G_1, kg) of each treatment was estimated and the average grain yield, *Y* ($kg \cdot ha^{-1}$), was computed as:

$$Y = k \times {G_1}/{20} \times 10000$$

where *k* is the ratio of grain dry weight to fresh ear weight for each treatment. To estimate the values of *k*, ten medium-sized ears were sampled from each experimental plot and their average fresh ear weight (G_2, kg) and average fresh grain weight (G_3, kg) were measured for each treatment The grain of Zhengdan 958 dries slowly before harvest. Extra moisture in maize grain should be removed before estimating the average maize grain yield. The

Table 1. Experimental treatments of maize using a 311-A optimized saturation design during 2006–2009 in Fuxin County.

Treatments	Code level			Application rate		
	I[†]	*N*	*P*	*I*	*N*	*P*
	X_1[‡]	X_2	X_3	$m^3 \cdot ha^{-1}$	$kg \cdot ha^{-1}$	$kg \cdot ha^{-1}$
1	2	0	0	900	225	135
2	−2	0	0	0	225	135
3	1	−1.414	−1.414	675	66	39.54
4	1	1.414	−1.414	675	384	39.54
5	1	−1.414	1.414	675	66	230.46
6	1	1.414	1.414	675	384	230.46
7	−1	2	0	225	450	135
8	−1	−2	0	225	0	135
9	−1	0	2	225	225	270
10	−1	0	−2	225	225	0
11	0	0	0	450	225	135

[†] *I*, *N* and *P* are abbreviations for irrigation, nitrogen fertilization and phosphorus fertilization, respectively.
[‡] X_1, X_2 and X_3 are non-dimensional linear code for irrigation, nitrogen fertilization and phosphorus fertilization, respectively.

average moisture content of fresh grain for each treatment (A%) was therefore determined using a PM-8188 Grain Moisture Tester (Japan). Then k was calculated as:

$$k = {G_3}/{G_2} \times {(100-A)}/{(100-18)}$$

A quadratic regression orthogonal design was used to construct the yield function in this study. Regression analysis was conducted by R software and the results were presented in graphs by SigmaPlot 10.0.

Results and Discussion

Yield Function

The average maize grain yield of each treatment from 2006 to 2009 varied between 8468.2 kg·ha^{-1} and 10478.2 kg·ha^{-1}. The best-fitted yield function, which quantified the relationship between maize grain yield (Y_{INP}) and the code levels of I, N and P (i.e., X_1, X_2, X_3), was expressed as:

$$\begin{aligned} Y_{INP} = & 10130.63^{**} + 232.55X_1^{**} + 363.44X_2^{**} + 151.46X_3^{*} \\ & -75.19X_1^2 - 436.54X_2^{2**} - 349.33X_3^{2**} \\ & +120.94X_1X_2^{*} + 55.11X_2X_3 - 98.17X_1X_3 \end{aligned} \quad (1)$$

where the coefficient of determination (R^2) for the regression is 0.87 at the significance level of $P<0.01$ (F-test). The reliability of coefficients for X_1, X_2 and X_3 are tested by a t-test. The symbol "**" stands for the significance at the level of $P<0.01$ and the symbol "*" for the significance at the level of $P<0.05$. In addition, the coefficients of X_1^2, X_2X_3 and X_1X_3 were significant at $P<0.20$, $P<0.40$ and $P<0.10$. Considering the importance of I, N and P on maize growth, the parameters in equation (1) were all taken into account when estimating maize grain yield in this study.

The code levels of I, N and P were proportional to their actual rates in each treatment. Thus, equation (1) appropriately described the combination effects of water spply and fertilizer input on maize grain yield in the semi-arid area examined in this study. It accounted for 87% of the variation in maize grain yield. Because of its accuracy and explicitness, equation (1) was used to analyze the main and individual effects as well as the interaction effects of I, N and P on maize grain yield (Text S1).

The main effects of I, N and P on maize grain yield were evaluated by comparing their corresponding coefficients in equation (1). The positive coefficients for X_1, X_2 and X_3 suggested that I, N and P all had positive effects on maize grain yield. The largest value was observed for the coefficient of X_2, indicating that N was the dominant factor influencing maximum maize grain yield. Similarly, I was recognized as a secondary factor determining maize grain yield, whereas P was a relatively minor factor. These results were partially consistent with findings from other studies around the world. Nitrogen input has a large effect on maize grain yield because maize production is an extractive process, with removal of maize equating to removal of nitrogen from the soil [44,45]. The significant effect of water supply on maize grain yield in arid and semi-arid regions is ubiquitous and easily understood. It should be noted that the removal of phosphorus from agricultural soils by maize harvesting was also apparent in this study. It is well-established that supplemental phosphorus can significantly improve maize grain yield as well [7,9,17,22]. In our study, the decreased response of maize grain yield to P compared to N and I may be due to the dry soil conditions which are characteristics of semi-arid areas in Northeast China.

To examine the individual effects of I, N and P on maize grain yield, each factor was selected as an independent variable with the other two factors fixed at 0 in equation (1) (Text S1). Then a subset of equations of yield function was derived respectively as:

$$Y_I = 10130.63 + 232.55X_1 - 75.19X_1^2 \quad (2)$$

$$Y_N = 10130.63 + 363.44X_2 - 436.54X_2^2 \quad (3)$$

$$Y_P = 10130.63 + 151.46X_3 - 349.33X_3^2 \quad (4)$$

The individual effects of I, N and P on maize grain yield derived from equations (2) ~ (4) are presented in Figure 2. The relationships between maize grain yield and I, N and P can be modeled using second-order parabolic equations. Apexes were observed when examining the trend of maize grain yield as the rates of I, N and P increased. Before the apex, maize grain yield increased as the rates of I, N and P increased. After the apex, maize grain yield decreased as the rates of I, N and P increased. These findings indicate that there must be optimum application rates of I, N and P when implementing agricultural management practices to improve maize grain yield.

The optimum application rates of I, N and P as individual influencing factors on maize grain yield were determined using marginal yield curves (Figure 3), which were the first-order differential analysis of equations (2) ~ (4). The marginal yield showed a monotonic descending trend with the increasing rates of I, N and P and had intersections with the x-axis. The intersection points revealed the optimum application rates of I, N and P. The code values at the intersecting points were +1.546, +0.416 and +0.217 for I, N and P, respectively. According to Table 1, they represented 797.9 m^3·ha^{-1} of I, 271.8 kg·ha^{-1} of N and 149.6 kg·ha^{-1} of P. In addition, levels of the majority of marginal yields before the intersections with the x-axis followed the sequence $N > P > I$. This suggests that the maize grain yield increased most with an increasing application rate of N compared to the increases of I and P. Maize grain yield was more sensitive to P than I.

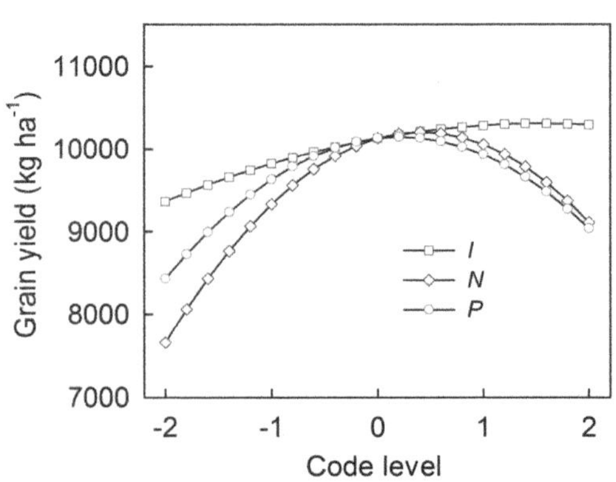

Figure 2. The individual effect of *I*, *N* and *P* on maize grain yield by fixing two factors at 0 level. *I*, *N* and *P* represent irrigation, nitrogen application and phosphorus application, respectively.

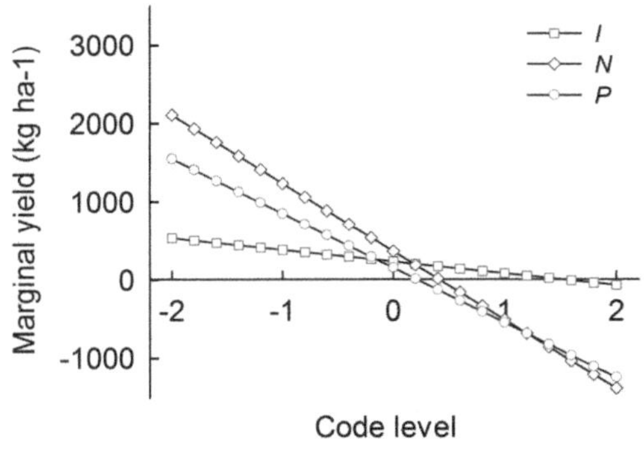

Figure 3. Marginal grain yield with respect to *I*, *N* and *P* through the first-order differential analysis of equations (2)~(4). *I*, *N* and *P* represent irrigation, nitrogen application and phosphorus application, respectively.

The interaction effects of *I*, *N* and *P* on maize grain yield were all less than their main effects in equation (1). By comparing the absolute values of the coefficients of X_1X_2, X_1X_3 and X_2X_3, the strength of interaction effects of *I*, *N* and *P* on maize grain yield followed the sequence *N*+*I* >*P*+*I*>*N*+*P*. The coefficient for X_1X_3 was positive, indicating that the interaction of *I* and *N* had a positive effect on maize grain yield. This finding is presented in Figure 4, which was constructed by fixing *P* at 0 level. In general, the maize grain yield increased as the application rates of *I* and *N* increased: an increase of any one input stimulates maize growth and creates a need for the other input. The optimum application rates of *I* and *N* for the highest grain yield (10,505.7 kg·ha^{-1}) were at code levels of +2.117 and +0.710, which corresponded to 926.3 m^3·ha^{-1} of *I* and 304.9 kg·ha^{-1} of *N*. The negative coefficient for X_1X_3 suggested that the interaction of *I* and *P* was antagonistic and had an inhibitory effect on maize grain yield. This relationship is presented in Figure 5, which was constructed by fixing *N* at 0 level. At code levels of +1.547 and −0.001, equal to 798.1 m^3·ha^{-1} of *I* and 134.9 kg·ha^{-1} of *P*, respectively, maize grain yield reached its highest value of 10,310.4 kg·ha^{-1}. Although the coefficient of X_2X_3 was positive, the interaction of *N* and *P* on maize grain yield was similar to that of *I* and *N*, but weaker and not significant at a relatively high level. This partly reveals the importance of water supply in maize growth in semi-arid areas. No contour plot of this interaction is presented here, although the interaction between *N* and *P* on maize grain yield was used to estimate the optimum schemes in next part.

Optimum Schemes

The findings presented above indicate the complex and sometimes antagonistic interactions of water supply and fertilizer input on maize productivity in semi-arid areas of Northeast China. The maize grain yield does not always increase as the additions of *I*, *N* and *P* improve. The optimum application rates of *I*, *N* and *P* were estimated to obtain the maximum grain yield and best economic profit.

To maximize maize grain yield, we set the first-order partial derivatives of equation (1) to zero (Text S1):

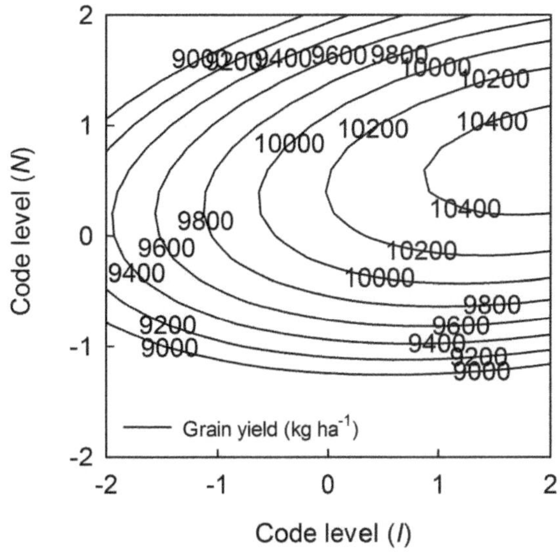

Figure 4. Interaction effects of *I* and *N* on maize grain yield by fixing *P* at 0 level. *I* and *N* represent irrigation and nitrogen application, respectively.

$$\begin{cases} \dfrac{\partial Y}{\partial X_1} = 232.55 - 150.38X_1 + 120.94X_2 - 98.17X_3 = 0 \\ \dfrac{\partial Y}{\partial X_2} = 363.44 + 120.94X_1 - 873.08X_2 + 55.11X_3 = 0 \\ \dfrac{\partial Y}{\partial X_3} = 151.46 - 98.17X_1 + 55.11X_2 - 698.66X_3 = 0 \end{cases}$$

By solving the above equations, the maximum maize grain yield, 10506.0 kg·ha^{-1}, can be obtained when the code levels of *I*, *N* and *P* were at +2.135, +0.710 and −0.027, respectively. According to Table 1, the optimum rates of *I*, *N* and *P* for the

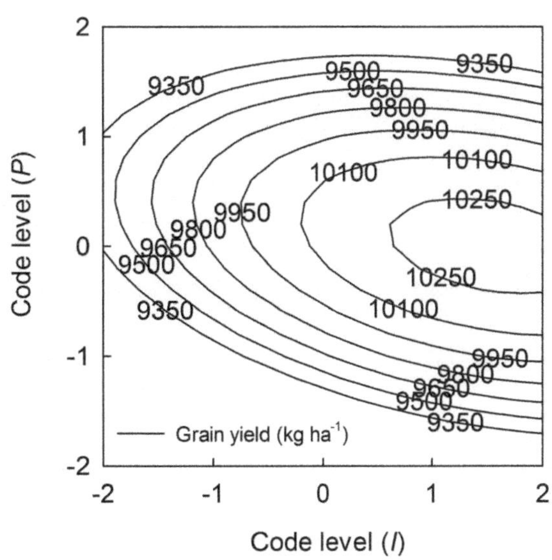

Figure 5. Interaction effects of *I* and *P* on maize grain yield by fixing *N* at 0 level. *I* and *P* represent irrigation and phosphorus application, respectively.

maximum maize grain yield were 930.4 $m^3 \cdot ha^{-1}$, 304.9 $kg \cdot ha^{-1}$ and 133.2 $kg \cdot ha^{-1}$, respectively.

Compared to the optimum application rates of *I*, *N* and *P* based on single-factor and binary combination effects, the optimum application rates based on the ternary combination effect have higher demand for *I* and *N* and less demand for *P*. This should be attributed to the effects of complex interactions of *I*, *N* and *P* on maize grain yield. In addition, maize grain yield showed different sensitivities to *I*, *N* and *P* in Figure 3. In the case of the ternary combination, these factors were considered holistically when estimating the maximum of the yield function. The optimum scheme indicated by the ternary combination enabled farmers to maximize maize grain yield by adjusting management practices without considering the cost of maize growing.

The increase of water supply and fertilizer input may increase maize grain yield as well as the economic profit. The economic profit of growing maize, *EP*, is decided by the relationship between outputs (i.e., the maize grain yield times the price of maize grain) and inputs (i.e., cost of irrigation, fertilization and seed). This relationship can be expressed as:

$$EP = p \cdot Y_{INP} - p_1 \cdot (225X_1 + 450) - p_2 \cdot (112.5X_2 + 225) - p_3 \cdot (67.5X_3 + 135) - C_{Seed} \quad (5)$$

where p stands for the price of maize grain (1.3 $CNY \cdot kg^{-1}$, CNY, Chinese Yuan), and p_1, p_2 and p_3 for the prices of water (0.750 $CNY \cdot m^{-3}$, i.e., electricity and labor fees for irrigation), urea (3.913 $CNY \cdot kg^{-1}$) and calcium superphosphate (5.769 $CNY \cdot kg^{-1}$), respectively. Note that the prices of urea and calcium superphosphate per kilogram were calculated from the cost of them to provide a kilogram of pure total nitrogen and P_2O_5. C_{Seed} represents the cost of maize seeds for each treatment in this study (450 $CNY \cdot ha^{-1}$). To obtain the maximum economic profit, the highest value of equation (5) was solved using the method of first-order partial derivatives as well. The results show that the highest economic profit (10827.3 $CNY \cdot ha^{-1}$) was obtained when *I*, *N* and *P* were at code levels of 1.033, 0.142 and −0.345, respectively. This optimum scheme consisted of a ternary combination of *I*, *N* and *P* at rates of 682.4 $m^3 \cdot ha^{-1}$, 241.0 $kg \cdot ha^{-1}$ and 111.7 $kg \cdot ha^{-1}$, respectively.

To further discern the results, we compared the two optimum schemes based on the maximum grain yield and the best economic profit. The economic profit of the optimum scheme for maximum grain yield (10506.0 $kg \cdot ha^{-1}$) was 10548.4 $CNY \cdot ha^{-1}$ while the grain yield of the optimum scheme for the best economic profit (10827.3 $CNY \cdot ha^{-1}$) was 10289.5 $kg \cdot ha^{-1}$. Notwithstanding a little higher yield, the optimum scheme for maximum grain yield produced less economic profit than that of the optimum scheme for the best economic profit. In addition, the optimum scheme for maximum grain yield consumed more water and fertilizer than the optimum scheme for the best economic profit. This apparently will increase demand on water and mineral resources and result in leaching of surplus nitrogen and phosphorus into soil and water environments. In contrast, the optimum scheme for the best economic profit produced both relatively high grain yield and economic profit while consumed relatively less water and fertilizer. Considering the water resource and soil conditions, the optimum scheme for the best economic profit was therefore more acceptable and should be recommended in the study area.

Conclusions

Yield response of spring maize (Zhengdan 958, *Zea mays* L) to water supply and fertilizer input in a semi-arid area of Northeast China was studied. In this field experiment, *I*, *N* and P as well as their interaction effects had significant influences on maize grain yield. The yield function derived by the ternary combination was able to describe these influences holistically. To obtain maximum maize grain yield (10506.0 $kg \cdot ha^{-1}$) in the semi-arid areas examined in this study, the optimum application rates of *I*, *N* and *P* based on the present findings were 930.4 $m^3 \cdot ha^{-1}$, 304.9 $kg \cdot ha^{-1}$ and 133.2 $kg \cdot ha^{-1}$, respectively. Alternately, to obtain the best economic profit (10827.3 $CNY \cdot ha^{-1}$), the optimum application rates of *I*, *N* and *P* were 682.4 $m^3 \cdot ha^{-1}$, 241.0 $kg \cdot ha^{-1}$ and 111.7 $kg \cdot ha^{-1}$, respectively. The latter scheme is recommended in the study area because of its relatively high grain yield and economic profit performance.

Author Contributions

Conceived and designed the experiments: GHY FSZ ZXL. Performed the experiments: GHY FSZ JG LH PFC. Analyzed the data: GHY FSZ JG. Contributed reagents/materials/analysis tools: GHY FSZ. Wrote the paper: GHY FSZ.

References

1. Li J (2009) Production, breeding and process of maize in China. In: Bennetzen JL, Hake SC, editors. Handbook of maize: Its biology. New York: Springer. pp. 563–576.
2. Ray DK, Mueller ND, West PC, Foley JA (2013) Yield trends are insufficient to double global crop production by 2050. PLoS ONE 8: e66428.
3. Wang H, Zhang M, Cai Y (2009) Problems, challenges, and strategic options of grain security in China. Advances in Agronomy 103: 101–147.
4. Wang T, Ma X, Li Y, Bai D, Liu C, et al. (2011) Changes in yield and yield components of single-cross maize hybrids released in China between 1964 and 2001. Crop Science 51: 512–525.
5. Ci X, Li M, Xu J, Lu Z, Bai P, et al. (2012) Trends of grain yield and plant traits in Chinese maize cultivars from the 1950s to the 2000s. Euphytica 183: 1–12.
6. Zou C, Gao X, Shi R, Fan X, Zhang F (2008) Micronutrient deficiencies in crop production in China. In: Alloway BJ, editor. Micronutrient deficiencies in global crop production. The Netherlands: Springer. pp. 127–148.
7. Wang CT, Li SK (2010) Assessment of limiting factors and techniques prioritization for maize production in China. Scientia Agricultura Sinica 43: 1136–1146. (in Chinese).
8. Zhang F, Yin G, Wang Z, McLaughlin N, Geng X, et al. (2013) Quantifying spatial variability of selected soil trace elements and their scaling relationships using multifractal techniques. PLoS ONE 8: e69326.9.
9. Fan T, Stewart BA, Wang Y, Luo J, Zhou G (2005) Long-term fertilization effects on grain yield, water-use efficiency and soil fertility in the dryland of Loess Plateau in China. Agriculture, Ecosystems & Environment 106: 313–329.
10. Khan S, Hanjra MA, Mu J (2009) Water management and crop production for food security in China: A review. Agricultural Water Management 96: 349–360.
11. Neumann K, Verburg PH, Stehfest E, Müller C (2010) The yield gap of global grain production: A spatial analysis. Agricultural Systems 103: 316–326.
12. Meng Q, Hou P, Wu L, Chen X, Cui Z, et al. (2013) Understanding production potentials and yield gaps in intensive maize production in China. Field Crops Research 143: 91–97.
13. Liu Z, Gao P (2009) Application basis and technology of water-saving agriculture in semi-arid areas of Northeast China. Beijing: Science Press. p. 314.
14. Fang Q, Ma L, Yu Q, Ahuja LR, Malone RW, et al (2010) Irrigation strategies to improve the water use efficiency of wheat–maize double cropping systems in North China Plain. Agricultural Water Management 97: 1165–1174.
15. Bernacchi CJ, Hollinger SE, Meyers T (2005) The conversion of the corn/soybean ecosystem to no-till agriculture may result in a carbon sink. Global Change Biology 11: 1867–1872.
16. Wang X, Willms WD, Hao X, Zhao M, Han G (2010) Cultivation and reseeding effects on soil organic matter in the mixed prairie. Soil Science Society of America Journal 74: 1348–1355.

17. Ma W, Ma L, Li J, Wang F, Sisák I, et al. (2011) Phosphorus flows and use efficiencies in production and consumption of wheat, rice, and maize in China. Chemosphere 84: 814–821.
18. Wang X, Wang J, Zhang J. (2012) Comparisons of three methods for organic and inorganic carbon in calcareous soils of northwestern China. PLoS ONE 7: e44334.
19. Wang X, Gan YT, Hamel C, Lemke RL, McDonald CL (2012) Water use profiles across the rooting zones of various pulse crops. Field Crops Research 134: 130–137.
20. Cui Z, Yue S, Wang G, Meng Q, Wu L, et al. (2013) Closing the yield gap could reduce projected greenhouse gas emissions: a case study of maize production in China. Global Change Biology 19: 2467–2477.
21. Jokela WE (1992) Nitrogen fertilizer and dairy manure effects on corn yield and soil nitrate. Soil Science Society of America Journal 56: 148–154.
22. Wortmann CS, Dobermann AR, Ferguson RB, Hergert GW, Shapiro CA, et al. (2009) High-yielding corn response to applied phosphorus, potassium, and sulfur in Nebraska. Agronomy Journal 101: 546–555.
23. Ju X, Christie P (2011) Calculation of theoretical nitrogen rate for simple nitrogen recommendations in intensive cropping systems: A case study on the North China Plain. Field Crops Research 124: 450–458.
24. Hou P, Gao Q, Xie R, Li S, Meng Q, et al. (2012) Grain yields in relation to N requirement: Optimizing nitrogen management for spring maize grown in China. Field Crops Research 129: 1–6.
25. Jin L, Cui H, Li B, Zhang J, Dong S, et al. (2012) Effects of integrated agronomic management practices on yield and nitrogen efficiency of summer maize in North China. Field Crops Research 134: 30–35.
26. Hu H, Ning T, Li Z, Han H, Zhang Z, et al. (2013) Coupling effects of urea types and subsoiling on nitrogen–water use and yield of different varieties of maize in northern China. Field Crops Research 142: 85–94.
27. Dang TH, Cai GX, Guo SL, Hao MD, Heng LK (2006) Effect of nitrogen management on yield and water use efficiency of rainfed wheat and maize in Northwest China. Pedosphere 16: 495–504.
28. Liu WZ, Zhang XC (2007) Optimizing water and fertilizer input using an elasticity index: a case study with maize in the loess plateau of china. Field Crops Research 100: 302–310.
29. El-Hendawy SE, Schmidhalter U (2010) Optimal coupling combinations between irrigation frequency and rate for drip-irrigated maize grown on sandy soil. Agricultural Water Management 97: 439–448.
30. Wang X, Dai K, Zhang D, Zhang X, Wang Y, et al. (2011) Dryland maize yields and water use efficiency in response to tillage/crop stubble and nutrient management practices in China. Field Crops Research 120: 47–57.
31. Grassini P, Cassman KG (2012) High-yield maize with large net energy yield and small global warming intensity. Proceedings of the National Academy of Sciences 109: 1074–1079.
32. Mulwa R, Emrouznejad A, Muhammad L (2009) Economic efficiency of smallholder maize producers in Western Kenya: a DEA meta-frontier analysis. International Journal of Operational Research 4: 250–267.
33. Rahman S, Rahman MS, Rahman MH (2012) Joint determination of the choice of growing season and economic efficiency of maize in Bangladesh. Journal of the Asia Pacific Economy 17: 138–150.
34. Sharma AR, Singh R, Dhyani SK, Dube RK (2011) Agronomic and economic evaluation of mulching in rainfed maize–wheat cropping system in the Western Himalayan region of India. Journal of Crop Improvement 25: 392–408.
35. Guo R, Lin Z, Mo X, Yang C (2010) Responses of crop yield and water use efficiency to climate change in the North China Plain. Agricultural Water Management 97: 1185–1194.
36. Xiong W, Holman I, Lin E, Conway D, Jiang J, et al. (2010) Climate change, water availability and future cereal production in China. Agriculture, Ecosystems & Environment 135: 58–69.
37. Piao S, Ciais P, Huang Y, Shen Z, Peng S, et al. (2010) The impacts of climate change on water resources and agriculture in China. Nature 467: 43–51.
38. Li X, Takahashi T, Suzuki N, Kaiser HM (2011) The impact of climate change on maize yields in the United States and China. Agricultural Systems 104: 348–353.
39. Tao F, Zhang Z (2011) Impacts of climate change as a function of global mean temperature: maize productivity and water use in China. Climatic Change 105: 409–432.
40. Tong C, Hall CAS, Wang H (2003) Land use change in rice, wheat and maize production in China (1961–1998). Agriculture, Ecosystems & Environment 95: 523–536.
41. Owombo PT, Akinola AA, Ayodele OO, Koledoye GF (2012) Economic impact of agricultural mechanization adoption: Evidence from maize farmers in Ondo State, Nigeria. Journal of Agriculture and Biodiversity Research 1: 25–32.
42. Ding XQ (1986) The applied regression design of the field experiment. Jilin Science&Technology Press. 199–203. (in Chinese).
43. Jin X, Fu Z, Ding D, Li W, Liu Z, et al. (2013) Proteomic identification of genes associated with maize grain-filling rate. PLoS ONE 8: e59353.
44. Lawlor DW, Lemaire G, Gastal F (2001) Nitrogen, plant growth and crop yield. In: Lea PJ, Morot-Gaudry JF, editors. Plant nitrogen. Berlin Heidelberg: Springer. pp. 343–367.
45. Peng Y, Li X, Li C (2012) Temporal and spatial profiling of root growth revealed novel response of maize roots under various nitrogen supplies in the field. PLoS ONE 7: e37726.

17

Validating and Improving Interrill Erosion Equations

Feng-Bao Zhang[1,2], Zhan-Li Wang[1,2]*, Ming-Yi Yang[1,2]

1 State Key Laboratory of Soil Erosion and Dryland Farming on the Loess Plateau, Institute of Soil and Water Conservation, Northwest A&F University, Yangling, P. R. China,
2 Institute of Soil and Water Conservation, Chinese Academy of Science and Ministry of Water Resources, Yangling, P. R. China

Abstract

Existing interrill erosion equations based on mini-plot experiments have largely ignored the effects of slope length and plot size on interrill erosion rate. This paper describes a series of simulated rainfall experiments which were conducted according to a randomized factorial design for five slope lengths (0.4, 0.8, 1.2, 1.6, and 2 m) at a width of 0.4 m, five slope gradients (17%, 27%, 36%, 47%, and 58%), and five rainfall intensities (48, 62.4, 102, 149, and 170 mm h^{-1}) to perform a systematic validation of existing interrill erosion equations based on mini-plots. The results indicated that the existing interrill erosion equations do not adequately describe the relationships between interrill erosion rate and its influencing factors with increasing slope length and rainfall intensity. Univariate analysis of variance showed that runoff rate, rainfall intensity, slope gradient, and slope length had significant effects on interrill erosion rate and that their interactions were significant at $p = 0.01$. An improved interrill erosion equation was constructed by analyzing the relationships of sediment concentration with rainfall intensity, slope length, and slope gradient. In the improved interrill erosion equation, the runoff rate and slope factor are the same as in the interrill erosion equation in the Water Erosion Prediction Project (WEPP), with the weight of rainfall intensity adjusted by an exponent of 0.22 and a slope length term added with an exponent of −0.25. Using experimental data from WEPP cropland soil field interrill erodibility experiments, it has been shown that the improved interrill erosion equation describes the relationship between interrill erosion rate and runoff rate, rainfall intensity, slope gradient, and slope length reasonably well and better than existing interrill erosion equations.

Editor: Juan A. Añel, University of Oxford, United Kingdom

Funding: Financial support for this research was provided by the National Natural Science Foundation of China (41171227, 40901127, 40971172), Natural Science Foundation of Shannxi Province (S2012JC6612), and the special funds of Northwest A&F University for the operational costs of basic scientific research (QN2011072). The funders had no role in study design, data collection and analysis, decision to publish, or preparation of the manuscript.

Competing Interests: The authors have declared that no competing interests exist.

* E-mail: zwang@ms.iswc.ac.cn

Introduction

Soil erosion on a hill slope consists of two major components: interrill and rill erosion [1]. Interrill erosion is the detachment and transport of soil material from the surface of the soil matrix by raindrop impact and overland flow [2]. More recent process-based models, e.g., WEPP [3], EUROSEM [4], LISEM [5], and PSEM-2D [6], make an explicit distinction between rill and interrill erosion processes and estimate interrill erosion rate using a function of various factors, including interrill erodibility parameter which expresses the soil's resistance to interrill erosion within interrill areas, rainfall erosivity, slope characteristics, hydraulic factors, vegetation, and land use [4–10].

It is difficult to describe and quantify interrill and rill erosion processes simultaneously during a rainfall event, especially to measure the rate of interrill erosion because it is less visible. The mini-plot, which eliminates the effect of rill erosion, has been an important means to study interrill erosion and to establish interrill erosion equations. The effects of rainfall intensity, slope length, slope gradient, runoff rate, and soil type on interrill erosion rate have been described. Power functions were used to describe the effect of rainfall intensity on interrill erosion. The rainfall intensity exponent varied from 1.6 to 2.3 [11] and from 1.36 to 2.54 [12]. The exponent decreased as the clay content of the soil increased [11]. Therefore, interrill erosion rate has been shown to be approximately proportional to the square of the rainfall intensity for a given mini-plot if runoff is not considered [12–15]. To describe the effect of slope gradient on interrill erosion, several formulas of slope factors have been proposed [16–19]. The role of runoff was initially considered as sediment transport for interrill erosion [20] and was not included in interrill erosion equations [3], but some researchers found that model predictability was improved when a separate runoff parameter was included in interill erosion equations [21,22]. In general, interrill erosion rate increased gradually with increasing runoff. For the effect of slope length on interrill erosion, there have been inconsistent results, which will be described detailedly in the following discussion.

According to the relationships between interrill erosion rate and its influencing factors, interrill erosion equations have been proposed on the basis of results from mini-plot experiments. In the WEPP model, the following equation was first used to estimate interrill erosion rates [3]:

$$D_i = K_i I^2 S_f \tag{1}$$

where D_i is interrill erosion rate (kg m^{-2} s^{-1}), K_i is interrill erodibility, I is rainfall intensity (m s^{-1}), and S_f is slope factor that can be calculated using the following equation [16]:

$$S_f = 1.05 - 0.85\exp(-4\sin\theta) \tag{2}$$

where θ is the slope gradient (°).

As described above, previous studies have shown that model predictability was improved when a separate runoff parameter was included in interrill erosion equations [21,22]. Therefore, Eq. (1) was modified by adding the factor of overland flow [22] and was then used in the WEPP model in 1995 as the following formula [23]:

$$D_i = K_i R I S_f \quad (3)$$

where R is runoff rate (m s^{-1}). In addition, to describe interrill erosion processes more accurately and to develop a simple lumped parameter model, the following equation was proposed by analyzing simulated rainfall data [19]:

$$D_i = K_i q^{1/2} I S^{2/3} = K_i R^{1/2} I S^{2/3} L^{1/2} \quad (4)$$

where q is flow discharge per unit width (m^2 s^{-1}) and S is slope inclination (m m^{-1}). This equation thought that the rainfall intensity term, I, represented the detachment of soil by raindrop impacts and the enhancement of the transport capacity of overland flow and that the term $q^{1/2}S^{2/3}$ represented sediment transport by thin overland flow. Compared to Eq. (3), the effect of runoff on interrill erosion rate was weakened, and the positive effect of slope length on interrill erosion rate was added to Eq. (4).

Table 1. Interrill erodibilities (K_i) and coefficients of determination (R^2) for Eqs (1), (3) and (4) for different slope lengths.

L (m)		Eq (1)	Eq (3)	Eq (4)
0.4	K_i	1,773,995	3,734,418	43,430
	R^2	0.90	0.90	0.96
0.8	K_i	2,026,011	3,420,684	31,387
	R^2	0.77	0.78	0.88
1.2	K_i	1,815,288	2,927,228	22,287
	R^2	0.79	0.80	0.84
1.6	K_i	1,531,958	2,435,898	16,294
	R^2	0.71	0.70	0.82
2.0	K_i	1,571,555	2,437,679	15,029
	R^2	0.64	0.64	0.77
Mean	K_i	1,743,761	2,991,181	25,686
CV		11%	19%	46%

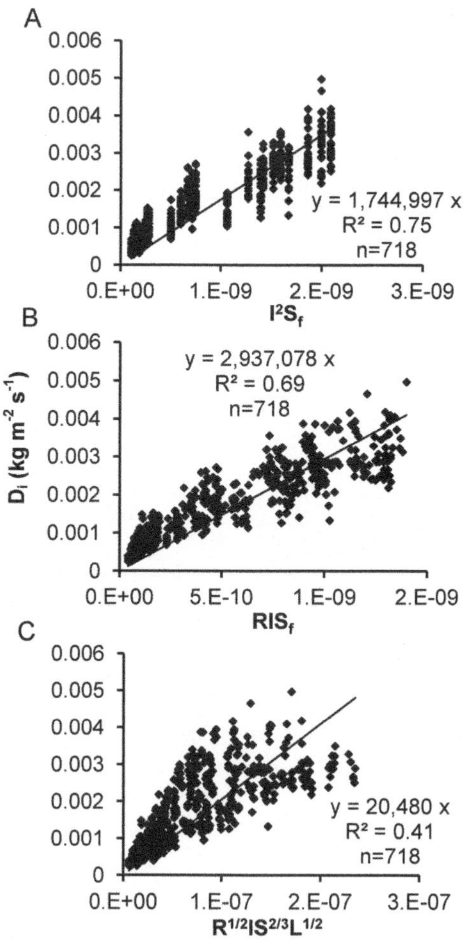

Figure 1. Linear regressions with zero intercept. Using linear regression with zero intercept between interrill erosion rate and the scaling factors I^2S_f (A), RIS_f (B), and $IR^{1/2}S^{2/3}L^{1/2}$ (C) to assess the previous interrill equations based on the mini plots.

Parameters for soil type (K_i), slope gradient, rainfall intensity, and overland flow are included in Eqs. (1), (3), and (4). The parameter for slope length is included only in Eq. (4). However, inconsistent results have been reported for the effect of slope length on interrill erosion. Running the similar experiments which simulated slope length using inflow rates, some experiments indicated consistently greater interrill erosion from increased overland flow [24], but other experiments showed that greater discharge rates resulted in decreased soil loss [25,26]. The authors thought that interrill erosion at a particular downslope distance is dictated by the soil detachment or sediment transport capacity characteristics existing at that particular location [25,26]. The effects of row-sideslope length within a range of 0.15 m to 0.6 m on sideslope erosion for four soils at several rain intensities demonstrated that slope length affected erosion very little until the soils began to rill, but had a major effect once rilling occurred [27]. However, the inconsistent viewpoint, which was that the effect of slope length on interrill erosion might have been due to a change from erosion which is dominated by raindrop-induced flow transport (RIFT) to erosion which is dominated by raindrop detachment-flow transport (RD-FT) except for the effect of slope length on rilling, was concluded by using the same data [28]. Other experiments indicated that soil losses were significantly correlated with rainfall intensity ($r = 0.89$; $P < 0.001$) and slope length ($r = 0.43$; $P < 0.001$), and suggested that careless scale transfer of erosion data may lead to erroneous conclusions [29]. In addition, some researchers have attributed the effect of slope length on interrill erosion to rainfall intensity and soil properties [27,30] and have assumed that slope length has little or no effect on interrill erosion per unit of area [31,32]. In short, although numerous studies were reported, the effect of slope length on interrill erosion has not been clearly established.

In addition, Eqs. (1), (3), and (4) were developed for smaller plot sizes (less than 0.5 m^2), shorter slope lengths (less than 0.8 m), and small slope gradients, and the different researchers who developed these equations used different plot sizes and slope lengths [3,19,22,27]. They suggested that the applicability of their model to longer slopes required further testing [19]. To date, no

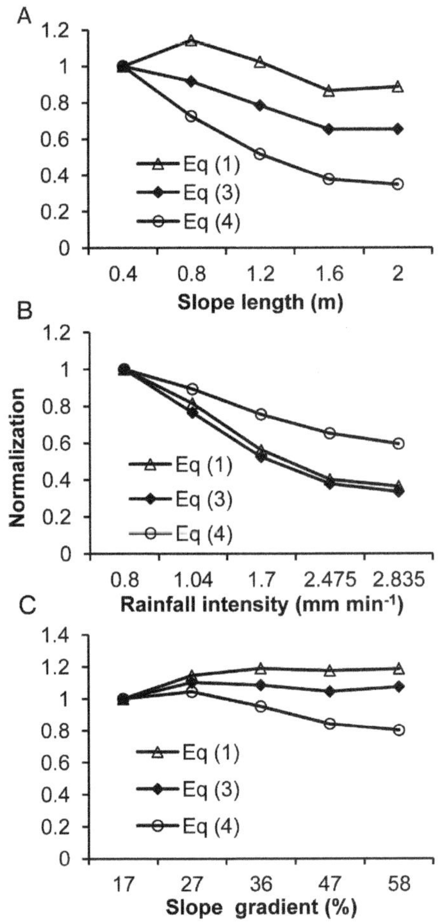

Figure 2. Normalized interrill erodibilities. Variations in interrill erodibility for the three equations for different slope lengths (A), rainfall intensities (B), and slope gradients (C), normalized to the lowest values of the independent factors.

systematic experiments based on mini-plot experiments have been conducted to validate and compare the existing interrill erosion equations for different slope lengths or plot sizes and a wider range of slope gradients and rainfall intensities. The objective of the present study is to validate the existing interrill erosion equations under different controlled experimental conditions and to describe interrill erosion rate accurately by adding or adjusting factors in the interrill erosion equations.

Materials and Methods

Erosion tests were conducted at the State Key Laboratory of Soil Erosion and Dryland Farming on the Loess Plateau in Yangling, China. Artificial rainfall was produced using a rain-making machine with an effective rainfall area of 2 m×3 m at a height of 8.67 m, which produced a simulated rainstorm at a controllable intensity with distribution uniformity greater than 85% and a median raindrop diameter of 2.2 mm, comparable to natural rainfall.

The soil used in the tests was cultivated Huangmian soil (Calcaric Cambisols, FAO) from Ansai County in Shaanxi Province, located in the northern part of the Loess Plateau. The cultivated land sampled in this study is publicly owned and managed by the Ansai Research Station of Soil and Water Conservation, Chinese Academy of Science. No specific permits were required for the field studies described. The soil contained 38.72% sand, 45.59% silt, 15.69% clay, and 0.53% organic matter. The soil was collected from the top 0.25 m layer in cultivated land, air-dried, crushed to pass through a 4-mm sieve, and thoroughly mixed. The soil moisture content was held constant at 14% for all experiments.

Perforated metal flumes with different lengths (0.4, 0.8, 1.2, 1.6, and 2 m), 0.4 m in width and 0.25 m in depth, were manufactured for use in the tests. The slope gradient of each flume could easily be adjusted from 0% to 84%. A trough was placed at the lower edge of the flume to collect runoff and sediment samples. The soil was packed uniformly into the flumes in four 5.5-cm layers to a total depth of 22 cm with a bulk density of 1.3 g cm^{-3}. To reduce discontinuities between layers, the surface of each soil

Table 2. Interrill erodibilities (K_i) and coefficients of determination (R^2) for Eqs. (1), (3), and (4) for different rainfall intensities.

I (mm min^{-1})		Eq (1)	Eq (3)	Eq (4)
0.8	K_i	4,386,185	7,967,808	29,897
	R^2	0.26	0.49	0[a]
1.04	K_i	3,572,441	6,097,944	27,063
	R^2	0.39	0.45	0
1.7	K_i	2,465,436	4,171,479	23,615
	R^2	0.4	0.37	0
2.475	K_i	1,761,220	3,009,673	20,424
	R^2	0.6	0.39	0
2.835	K_i	1,596,803	2,662,866	19,345
	R^2	0.46	0.17	0
Mean	K_i	2,756,417	4,781,954	24,069
CV		43%	47%	18%

[a]R^2 was negative for the linear regression with zero intercept between interrill erosion rate and $Iq^{1/2}S^{2/3}$ in Eq (4); therefore, it was replaced by zero.

Table 3. Interrill erodibilities (K_i) and coefficients of determination (R^2) for Eqs. (1), (3), and (4) for different slope gradients.

S (%)		Eq (1)	Eq (3)	Eq (4)
17	K_i	1,503,203	2,753,304	23,327
	R^2	0.71	0.7	0.47
27	K_i	1,722,260	3,036,752	23,789
	R^2	0.74	0.68	0.44
36	K_i	1,788,967	2,989,530	21,976
	R^2	0.72	0.65	0.37
47	K_i	1,766,816	2,875,945	19,512
	R^2	0.71	0.64	0.34
58	K_i	1,784,369	2,958,936	18,592
	R^2	0.75	0.67	0.34
Mean	K_i	1,713,123	2,922,893	21,439
CV		7%	4%	11%

Table 4. Univariate Analysis of Variance for selected parameters with interrill erosion rate as a dependent variable using all effective steady-state values.

Source	Type III Sum of Squares	df	Mean Square	F Value	Sig.
Corrected Model	.001[a]	125	5.405E-06	115.591	.000
Intercept	2.797E-06	1	2.797E-06	59.824	.000
Runoff (R)	2.747E-07	1	2.747E-07	5.874	.016
Intensity (I)	3.518E-06	4	8.794E-07	18.808	.000
Slope length (L)	1.939E-05	4	4.847E-06	103.660	.000
Slope gradient (S)	3.017E-05	4	7.542E-06	161.305	.000
I * L	1.104E-05	16	6.901E-07	14.760	.000
I * S	1.507E-05	16	9.419E-07	20.144	.000
L * S	8.892E-06	16	5.558E-07	11.886	.000
I * L * S	8.576E-06	64	1.340E-07	2.866	.000
Error	2.768E-05	592	4.676E-08		
Total	.003	718			
Corrected Total	.001	717			

[a]$R^2 = 0.961$ (Adjusted $R^2 = 0.952$).

layer was gently scored. The surface of the top layer was smoothed to minimize microtopographic effects.

Two replications of all combinations of one soil type, five levels of slope length (0.4, 0.8, 1.2, 1.6, and 2 m), five levels of slope gradient (17%, 27%, 36%, 47%, and 58%), and five levels of rainfall intensity (48, 62.4, 102, 149, and 170 mm h^{-1}) were tested using a randomized factorial design. A total of 250 runs were conducted. Replicate means were used in the data analyses.

Simulated rainfall was generally applied for 60 min for each rainfall event and was terminated if rill erosion occurred during the 60 min. Sediment and runoff samples were collected continuously throughout the rainfall event. Samples were collected at 1, 2, 3, 4, and 5 min within the first 15 minutes after runoff production and then every 5 min thereafter. Sediment and runoff were measured gravimetrically. In general, steady states of soil loss and runoff rate were reached within 30 min after runoff commenced. To minimize the effect of surface condition, steady-state values were used in the analysis [3,19,27]. Values identified as outliers were not included [3]. In total, 36 of 1740 experimental values and 22 of 740 steady-state values were rejected. Statistical analyses were performed using the SPSS PASW Statistics (Version 18.0) software.

Results and Discussion

Validation of existing interrill erosion equations

Interrill erodibility based on mini-plot experiments for a given soil type was defined as a constant for each equation. This constant was determined from linear regression between the interrill erosion rate D_i and the scaling factors I^2S_f in Eq. (1), RIS_f in Eq. (3), and $IR^{1/2}S^{2/3}L^{1/2}$ in Eq. (4) for all effective steady-state values. A zero intercept was used in the regression, and the slope of the regression line was assumed to be the interrill erodibility (K_i). Figure 1 presents the results of linear regression with zero intercept for Eqs. (1), (3), and (4). Good linear regressions could not be gained for the three interrill erosion equations for all effective steady-state values, which indicated that in some cases the existing interrill erosion equations did not adequately describe the relationship between interrill erosion rate and its influencing factors.

To find further key factors influencing the equation's predictive power, interrill erodibility was also calculated using linear regression with a zero intercept under various controlled experimental conditions. Table 1 presents interrill erodibility (K_i) and coefficient of determination (R^2) for different slope lengths for the three existing equations. When rainfall intensity and slope gradient were varied and slope length was increased from 0.4 m to 2 m, the interrill erodibility (K_i) from Eq. (1) varied slightly, with a coefficient of variation (CV) of 11%, whereas that from Eq. (3) decreased by approximately 1.5 times, with a CV of 19%, and that from Eq. (4) varied widely and decreased by approximately 2.9 times, with a CV of 46% (Fig. 2a). These results indicated that slope length plays a key role in Eqs. (3) and (4) and should be included in these equations. Compared with Eq. (1), the wide variations in interrill erodibility for Eqs. (3) and (4) might imply interactions between runoff and slope length. The change in the coefficient of determination (R^2) corresponding to the change in slope length from 0.4 m to 2 m demonstrated that the three existing interrill equations could describe interrill erosion processes effectively for plots with shorter slope lengths, especially those experiments of approximately 0.4 m. This finding is consistent with existing interrill erosion equations established on the basis of data from mini-plots with shorter slope lengths in previous works [3,12,27].

Table 2 presents the interrill erodibility (K_i) values and coefficients of determination (R^2) obtained for different rainfall intensities using the three existing equations. Interrill erodibility (K_i) for the three equations decreased as rainfall intensity increased from 0.8 to 2.835 mm min^{-1} (Fig. 2b). Interrill erodibility (K_i) varied obviously for Eqs. (1) and (3), with a CV of approximately 45%, but varied relatively little for Eq. (4), with a CV of 18% because of the decrease in the weight of runoff rate in the equation. These results indicated that the effects of rainfall intensity or runoff rate are overestimated by the three existing interrill erosion equations. When the weight of runoff or rainfall intensity in the interrill erosion equations was decreased, the equations' predictive power was improved. Therefore, an adjustment to these weights should be considered.

Table 3 presents the interrill erodibility (K_i) values and coefficients of determination (R^2) obtained for different slope gradients for the three existing equations. Only for Eq. (4) did interrill erodibility decrease with an increase in slope gradient (Fig. 2c). In general, interrill erodibility from these three equations varied slightly with slope gradient. These results suggested that the effect of slope gradient on interrill erosion rate is almost completely expressed by the existing equations, but that use of the S_f term in Eqs. (1) and (3) yields better results than use of the $S^{2/3}$ term in Eq. (4). The lower coefficient of determination (R^2) values shown in Tables 2 and 3 also implied that slope length is the most important factor affecting interrill erosion processes and results in lower coefficients of determination (R^2).

In summary, the results from the analysis described above suggested that the three existing interrill erosion equations considered do not completely describe the relationship between interrill erosion rate and its influencing factors, especially slope length and rainfall intensity or runoff. In addition, univariate analysis of variance was conducted on the assumption that runoff rate was the covariant and that rainfall intensity, slope gradient, and slope length were the fixed factors. The results of this analysis showed that runoff rate, rainfall intensity, slope gradient, and slope length had significant effects on interrill erosion rate and that their interactions were significant at $p<0.01$ (Table 4). These results

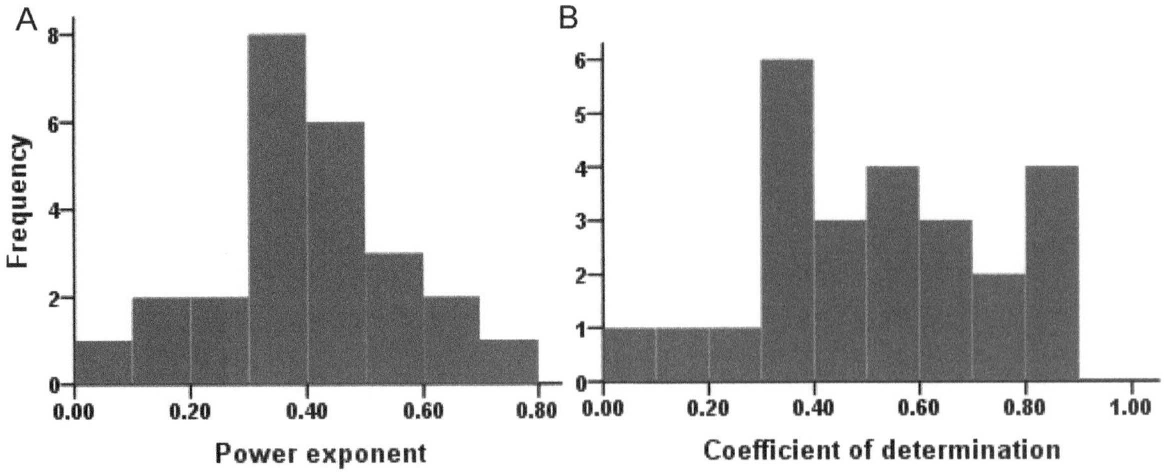

Figure 3. Regression parameters of power function. Power exponents (A) and coefficients of determination (B) for the relationship between sediment concentration and slope gradient for all tests.

also implied that the effect of slope length on interrill erosion rate should be considered in the interrill erosion equations. Therefore, an interrill erosion equation is needed that includes a factor for slope length and weight adjustments for the other factors.

Improved interrill erosion equation

According to Eqs. (1), (3), and (4) and the results of the analysis described above, the interrill erosion rate for a given soil type can be described by a function of the following general form:

$$D_i = f(R,I,S,L) \quad (5)$$

In this function, slope gradient (S), rainfall intensity (I), and slope length (L) are three independent variables that directly affect runoff rate (R) and directly or indirectly affect interrill erosion rate (D_i). Runoff rate also affects interrill erosion rate. In fact, these factors are interdependent and interactive. Multivariate regression does obtain the best-fit expression, but does not reflect the physical interpretation of the interaction between interrill erosion rate and its influencing factors. It is difficult to eliminate the interdependence and interaction among these factors and to express accurately the relationship between interrill erosion rate and its influencing factors using multivariate regression. Therefore, before structuring the interrill erosion equation, it is necessary to analyze individually the relationships between interrill erosion rate and each influencing factor and to understand their physical interpretation under controlled conditions.

Sediment concentration (C) is a composite indicator whose variations reflect the dynamic variations of runoff and soil loss. Transport by interrill overland flow is the predominant process in interrill erosion [2], and the detachment process can be considered negligible [20]. Based on the theory of erosion by rainfall-disturbed flow [33,34], sediment concentration was determined by soil factors, rainfall intensity, kinetic energy per unit quantity of rain, slope length, and slope gradient [28]. Because the kinetic energy of rain is an expression of rainfall intensity [35], the sediment concentration for a given soil type can also be expressed by a function of the following general form:

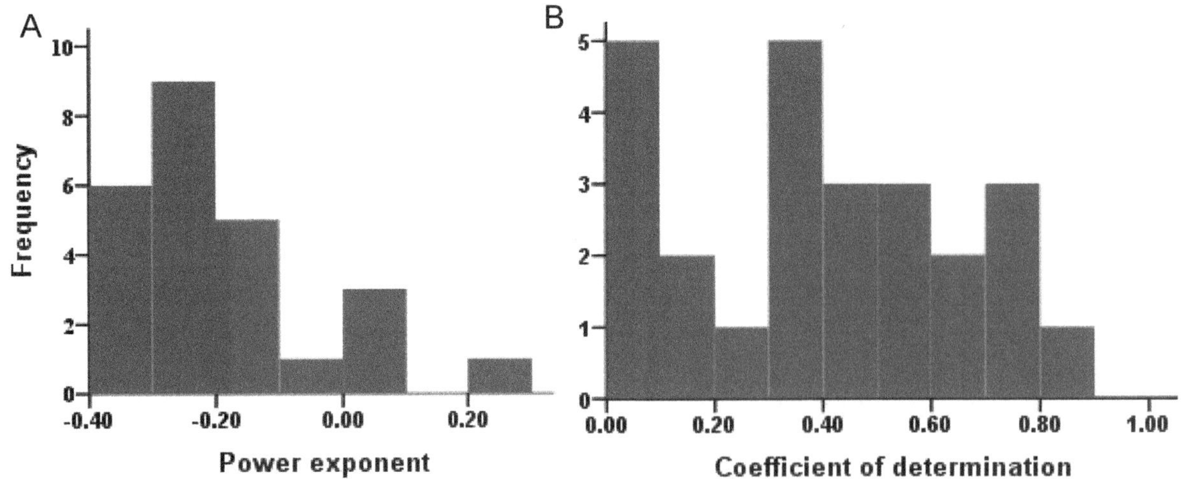

Figure 4. Regression parameters of power function. Power exponents (A) and coefficients of determination (B) for the relationship between sediment concentration and slope length for all tests.

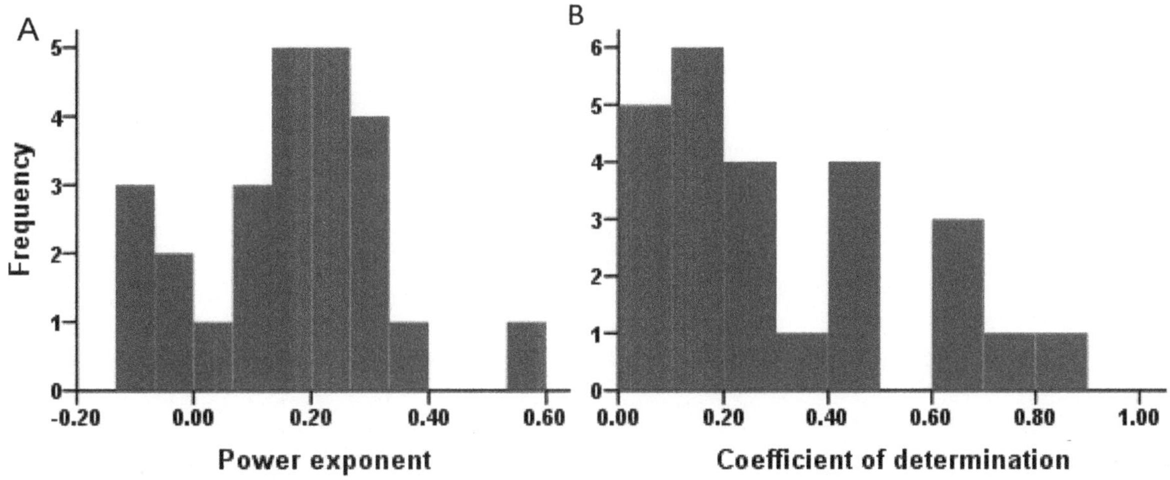

Figure 5. Regression parameters of power function. Power exponents (A) and coefficients of determination (B) for the relationship between sediment concentration and rainfall intensity for all tests.

$$C = f(I,S,L) \quad (6)$$

Compared with Eq. (5), there is one dependent parameter (C) in Eq. (6), and the effects of the interdependence between runoff rate and rainfall intensity, slope gradient, and slope length in Eq. (5) are not considered in Eq. (6), while interactions among rainfall intensity, slope gradient, and slope length may exist in Eq. (6).

Sediment concentration (C) can also be calculated by the following formula:

$$C = D_i / R \quad (7)$$

Therefore, if the relationships between sediment concentration and slope gradient, rainfall intensity, and slope length can be determined, the relationships between interrill erosion rate and runoff rate, slope gradient, rainfall intensity, and slope length can also be determined. The relationships between sediment concentrations and each influencing factor are quantified in the following discussion.

For the same slope length and rainfall intensity, highly significant power-function relationships ($p<0.05$) between sediment concentration and slope gradient were found except for one combination test, although the power exponents were different for different combinations of slope length and rainfall intensity (Fig. 3). In general, the change in sediment concentration was less than linearly related to the change in slope gradient, with a mean power exponent of 0.4 and a CV of 0.41. The power exponents and the coefficients of determination (R^2) decreased with increasing slope length. This suggested that the effect of slope gradient on sediment concentration decreases with increasing slope length. The variability of the power exponent was less for higher rainfall intensities than for lower rainfall intensities.

For the same slope gradient and rainfall intensity, significant negative power-function relationships ($p<0.05$) between sediment concentration and slope length were detected in 19 out of 25 combination tests, and one had a negative power-function relationship ($p>0.05$). Figure 4 presents the frequencies of the power exponents and the coefficients of determination of the relationship between sediment concentration and slope length for all tests. For five combination tests with positive exponents, four tests were insignificant ($p>0.05$), and one had a significantly positive correlation ($p<0.05$). The positive power exponents

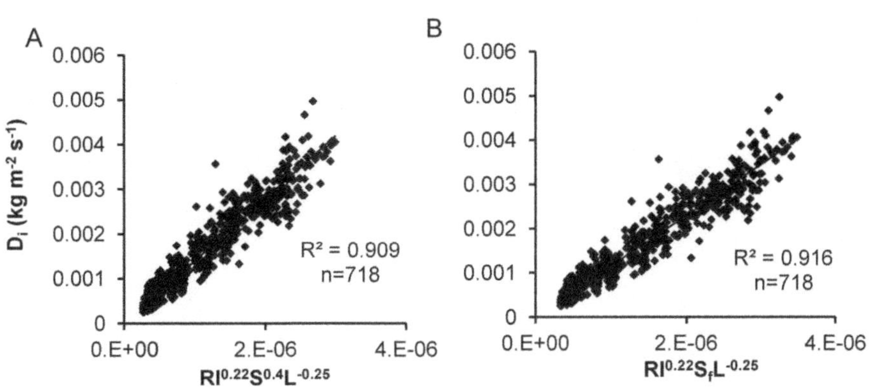

Figure 6. Linear regressions with zero intercept. Linear regression with zero intercept between the interrill erosion rate and the scaling factors $R\,I^{0.22}S^{0.4}L^{-0.25}$ (A) and $R\,I^{0.22}\,S_f\,L^{-0.25}$ (B) to assess the slope factor of $S^{0.4}$ and S_f.

Table 5. Slopes of the linear regression (K_i) and coefficients of determination (R^2) from Eq (10) for different experimental conditions.

L (m)		Eq (10)	I (mm min^{-1})		Eq (10)	S (%)		Eq (10)
0.4	K_i	1,132	0.8	K_i	1,271	17	K_i	1,082
	R^2	0.95		R^2	0.5		R^2	0.86
0.8	K_i	1,234	1.04	K_i	1,190	27	K_i	1,197
	R^2	0.95		R^2	0.49		R^2	0.89
1.2	K_i	1,163	1.7	K_i	1,192	36	K_i	1,182
	R^2	0.93		R^2	0.57		R^2	0.92
1.6	K_i	1,050	2.475	K_i	1,148	47	K_i	1,134
	R^2	0.90		R^2	0.72		R^2	0.9
2.0	K_i	1,153	2.835	K_i	1,134	58	K_i	1,140
	R^2	0.86		R^2	0.67		R^2	0.94
Mean	K_i	1,146	Mean	K_i	1,187	Mean	K_i	1,147
CV		5.7%	CV		4.5%	CV		3.9%

between sediment concentration and slope length might have resulted from poor control of the experimental conditions or from other random factors. The mean power exponent of the negative correlation was −0.25, with a CV of 0.33. In general, the power exponents decreased as slope gradient and rainfall intensity increased. The negative correlation between sediment concentration and slope length might indicate that the role of raindrop detachment on detached soil particles from the soil surface weakens with increasing slope length and that transport-limited erosion processes transfer to detachment-limited erosion with increasing slope length.

A power-function relationship between sediment concentration and rainfall intensity was derived for controlled conditions of fixed slope length and gradient. Among the 25 combination tests, 18 were significant positive power-function relationships ($p<0.05$), 2 were insignificant positive power-function relationships ($p>0.05$), and 5 were insignificant negative power-function relationships ($p>0.05$). These results indicate that positive power-function relationships exist between sediment concentration and rainfall intensity, while for a few tests, an insignificant negative correlation was detected, and that the power exponents are different for different tests (Fig. 5). The mean power exponent of the positive correlations was 0.22, with a CV of 0.55. In general, the power exponents decreased as slope length increased, which indicated that the effect of rainfall intensity on sediment concentration weakens due to the increase of flow depth and the reduction in raindrop detachment with increasing slope length and plot size.

Based on the results of the analysis described above, power-function relationships were derived between sediment concentration (C) and slope gradient (S), rainfall intensity (I), and slope length (L) for all combination tests. The averaged power exponents were used as a basis for describing the relationship between C and I, S, and L in Eq. (6). As a result, Eq. (6) can be expressed as:

$$C = AI^{0.22}S^{0.4}L^{-0.25} \quad (8)$$

In Eq. (8), A is a constant for a given soil type and is equivalent to the interrill erodibility (Ki) of that soil type.

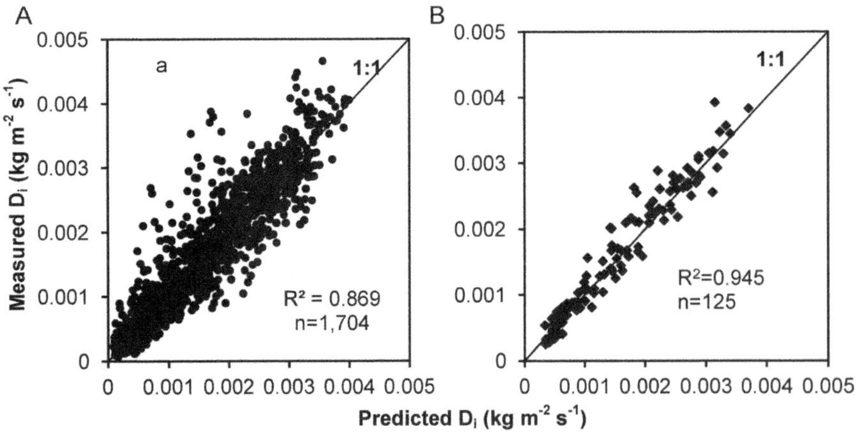

Figure 7. Measured vs. predicted interrill erosion rates. Measured and predicted interrill erosion rates for Eq. (10) for the erosion processes during a rainfall event (A) and the average of each rainfall event (B).

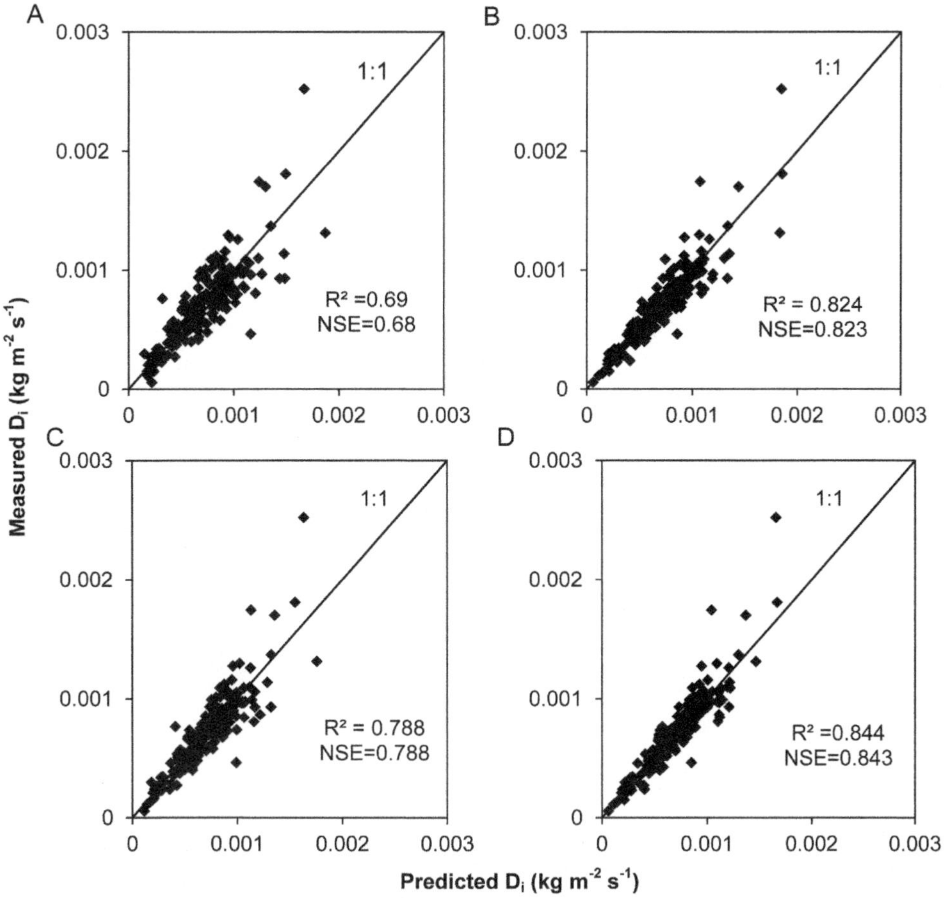

Figure 8. Measured vs. predicted interrill erosion rates. Measured and predicted interrill erosion rates for Eq. (1) (A), Eq. (3) (B), Eq. (4) (C), and Eq. (10) (D) using experimental data from WEPP cropland soil field interrill erodibility experiments.

Combining Eqs. (7) and (8), the interrill erosion rate, D_i (kg m^{-2} s^{-1}), can be expressed by the following equation:

$$D_i = K_i RI^{0.22} S^{0.4} L^{-0.25} \tag{9}$$

However, as the previous work noted [19], the exponents of rainfall intensity, slope gradient, and slope length can be slightly altered to account for interactions among soil type, gradient, slope length, and rainfall intensity.

Figure 6 shows linear regressions with zero intercept between the interrill erosion rate (D_i) and the scaling factor $RI^{0.22}$ $S^{0.4}$ $L^{-0.25}$ and the scaling factor $RI^{0.22}$ S_f $L^{-0.25}$. Although the coefficient of determination (R^2) values using $S^{0.4}$ and S_f were not significantly different, using S_f to describe the effect of gradient on interrill erosion yielded slightly better results than using $S^{0.4}$. Moreover, using S_f was consistent with results obtained by other researchers [16]. Equation (9) could therefore be transformed as follows:

$$D_i = K_i RI^{0.22} S_f L^{-0.25} \tag{10}$$

To assess the performance of Eq. (10), table 5 presents the linear regression slopes (K_i) and coefficients of determination (R^2) obtained from linear regressions with zero intercept between the interrill erosion rate (D_i) and the scaling factor $RI^{0.22}$ S_f $L^{-0.25}$ for different experimental conditions. The higher coefficients of determination (R^2) and the lower CVs of interrill erodibility (K_i) indicated that Eq. (10) is better than previously developed equations for interrill erosion and adequately predicts interrill erosion for a wide range of rainfall intensities, slope gradients, and slope lengths.

As stated previously, Eq. (10) was developed for steady-state overland flow and interrill erosion rate. Equation (10) must be validated for non-steady-state conditions. Equation (10) was therefore used to predict interrill erosion rates during a rainfall event and the averaged interrill erosion rate of each rainfall event for all rainfall events in this study. The measured and predicted interrill erosion rates are presented in Fig. 7. Correlation coefficients of 0.932 and 0.972 between the measured and predicted interrill erosion rates were obtained, and Nash-Sutcliffe coefficients of 0.865 and 0.942 were obtained for the interrill erosion rates of erosion processes and the average of each rainfall event respectively. These results suggested that Eq. (10) could predict interrill erosion rates reasonably well, despite a few predicted values that were considerably different from the corresponding measured values.

Validation of the improved interrill erosion equation

To evaluate further the effectiveness of the improved interrill erosion equation, experimental data from WEPP cropland soil field interrill erodibility experiments were used for validation.

Table 6. Interrill erodibility and its CV from the different interrill erosion equations for the experimental data from the WEPP cropland soil field interrill erodibility experiments.

Soil type	No. of plots	$D_i = K_i I^2 S_f$		$D_i = K_i R I S_f$		$D_i = K_i q^{1/2} I S^{2/3}$		$D_i = K_i R I^{0.22} S_f L^{-0.25}$	
		K_i	CV (%)	K_i	CV (%)	K_i	CV (%)	K_i	CV (%)
Sharpsburg	6	1,917,372	24.05	2,457,363	13.22	23,850	17.18	323	13.03
Hersh	6	4,153,649	37.64	6,801,244	31.43	58,530	31.28	920	26.69
Keith	6	3,403,141	23.66	4,229,851	24.27	44,048	21.04	597	19.94
Amarillo	6	4,213,438	31.33	6,235,421	33.63	65,118	31.16	865	34.25
Woodward	6	4,592,143	27.85	7,568,362	20.60	74,617	28.95	1,049	27.73
Heiden	6	1,336,998	24.65	1,638,827	30.53	17,529	25.49	211	30.66
Whitney	6	3,101,189	20.41	4,367,166	16.02	48,903	18.38	590	14.56
Academy	6	2,994,017	25.59	3,932,019	11.12	40,099	16.97	514	9.04
Los Banos	6	2,405,840	14.00	3,262,543	17.95	30,191	10.92	414	10.58
Portneuf	6	1,382,542	16.90	2,314,773	14.54	20,975	13.61	280	12.42
Nansene	6	3,206,879	22.89	4,936,611	16.09	44,005	15.82	682	11.03
Palouse	6	4,003,363	18.95	5,201,600	12.97	48,338	13.65	682	9.57
Zahl	6	3,267,617	29.70	3,840,678	20.37	42,288	21.38	512	14.65
Pierre	6	2,300,664	17.06	2,719,824	7.11	27,829	11.27	367	6.60
Williams	6	2,684,238	24.36	3,548,762	11.90	38,403	10.74	508	10.17
Barnes - ND	6	1,876,045	10.85	3,056,669	10.09	31,502	8.41	474	8.12
Sverdrup	6	2,362,425	17.69	3,958,183	14.65	34,180	13.69	533	14.54
Barnes - MN	6	1,503,947	21.63	2,576,945	20.84	24,530	18.80	409	17.40
Mexico	6	2,970,456	16.24	4,153,863	15.04	41,848	13.23	570	12.10
Grenada	6	2,633,592	19.66	3,372,939	10.67	35,349	13.24	462	10.24
Tifton	6	652,793	54.22	1,603,183	27.15	12,053	36.29	221	23.88
Bonifay	6	871,497	73.08	4,226,263	15.32	22,179	47.29	581	21.05
Cecil	6	1,858,964	25.03	2,282,079	18.68	24,807	21.10	319	19.42
Hiwassee	6	1,873,194	18.16	2,243,785	17.27	25,030	15.54	320	13.94
Gaston	6	2,036,993	13.15	2,205,358	16.98	24,961	14.49	297	16.56
Opequon	6	3,195,673	16.05	3,398,535	9.63	39,643	11.97	474	7.20
Frederick	6	2,477,747	29.91	3,331,759	15.01	35,043	21.67	476	14.22
Manor	6	2,694,182	23.21	3,563,361	12.63	38,111	17.67	515	13.31
Caribou	6	1,551,246	11.13	1,845,618	7.78	20,635	8.21	263	7.82
Collamer	6	3,456,598	16.26	4,121,058	9.61	44,942	10.17	566	6.60
Miamian	6	1,591,767	22.99	2284814	18.64	22548	20.61	310	19.44
Lewisburg	6	2,257,824	20.43	2,803,097	15.93	29,807	16.97	381	14.62
Miami	6	1,970,749	26.84	2,579,164	18.97	26,927	20.07	356	15.01

These experiments were part of the USDA-ARS WEPP project, which included thirty-three soils from all areas of the United States in ridge plots. Flat plots were excluded from the validation because their slope gradients were much less than the lowest slope gradient in these experiments and some soil types were not represented in flat plots. The detailed information about the experimental materials and procedures were described [3]. Based on their data analysis techniques, the means of the last four erosion rates and flow discharges were selected as the erosion rate and flow discharge for a given plot in this validation. If there was an outlier value within the last four, then it was not considered, and the four most consistent of the last five values were used [3]. Although the fixed slope length was an obstacle to validating the improved interrill erosion equation, these experimental data could be used to judge the quality of different interrill erosion equations.

The "intrinsic" interrill erodibility for a given soil is independent of runoff rate, rainfall intensity, infiltration, slope length, and gradient. If an interrill erosion equation adequately describes the relationship between interrill erosion rate and its influencing factors, the interrill erodibility, as estimated by an interrill erosion equation for a given soil, should change little over a wide range of conditions. Therefore, the CV of interrill erodibility estimated from an interrill erosion equation for a given soil under a wide range of conditions can be used to judge the quality of the interrill erosion equation. Interrill erodibility values, as estimated from the different interrill erosion equations discussed in this paper, along with their CV for 33 soils in the United States, are given in Table 6.

In general, there were common trends in the interrill erodibilities predicted by the various interrill erosion equations. For all soils, very strong positive correlations ($p<0.05$) were found between the interrill erodibility values predicted from the various interrill erosion equations. Although the trends showed that interrill erodibility values predicted from the different interrill erosion equations reflected the strength of soil resistance to interrill erosion, the precision of the predictions was different among the equations. The CV of interrill erodibility from Eq. (10) was less than that from Eq. (1) for 30 soils, that from Eq. (10) was less than that from Eq. (3) for 25 soils, and that from Eq. (10) was less than that from Eq. (4) for 29 soils. For 23 soils, the CV of interrill erosion from Eq. (10) was the lowest among those from the four equations. Figure 8 shows the observed and predicted values for the various equations. The correlation coefficients and Nash-Sutcliffe coefficients between the measured and predicted values for Eq. (10) were the highest among all the equations, which suggested that Eq. (10) described the relationship between interrill erosion rate and its influencing factors better than previously proposed interrill erosion equations, even though the slope length of the plots in the experiments used for validation was fixed.

The CV of interrill erodibility from Eqs. (3) and (10) was generally less than that from Eq. (1), indicating that rainfall intensity or flow discharge alone cannot completely express interrill erosion processes. Compared with Eq. (3), the reduced proportion of flow discharge in Eq. (4) resulted in an increase in the CV of interrill erodibility, but the reduced proportion of rainfall intensity in Eq. (10) resulted in a decrease in the CV of interrill erodibility. These results indicated that flow discharge is still the most important transport agent for interrill erosion processes, even though rainfall intensity has two effects, i.e., detachment of soil and enhancement of the transport capacity of thin overland flow. The previously proposed interrill erosion equations overemphasized the effect of rainfall intensity on interrill erosion when slope length was relatively longer. The CVs of interrill erodibility from Eq. (3) and Eq. (10) were closer, perhaps because slope length was fixed in the WEPP cropland soil field interrill erodibility experiments.

Conclusions

Previously proposed interrill erosion equations based on mini-plot experiments largely ignored the effect of slope length and plot size on interrill erosion rate. A series of simulated rainfall experiments was conducted to validate the previously proposed interrill erosion equations. The results indicated that slope gradient almost adequately predicted interrill erosion in the previously proposed equations, but that the interrill erodibilities calculated from previously proposed equations generally decreased with increasing slope length and rainfall intensity. This result suggested that adding a factor for slope length and adjusting the factor for rainfall intensity are necessary to improve prediction of interrill erosion rate. An improved interrill erosion equation, $D_i = K_i RI^{0.22} S_f L^{-0.25}$, was developed by analyzing the relationships between sediment concentration and rainfall intensity, slope length, and slope gradient. To evaluate the effectiveness of the improved interrill erosion equation, part experimental data from WEPP cropland soil field interrill erodibility experiments were used for validation. The results indicated that the improved interrill erosion equation predicted interrill erosion rate better than previously proposed interrill erosion equations, even though the slope length of the plots in the WEPP experiments was fixed.

Author Contributions

Conceived and designed the experiments: ZW FZ MY. Performed the experiments: ZW FZ. Analyzed the data: FZ ZW MY. Contributed reagents/materials/analysis tools: ZW. Wrote the paper: FZ ZW MY.

References

1. Meyer LD, Foster GR, Romkens MJM (1975) Source of soil eroded by water from upland slopes. In: Present and prospective technology for predicting sediment yields and sources, Washington, USDA-Agricultural Research: 177–189.
2. Kinnell PIA (1991) The effect of flow depth on sediment transport induced by raindrops impacting shallow flows. Transactions of the ASAE 34: 161–168.
3. Elliot WJ, Liebenow AM, Laflen JM, Kohl KD (1989) A Compendium of Soil Erodibility Data from WEPP Cropland Soil Field Erodibility Experiments 1987 and 1988. NSERL Report No 3, The Ohio State University, and USDA Agricultural Research Service.
4. Morgan RPC, Quinton JN, Smith RE, Govers G, Poesen JWA, et al. (1998) The European Soil Erosion Model (EUROSEM): a dynamic approach for predicting sediment transport from fields and small catchments. Earth Surf Proc Land 23: 527–544.
5. De Roo APJ, Wesseling CG, Ritsema CJ (1996) LISEM: A single-event physically based hydrological and soil erosion model for drainage basins I: theory, input and output. Hydrol Process 10: 1107–1117.
6. Nord G, Esteves M (2005) PSEM_2D: A physically based model of erosion processes at the plot scale. Water Resour Res 41: W08407.
7. Valmis S, Dimoyiannis D, Danalatos NG (2005) Assessing interrill erosion rate from soil aggregate instability index, rainfall intensity and slope angle on cultivated soils in central Greece. Soil Till Res 80: 139–147.
8. Yan FL, Shi ZH, Li ZX, Cai CF (2008) Estimating interrill soil erosion from aggregate stability of Ultisols in subtropical China. Soil Till Res 100: 34–41.
9. Ascough JC, Baffaut C, Nearing MA, Liu BY (1997) The WEPP watershed model: I. Hydrology and erosion. Transactions of the ASAE 40: 921–933.
10. Wei H, Nearing MA, Stone JJ, Guertin DP, Spaeth KE, et al. (2009) A New Splash and Sheet Erosion Equation for Rangelands. Soil Sci Soc Am J 73: 1754–1754.
11. Meyer LD (1981) How Rain Intensity Affects Interrill Erosion. Transactions of the ASAE 24: 1472–1475.
12. Watson DA, Laflen JM (1986) Soil strength, slope, and rainfall intensity effects on interrill erosion. Transactions of the ASAE 29: 98–102.
13. Foster GR, Meyer LD, Onstad CA (1977) An Erosion Equation Derived from Basic Erosion Principles. Transactions of the ASAE 20: 678–682.
14. Meyer LD, Harmon WC (1984) Susceptibility of agricultural soils to interrill erosion. Soil Sci Soc Am J 48: 1152–1157.
15. Guy BT, Dickinson WT, Rudra RP (1987) The Roles of Rainfall and Runoff in the sediment transport capacity of interrill flow. Transactions of the ASAE 30: 1378–1386.
16. Elliot WJ, Laflen JM, Kohl KD (1989) Effect of soil properties on soil erodibility. Paper no 89-2150 Transactions of the ASAE, St Joseph, MI.
17. Mccool DK, Brown LC, Foster GR, Mutchler CK, Meyer LD (1987) Revised slope steepness factor for the universal soil loss equation. Transactions of the ASAE 30: 1387–1396.
18. Neal JH (1938) The Effect of the Degree of Slope and Rainfall Characteristics on Runoff and Soil Erosion. Soil Sci Soc Am J 2: 525–532.
19. Zhang XC, Nearing MA, Miller WP, Norton LD, West LT (1998) Modelling interrill sediment delivery. Soil Sci Soc Am J 62: 438–444.
20. Young RA, Wiersma JL (1973) The role of rainfall impact in soil detachment and transport. Water Resour Res 9: 1629–1636.
21. Truman CC, Bradford JM (1995) LABORATORY DETERMINATION OF INTERRILL SOIL ERODIBILITY. Soil Sci Soc Am J 59: 519–526.
22. Kinnell PIA (1993) Runoff as factor influencing experimentally determined interrill erodiblities. Aust J Soil Res 31: 333–342.
23. Flanagan DC, Nearing MA (1995) USDA-Water Erosion Prediction Project: Technical documentation. NSERL Rep no 10 Natl Soil Erosion Res Lab, West Lafayette, IN.
24. Monke EJ, Marelli HJ, Meyer LD, DeJong JF (1977) Runoff, erosion and nutrient movement from interrill areas. Transactions of the ASAE 20: 58–61.
25. Gilley JE, Woolhiser DA, McWhorter DB (1985) Interrill soil erosion - Part II: Testing and use of model equations. Transactions of the ASAE 28: 154–159.
26. Gilley JE, Woolhiser DA, McWhorter DB (1985) Interrill soil erosion-Part I. Development of model equations. Transactions of the ASAE 28: 147–153, 159.
27. Meyer LD, Harmon WC (1989) How Row-Sideslope Length and Steepness Affect sideslope erosion. Transactions of the ASAE 32: 639–644.
28. Kinnell PIA (2000) The effect of slope length on sediment concentrations associated with side-slope erosion. Soil Sci Soc Am J 64: 1004–1008.
29. Chaplot VAM, Le Bissonnais Y (2003) Runoff features for interrill erosion at different rainfall intensities, slope lengths, and gradients in an agricultural loessial hillslope. Soil Sci Soc Am J 67: 844–851.

30. Gabriels D (1999) The effect of slope length on the amount and size distribution of eroded silt loam soils: short slope laboratory experiments on interrill erosion. Geomorphology 28: 169–172.
31. Foster GR (1982) Modeling the erosion process. In Hydrologic modeling of small watersheds C T Haan et al, eds Chapter 8: 297–380. ASAE Monograph No. 5, St. Joseph, MI: ASAE.
32. Renard K, Foster GR, Weesies G, McCool DK, Yoder D (1997) Predicting Soil Erosion by Water; Guide to Conservation Planning with the Revised Universal Soil Loss Equation (RUSLE). ARS, USDA, Agricultural Handbook No 703.
33. Kinnell PIA (1993) Sediment concentrations resulting flow depth drop size interactions in shallow overland flow. Transactions of the ASAE 36: 1099–1103.
34. Kinnell P (1994) The effect of pre-detached particles on soil erodibilities associated with erosion by rain-impacted flows. Soil Res 32: 127–142.
35. Wischmeier WH (1958) Rainfall energy and its relationship to soil loss. Trans AGU 39: 285–291.

A Multi-Criteria Index for Ecological Evaluation of Tropical Agriculture in Southeastern Mexico

Esperanza Huerta[1]*, Christian Kampichler[2,3], Susana Ochoa-Gaona[4], Ben De Jong[4], Salvador Hernandez-Daumas[1], Violette Geissen[5,6]

1 El Colegio de la Frontera Sur, Unidad Campeche, Dpto. Agroecología, Campeche, México, **2** Universidad Juárez Autónoma de Tabasco, División de Ciencias Biológicas, Villahermosa, Tabasco, México, **3** Sovon Dutch Centre for Field Ornithology, Natuurplaza (Mercator 3), Nijmegen, The Netherlands, **4** El Colegio de la Frontera Sur, Unidad Campeche, Dpto. Sustainability Sciences, Campeche, México, **5** University of Bonn - INRES, Bonn, Germany, **6** Wageningen University and Research Center – Alterra, Wageningen, Gelderland, Netherlands

Abstract

The aim of this study was to generate an easy to use index to evaluate the ecological state of agricultural land from a sustainability perspective. We selected environmental indicators, such as the use of organic soil amendments (green manure) versus chemical fertilizers, plant biodiversity (including crop associations), variables which characterize soil conservation of conventional agricultural systems, pesticide use, method and frequency of tillage. We monitored the ecological state of 52 agricultural plots to test the performance of the index. The variables were hierarchically aggregated with simple mathematical algorithms, if-then rules, and rule-based fuzzy models, yielding the final multi-criteria index with values from 0 (worst) to 1 (best conditions). We validated the model through independent evaluation by experts, and we obtained a linear regression with an $r^2 = 0.61$ ($p = 2.4e\text{-}06$, $d.f. = 49$) between index output and the experts' evaluation.

Editor: Yong Deng, Southwest University, China

Funding: Financial support was obtained from the Conacyt-SEMARNAT Project "Uso sustentable de los recursos naturales en la frontera sur de México" (Sustainable use of natural resources on the southern border of Mexico) (code SEMARNAT-2002-C01-1109). El Colegio de la Frontera Sur provided infrastructural resources. The funders had no role in study design, data collection and analysis, decision to publish, or preparation of the manuscript.

Competing Interests: The authors have declared that no competing interests exist.

* Email: ehuerta@ecosur.mx

Introduction

In the past 60 years, degradation and deforestation of tropical forests worldwide have occurred much faster and more extensively than in any other period in history [1], [2]. Furthermore, countries like Mexico have been undergoing drastic land use changes. In the tropics of Mexico large parts of its lowland rainforest areas have been converted into pasture and cropland. In the state of Tabasco, for example, only 3.4% of the state is covered with original forest [3], whereas 76.4% of the surface was used for cattle production in 2000 [4] and 15.6% was used for agriculture, principally sugarcane and fruit plantations [3]. The ecological consequences of these land-use changes in Tabasco are well documented. Soil losses in hilly regions are very high up to 200 t ha^{-1} $year^{-1}$ [5]; high pesticide and fertilizer inputs to crops that have replaced forests have caused considerable environmental contamination [6], [7], and soil fertility is decreasing [8], [5], [9].

It is of the utmost importance to identify sustainable land use strategies which are economically attractive for the region's farmers and which may also reconcile the need for food production with that of soil conservation. In order to assist a variety of stakeholders at the local and regional level in making land use decisions, simple evaluation tools are needed. This is even more needed since a high percentage of the population consists of immigrants from other Mexican states, who are unfamiliar with the conditions of the humid tropics and use intensive techniques for farming the land. This is mostly due to large-scale agricultural development projects, such as "Plan Balancán-Tenosique," named after the two municipalities involved, in which, in the 1970 s, over 1100 km^2 of lowland rain forest was destroyed and converted into crop and pasture land. Ecological values of the 1970 s were very different from those of the present, and government representatives were willing to deforest in order to grant land to farmers [10].

The direct impact of farming is difficult to measure due to methodological difficulties (impossibility of measurement, complexity of the system) or practical reasons (time, costs) [11]. Therefore, the use of indicators appears to be an alternative way of guiding land use decisions [12], [13]. However, the "indicator explosion" [14], that is, the use of an exaggerated number of indicators aimed at assessing environmental impacts of agricultural activities, has been of little use to local decision-makers. Particularly in the tropics, land use decisions are still based on the informal opinion of local experts rather than on implementation of Decision Support Systems for environmentally sound resource management [15]. On the one hand, this is due to farmers' restricted access to modern communications and information technologies; on the other, application of indicators is often beyond the capability of local farmers. For example, the agricultural sustainability index proposed by Nambiar et al. [16], which aims to measure agricultural sustainability as a function of biophysical, chemical, economic, and social indicators, would require considerable training of government stakeholders and farmers in order to be applied, and such training is rarely available.

Thus, any tool which local farmers or regional decision makers may use to support their decision making must be as simple as possible. Agroecosystems (like any other ecosystem) are too complex to be precisely measured and evaluated [17]. We agree with the view of Darnhofer et al. [18] in favour of developing less precise rules of thumb which may be used by farmers as well as to guide local land-use decisions toward a more environmentally friendly system of agriculture.

This index may provide farmers and others involved with a tool to evaluate the sustainability of their management of crop land which is oriented toward diminishing soil damage and conserving soil fertility. The index, exclusively based on terms which describe the environmental conditions of the crop system, is accessible to most farmers, and may be calculated using a simple internet application. In this paper we present a simple, easy-to-use index in order to evaluate the ecological state of farms in south-eastern Mexico. We applied the indicator system to 52 crop production systems in south-eastern Mexico and compared the results of the indicator system with expert opinions.

Human knowledge of how to efficiently and sustainably manage complex systems (including agricultural systems) is incomplete and much of what is thought to be known about this topic is actually incorrect. Yet, decisions must be made by policy makers, agricultural extension agents, and farmers despite uncertainty and knowledge gaps [19]. Therefore, tools to support local decision-makers must be flexible, should not enter into too much detail or precision, and should allow for an adaptive strategy which promotes "learning through management" [19]. Consequently, our rationale for developing an index which aids farmers in making environmentally friendly land-use decisions is based on basic, simplified ecological concepts, i.e. the presence of trees, since trees within an agroecosystem enhance soil microclimate in terms of radiation partitioning (shading), evapotranspiration partitioning, and rain interception/redistribution [20]. These factors all help to retain soil moisture. Branches, bark, roots, and living and dead leaf surfaces provide shelter [21] for soil micro-, meso-, and macro-invertebrates. Tree cover for instance, enhances above- and below-ground diversity, serving to support agricultural sustainability [22].

Materials and Methods

1. Rationale of index composition

We define conventional agriculture as a cropping system, typically promoted by government development programs, that is "capital-intensive, large-scale, highly mechanized agriculture with monocropping and extensive use of synthetic fertilizers, and pesticides" [23], [24]. Furthermore, we acknowledge that farming systems are sustainable only if "they minimize the use of external inputs and maximize the use of internal inputs already present on the farm" [25], [26]. The strategy most frequently linked to sustainability is reduction or elimination of agrochemicals, particularly chemical fertilizers and pesticides [25], [27], [28], [29], [30], [31]. Another key to sustained productivity of agricultural systems is the maintenance of soil functions, such as organic matter and nutrient cycling [32], based on organic inputs [33], above-and below-ground biodiversity [22], and diversifying crop systems with nitrogen-fixing legumes [34]. The principal role of the index we propose is to characterize methods of tillage, external inputs, and crop structure.

2. Primary indicators

We chose 12 field variables as primary indicators related to the above mentioned aspects of ecologically sound agricultural land use based on farmer's practices. These are easy to evaluate in the field and characterize plot structure (primary indicators: tree cover, tree density, tree diversity), crop structure and crop conditions (primary indicators: crop type, crop rotation, crop density, crop colour), tillage (primary indicators: type of tillage, timing of tillage), the use of fertilizers (chemical versus organic) and pesticide application.

2.1 Tree cover. Tree cover is defined as the canopy of trees, measured in the field, and recorded as percentage classes of tree cover in three height classes (trees >15 m, 10–15 m, <10 m). Thus, this variable characterized one aspect of agroecosystem management: the farmers' decision to maintain the canopy of the trees in his or her agroecosystem.

2.2 Tree density. Tree density is defined as the number of trees per area. To measure this, we distinguished three categories of tree density: high density (abundant), medium density, and low density (isolated or no trees). A high number of trees per area guarantee carbon sequestration [35] while a stable microclimate is maintained. This variable, measured in the field, is one the variables that characterize the effect of trees in the agroecosystem.

2.3 Tree diversity. This variable was measured in the field by counting the number of trees species within the plot. In agroecosystems, biodiversity may; (i) contribute to constant biomass production and reduce the risk of crop failure in unpredictable environments, (ii) restore disturbed ecosystem services such as water and nutrient cycling, and (iii) reduce risks of pests and diseases through enhanced biological control or direct pest control [36], [20], [22].

2.4 Crop type. This variable indicates whether the crop is annual, seasonal, or perennial. We obtained this information by observing the type of crop. Annual crops in general have higher environmental impacts, ie: greenhouse emissions, and nutrient leaching, than perennial crops [37].

2.5 Crop rotation. In sustainable farm systems leguminous crops are increasingly used in crop rotations as a source of nutrients, particularly nitrogen for crop growth [38], [39], nitrogen-fixing legumes, contribute to maintaining biodiversity above and in the soil, contribute nitrogen to the soil/plant system, and help avoid the build-up of pest populations [34]. In this study, we asked farmers whether they planted another crop before planting the main crop and whether they practice crop rotation, as crop rotation may assist with weed and pest control [40], [34]. According to Bellon [41], an activity which leads into the maintenance or increase of renewable resources in agroecosystems, is considered as an ecological technology. This variable helped us to characterize the technology used in the agroecosystem.

2.6 Crop density. Crop density is defined as the number of plants (individuals) per area. Three categories were recorded in the field: abundant (high density: 3,000 plants/ha), medium density (1000–1600 plants/ha), and sparse (<1000 plants/ha). This variable also indicated the level of technology applied to the crop, as less intensive techniques typically yield lower densities [42].

2.7 Crop colour. The colour of a crop indicates the nutritional status of the plants; green plants generally have sufficient nutrients, while yellow plants lack nitrogen [43].

2.8 Type of tillage. Type of tillage was categorized into no tillage, manual tillage, and mechanical tillage using machinery, the latter of which generally indicates high disturbance of the soil surface and rapid loss of soil organic carbon and other nutrients

[44]. In the field, we asked the farmers how they prepared the land for crop planting.

2.9 Frequency of tillage. This variable indicates frequency of tillage - every year or every 2 years. With this information, it was possible to estimate the frequency of soil disturbance due to tillage.

2.10 Chemical input. Searching for sustainable productions, it is recommended no or low or use of inorganic fertilizers and pesticides [45], [46]. Long term use of some pesticides, as glifosate can decrease earthworm species number, density and biomass [47], [48]. This information allowed us to estimate the amount of chemical fertilizers and pesticides applied within a given area. These variables (chemical fertilizers and pesticides) are included in the indicator for chemical disturbance.

2.11 Green manure. Is defined as the presence or absence of leguminous crops mixed with the principal crop, generally used to increase total soil nitrogen content. Green manures should always be intercropped, as it has been proven that growing legumes with cereal crops decreases N_20 emissions [49], and therefore is a sustainable, environmentally friendly practice. Examples of green manure use have been observed in traditional Mesoamerican cultures, for example, intercropping beans, as well as other edible plants, within the *milpa* or traditional maize cropping system [50].

3. Index development

Primary indicators were hierarchically aggregated into higher levels, forming intermediate variables, which in turn are structured into a single index that evaluates the ecological condition of a given plot on a scale from 0 (worst) to 1 (best). An index close to zero either would mean that a more environmentally friendly farming techniques need to be implemented or that the plot should be subjected to a fundamental change in land-use, e.g., reforestation, in order to return to a more ecologically sound state. An index close to one, on the other hand, would indicate an ecologically sound land-use. Methods applied in indicator aggregation were (i) simple mathematical operations, ii) sets of if-then rules, and (iii) sets of if -then rules combined with fuzzy logic.

3.1. Aggregation through mathematical operations. Mathematical operations include calculating averages, weighed averages, minimum values, and maximum values. For example, if primary indicator A has the value a and primary indicator B has the value b, then the value x of the intermediate variable X is determined as $x = (a+b)/2$, $x = (w_1*a+w_2*b)$, where $w1$ and $w2$ are weights, or $x = \min(a, b)$ or $\max(a, b)$.

3.2. Aggregation by IF-THEN rules. If primary indicator A can have the discrete values a_1, a_2, and a_3, and primary indicator B can have the discrete values b_1, b_2, and b_3, then the value x of the intermediate variable X is determined by a set of nine (number of levels of A x number of levels of B) rules. For example, IF $A = a_1$ and $B = b_1$ THEN $X = x_1$.

3.3. Aggregation by IF-THEN rules and fuzzy logic. We used small fuzzy rule-based models for aggregation in the case of non-linear interactions among indicators using a continuous numerical scale or an ordinal scale with a large number of possible values. In classic set theory, an object can either be a member (membership = 1) or not (membership = 0) of a given set. The central idea of fuzzy set theory is that an object may have a partial membership of a set, which consequently may possess all possible values between 0 and 1. The closer an element is to 1, the more it belongs to the set; the closer the element is to 0, the less it belongs to the set. To apply the fuzzy set theory, three steps are involved in calculating the model's output. First, for any observed value of the primary indicators, its corresponding membership value in the fuzzy set domain is calculated (*fuzzification*); second, the memberships of the intermediate variable X are calculated, applying the rules in the fuzzy set theory (*fuzzy inference*); third, the fuzzy results are converted into a discrete numerical output (*defuzzification*; see Wieland 2008 for an introduction to fuzzy models). Fuzzy rule-based models have become popular in ecological modelling [51], [52], and several examples exist of its usefulness in the context of ecosystem evaluation, bioindication, and sustainable management [15], [53], [54]. Here, if both primary indicators A and B can have numerical values from 0 to 1, then the value x of the intermediate variable X is determined by a series of fuzzy set rules representing the linguistic variables "low A", "medium A", "high A", and "low B", "medium B", and "high B", as well as the output "low X", "medium X", and "high X". The value of the intermediate variable X is determined by nine levels of A x the number of levels of B; for example: If A = low and B = low Then X = [low, medium, or high].

To maintain the number of rules as well as their complexity as low as possible, we aggregated only two variables at a time. A simple example shows the reasoning behind this decision; if there are three primary variables, A, B, and C with three categories for each variable, a single rule node requires 3*3*3 = 27 If-Then rules of the type "If $A = a_1$, a_2, or a_3 and If $B = b_1$, b_2, or b_3 and if $C = c_1$, c_2, or c_3 Then…", whereas if an additional intermediate variable X is introduced to the model, only 18 rules are needed: 3*3 = 9 rules to aggregate A and B to X (If $A = a_1$, a_2, or a_3 and If $B = b_1$, b_2, or b_3), and 3*3 = 9 rules to aggregate X and C (If $X = x_1$, x_2, or x_3 and If $C = c_1$, c_2, or c_3).

4. Study area and application of the index

The state of Tabasco in south-eastern Mexico is characterized by a humid tropical climate with a mean annual rainfall between 1200 and 4000 mm and a mean annual temperature of 27°C [55]. Predominant soils are Gleysols and Fluvisols over alluvial sediments in the plains, Vertisols, Cambisols, Luvisols, and Acrisols over Miocene or Oligocene sediments, and Leptosols and Regosols over limestone mountains [9], [56]. We chose the municipalities Balancán and Tenosique in western Tabasco (17°81′50′′–18°81′00′′ N, 91°80′10′′–91°84′60′′ W) as a study area (Figure 1), we worked with private, *ejidal* (multipurpose land, where owners can or cannot sell the property according to their legal status in the National Agrarian File), and communal lands (coordinates of each plot are shown in Table 1). The region is mainly a plain towards the North (67% of area has an elevation <20 m. a. s. l.) with hills (29%, 20–200 m. a. s. l.), and mountains (4%, max. 640 m. a. s. l.) in the South, comprising a total area of 5474 km^2. These municipalities have undergone a high degree of land use change over the past 40 years. Until the early 1970 s, this region was still covered by lowland rain forest. The principal form of land use is pastureland, and covers 60% of the land [57]. An additional 30% is cropland, mainly cultivated under small-medium holder systems with seasonal conventional agriculture using high levels of agrochemical inputs (Table 2 & 3). Common crops are maize (*Zea mays*), a variety of hot peppers (*Capsicum* sp), cucumbers (*Cucurbita argyrosperma*), watermelon (*Citrullus lanatus*), perennial fruit crops such as papaya (*Carica papaya*), and biannual crops such as sugar cane (*Saccharum officinarum*) [58].

We chose 52 farms in the study area (Table 1), and selected those farms whose main economic activity (100%) is agriculture and chose one agricultural plot from each farm for evaluation. There were annual and biannual crops with or without trees (Table 2). Average plot size was 32.4±55.1 ha, and the average time that the plot had been used for a given crop system was 2.5±3.0 years.

Table 1. Characterization of plots.

Plot #	Municip	Lat	Long	Altitud (mls)	Plot size (ha)	Type of property	Original-vegetation
1	Bal	2007880	652445	43	18	Ejidal	A
2	Bal	1964368	642505	10	0.5	Private	TRF
3	Bal	1970936	647701	11	160	Private	TRF
4	Bal	1957459	667496	14	2	Ejidal	TRF
5	Bal	1964107	660656	14	7	Private	PT
6	Bal	1960297	658989	22	15	Private	TRF
7	Bal	1976300	676424	40	30	Private	TRF
8	Bal	1990326	665947	40	28	Private	TRF
9	Bal	1964590	667264	11	3	Private	RV
10	Bal	1972260	672382	28	2.5	Ejidal	TRF
11	Bal	1950339	666409	49	40	Private	TRF
12	Bal	1966112	682994	40	80	Private	PT
13	Ten	1931754	704474	65	14	Private	TRF
14	Ten	1912951	690381	130	2	Communal	TRF
15	Ten	1932644	665521	15	3	Ejidal	TRF
16	Ten	1926163	678608	56	68	Private	TRF
17	Ten	1933663	674227	59	35	Private	A
18	Ten	1926570	669486	35	8.5	Ejidal	TRF
19	Ten	1942431	663885	21	1.5	Ejidal	TRF
20	Ten	1942068	671156	31	7.25	Ejidal	TRF
21	Ten	1943167	650736	82	14	Ejidal	A
22	Ten	1923838	669841	118	0. 25	Private	TRF
23	Bal	1973301	695548	105	40	Ejidal	TRF
24	Bal	1982392	701383	74	20	Ejidal	A
25	Bal	1965354	709053	44	11	Ejidal	A
26	Bal	1960496	706464	51	44	Ejidal	TRF
27	Bal	1942484	709731	66	10	Ejidal	A
28	Ten	1944148	673673	16	320	Ejidal	TRF
29	Ten	1930350	659650	29	20	Private	TRF
30	Ten	1924549	656910	41	53	Ejidal	TRF
31	Ten	1926226	652978	65	3	Ejidal	TRF
32	Bal	1955194	688843	45	44	Private	TRF
33	Bal	1960980	679265	49	29	Ejidal	TRF
34	Bal	1942552	696222	48	10	Private	TRF
35	Ten	1938120	686470	49	5	Private	TRF

Table 1. Cont.

Plot #	Municip	Lat	Long	Altitud (mls)	Plot size (ha)	Type of property	Original-vegetation
36	Bal	1968750	695595	60	2	Private	A
37	Bal	1977546	699417	53	200	Ejidal	TRF
38	Bal	1949770	708854	53	23	Ejidal	TRF
39	Bal	1951216	697890	46	2	Ejidal	TRF
40	Ten	1931041	667544	31	17	Private	TRF
41	Ten	1927447	664223	3	3	Ejidal	TRF
42	Ten	1928226	664551	31	3	Ejidal	TRF
43	Ten	1930579	691785	54	15	Communal	TRF
44	Ten	1931219	677895	59	63	Private	TRF
45	Ten	1931525	676500	60	6.75	Ejidal	TRF
46	Ten	1908445	675040	140	40	Ejidal	TRF
47	Ten	1910875	669750	222	45	Ejidal	TRF
48	Ten	1924222	704702	42	30	Ejidal	A
49	Ten	1935076	712227	88	20	Ejidal	TRF
50	Ten	1914876	686285	113	4	Ejidal	TRF
51	Ten	1911365	684878	389	22	Ejidal	TRF
52	Ten	1911503	698606	139	10	Ejidal	TRF

Municip: municipality, Bal: Balancan, Ten: Tenosique. TRF: tropical rain forest, PA: pasture, A: acahual (secondary vegetation). Ejidal: multipurpose land, where owners can or cannot sell the property according to their legal status in the National Agrarian File.

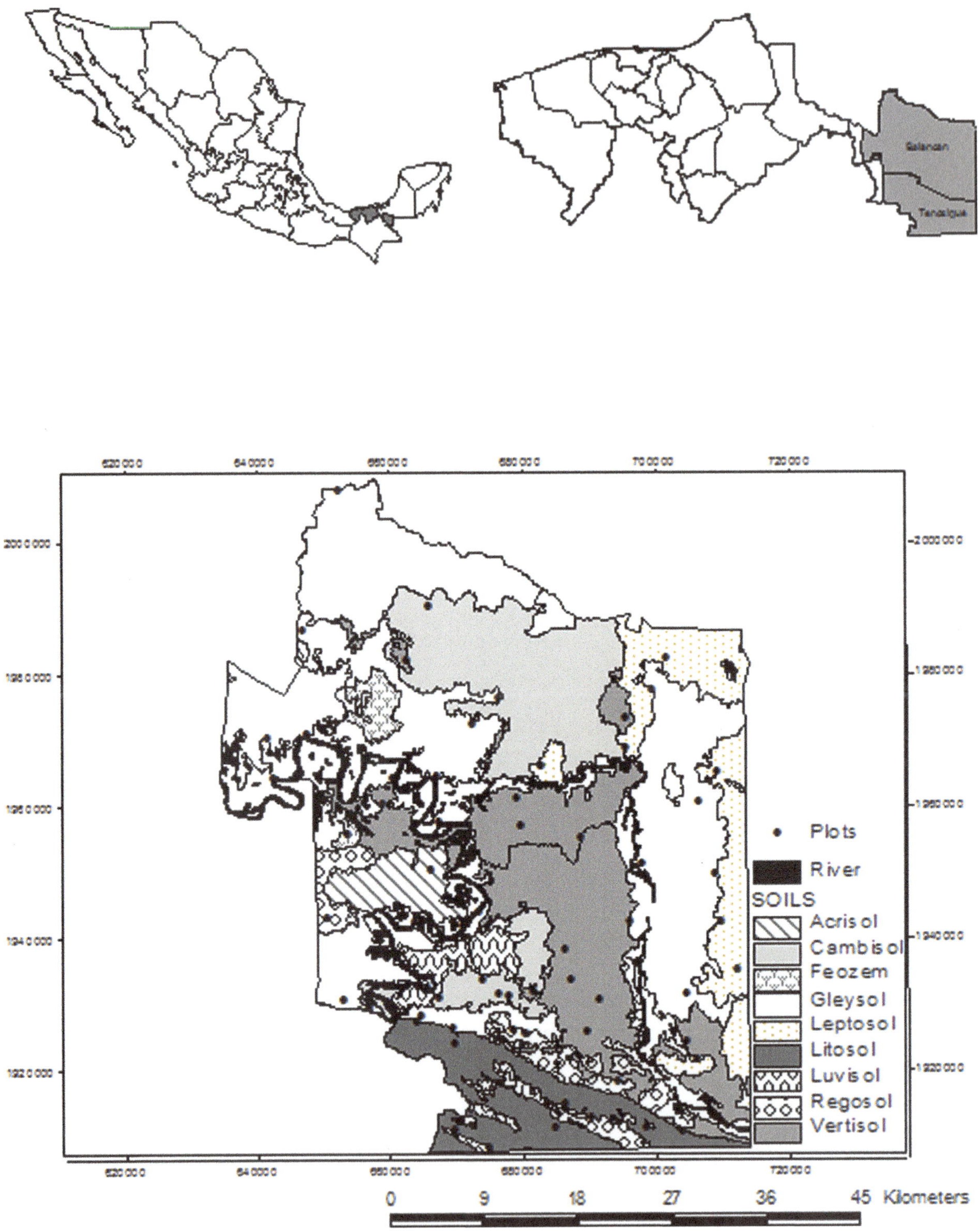

Figure 1. Distribution of 52 evaluated agricultural plots in tropical South-East Mexico.

We questioned farmers as to frequency, amount and type of chemical fertilizers and pesticides applied per area of cropland. After their verbal consent, farmers gave us their complete name and signed next to their name on the record sheet, validating all the information. Due to the fact that we only asked about the use of fertilizers and management of the land, the procedure approval from the Ethics Committee was not required.

Between March and October 2004, for each plot, the values of the primary indicators were determined and the index of ecological condition was calculated. Prior to index calculation, the plots were also evaluated by experts (2 scientists, each with

Table 2. Crop characterization.

Plot	Cycle	Main seasonal or perennial crop	Trees
1	s-s	*Zm, Pv, Cs, Cl*	*Tr, Pa, Cpa*
2	s-s	*Zm, Pv, Cs, Cl*	*Sh, Dg*
3	w-s	*Sv*	*Sa*
4	a-w	*Zm, Pv, Cpa*	*0*
5	Y	*Zm, Pv, Cl, Cm*	*0*
6	Y	*Zm*	*Cpa, Eg, Tr*
7	s-a	Ca	Tr
8	Y	*Cp*	*Cp*
9	Y	*Zm*	*0*
10	Y	*Zm*	*0*
11	w-s	*Zm, Pv, Cl*	*Sm, Cpa*
12	Y	*Co*	*Co*
13	s-s	*Zm*	*0*
14	s-s	*Le, Se*	*0*
		Cpa, Zm, Pv	
15	s-s	*Cl, Ta, Mp,*	*Fruit trees*
16	Y	*Jj*	*Gs, Hc, Cc, Tr*
17	Y	*So*	*0*
18	Y	*So*	*0*
19	Y	*So*	*0*
20	Y	*Zm*	*Mi*
21	s-s	*Zm*	*0*
22	Y	*Csi, Cli, Js*	*Co, Pa, Bc, Cn, Mi*
23	s-s	*Zm, Cpa*	*Cpe*
24	s-s	*Zm*	*Cpa*
25	s-s	*Zm*	*0*
26	s-s	*Zm, Cpa*	*0*
27	s-s	*Zm, Ca*	*Sm*
28	Y	*So*	*0*
29	s-s	*Zm, Pv, Cs, Cl*	*0*
30	Y	*Zm, Pv, Cs, Cl, Ta*	*0*
31	Y	*Zm, Pv, Cs*	*0*
		Cl, Os, Le	
32	s-s	*Zm, Pv, Cpa, Cl, Os*	*Tr*
33	Y	*Zm, Pv*	*Sm, Bg*
34	s-s	*Zm, Pv, Os*	*Cpa*
			Tr
35	Y	*Zm, Cl*	*Tr*
36	Y	*Zm, Pv, Cpa, Cl, Ca*	*0*
37	s-s	*Zm, Pv, Ca*	*Sm, Pa, Sh*
38	Y	*Zm, Pv, Cpa, Cl, Ta*	*Co, Sh*
39	Y	*Zm, Pv, Cpa, Cl, Ca*	
		Cs, Cm	
40	Y	*So*	*Tr, Cpe*
			Co, D a
41	Y	*So*	*Tr, Cpe*
			Co, D a
42	Y	*So*	*Tr, Cpe*
			Co, D a

Table 2. Cont.

Plot	Cycle	Main seasonal or perennial crop	Trees
43	Y	*So*	*Tr, Cpe*
			Co, Da
44	Y	*So*	*Tr, Cpe*
			Co, Da
45	Y	*So*	*Tr, Cpe*
			Co, D a
46	Y	*Zm, Pv, Cpa, Me*	*Sm*
47	Y	*Zm, Pv, Me, Ib*	*Bc, Mi, Dg, Cs, Ll*
48	Y	*Zm, Pv, Cpa, Cl*	*0*
49	Y	*Zm, Pv, Ca*	*Tr, Gs*
50	Y	*Zm, Pv, Cpa, Cl*	*Sm, Co, Ma*
		Cs, Ib, Me	0
51	Y	*Zm, Pv, Cpa, Cs*	*Fruit trees*
52	Y	*Zm, Pv, Cpa, Cl, Cs*	*Co, Sh, Bc*
		Mp, Ca, Me	0

Cycle: s-s: spring-summer, w-s: winter-summer, s-a: summer-autumn, Y: all the year. **Main seasonal or perennial crop**: Ca: *Cucurbita argyrosperma* (cushaw pumpkin); Cp: *Carica papaya* (papaya); Cl: *Citrullus lanatus* (watermelon); Cli: *Citrus limon* (lemon); Cm: *Cucumis melo* (muskmelon); Cpa: *Cucurbita pepo* (squash); Cs: *Capsicum* sp. (pepper); Csi: *Citrus sinensis* (orange tree); Ib: *Ipomoea batatas* (sweet potato); Jj: *Jarthropha jurcas* (oil palm); Js: *Jobo spondia* (plum); Le: *Lycopersicon esculentum* (tomato); Me: *Manihot esculenta* (cassava); Mp: *Musa paradisiaca* (banana); Os: *Oryza sativa* (rice); Pv: *Phaseolus vulgaris* (bean), So: *Saccharum officinarum* (sugarcane); Se: *Sechium edule* (pear squash); Sv: *Sorghum vulgare* (milo); Ta: *Triticum aestivum* (wheat); Zm: *Zea mays* (maize). **Trees**: *Bc: Byrsonima crassifolia* (nanche); *Cc: Crescentia cujete* (calabash tree); *Co: Cedrela odorata* (Mexican cedar); *Cpa: Carludovica palmate* (toquilla palm); *Cpe: Ceiba pentandra* (kapok); Cn: *Cocos nucifera* (Coconut); *Dg: Dialium guianense* (wild tamarind); Eg: *Eucalyptus grandis* (Eucalyptus); *Gs: Gliricidia sepium* (Cocoite, gliricidia, cacao de nance); *Ll: Leucaena leucocephala* (white lead tree, jumbay); *Ma: Mammea americana* (mamey apple); *Mi: Mangifera indica* (mango); *Pa: Persea americana* (avocado); *Sa: Sterculia apetala* (camoruco, manduvi tree); *Sh: Swietenia humilis* (small mahagoni); *Sm: Spondias mombin* (Yellow plum, bai makok); *Tr: Tabebuia rosea* (savannah oak, macuilis).

over 15 years of experience in agroecology) on a scale from 0 (poor condition) to 1 (good condition) in order to test the correlation between the experts' opinions and the index.

Data was normalized for carrying out multiple regression with plot size, altitude of plot site, or previous vegetation cover and the ecological index.

Finally we located the index in this public web address: http://201.116.78.102/~modelo/Index.html, where farmers in the near future can introduce independent data sets and evaluate or monitor their own agro ecosystems.

Results

1. Index structure

Ten nodes were used to aggregate primary indicators to the final index of ecological conditions of agricultural systems (Figure 2). Six of these used simple mathematical operations; two were based on rule sets, and two on fuzzy rule-based models (Table 4).

2. Index application

2.1. General characterization of sampled plots. A total of 67% of the plots were cultivated after cutting of primary lowland forest, 24% after cutting secondary forests of various ages, and 1.5% after cutting riparian vegetation, the rest conversion from pastureland (livestock production) to crop system. On 63% of the farms, maize, beans and pumpkins were cultivated, 13% sugar cane, 12% watermelon, 9% rice and 3% pepper. On 68.5% of the plots, trees were scattered among the crops. Conventional tillage was used on 55.5% of the plots, and pesticides were used on 79.4%, green manure was used on 21% of the plots and chemical fertilizers on 67%. All farmers understood the meaning of the variables evaluated for their plots (for a complete list of plots, see Table 5).

2.2. Plot evaluation with the index. Ecological condition of the plots ranges from 0.0 to 0.8125 (Figure 3; Table 5). One plot which was intercropped with timber and fruit trees presented the highest value. This site was characterized by an absence of agrochemical use, use of green manure, presence of annual crop rotation, and high tree diversity. We found 18 plots with index values of 0–0.25, 7 plots with values of 0.3–0.5, and 26 plots with values of 0.5–0.7. Thus, the majority of the plots evaluated in this study had an intermediate index value. 50% of the plots with this intermediate index are *ejidal* property, 38% private land, and 12% is under another type of ownership (smallholder or communal). 80% of these intermediate plots were lowland rain forest before being converted into agricultural land, 12% were secondary vegetation, and 8% were pastureland (Table 1). One might believe that a prior forest condition implies a more environmentally friendly agroecosystem that allows for preserving a more diverse system. However, plots with low index values were also lowland rain forest before being turned into agricultural land (Table 5). In this case, the land managers or owners decided to deforest the land to subsequently plant annual crops. 13 plots were larger than 40 ha (Table 1). All of these were previously covered with lowland rain forest; the ecological index ranges from 0.5–0.56 for 46% of these plots, and another 23% have an ecological index of 0.37–0.43. 10 plots fell under the smallest size category (0.25–3 ha) and had an ecological index of 0.5–0.68 (4 plots), 0 to 0.18 (5 plots), and 0.39 (1 plot).

Carrying out multiple regression with normalized data, we did not observe significant correlations between plot size, altitude of

Table 3. Crop land preparation and inputs characterization.

Plot	Tillage	Chemical Fertilization	Pesticides Use	Gm	Oa
1	Ma & Me	NPK 17:17:17 & U	Endosulfan**	0	0
2	Ma	0	Chlorpyrifos*	0	0
3	Me	Pholiar	0	0	c s m
4	Ma	0	0	0	0
5	Ma & Me	NPK 17:17:17	Chlorpyrifos**	Y	Y
6	Ma & Me	NPK 17:17:17	Chlorpyrifos*	0	0
7	Me	0	ID	0	0
8	Me	ID	ID	Y	0
9	Ma & Me	0	Carbofuran*	0	0
10	Me	NPK 18-46-0	ID	0	0
11	Me	NPK 18- 46- 0, U & P	ID	Y	0
12	Me	NPK 18-46-0	ID	0	0
13	Me	NPK 17:17:17 & U	Chlorothalonil**	0	0
			Chlorpyrifos*		
14	Ma	0	Endosulfan*	0	0
15	Me	0	0	Y	0
16	Ma	Urea	0	0	0
17	Me	NPK 19-19-19	ID	0	0
18	Me	NPK 19-19-19	ID	0	0
19	Me	Urea	ID	0	bg
20	Me	0	Methilic*	0	0
21	Me	NPK 17:17:17 & U	Zeta**	0	0
22	Ma	0	0	0	0
23	Me	NPK 17:17:17 & U	Chlorpyrifos**	0	lcf
24	Me	Urea & P	Zeta*	Y	lcf
25	Me	Urea	(2,4-D AMINA)*	Y	0
26	Me	NPK 17:17:17 & U	(Z)-(1R,3R)**	0	0
27	Me	0	Zeta*	N	0
28	Me	NPK 17:17:17	(RS)*	0	0
29	Me	NPK 17:17:17 & U	Chlorpyrifos**	0	0
30	Ma	0	Chlorpyrifos**	Y	0
31	Me	NPK 17:17:17 & U	Chlorpyrifos*	0	0
32	Me	0	0	Y	0
33	Me	0	0	0	0
34	Me	0	0	0	0
35	Me	NPK 17:17:17 & U	0	0	0
36	Me	NPK 17:17:17 & U	Chlorpyrifos*	Y	0
37	Me	0	Urea	0	0
38	Me	NPK 17:17:17 & U	Carbofuran*	Y	0
39	Me	NPK 17:17:17 & U	Chlorpyrifos**	0	0
40	Me	NPK 17:17:17	(Z)-(1R,3R)*	0	0
41	Me	NPK 17:17:17 & U	(Z)-(1R,3R)*	0	0
42	Me	NPK 17:17:17 & U	(Z)-(1R,3R)*	0	0
43	Me	NPK 17:17:17 & U	(Z)-(1R,3R)*	0	0
44	Me	NPK 17:17:17 & U	(Z)-(1R,3R)*	0	0
45	Me	NPK 17:17:17 & U	(Z)-(1R,3R)*	0	0
46	Ma	0	0	Y	0
47	Me	U	0	0	0
48	Ma &Sb	Urea	Chlorpyrifos**	0	0

Table 3. Cont.

Plot	Tillage	Chemical Fertilization	Pesticides Use	Gm	Oa
49	Me	Urea & P	Zeta*	0	0
50	Ma	Urea	Endosulfan*	0	0
51	Ma	0	ID	0	0
52	Me	NPK 17:17:17 & U	Chlorpyrifos*	0	0

Tillage: Ma: Manual; Me: Mechanized. **Pesticides**: Endosulfan: 6,7,8,9,10,10-Hexachlor- 1,5,5a,6,9,9a-Hexahidro-6,9-metane-2,4,3-benzodioxatiepin-3-oxide; Chlorpyrifos: 0,0-dimetil 0-(3,5,6-trichlore-2-piridinil) fosforotioate (33.8%), Permetrine: 3-fenoxibenzil (1RS)-cis, trans-3-(2,2 diclorovinil)-2,2 dimetil ciclopropane-carboxilate (4.8%); Carbofuran: 2,3 Dihidro-2,2-dimetil-7-benzofuranil metil carbamate; Chlorothalonil: Tetrachloroisoftalonitrile; Zeta: Zeta-cipermetrine a-ciano-3-(fenoxifenil) metil (±) cis-trans; (Z)-(1R,3R): (Z)-(1R,3R)-3-(2-chloro-3,3,3-trifluoroprop-1-enil)-2,2-dimetilcichlopropanecarboxilate de (S)-α-ciano-3-fenoxibencile & (Z)-(1S,3S)-3-(2-chloro-3,3,3-trifluoroprop-1-enil)-2,2-dimetilcichlopropanecarboxilate de (R)-α-ciano-3-fenoxibencile; (RS): (RS)- alfa- ciano-3-fenoxybencil(1RS)-cis-trans-3-(2,2-dichlorvinil)- 1,1-dimetilcichlopropanecasrboxilate. * Once per year, ** 2 per year. **Chemical Fertilization**: U: Urea, P: Phosphorus. **Gm**: Green manure, Y: Yes, 0: null application. **Oa**: Organic amendments: csm: cow, sheep manure, bg: burned grass, lcf: last crop fallow, 0: null application.

plot site, or previous vegetation cover and the ecological index (Kendall's Tau, T = 0.016 p = 0.98, T = 0.11 p = 0.90, T = −0.31 p = 0.75, respectively). Therefore, in this study, it seems that neither size nor location of the plot determines the type of plot management; rather, plot management is likely determined by government development programs and traditional farming techniques.

3. Correlation between index and expert opinion

We obtained a Pearson correlation coefficient of $r^2 = 0.61$ ($p = 2.4\text{e-}06$, $d.f. = 49$) between the index and the values determined by independent experts, indicating a satisfactory correspondence (Figure 4). However, the experts systematically awarded higher scores to the plots than did the index. Moreover, they suggested to include additional variables to the index which would yield better information regarding (i) type of organic inputs to the crop system, (ii) types of pest and disease control used, (iii) number of native plant species among the crop, (iv) origin of crop seeds, (v) vegetation surrounding the crop, (vi) presence of vertebrate fauna, and (vii) diversity of soil macroinvertebrates.

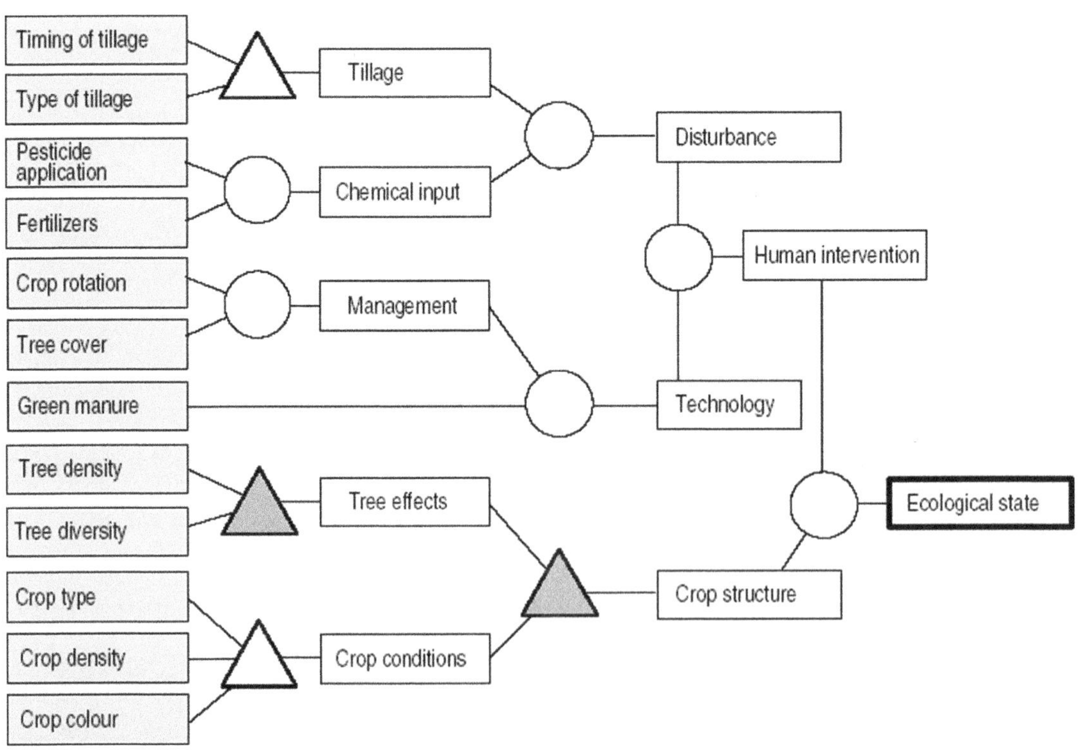

Figure 2. Structure of the Index of ecological condition of tropical agroecosystems. Primary indicators are shaded in grey. Circles represent simple mathematic algorithms, white triangles represent rule sets, and grey triangles represent rule sets, and grey triangles represent rule sets based on fuzzy logic. This index was presented together with other indexes within a frame of Indicators of environmentally sound land use in the humid tropics [15].

Table 4. Description of the measurement levels of the variables.

	Groupal variables	Field data variables	Best	Intermediate			Worst
				5 levels			
			1	0.75	0.5	0.25	0
	Management	Tree cover	all year	presence polyculture	present	rare	null
	Tree effect	Tree density	high	medium	low	isolated	null
				4 levels			
			1	0.66	0.33		0
		Tree diversity	>4 species	4 species	2 species		1 species
Crop structure	Crop conditions	Crop type	perennial polyculture	perennial monoculture	biannual		annual
				3 levels			
			1	0.33			0
	Plough (tillage)	Tillage time	null	frequent			constant
		Tillage form	null	manual			technical
Disturbance	Chemical inputs	Pesticides	null	frequent			constant
		Fertilizers	null	frequent			constant
	Management	Crop rotation	constant	frequent			null
Technology		Green manure	constant	frequent			null
				2 levels			
			1				0
	Crop conditions	Crop density	abundant				disperse
		Crop coleur	green				yellow

Table 5. Normalized Ecological Index for plots evaluated (52).

Plot	TT	TF	P	F	GM	RC	CA	DA	DivA	CT	CD	Ccol	Ev	Index
1	n	m	c	c	n	n	>1	l	>4	pm	a	g	0.24	0.56
2	n	n	c	n	n	n	>1	i	>4	a	a	g	0.35	0.5
3	c	m	n	c	c	n	>1	i	>4	a	a	g	0.26	0.56
4	n	n	n	n	n	n	n	n	1	a	a	g	0.01	0.37
5	n	n	c	c	c	c	>1	i	>4	a	a	g	0.09	0.62
6	n	m	c	c	n	n	>1	l	>4	a	a	g	0.22	0.5
7	c	t	c	n	n	n	>1	i	>4	a	d	b	0	0.15
8	c	t	c	c	c	c	n	n	1	pp	d	g	0.1	0.12
9	n	t	c	n	n	n	n	n	1	a	a	g	0.09	0.25
10	c	t	c	c	n	n	n	n	1	a	d	b	0	0
11	c	t	c	c	n	c	>1	i	>4	pm	ID	ID	0.03	0.21
12	c	t	c	c	n	n	>1	h	>4	ba	a	g	0.2	0.5
13	n	t	c	c	n	n	>1	i	>4	a	a	g	0.03	0.5
14	n	n	c	n	n	n	n	n	1	pp	a	g	0.29	0.37
15	n	t	n	n	n	c	>1	h	>4	pp	a	g	0.2	0.81
16	n	n	c	c	n	n	>1	l	>4	ba	a	g	0.25	0.5
17	c	t	c	c	n	n	n	n	1	a	a	g	0	0.25
18	c	t	c	c	n	n	n	n	1	a	d	g	0	0
19	c	t	c	c	c	n	n	n	1	a	d	g	0	0.06
20	n	t	c	n	n	n	>1	i	>4	a	a	g	0.1	0.5
21	c	t	c	c	n	n	n	n	1	a	a	g	0	0.25
22	n	n	n	n	n	n	>1	i	>4	pp	d	g	0.33	0.62
23	n	t	c	c	c	n	>1	i	>4	a	a	g	0.15	0.56
24	n	t	c	c	c	c	>1	l	>4	a	a	g	0.2	0.62
25	n	n	c	c	n	c	n	n	1	a	d	g	0.01	0.06
26	n	m	c	c	n	n	n	n	1	a	a	g	0.25	0.25
27	n	t	c	n	n	n	>1	i	>4	a	a	g	0.01	0.5
28	n	t	c	c	n	n	n	n	1	a	a	g	0	0.25
29	c	t	c	c	n	n	n	n	1	pm	a	g	0.01	0.37
30	n	m	c	n	n	c	n	n	1	pm	a	g	0.1	0.43
31	c	t	c	c	n	n	n	n	1	a	a	g	0.01	0.25
32	c	t	n	n	n	c	>1	i	>4	a	d	y	0.09	0.34
33	c	t	n	n	n	n	>1	h	>4	a	a	g	0.15	0.62
34	c	t	n	n	n	n	>1	i	>4	ba	a	g	0.13	0.62
35	c	t	n	c	n	n	>1	l	>4	a	a	g	0.01	0.5

Table 5. Cont.

Plot	TT	TF	P	F	GM	RC	CA	DA	DivA	CT	CD	Ccol	Ev	Index
36	c	t	c	c	n	c	n	n	1	a	a	g	0	0.31
37	c	t	c	n	n	n	>1	i	>4	pm	a	y	0.01	0.15
38	c	t	c	c	n	c	>1	h	>4	pm	a	g	0.2	0.68
39	c	t	c	c	n	n	>1	l	>4	a	a	b	0.06	0.18
40	c	t	c	c	n	n	>1	l	>4	a	a	g	0.19	0.5
41	c	t	c	c	n	n	>1	l	>4	a	a	g	0.19	0.5
42	c	t	c	c	n	n	>1	l	>4	a	a	g	0.17	0.5
43	c	t	c	c	n	n	>1	l	>4	a	a	g	0.2	0.5
44	c	t	c	c	n	n	>1	l	>4	a	a	g	0.2	0.5
45	c	t	c	c	n	n	>1	l	>4	a	a	g	0.2	0.5
46	n	n	n	n	n	c	>1	l	>4	pp	a	b	0.6	0.37
47	n	t	n	c	n	n	>1	l	>4	pp	d	g	0.18	0.5
48	c	n	c	c	n	n	n	n	1	a	a	b	0	0
49	c	t	c	c	n	n	>1	l	>4	a	a	b	0.07	0.18
50	c	n	c	c	n	n	>1	l	>4	pm	a	g	0.2	0.56
51	c	n	c	n	n	n	>1	l	>4	pm	a	g	0.15	0.56
52	c	t	c	c	n	n	>1	l	>4	pm	a	ID	0.24	0.18

TT: Tillage Time: c: >once per year; n: once per year; TF: Tillage Form: n: null, m: manual, t: technical; P. Pesticide use: c: constant, n: null, F: Use of chemical Fertilizers: c: constant, n: null; GM: Green Manure, c: constant, n: null; RC: Crop Rotation: c: constant, n: null; CA: tree cover: >1 year, n: null; DA: tree density: h: high, l: low, i: isolated, n:null; DivA: tree diversity: >4 species, 1 species, CT: crop type: a: annual, ba: biannual, pp: perennial polyculture, pm: perennial monoculture; CD: crop density: a: abundant, d: disperse; Ccol: crop colour: g: green, b: brown, y: yellow; EV: evaluation.

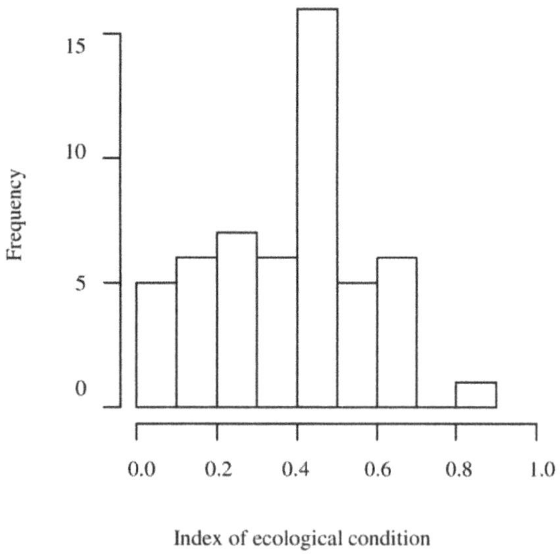

Figure 3. Frequency histogram of values of the index of ecological condition applied to 52 plots in South-East Mexico.

Discussion

Since the Rio Earth Summit, there has been a concerted effort to construct indicators to monitor progress toward sustainable development [59]. Most of these indicators have been developed in Europe (10) and Asia (2) [60]. In Latin America, sustainable indicators have been developed by Astier et al. [61]; this index mainly focuses on subsistence level agriculture, and evaluates the sustainability of a system.

Our index is geared toward small and medium scale producers who principally grow for market and have a fairly large crop area (32.4 ha on average). According to Bockstaller and Girardin [62], indicators must be elaborated according to a scientific approach, and one of the important steps in this elaboration is validation.

Our index was developed according to the consensus of a group of scientists, with knowledge in agroecology, and was validated independently by 2 scientists, each with over15 years of experience in agroecology. The evaluation included 3 important steps: design validation, output validation, and end use validation [62].

In previous studies, only seven indicators have been used to evaluate farm systems: crop diversity, crop succession, pesticide use, nitrogen level, phosphorus level, soil organic matter, and irrigation methods [63]. All of these practices depend on the farmer's decision, and to a large extent they impact the environment.

Our index is based on qualitative and quantitative concrete data and includes most of these 7 indicators, except nitrogen and phosphorus, both of which are observed indirectly via plant health, through the crop colour indicator (according to whether plants have a greenish or yellowish colour); organic matter, which is indirectly characterized by the technology applied in the system (green manure indicator, Figure 2); and irrigation, which in this study was not evaluated, given that all plots evaluated were only used for seasonal rainfed agriculture, according to local rainfall patterns.

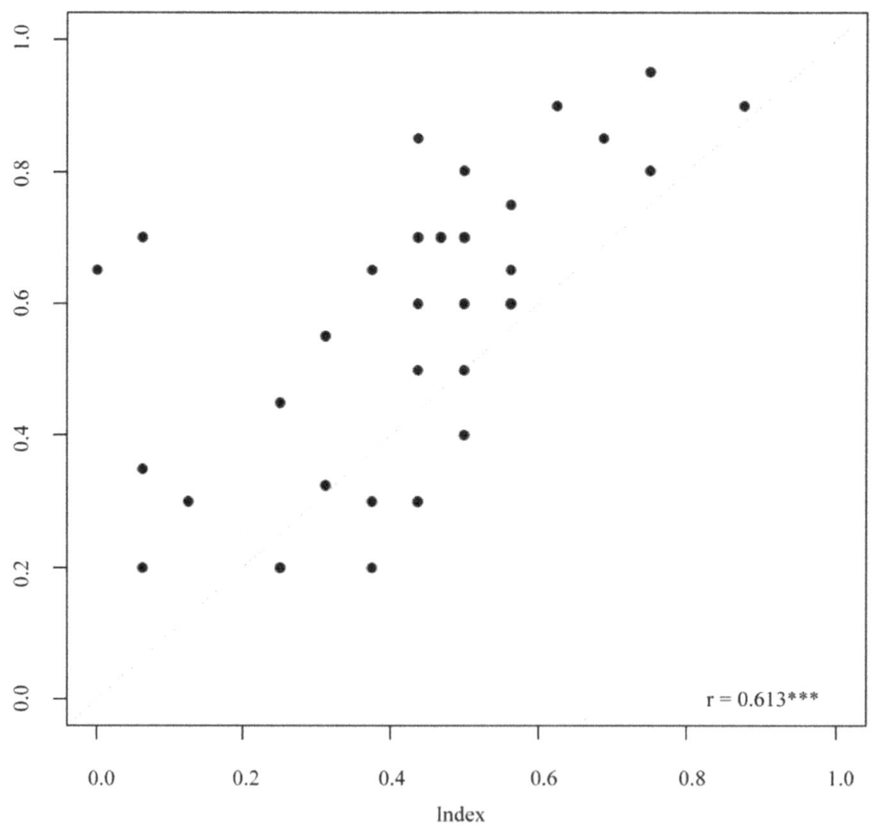

Figure 4. Scatterplot of the values of the index of ecological condition applied to 52 plots in South-East Mexico (x-axis) versus quality values between 0 (worst) and 1 (best) assigned by experts to the same plots (y-axis).

At the farm level there are indicators that evaluate the environmental impact of the agricultural practices and indicators that evaluate the effect of those practices at the local and global level [60].

Our index includes both types of indicators; evaluating agricultural practices: those variables taken in the field: type and frequency of tillage, pesticide application, fertilizer use, crop rotation, tree cover and density, green manure, crop density (see Table 2 & 3). Effect indicators used were disturbance (mainly soil disturbance) measured by tillage (frequency and type) and chemical inputs, technology used, and crop structure (see index, Figure 2). Some existing indexes focus on evaluation or evolution of environmental performance, thus encouraging environmentally sound practices [60], such as crop rotation, organic fertilizers, and no-tillage. Meanwhile, our index identifies the indicator that has the highest environmental impacts in each plot evaluated, with the idea that the farmer could potentially improve these with a given practice, ie. to monitor the soil ecological condition via the use of a soil macroinvertebrates index [64], where the lack of macroinvertebrates informs of a severe negative activity as pollution or conventional tillage.

In developing the variables to be included in our index, we reviewed bibliographic studies and carried out field work obtaining data which we hoped would reflect the negative and environmentally friendly practices commonly used in agro ecosystems of southeastern Mexico. The index can be used by farmers, using the following web address: http://201.116.78.102/~modelo/Index.html.

However, we did not evaluate certain indicators such as water use, and water quality, as did - for example - the index (monitoring tool) of ecological indicators used on a Flemish dairy farm [65]. Nor did we evaluate environmental impacts due to energy consumption [66]. Within a tropical framework, in south-eastern Mexico, the priorities were to identify those practices that were soil perturbing and environment polluting, practices that can be modified by the farmers, by an attitude changing. Nambiar et al. [16], proposed an agricultural sustainability index (ASI) to measure sustainability as a function of biophysical, chemical, economic, and social indicators, our index only measures the ecological state of the agroecosystem, and provides easy to use tools for improving those practices which negatively impact the environment. Van der Werf and Petit [60], state that indicators based on farmer practices cost less in data collection but do not allow for an actual evaluation of environmental impact. In the case of our study, the experts' evaluation correlated satisfactorily with the index, although the index rendered more penalizing scores than did the experts. Some improvements must be made to our index, relating quality and quantity of the applied inputs for instance; the index should specify the kind of manures used and then to evaluate their effect when added to the systems. The consulted experts found important to integrate this information into the index, they also found that the possible relations among different crops and environmental effects of using cow manure, vermicompost, or traditional compost have to be considered. Another variable which the experts suggested should be added to the index is the presence of natural vegetation surrounding the crop. Farmers see advantages of having crops surrounded by secondary forest, diversity in the agricultural area can be increased, ie when different pollinators arrive.

The advantages of using this index is that the common agricultural practices (mechanized land preparation and use of common pesticides ie. Carbofuran, Chlorpyrifos), evaluated in this study as indicators are used throughout the world; over time, through practice, the index may be improved. Farmers and other land owners may realize which of the practices they use are disturbing the environment, due to the fact that these practices generate a value in each of the evaluated indicators. The variables obtained in the field contribute to the information of each of the indicators, and the index is the compendium of all the indicators. Our index doesn't give a sustainability measure, because it does not include socioeconomic indicators of the farms. Further studies are required in order to observe the acceptance of this index by farmers in a regional scale.

Supporting Information

Figure S1 Indicators of environmentally sound land use in the humid tropics [15].

Acknowledgments

We thank Lorena Soto and Guillermo Jimenez for participating in the evaluation of the plots. We thank Simon Hernandez de la Cruz, Alejandra Sepulveda Lozada, Lauritania Ibarra Hernandez, and Marcelo Rodriguez Ricardes, who helped to collect data in the field. We are also thankful to Lorena Reyes for bibliographic support. El Colegio de la Frontera Sur provided infrastructural resources. Language revision was done by Ann Greenberg.

Author Contributions

Conceived and designed the experiments: EH CK SOG BDJ SHD VG. Performed the experiments: EH CK. Analyzed the data: EH CK. Contributed reagents/materials/analysis tools: EH CK SOG BDJ SHD VG. Contributed to the writing of the manuscript: EH CK SOG BDJ VG. Figures elaboration: CK SOG.

References

1. Houghton RA (1994) The world wide extent of land-use change. BioScience 44: 305–313.
2. Tilman D, Lehman C (2001) Human-cause environmental change: impacts on plant diversity and evolution. Proceedings of the National Academy of Sciences 98: 5433–5440.
3. INEGI (2008) Cuaderno Estadístico Municipal de Tenosique. Villahermosa, Tabasco: Gobierno del Estado de Tabasco.
4. Grande D, de Leon F, Nahed J, Perez-Gil F (2010) Importance and Function of Scattered Trees in Pastures in the Sierra Region of Tabasco, Mexico. Research Journal of Biological Sciences 5: 75–87.
5. Geissen V, Morales-Guzman G (2006) Fertility of tropical soils under different land use systems-a case study of soils in Tabasco, Mexico. Appl Soil Ecol 841: 1–10.
6. Melgar C, Geissen V, Cram S, Sokolov M, Bastidas P, et al. (2008) Pollutants in drainage channels following long-term application of Mancozeb to banana plantations in southeastern Mexico. Journal of Plant Nutrition and Soil Science 171: 597–604.
7. Aryal DR, Geissen V, Ponce-Mendoza A, Ramos-Reyes RR, Becker M (2012) Water quality under intensive banana production and extensive pastureland in tropical Mexico. Journal of Plant Nutrition and Soil Science 175(4): 553–559.
8. Ortiz SM, Anaya G, Estrada BW (1994) Evaluación, Cartografía y Políticas Preventivas de la Degradación de la Tierra. Chapingo, Mexico.
9. Geissen V, Sanchez-Hernandez R, Kampichler C, Ramos-Reyes R, Sepulveda-Lozada A, et al. (2009) Effects of land-use change on some properties of tropical soils – An example from Southeast Mexico. Geoderma 151: 87–97.
10. Moreno-Unda AA (2011) Environmental effects of the National Tree Clearing Program, Mexico, 1972–1982 Cologne: Cologne University of Applied Sciences 119 p.
11. Bockstaller CGL, Keichinger O, Girardin P, Galan MB, Gaillard G (2009) Comparison of methods to assess the sustainability of agricultural systems. A review. Agron Sustain Dev 29: 223–235.
12. Mitchell GMA, McDonald A (1995) PICABUE: a methodological framework for the development of indicators of sustainable development. Int J Sust Dev World 104–123.

13. Bockstaller C, Guichard L, Makowski D, Aveline A, Girardin P, et al. (2008) Agri-Environmental Indicators to Assess Cropping and Farming Systems: A Review Sustainable Agriculture. In: Lichtfouse E, Navarrete M, Debaeke P, Véronique S, Alberola C, editors. Springer Netherlands. 725–738.
14. Riley J (2001) The indicator explosion: local needs and International challenges. Agr Ecosyst Environ 87: 119–120.
15. Kampichler C, Hernández-Daumás S, Ochoa-Gaona S, Geissen V, Huerta-Lwanga E, et al. (2010) Indicators of environmentally sound land use in the humid tropics: The potential roles of expert opinion, knowledge engineering and knowledge discovery. Ecological Indicators 10: 320–329.
16. Nambiar KK, Gupta AP, Fu Qinglin, Li S (2001) Biophysical, chemical and socio-economic indicators for assessing agricultural sustainability in the Chinese coastal zone Agriculture Ecosystems and Environment 87: 209–214.
17. Cabell JF, Oelofse M (2012) An indicator framework for assessing agroecosystem resilience. Ecology and Society 17: 18 http://dx.doi.org/10.5751/ES-04666-170118.
18. Darnhofer I, Bellon S, Dedieu B, Milestad R (2010) Adaptiveness to enhance the sustainability of farming systems. A review. Agronomy for Sustainable Development 30: 545–555.
19. Allen CR, Fontaine JJ, Pope KL, Garmestani AS (2011) Adaptive management for a turbulent future. Journal of Environmental Management 92: 1339–1345.
20. Malezieux E, Crozat Y, Dupraz C, Laurans M, Makowski D, et al. (2009) Mixing plant species in cropping systems: concepts, tools and models. A review. Agron Sustain Dev 29: 43–62.
21. Jones CG, Lawton J, Shachak M (1997) Positive and negative effects of organisms as physical ecosystem engineers. Ecology and Society 78: 1946–1957.
22. Brussaard L, Caron P, Campbell B, Lipper L, Mainka S, et al. (2010) Reconciling biodiversity conservation and food security: scientific challenges for a new agriculture. Current opinion in Environmental sustainability 2: 34–42.
23. Knorr D, Watkins TR (1984) Alterations in Food Production Knorr DW, Watkins, T.R., editor. New York: Van Nostrand Reinhold.
24. Seufert V, Ramankutty N, Foley JA (2012) Comparing the yields of organic and conventional agriculture. Nature 485: 229–232.
25. Carter H (1989) Agricultural sustainability: an overview and research assessment. Calif Agric 43: 1618–1637.
26. Tellarini V, Caporali F (2000) An input/output methodology to evaluate farms as sustainable agroecosystems: an application of indicators to farms in central Italy. Agriculture, Ecosystems and Environment 77: 111–123.
27. Stinner BR, House GJ (1987) Role of ecology in lower-input, sustainable agriculture: an introduction. Am J Alternative Agric 2: 146–147.
28. Lockeretz W (1988) Open questions in sustainable agriculture. Am J Alternative Agric 3: 174–181.
29. Hauptli H, Katz D, Thomas BR, Goodman RM (1990) Biotechnology and crop breeding for sustainable agriculture. In: Edwards CA, Lal R, Madden P, Miller RH, House G, editors. Sustainable Agricultural Systems Soil and Water Conservation Society: Ankeny, Iowa. 141–156.
30. Madden P (1990) The economics of sustainable low-input farming systems. In: Francis CA, Flora CB, King LD, editors. Sustainable Agriculture in Temperate Zones. New York: John Wiley & Sons. 315–341.
31. Dobbs TL, Becker DL, Taylor DC (1991) Sustainable agriculture policy analyses: South Dakota on-farm case studies. J Farming Systems ResExt 2: 109–124.
32. Blair GJ, Lefroy RD, Lisle L (1995) Soil carbon fractions based on their degree of oxidation, and the development of a carbon management index for agricultural systems. Aust J Soil Res 46: 1459–1466.
33. Ouédraogo E, Mando A, Zombré NP (2001) Use of compost to improve soil properties and crop productivity under low input agricultural system in West Africa. Agriculture, Ecosystems and Environment 84: 259–266.
34. Pretty J, Toulmin C, Williams S (2011) Sustainable intensification in African agriculture. International Journal of Agricultural Sustainability 9: 5–24.
35. Lal R (2004) Soil carbon sequestration impacts on global climate change and food security. Science 304: 1623–1627.
36. Gurr GM, Wratten SD, Luna JM (2003) Multi-function agricultural biodiversity: pest management and other benefits. Basic Appl Ecol 4: 107–116.
37. Börjesson P (1999) Environmental effects of energy crop cultivation in Sweden I: Identification and quantification. Biomass and Bioenergy 16: 137–154.
38. Rommelse R (2001) Economic assessment of biomass transfer and improved fallow trials in western Kenya In: ICRAF Natural Resource Problems PaPP, editor. Natural Resource Problems. Nairobi (Kenya): International Centre for Research in Agroforestry (ICRAF).
39. Nyende P, Delve R (2004) Farmer participatory evaluation of legume cover crop and biomass transfer technologies for soil fertility improvement using farmer criteria, preference ranking and logit regression analysis. Exp Agric 40: 77–88.
40. Koocheki A, Nassiri M, Alimoradi L, Ghorbani R (2009) Effect of cropping systems and crop rotations on weeds. Agronomy for Sustainable Development 29: 401–408.
41. Bellon M (1995) Farmers 'Knowledge and Sustainable Agroecosystem Management: An Operational Definition and an Example from Chiapas, Mexico. Human Organization 54: 263–272.
42. Belalcázar S, Espinosa J (2000) Effect of Plant Density and Nutrient Management on Plantain Yield. Better Crops International 14: 12–15.
43. Shaahan MM, El-Sayed AA, Abou El-Nour EAA (1999) Predicting nitrogen, magnesium and iron nutritional status in some perennial crops using a portable chlorophyll meter. Scientia Horticulturae 82: 339–348.
44. Agbede TM (2008) Nutrient availability and cocoyam yield under different tillage practices. Soil Tillage Research 99: 49–57.
45. Edwards CA (1989) The Importance of Integration in Sustainable Agricultural Systems. Agriculture, Ecosystems and Environment 27: 25–35.
46. Edwards CA, Grove TL, Harwood RR, Pierce Colfer CJ (1993) The role of agroecology and integrated farming systems in agricultural sustainability. Agriculture, Ecosystems and Environment 46: 99–121.
47. García-Pérez JA, Alarcón-Gutiérrez E, Perroni Y, Barois I (2014) Earthworm communities and soil properties in shaded coffee plantations with and without application of glyphosate. Applied Soil Ecology 83: 230–237.
48. Correia FV, Moreira JC (2010) Effects of Glyphosate and 2,4-D on Earthworms (*Eisenia foetida*) in Laboratory Tests. Bull Environ Contam Toxicol 85: 264–268.
49. Dick J, Kaya B, Soutoura M, Skiba U, Smith R, et al. (2008) The contribution of agricultural practices to nitrous oxide emissions in semi-arid Mali Soil. Use and Management 24: 292–301.
50. Morales H, Perfecto I (2000) Traditional knowledge and pest management in the Guatemalan highlands. Agriculture and Human Values 17: 49–63.
51. Li BL, Rykiel EJ (1996) Introduction. Ecological Modelling 90: 109–110.
52. Salski A (1996) Introduction Fuzzy Logic in Ecological Modelling. Ecological Modelling 85: 1–2.
53. Mendoza GA, Prabhu R (2003) Fuzzy methods for assessing criteria and indicators of sustainable forest management. Ecological Indicators 3: 227–236.
54. Kampichler C, Platen R (2004) Ground beetle occurrence and moor degradation: modelling a bioindication system by automated decision-tree induction and fuzzy logic. Ecological Indicators 4: 99–109.
55. INEGI (2000) Cuaderno Estadístico Municipal de Tenosique. Gobierno del Estado de Tabasco; INEGI, editor. Villahermosa, Tabasco.
56. INEGI (1985) Carta Edafológica, Villahermosa In: E15-8, editor. 1: 250,000. Aguascalientes, México: Instituto Nacional de Estadística, Geografía e Informática.
57. Manjarrez-Muñoz B (2008) Ordenamiento territorial de la ganadería bovina en Balancán y Tenosique, Tabasco. Villahermosa Tabasco: El Colegio de la Frontera Sur. 105 p.
58. Isaac-Márquez R (2008) Análisis del Cambio de Uso y Cobertura del Suelo en los Municipios de Balancán y Tenosique, Tabasco, México. Villahermosa, Tabasco, México: El Colegio de la Frontera Sur.
59. Rigby D, Woodhouse P, Young T, Burton M (2001) Constructing a farm level indicator of sustainable agricultural practice. Ecological Economics 39: 463–478.
60. van der Werf HMG, Petit J (2002) Evaluation of the environmental impact of agriculture at the farm level: a comparison and analysis of 12 indicator-based methods. Agriculture, Ecosystems and Environment 93: 131–145.
61. Astier M, Speelman S, López-Ridaura S, Masera O, Gonzalez-Esquivel CE (2011) Sustainability indicators, alternative strategies and trade-offs in peasant agroecosystems: analysing 15 case studies from Latin America. International Journal of Agricultural Sustainability 9: 409–422.
62. Bockstaller C, Girardin P (2003) How to validate environmental indicators. Agr Syst 76: 639–653.
63. Bockstaller C, Girardin P, Van der Werf HGM (1997) Use of agroecological indicators for the evaluation of farming systems. Eur J Agron 7: 261–270.
64. Huerta E, Kampichler C, Geissen V, Ochoa-Gaona S, de Jong B, et al. (2009) Towards an ecological index for tropical soil quality based on soil macrofauna. Pesq agropec bras 44 (8): 1056–1062.
65. Meul M, Nevens F, Reheul D (2009) Validating sustainability indicators: Focus on ecological aspects of Flemish dairy farms. Ecological Indicators 9: 284–295.
66. Pervanchon F, Bockstaller C, Girardin P (2002) Assessment of energy use in arable farming systems by means of an agro-ecological indicator: the energy indicator. Agricultural Systems 72: 149–172.

Effects of Soil Data and Simulation Unit Resolution on Quantifying Changes of Soil Organic Carbon at Regional Scale with a Biogeochemical Process Model

Liming Zhang[1,2], Dongsheng Yu[2]*, Xuezheng Shi[2], Shengxiang Xu[2], Shihe Xing[1]*, Yongcong Zhao[2]

1 College of Resource and Environment, Fujian Agriculture and Forestry University, Fuzhou, China, **2** State Key Laboratory of Soil and Sustainable Agriculture, Institute of Soil Science, Chinese Academy of Sciences, Nanjing, China

Abstract

Soil organic carbon (SOC) models were often applied to regions with high heterogeneity, but limited spatially differentiated soil information and simulation unit resolution. This study, carried out in the Tai-Lake region of China, defined the uncertainty derived from application of the DeNitrification-DeComposition (DNDC) biogeochemical model in an area with heterogeneous soil properties and different simulation units. Three different resolution soil attribute databases, a polygonal capture of mapping units at 1:50,000 (P5), a county-based database of 1:50,000 (C5) and county-based database of 1:14,000,000 (C14), were used as inputs for regional DNDC simulation. The P5 and C5 databases were combined with the 1:50,000 digital soil map, which is the most detailed soil database for the Tai-Lake region. The C14 database was combined with 1:14,000,000 digital soil map, which is a coarse database and is often used for modeling at a national or regional scale in China. The soil polygons of P5 database and county boundaries of C5 and C14 databases were used as basic simulation units. Results project that from 1982 to 2000, total SOC change in the top layer (0–30 cm) of the 2.3 M ha of paddy soil in the Tai-Lake region was +1.48 Tg C, −3.99 Tg C and −15.38 Tg C based on P5, C5 and C14 databases, respectively. With the total SOC change as modeled with P5 inputs as the baseline, which is the advantages of using detailed, polygon-based soil dataset, the relative deviation of C5 and C14 were 368% and 1126%, respectively. The comparison illustrates that DNDC simulation is strongly influenced by choice of fundamental geographic resolution as well as input soil attribute detail. The results also indicate that improving the framework of DNDC is essential in creating accurate models of the soil carbon cycle.

Editor: Jose Luis Balcazar, Catalan Institute for Water Research (ICRA), Spain

Funding: The work was funded by National Natural Science Foundation of China (No. 41001126), and the National Basic Research Program of China (973 Program) (2010CB950702). The funders had no role in study design, data collection and analysis, decision to publish, or preparation of the manuscript.

Competing Interests: The authors have declared that no competing interests exist.

* E-mail: dshyu@issas.ac.cn (DY); fafuxsh@126.com (SX)

Introduction

An estimated 1500 Pg of C is held in the form of soil organic carbon (SOC), representing 2/3 of the global terrestrial organic carbon pool [1–3]. SOC plays a vital role in the global carbon cycle, where a slight alteration of the soil carbon pool can cause profound changes in atmospheric CO_2 concentrations. Agro-ecosystems, accounting for 10% of the total terrestrial area, are one of the most sensitive terrestrial ecosystems subject to heavy human activity [3]. Increasing agricultural soil C sequestration is recognized as one strategy for achieving food security and improving soil quality.

Paddy soil is a major cultivated soil in China, and a unique type of anthropogenic soil recognized by Chinese Soil Taxonomy [3–5]. The total area of paddy soils is 45.7 M ha, which accounts for 34% of the total cultivated land in China [6]. This area also accounts for 22% of the total waterlogged farming area worldwide and produces about 44% of all grain in China [4]. Therefore, accurate estimation of paddy soil SOC change in China is vitally important for a comprehensive understanding of SOC dynamics and agro-ecosystem sustainability.

Recently, scientists have applied modeling to estimate SOC change in cropping systems [7–14]. The DeNitrification-DeComposition (DNDC) model, developed by Li et al. [15,16], is a process-based model focused on agrosystem carbon and nitrogen cycling and has been widely used for regional studies in the USA [17], China [11], India [18] and Europe [19]. Recently the DNDC model was determined to be one of the well performing models based on seven long-term experiments selected by the Global Change and Terrestrial Ecosystems Soil Organic Matter Network (GCTE SOMNET), which evaluated model performance using three different land uses, a range of climatic conditions within the temperate region, and different treatments [11,14].

In China, scientists have studied SOC change using the DNDC model for many years. At the regional scale, Tang et al. [11] simulated SOC changes for cropland in China for 1998 using the DNDC model, and they found that SOC would be lost at a rate of 78.89 Tg C year^{-1}. Zhang et al. [20] linked the DNDC model and 1: 14,000,000 soil database to estimate SOC stock changes for the year 2000 in Northwest China, revealing a decline in SOC stock. At the field scale, Wang et al. [21] tested DNDC uncertainty based on six long-term (10–20 year) SOC datasets from the Northeast, North, Northwest, Central South, East, and Southwest China. Results from the six validation tests supported the previous

Figure 1. Geographical location of the study area in China.

conclusions that the DNDC model was capable of quantifying SOC change in the agroecosystems across the entire area of China.

To date, the county boundary was used as the basic simulation unit in most DNDC simulations conducted at regional scale [11,20]. As a result, these simulations are often subject to great uncertainties since the soil property data were averaged for the area, which greatly ignore the impacts of soil heterogeneity therein [18,22]. Moreover, many researchers used coarse soil attribute data obtained from the books such as Soil in China (Vol. 1–6) and 1: 14,000,000 soil maps at national or a regional scale in China [11,20]. However, studies have already pointed out that the effect of soil heterogeneity on SOC change estimation is a major source of uncertainty when using the DNDC model at the regional scale [18,22,23].

This study, which was carried out in the rice-dominated Tai-Lake Region of China, provides a chance to test the uncertainty of the DNDC model caused by different precisions of soil data and basic simulation unit. The goals of this study were to: (1) compare SOC changes modeled with different resolutions of soil databases and varied basic simulation units, (2) assess the uncertainty derived from these soil databases with different resolutions and basic simulation units, and (3) give some suggestions for improving the performance of the biogeochemical DNDC model applied at the regional scale.

Materials and Methods

Study area

The Tai-Lake region (118°50′-121°54′E, 29°56′-32°16′N), an area of intensive rice cultivation, is located in the middle and lower reaches of the Yangtze River paddy soil region of China. The region includes the entire Shanghai City administrative area and a part of Jiangsu and Zhejiang provinces, and covers a total area of 36,500 km^2 (**Fig. 1**) [4]. The Tai-Lake region mainly consists of plains formed on deltas with numerous rivers and lakes. The climate is warm and moist with abundant sunshine and a long growing season. Annual rainfall is 1,100–1,400 mm, with a mean temperature of 16°C, and average annual sunshine of 1,870–2,225 hours. The frost-free period is over 230 days. The study area is one of the oldest agricultural regions in China, with a long history of rice cultivation spanning several centuries. Most cropland in the region is managed as a rice and winter wheat rotation. Rice is planted in June and harvested in October and wheat is planted in November and harvested in May [24].

Approximately 66% of the total land area is covered with paddy soils [24]. Paddy soils in the Tai Lake area are derived mostly from loess, alluvium, and lacustrine deposits, and are classified into 6 soil subgroups according to the Genetic Soil Classification of China (GSCC) system which are represented in the 1:50,000 digital soil map (**Table 1**). As map scale decreased, the soil subgroups of submergenic, bleached, percogenic and degleyed on the 1:50,000 soil map was eliminated and emerged into the soil subgroups of degleyed and hydromorphic in the 1:14,000,000 soil map. Therefore, those were only two paddy soil subgroups of degleyed and hydromorphic in the 1:14,000,000 soil map. The GSCC nomenclature as well as the subgroup's reference name in US Soil Taxonomy (ST) include; Hydromorphic (Typic Epiaquepts), Submergenic (Typic Endoaquepts), Bleached (Typic Epiaquepts), Gleyed (Typic Endoaquepts), Percogenic (Typic Epiaquepts), and Degleyed (Typic Endoaquepts) [25,26].

Description of the DNDC model

The DNDC model (Version 9.1) is a process-based soil biogeochemical research tool that was developed to estimate the impact of management strategies on the fate of nitrogen (N) and carbon (C) in agroecosystems. It integrates crop growth and soil biogeochemical processes on a daily time step and simulates N and C cycles in plant-soil systems.

The model contains six interacting sub-models which describe the generation, decomposition, and transformation of organic matter, and outputs the dynamic components of SOC and greenhouse gas fluxes. The six sub-models include: 1) a soil climate component which use soil physical properties, air temperature, and precipitation data to calculate soil temperature, moisture, and redox potential (Eh) profiles and soil water fluxes through time. The results of the calculation are then fed to the other sub-models; 2) a nitrification component; 3) a denitrification module, which calculates hourly denitrification rates and N_2O, NO, and N_2 production during periods when the soil Eh decreases due to rainfall, irrigation, flooding, or soil freezing; 4) simulation of SOC decomposition and CO_2 production through soil microbial respiration; 5) a plant growth component, which calculates daily

Table 1. The subgroups of paddy soil in the Tai-Lake region, China.

Subgroups	Horizonation*	Descriptions
Bleached	A-P-E-C	Mainly distributed in foothills, usually no underground water, impervious layer at 60 cm depth, soil reaction close to neutral or slightly acid.
Gleyed	Aa-Ap-G-C	Mainly distributed in depressional areas, high underground water level, poorly drained, distinct gleyization, soil reaction was slightly acid.
Percogenic	Aa-Ap-C	Mainly distributed on gentle hill slopes, no underground water, associated with rain-fed paddy fields, soil reaction was neutral to slightly acid.
Degleyed	Aa-Ap-Gw-G	Same distribution area as Gleyed paddy soils, after man-made drainage the underground water level decreases leading to degley processes, soil reaction was slightly acid.
Submergenic	A-Ap-P-C	Mainly distributed in alluvial plain or low flat ground, moderate drainage, underground water level was below 60 cm, soil reaction was neutral.
Hydromophic	Aa-Ap-P-W-G-C	Mainly distributed in floodplain, long cultivation history, well-drained, underground water level was below 90 cm, soil reaction was neutral.

*According to GSCC (Genetic Soil Classification of China), Aa means arable layer, Ap plow pan, C undeveloped parent material, Ds fragmental deposit horizon, E bleached horizon, G gley horizon, Gw degley horizon, P percogenic horizon, W waterlogogenic horizon.

root respiration, water, and N uptake by plants, and plant growth; and 6) a fermentation module, which calculates daily methane (CH_4) production and oxidation. The DNDC model can simulate C and N biogeochemical cycles in paddy rice ecosystems, as the model has been modified by adding a series of anaerobic processes [15,16,22,23,27,28,29,30].

At present, the DNDC model has been utilized by scientists in many countries, for example, the model is applied to simulate the carbon cycle in paddy field in Italy, China and Germany, in wheat fields in Canada, and it has been used to simulate the dynamics of soil organic matter in a 100 year experimental field in Rothamsted Experimental Station in England [14,31]. At the international conference on global change in Asia-Pacific areas in 2000, the DNDC model was recommended as the primary method for SOC studies in the in the Asia-Pacific region [31].

Database development

A major challenge for using an ecosystem model at regional scale is to assemble adequate datasets required to initialize and run the model. We examined the influence of database choices by executing simulation runs with different input sets using individual or combinations of databases. The geographic resolution or fundamental simulation unit could be represented by any of three assessment unit format datasets, polygon-based database of 1:50,000 (P5), county-based database of 1:50,000 (C5), and county-based database of 1:14,000,000 (C14). The three soil datasets covered 37 counties in Tai-Lake region.

The polygon-based database of 1:50,000 (P5) was linked a digital soil map (1:50,000), the most detailed of the three databases, in the Tai-Lake region contains 52,034 paddy soil polygons (**Table 2**). The polygons were derived from 1,107 soil profiles extracted from the latest national soil map (1:50,000), the Second National Soil Survey of China in the 1980s-1990s, with attribute assignment using the Pedological Knowledge Based (PKB) method based on GSCC [32]. The 1:50,000 digital soil database consists of many soil attributes, such as soil name, horizon thickness, bulk density, organic carbon content, clay content, pH, etc.

Soil parameters in C5 were derived from the 1:50,000 digital soil map (**Fig. 2 and Table 2**). However the attributes for C14 were derived from different sources than C5, primarily the 1:14,000,000 national soil map [33,34] (**Fig. 2**). C14 was widely used when the DNDC model was applied to national or regional scale in China [11,20]. The C14 in the Tai-Lake region contained 8 polygons of paddy soils representing 49 paddy soil profiles, and was also compiled via the Pedological Knowledge Based (PKB) method based on GSCC [32].

The C5 and C14 were built from the default method developed for DNDC, in which the maximum and minimum values of soil texture, pH; bulk density, and organic carbon content were recorded for each county (**Fig. 2**). So, the DNDC modeling of C5 and C14 methods conducted have used counties as the basic simulation unit in the Tai-Lake region (**Fig. 2**). After regional runs with C5 and C14 database, the DNDC model produced two SOC

Table 2. Characteristics of different resolution soil attribute databases of paddy soils in GSCC in the Tai-Lake region, China.

Soil database	Map scale	Source of soil maps	Source of soil data	Basic map units	Number of soil profiles	Number of polygons	Simulation unit
P5	1:50,000	Soil Survey Office of County in Jiangsu Province, Zhejiang Province and Shanghai City	Soil Series of County in Jiangsu Province, Zhejiang Province and Shanghai City	Soil Species	1,107	52,034	polygon
C5	1:50,000	Soil Survey Office of County in Jiangsu Province, Zhejiang Province and Shanghai City	Soil Series of County in Jiangsu Province, Zhejiang Province and Shanghai City	Soil Species	1,107	52,034	county
C14	1:14,000,000	Institute of Soil Science, Chinese Academy of Sciences	Soil Series of China	Subgroups	49	8	county

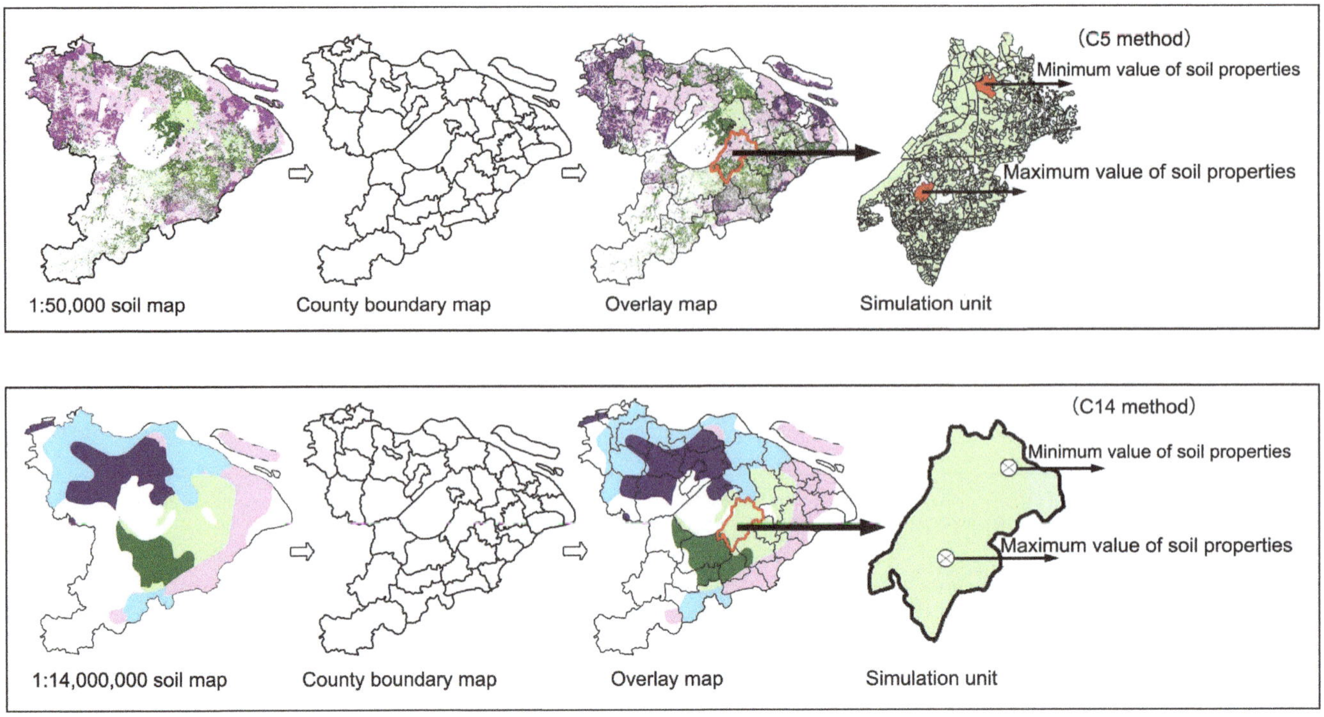

Figure 2. Description of C5 and C14 methods in the Tai-Lake region of China.

change (0–30 cm) resulting from two runs with the maximum and minimum soil values in each county. In this paper we present the mean results (average of maximum and minimum estimates) [11]. The DNDC modeling of P5 method conducted has used polygon as the basic simulation unit in the Tai-Lake region (**Table 2**). Therefore, the DNDC model runs with P5 database produced a single annual SOC change (0–30 cm) for each polygon. The total

Figure 3. Geographical location of weather stations across or near the Tai-Lake region, China.

SOC change of each county in the P5 was calculated by summing the SOC change of all polygons in a county. For a more complete description of P5 method see Zhang et al [35,36] and Xu et al [37].

For comparison in this study, both the polygon-based (P5) and county-based (C5 and C14) soil databases in the Tai-Lake region were run concurrently so the DNDC model could generalize regional SOC change from 1982 to 2000. The results simulated by DNDC with the two types of databases were compared to assess the advantages of using detailed, polygon-based 1:50,000 soil dataset (P5) [35,36,38,39].

In this study, the crop dataset included physiological data for summer rice and winter wheat in the Tai-Lake region. The crop parameters were obtained from thorough testing with that reflected the typical conditions of Tai-Lake region, which were founded on a wide range of information form Chinese literature published during the past decade and a publication of Gou et al [40,41].

Daily meteorological data (precipitation, maximum and minimum air temperature) for 1982–2000 from 13 weather stations across and near the Tai-Lake region were acquired from the National Meteorological Information Center, China Meteorological Administration (**Fig. 3**) [42]. Each county in the simulation was assigned to the nearest weather station [11,20,31].

The agricultural management dataset included sowing acreage, nitrogen fertilizer application rates, livestock, planting and harvest dates, and agricultural population at the county level from 1982 to 2000 in three resolution databases. The crop management practices of different counties were almost the same because the Tai-Lake region was a plain in topography. The main measures of farming management in the study area included: (1) fertilizer application: nitrogen synthetic fertilizer was applied for 6 times in the basal, tillering and heading stage for rice, and in the basal, jointing and heading stage for wheat; and organic manure (20% of livestock wastes and 10% of human wastes) was applied twice as base fertilizer for rice and wheat at the rates calculated based on the local livestock numbers (866, 44, 95, and 23 kg C head^{-1} yr^{-1} for cattle, sheep, swine and human, respectively); and N concentration in rainfall was 2.07 ppm; (2) crop residue management: 15% of aboveground crop residue was returned to the soil; (3) water management: one time of midseason and 5 time of shallow flooding (from June 17 to July 23, from July 28 to August 12, from August 24 to September 11, from September 18 to September 25, and from September 27 to October 2, respectively) were applied at summer rice; (4) tillage: twice at the 20 cm tilling depth for rice and 10 cm for wheat on the planting dates before 1990; and no-till applied for wheat after 1990; (5) growing period: rice is planted in June and harvested in October and wheat is planted in November and harvested in May; (6) optimum yield: rice is 7500 kg dry matter ha^{-1} and wheat is 3750 kg dry matter ha^{-1} [11,14,24,35,41,43]. All simulation methods within a certain county have the same feature input value such as crops, agricultural management, and climate, except soil feature [38,39].

Evaluation of simulation accuracy in three resolution databases

In order to evaluate the accuracy in three resolution databases, the simulated results of DNDC model were tested against measured data from paddy soils of the Tai-Lake region, which is the same area examined here.

From the perspective of previous studies, most dynamic models were only tested or validated with static long-term field-scale observations due to a lack of available soil data with temporal and spatial variation. Since these models have not yet been validated by regional scale data, uncertainty concerning their accuracy exists when they were applied to larger area dynamic SOC simulation [3].This study compared simulation results with the spatial distribution of SOC measurements from 1033 paddy soil sampling sites acquired in 2000, to validate and assess model performance in different simulation methods (P5, C5, and C14). The bias in the total difference between simulation and measurement were determined by calculating the correlation coefficient (r), the relative error (E), the mean absolute error (MAE) and the root mean square error (RMSE), as follows: [8,44].

$$r=\frac{\sum_{i=1}^{n}\left(V_{oi}-\overline{V_{oi}}\right)\left(V_{Si}-\overline{V_{Si}}\right)}{\left[\sum_{i=1}^{n}\left(V_{oi}-\overline{V_{oi}}\right)^2\right]^{1/2}\left[\sum_{i=1}^{n}\left(V_{Si}-\overline{V_{Si}}\right)^2\right]^{1/2}} \quad (1)$$

$$E=\frac{100}{n}\times\sum_{i=1}^{n}\frac{V_{oi}-V_{Si}}{V_{oi}} \quad (2)$$

$$\mathrm{RMSE}=\sqrt{\frac{1}{n}\sum_{i=1}^{n}(V_{oi}-V_{Si})^2} \quad (3)$$

$$MAE=\frac{1}{n}\sum_{i=1}^{n}ABS(V_{oi}-V_{Si}) \quad (4)$$

Where V_{oi} are the observed values, $\overline{V_{oi}}$ is the mean of the observed data, V_{Si} is the simulated value, $\overline{V_{Si}}$ is the mean of the simulated value, $Si\in(P5,C5,C14)$, and n is the number in the sequence of the data pairs. If E is less than 5% or between 5% and 10%, the simulation is satisfactory or acceptable, respectively; otherwise, it is unacceptable [44]. The greater r value is and the smaller RMSE or MAE value is, the greater prediction accuracy is. Conversely, a lower r value and more elevated RMSE or MAE value, the lower prediction accuracy is.

Data comparison and analysis

SOC change as quantified by DNDC modeling with the P5 assessment unit data set are recognized as a benchmark for comparison with the results of the DNDC model runs with the other two assessment unit data sets as input. The P5 are thought

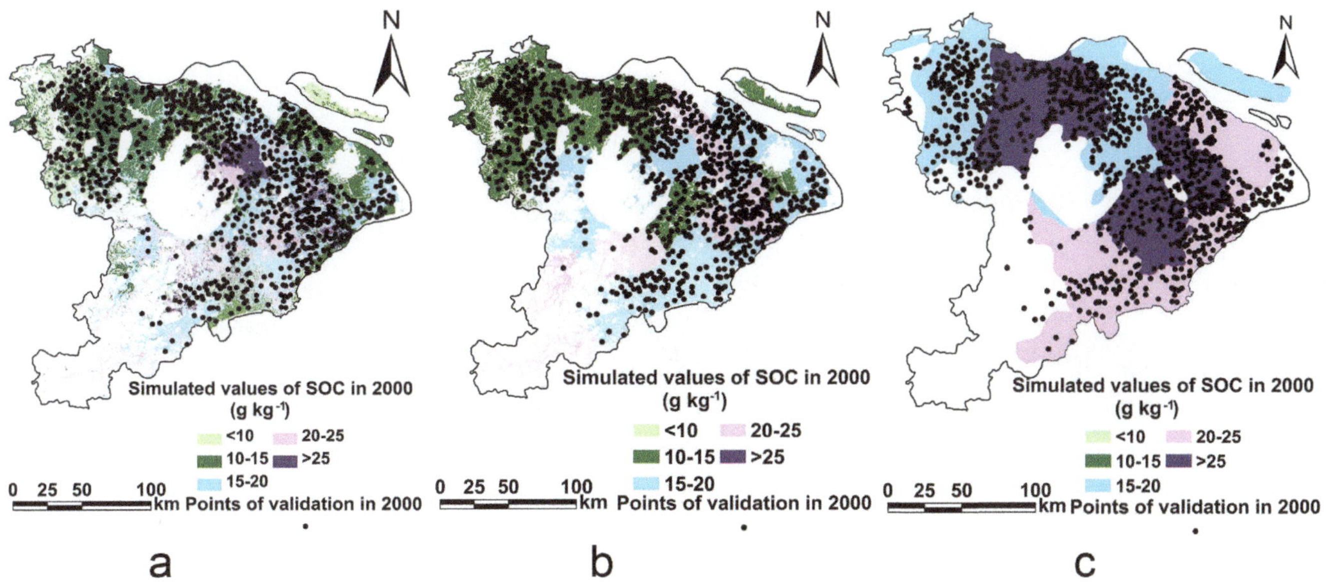

Figure 4. Spatial distribution of validation points and simulated SOC values from different simulation methods for the Tai-Lake region for 2000 (a: P5, b: C5, and c: C14).

theoretically to be more accurate than the C5 and C14 because of their relative greater detail and accuracy [35,36,38,39]. Relative variation of an index value (VIV, %) of C5 and C14 methods is calculated as the formula (1). The index values (IV) were quantified from the P5 data set (IV_{P5}) and other data sets (IV_{Ci}) to support data set comparison [23,38,39].

$$VIV(\%) = ABS(100 \times (IV_{P5} - - IV_{Ci})/IV_{P5}) \quad (5)$$

Where ABS is the absolute value function, IV_{P5} is the total SOC change with P5, and IV_{Ci} is the total SOC change produced by C5 (or C14).

Previous results from the sensitivity tests of the DNDC model indicated that the spatial heterogeneity of soil properties (e.g. texture, SOC content, bulk density, and pH) are the major sources of uncertainty for simulating SOC changes under specific management conditions at regional scale [18,20,22,23]. In order to test the most sensitive soil properties factor, the correlation of soil properties and average annual SOC changes were determined by step-wise regression analysis by using SPSS statistical software [37,45]. The step-wise regression is useful in checking how entering each variable affects the overall regression model, which begins by entering the variable with the largest partial statistic and checking the importance of the coefficient of the variable [45,46]. This method keeps adding more variables, each time recalculating the coefficients. During the incorporation of a variable into the model, the partial statistic of the already entered variable changes and might cause it to be unimportant. The operation stops when the model has incorporated the variables with the most significant contribution and discarded the least significant ones [47].

Results and Discussion

Difference of simulation accuracy in three resolution databases

Three maps of average SOC content for paddy soils at surface layers (0–15 cm) in the study area in 2000 were constructed on the basis of simulated data in different simulation methods (P5, C5, and C14) (**Fig. 4**). Also, corresponding SOC validation points were constructed from measurements of the surface layer (0–15 cm) of 1033 paddy soil samples taken in the study area in 2000. Fig. 4 demonstrates that the observed SOC in 2000 varied from 1.9 g kg^{-1} to 36 g kg^{-1}. By comparison, Fig. 4 also illustrates that simulated SOC in 2000 varied from 5.1 g kg^{-1} to 34 g kg^{-1} in P5, from 11 g kg^{-1} to 24 g kg^{-1} in C5, and from 17 g kg^{-1} to 28 g kg^{-1} in C14; where 99.6%, 84.1% and 57.1% of simulated paddy soil samples in P5, C5 and C14 were within the ranges produced by the observed SOC data. Furthermore, the relative errors (E) of P5 and C5 were 6.4% and 5.0%, respectively; and within the range of 5%–10%, demonstrating that the DNDC model in P5 and C5 were acceptable for modeling SOC of paddy soils in the Tai-Lake region according to the evaluation criteria described earlier (**Fig. 5a and b**) [8,44]. Moreover, the small values of MAE (4.0 g kg^{-1}) and RMSE (5.0 g kg^{-1}) in P5 and C5 also indicated that the modeled results were encouragingly consistent with observations in the Tai-Lake region (**Fig. 5a and b**). However, the E, MAE and RMSE of C14 reached −33%, 6.0 g kg^{-1} and 7.0 g kg^{-1}, respectively, suggesting that the simulated results of C14 were not suitable for simulating paddy soils in the Tai-Lake region (**Fig. 5c**).

Overall, though the values of E, MAE and RMSE between P5 and C5 had no significant differences, P5 was recognized better due to high correlation coefficient (0.5) and accurate simulation range (99.6%) (**Fig. 5a and b**). Furthermore, the simulation of P5 can differentiate the difference of paddy soil type within a county. Some studies showed that SOC content spatial variability was correlated with soil type spatial variability(**Fig. 5a**) [32,48,49]. Compared to the SOC validation of DNDC model in cropland by other scientists, accurate simulation of P5 ($r = 0.50^{**}$; E = 6.4%; MAE = 4.0 g kg^{-1}; RMSE = 5.0 g kg^{-1} and n = 1033) and C5 ($r = 0.40^{**}$; E = 5.0%; MAE = 4.0 g kg^{-1}; RMSE = 5.0 g kg^{-1} and n = 1033) are higher than those of Liu et al [50] (r = 0.25–0.66 and n = 68), Liu et al [51] (E = 27.6% and n = 49), and Xu et al [52] ($r = 0.22^{**}$; RMSE = 4.4 g kg^{-1} and n = 243); and are almost similar to that of the Xu et al [52] ($r = 0.52^{**}$; RMSE = 4.1 g kg^{-1} and n = 1385). However, the SOC accurate simulation of DNDC model in P5, C5 and C14 are lower than those of Studdert et al [53] ($r = 0.73^{**}$ and n = 286) by using the RothC model and Yu et al [54] ($r = 0.98^{**}$ and n = 349) by using the Agro-C model. Therefore, the results mentioned above suggest that modification of the DNDC model is necessary to better simulate SOC change from cropping systems. With continued modification, DNDC model could become a powerful tool for estimating SOC change at regional and national scales.

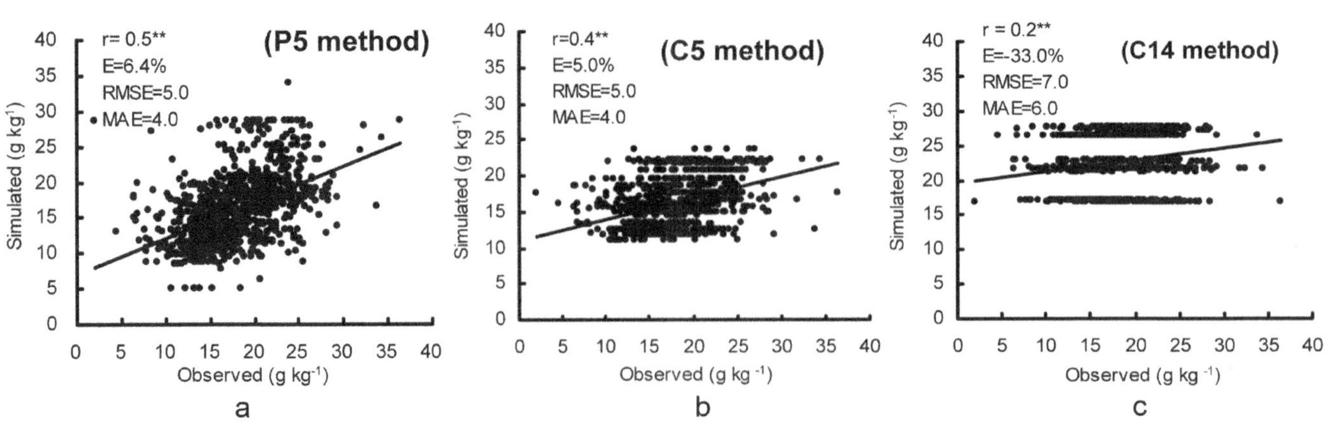

Figure 5. Comparison between simulated and observed SOC values from different simulation methods of the Tai-Lake region for 2000 (a: P5, b: C5, and c: C14).

Table 3. Soil properties at three resolution soil attribute databases contributing to the variability of average annual SOC change in Tai-Lake region paddy soils from 1982 to 2000.

Soil database	Number of simulation units	ΔR^{2a}				Adjusted R^2
		Initial SOC (g kg^{-1})	Clay(%)	pH	Bulk density (g cm^{-3})	
P5	52,034	0.778***	0.066***	0.009***	0.025***	0.878***
C5	37	0.881***			0.062***	0.939***
C14	37	0.185***		0.757***		0.938***

***significant at 0.001 probability levels, respectively.
[a]The change in the R^2 statistic is produced by adding a soil property into stepwise multiple regressions.

Variation of soil properties derived as input for DNDC modeling in three resolution databases in Tai-Lake region

Results of the contribution of soil properties to the variability of average annual SOC change are given in Table 3. All variables (i.e., initial SOC content, pH, bulk density, and clay content) were included in the step-wise regression analysis. For the P5 and C5 resolution databases, initial SOC content accounted for 77.8%–88.1% of the difference of average annual SOC change for paddy soils from 1982 to 2000, while other soil parameters only accounted for less than 6.6% of the difference. For the C14 resolution database, initial SOC content accounted for 18.5% of the difference of average annual SOC change for paddy soils from 1982 to 2000, and soil pH accounted for 75.7% of the difference. Therefore, it could be inferred that the differences in SOC change modeled with the three resolution databases were primarily due to the differences in initial SOC content and pH.

Table 4 shows the initial SOC content (0–5 cm), clay content (0–10 cm), pH (0–10 cm), and bulk density (0–10 cm) derived as input for DNDC modeling, from P5, C5 and C14 for the Tai-Lake region. As for the entire Tai-Lake region, the average initial SOC values sourced from P5 was lower than that from C5 and C14. Another difference is that the average values of clay content and pH sourced from C14 were also higher than those from P5 and C5. The average bulk density sourced from C5 was higher than that from P5 and C14.

The differentiation of soil properties was also shown at the county scale in the Tai-Lake region (**Table 4**). The average values of initial SOC content and bulk density sourced from C5 for 24 counties were higher than those from P5; the other was that the average values of clay content for 24 counties and pH for 20 counties in C5 were lower than those from P5. Although the average clay content sourced from P5 for 25 counties was slightly lower than that from C14, but the average initial SOC content sourced from C14 for 34 counties was obviously higher than that of P5. According to statistics describing the 1:50,000 digital soil database of the Tai-Lake region, initial SOC content of six paddy soil subgroups, namely submergenic, bleached, percogenic, hydromorphic, degleyed and gleyed, were 10 g kg^{-1}, 10 g kg^{-1}, 11 g kg^{-1}, 15 g kg^{-1}, 19 g kg^{-1}, and 25 g kg^{-1}, respectively. As map scale decreased from 1:50,000 to 1:14,000,000, the submergenic, bleached, percogenic and degleyed subgroups on the 1:50,000 digital soil map were eliminated and merged into the hydromorphic and degleyed subgroups in the 1:14,000,000 digital soil map [32,39]. The initial SOC content of the hydromorphic and gleyed subgroups in the 1:14,000,000 digital soil database were 17 g kg^{-1} and 28 g kg^{-1}, respectively, which were higher than most paddy soil subgroups in the 1:50,000 digital soil database. Therefore, the average initial SOC content of most counties in C14 was significantly higher than that from P5, while the average values of bulk density for 20 counties and pH for 24 counties in C14 was lower than those from P5. The results demonstrated that the soil properties (i.e., texture, SOC content, bulk density, and pH) in three resolution databases methods had large differences in the Tai-Lake region. Many studies have showed that SOC spatial variability is expressed by map delineations and map unit composition which varied with scales, resulting in the assignment of different soil properties at each scale of aggregation [32,48,49]. As such, an improper of soil map scales and simulation unit may lead to SOC estimation inaccuracy.

Variation of the average annual-, total SOC change modeled with the three resolution databases in Tai-Lake region

Similar trends can be observed in estimates of average annual-, total SOC change over the 19 year study period for three resolution databases decreased from P5 to C14 (**Fig. 6**). Simulation results demonstrate that total SOC change of P5 in the top layer (0–30 cm) of the 2.3 M ha of paddy rice fields in the Tai-Lake region was +1.48 Tg C from 1982 to 2000, with the annual SOC change ranging from -45 kg C ha^{-1} yr^{-1} to 92 kg C ha^{-1} yr^{-1} (**Fig. 6**). From 1982 to 1988, the SOC change modeled with P5 inputs was almost negative with annual changes ranging from -3.2 kg C ha^{-1} yr^{-1} to -45 kg C ha^{-1} yr^{-1}. According to agricultural statistical data, chemical fertilizer application rate ranged from 180 kg N ha^{-1} yr^{-1} to 350 kg N ha^{-1} yr^{-1}, which is a relatively low value. Low fertilizer application rates often result in reduced SOC sequestration [31,55]. From 1989 to 2000, rural economic development led to increased fertilizer application from 350 kg N ha^{-1} yr^{-1} to 400 kg N ha^{-1} yr^{-1}. Increasing fertilizer application results in enhanced crop production and residue accumulation, and the latter leads to an increase of SOC. Further, much of the region has been utilizing no-tillage practices in planting wheat since 1991, which contribute to reduced SOC decomposition [35].

Although three resolution databases within a certain county have the same feature input value such as crops, agricultural management, and climate; SOC balance of C5 (or C14) in the Tai-Lake region was almost negative with annual changes ranging from 86 kg C ha^{-1} yr^{-1} to -205 kg C ha^{-1} yr^{-1} (or -185 kg C ha^{-1} yr^{-1} to -693 kg C ha^{-1} yr^{-1}) from 1982 to 2000 (**Fig. 6**). The total SOC changes of C5 and C14 in the Tai-Lake region were −3.99 Tg C and −15.38 Tg C, respectively, from 1982 to 2000. With the total SOC change as modeled with P5 inputs as the baseline, the relative deviation of C5 and C14 were 368% and 1126%, respectively.

Table 4. Statistics for soil properties derived as input for DNDC modeling in different counties, from P5, C5 and C14 for the Tai-Lake region.

County	P5				C5								C14							
	SOC	Clay	BD	pH	SOC		Clay		BD		pH		SOC		Clay		BD		pH	
	-----WA------				Range	Ave	Range	Ave	Range	Ave	Range	Ave	Range	Ave	Range	Ave	Range	Ave	Range	Ave
Zhangjiagang	14	28	1.22	7.7	10–17	14	5–32	19	1.16–1.33	1.25	7.4–8.0	7.7	12–21	17	24–31	28	1.19–1.23	1.21	6.0–7.4	6.7
Changshu	17	25	1.23	7.0	9–38	24	9–34	22	1.05–1.46	1.26	5.5–8.1	6.8	12–21	17	24–31	28	1.19–1.23	1.21	6.0–7.4	6.7
Taicnang	14	33	1.20	7.7	9–20	15	23–42	33	1.11–1.38	1.25	7.4–8.6	8.0	11–31	21	18–48	33	1.12–1.27	1.20	5.5–7.4	6.5
Kunshan	19	35	1.15	7.1	11–34	18	22–44	33	0.94–1.40	1.17	6.4–7.6	7.0	23–33	28	32–56	44	1.14–1.14	1.14	6.2–6.9	6.6
Wuxian	24	41	1.08	6.6	6–30	23	26–47	37	0.97–1.47	1.22	3.4–7.4	5.6	12–21	17	24–31	28	1.19–1.23	1.21	6.0–7.4	6.7
Wujiang	17	36	1.06	5.9	3–26	15	17–58	38	0.89–1.67	1.28	4.9–6.9	5.9	23–33	28	32–56	44	1.14–1.14	1.14	6.2–6.9	6.6
Wuxi	14	28	1.16	6.7	4–17	11	14–34	24	1.09–1.39	1.24	5.3–7.2	6.3	21–33	27	25–31	28	1.13–1.23	1.18	6.0–6.3	6.1
Jiangyin	13	13	1.28	6.2	6–17	12	8–25	17	0.99–1.54	1.27	5.4–8.0	6.7	21–33	27	25–31	28	1.13–1.23	1.18	6.0–6.3	6.1
Wujin	12	9	1.22	6.8	7–18	13	4–13	9	1.08–1.51	1.30	6.2–7.9	7.1	21–33	27	25–31	28	1.13–1.23	1.18	6.0–6.3	6.1
Jintan	10	9	1.32	6.8	7–14	11	4–13	9	1.17–1.58	1.38	5.5–7.6	6.6	12–21	17	24–31	28	1.19–1.23	1.21	6.0–7.4	6.7
Liyang	10	10	1.23	6.2	6–17	12	7–12	10	1.05–1.37	1.21	6.0–7.2	6.6	12–21	17	24–31	28	1.19–1.23	1.21	6.0–7.4	6.7
Yixing	13	27	1.17	6.0	3–30	17	10–53	32	1.11–1.58	1.35	4.4–8.5	6.5	21–33	27	25–31	28	1.13–1.23	1.18	6.0–6.3	6.1
Dantu	7	36	1.25	6.6	2–19	11	12–49	31	1.07–1.39	1.23	5.8–8.0	6.9	12–21	17	24–31	28	1.19–1.23	1.21	6.0–7.4	6.7
Jurong	10	30	1.23	5.4	6–13	10	15–38	27	1.10–1.29	1.20	5.1–7.4	6.3	12–21	17	24–31	28	1.19–1.23	1.21	6.0–7.4	6.7
Danyang	12	29	1.24	6.7	8–16	12	16–53	35	1.07–1.36	1.22	5.8–7.8	6.8	12–21	17	24–31	28	1.19–1.23	1.21	6.0–7.4	6.7
Jiaxing	19	34	1.19	6.5	10–26	18	20–56	38	0.98–1.34	1.16	5.8–7.6	6.7	23–33	28	32–56	44	1.14–1.14	1.14	6.2–6.9	6.6
Jiashan	21	39	1.23	6.2	15–27	21	22–44	33	1.03–1.34	1.19	5.7–7.0	6.4	23–33	28	32–56	44	1.14–1.14	1.14	6.2–6.9	6.6
Pinghu	15	35	1.10	6.6	9–24	17	22–43	33	0.92–1.48	1.20	6.3–7.2	6.8	11–31	21	18–48	33	1.12–1.27	1.20	5.5–7.4	6.5
Haiyan	17	40	1.17	6.7	7–25	16	22–52	37	0.92–1.51	1.22	5.7–7.3	6.5	11–31	21	18–48	33	1.12–1.27	1.20	5.5–7.4	6.5
Haining	14	36	1.19	6.6	7–24	16	19–52	36	0.92–1.51	1.22	6.0–7.5	6.8	11–31	21	18–48	33	1.12–1.27	1.20	5.5–7.4	6.5
Tongxiang	14	30	1.05	6.5	7–29	18	20–52	36	0.91–1.34	1.13	6.0–7.4	6.7	14–33	24	25–34	30	1.12–1.13	1.13	6.3–6.6	6.5
Huzhou	23	30	1.10	6.2	13–37	25	7–42	25	0.99–1.37	1.18	5.6–6.7	6.2	14–33	24	25–34	30	1.12–1.13	1.13	6.3–6.6	6.5
Changxing	17	31	1.14	5.8	6–31	19	9–47	28	0.84–1.53	1.19	3.6–7.1	5.4	14–33	24	25–34	30	1.12–1.13	1.13	6.3–6.6	6.5
Anji	18	22	1.16	6.0	12–34	23	12–42	27	0.84–1.44	1.14	5.4–6.7	6.1	14–33	24	24–35	30	1.12–1.13	1.13	6.3–6.6	6.5
Deqing	19	32	1.12	6.3	7–26	17	18–38	28	0.87–1.53	1.20	5.2–7.2	6.2	14–33	24	24–35	30	1.12–1.13	1.13	6.3–6.6	6.5
Yuhang	15	5	1.16	6.6	9–21	15	16–48	32	0.95–1.34	1.15	5.9–7.3	6.6	11–31	21	18–48	33	1.12–1.27	1.20	5.5–7.4	6.5
Linan	22	22	1.09	6.2	18–27	23	8–29	19	0.91–1.14	1.03	5.5–7.8	6.7	11–31	21	18–48	33	1.12–1.27	1.20	5.5–7.4	6.5
Minhang	13	26	1.18	7.6	10–18	14	17–46	32	1.11–1.30	1.21	6.4–8.0	7.2	11–31	21	18–48	33	1.12–1.27	1.20	5.5–7.4	6.5
Jiading	13	28	1.10	7.6	9–20	15	13–44	29	0.94–1.24	1.09	6.5–8.1	7.3	11–31	21	18–48	33	1.12–1.27	1.20	5.5–7.4	6.5
Chuangsha	12	29	1.15	7.6	9–20	15	17–36	27	1.06–1.33	1.20	7.3–8.0	7.7	11–31	21	18–48	33	1.12–1.27	1.20	5.5–7.4	6.5
Nanhui	16	31	1.18	7.4	13–22	18	8–35	22	1.11–1.21	1.16	6.5–8.1	7.3	11–31	21	18–48	33	1.12–1.27	1.20	5.5–7.4	6.5
Qingpu	21	27	1.15	7.1	7–33	20	11–36	24	0.94–1.53	1.24	5.6–8.3	7.0	23–33	28	32–56	44	1.14–1.14	1.14	6.2–6.9	6.6
Songjiang	23	26	1.20	6.8	10–33	22	8–37	23	1.03–1.47	1.25	5.6–8.1	6.9	23–33	28	32–56	44	1.14–1.14	1.14	6.2–6.9	6.6
Jinshan	20	29	1.22	7.0	11–37	24	18–36	27	1.11–1.47	1.29	4.6–8.3	6.5	11–31	21	18–48	33	1.12–1.27	1.20	5.5–7.4	6.5
Fengxian	15	25	1.20	7.4	12–18	15	19–39	29	1.11–1.49	1.30	6.9–8.1	7.5	11–31	21	18–48	33	1.12–1.27	1.20	5.5–7.4	6.5
Baoshan	11	23	1.21	7.9	9–19	14	8–44	26	1.11–1.28	1.20	7.2–8.2	7.7	11–31	21	18–48	33	1.12–1.27	1.20	5.5–7.4	6.5
Chongming	10	17	1.11	8.1	9–13	11	15–29	22	1.11–1.21	1.12	7.8–8.1	8.0	12–16	14	24–39	31	1.17–1.27	1.22	7.3–7.4	7.5
Tai-Lake region	**15**	**26**	**1.18**	**6.7**	**2–38**	**16**	**4–58**	**27**	**0.84–1.54**	**1.23**	**3.4–8.3**	**6.7**	**11–33**	**22**	**18–56**	**32**	**1.12–1.27**	**1.18**	**5.5–7.4**	**6.5**

WA = Weighted average of soil properties by the area of each polygon; SOC = Initial SOC content (g kg^{-1}); Clay = Clay content (%); BD = Bulk Density (g cm^{-3}); Range = Range of maximum and minimum soil properties; Ave = Average of maximum and minimum soil properties.

As Table 3 illustrated, initial SOC content was the most sensitive parameter controlling SOC change among all soil factors in P5 and C5 [20,22]. The average initial SOC value of P5 and C5 were 15 g kg^{-1} and 16 g kg^{-1} for the entire Tai-Lake region, respectively. Furthermore, the average initial SOC content sourced from P5 for 24 counties was lower than that from C5, while the average clay content sourced from P5 for 24 counties was also higher than that from C5. Many previous studies showed that soils with lower initial organic carbon and higher clay content tended to sequester C [20,22,35]. The high SOC sequestration

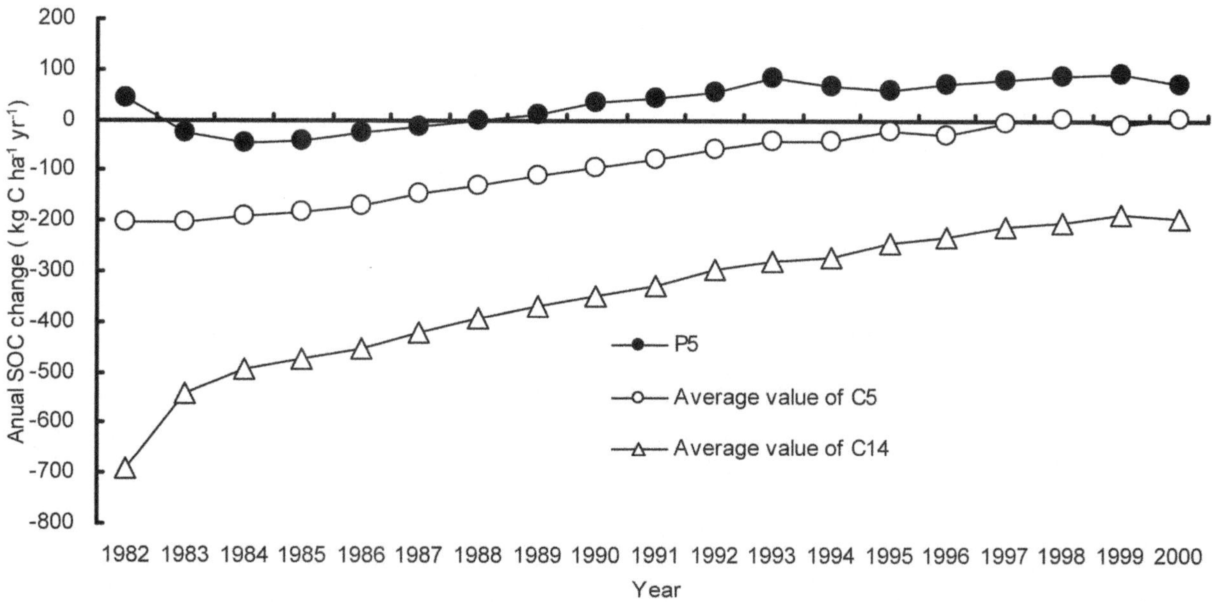

Figure 6. Temporal distribution of average annual SOC change modeled with P5, C5 and C14 from 1982 to 2000 in the Tai-Lake region, China.

rate (34 kg C ha^{-1} yr^{-1}) was thus associated with P5 (**Fig. 7a**). Conversely, the high SOC losses rate (-91 kg C ha^{-1} yr^{-1}) was associated with C5 (**Fig. 7b**). The SOC losses rate (-349 kg C ha^{-1} yr^{-1}) in C14 was the highest in the three resolution databases (**Fig. 7c**). Table 3 demonstrates that pH and initial SOC content are the most sensitive parameters controlling SOC change among all soil factors in C14. The average initial SOC value (22 g kg^{-1}) of C14 was significantly higher than that of P5 (15 g kg^{-1}) and C5 (16 g kg^{-1}) for the entire Tai-Lake region. In addition, the average pH value of C14 for 34 counties was close to neutral (6.5–7.5), and the average initial SOC contents of C14 for 28 counties were higher than 20 g kg^{-1}. Some studies showed that soils with neutral pH value and higher organic carbon content were favorable for CO_2 production by providing more substrates and better living environment for microbes [22,56].

The comparison illustrates that using different basic simulation units and soil data sources will produce different conclusions as to C sequestration or C liberation in the same study area. The implication is that more precise soil data and high resolution simulation units were necessary for better simulating regional scale SOC dynamics. The simulation outcome can be attributed to how the databases represent soil types and spatial heterogeneity, which is more precisely done with larger scale soil data and high resolution simulation units (e.g., 1:50,000 soil database).

Distribution of the average annual-, total SOC change modeled with the three resolution databases in different counties

The differentiation of the average annual-, total SOC change in P5, C5 and C14 was also shown at the county scale in the Tai-Lake region (**Table 5 and Fig. 7**). In the modeled domain, there were 26 counties that gained SOC and 11 counties that lost SOC from 1982 to 2000 in P5. The highest SOC sequestration rate of P5 were in Dantu, Jurong, Jiading and Baoshan counties which was higher than 200 kg C ha^{-1} yr^{-1}, due to the low initial SOC content (7.1 g kg^{-1}, 9.5 g kg^{-1}, 13 g kg^{-1} and 11 g kg^{-1}, respectively). In addition, the clay content of P5 in Dantu and Jurong counties were 36% and 30%, respectively. High clay content is associated with high SOC sequestration [22,57,58]. By contrast, the greatest SOC loss rate of P5 in the Huzhou, Songjiang, Linan and Wuxian county was more than 170 kg C ha^{-1} yr^{-1}, due to the high initial SOC content (23 g kg^{-1}, 23 g kg^{-1}, 22 g kg^{-1} and 24 g kg^{-1}, respectively). Moreover, the clay content of P5 in Linan and Songjiang counties were only 22% and 26%, respectively. Low clay content is linked to high CO_2 emissions [22].

However, under the same agricultural practice, there were only 14 counties that gained SOC and 23 counties that lost SOC from 1982 to 2000 in C5. The highest SOC sequestration rate of C5 were in Dantu, Jurong, Jintan, Chongming, and Baoshan counties which was higher than 150 kg C ha^{-1} yr^{-1}. The main reason was that the initial SOC content of C5 in Dantu, Jurong, Jintan, Chongming, and Baoshan counties were 11 g kg^{-1}, 10 g kg^{-1}, 11 g kg^{-1}, 11 g kg^{-1} and 14 g kg^{-1}, respectively; the other was that the average clay content of C5 in Dantu, Jurong, and Baoshan counties ranged from 26% to 31%. Some studies showed that low initial SOC value and high clay content were linked to low CO_2 emissions [22,57,58]. In contrast, the greatest SOC loss rate of C5 in Jinshan, Changshu, Huzhou, Anji, and Kunshan county were more than 400 kg C ha^{-1} yr^{-1}, which possessed high initial SOC and low bulk density [22,35]. Compared with the P5 resolution database, the average annual-, total SOC change modeled with C5 for 28 counties was lower than that from P5. With the total SOC change as modeled with P5 inputs as the baseline, the relative deviations of counties in Jiangyin, Zhangjiagang and Kunshan were relatively high (>1000%). The relative deviations ranged from 50% to 250% in most counties. Only fifteen counties (Wuxian, Wujin, Jintan, Liyang, Dantu, Jurong, Huzhou, Yuhang, Linan, Minhang, Jiading, Chuangsha, Songjiang, Baoshan, and Chongming) had relatively low value of relative deviation (<100%). The SOC changes for the two resolution databases are almost in agreement with the soil feature across the 37 simulated counties (**Table 4 and Table 5**). The average initial SOC content sourced from C5 for 24 counties was higher than

Table 5. Distribution of the average annual SOC change (kg C ha^{-1} yr^{-1}) and the total SOC change (Gg C) in different counties of the Tai-Lake region, China modeled with P5, C5 and C14 from 1982 to 2000.

County	Area 10^4ha	P5		C5						C14					
		ASC	TSC	ASC			TSC			ASC			TSC		
		WA		Max	Min	AVE	Max	Min	AVE	Max	Min	AVE	Max	Min	AVE
Zhangjiagang	2.54	2	1	138	−50	44	66	−24	21	57	−418	−181	27	−202	−87
Changshu	7.55	−64	−92	246	−1438	−596	353	−2063	−855	92	−404	−156	132	−580	−224
Taicnang	6.14	43	50	241	−360	−60	281	−420	−70	192	−993	−400	224	−1158	−467
Kunshan	7.57	38	55	252	−1057	−402	362	−1520	−579	−75	−838	−456	−108	−1205	−656
Wuxian	14.78	−172	−483	356	−876	−260	998	−2458	−730	72	−403	−166	201	−1132	−466
Wujiang	9.79	53	99	586	−717	−65	1091	−1333	−121	−59	−792	−425	−110	−1472	−791
Wuxi	9.77	49	91	478	−196	141	888	−364	262	−256	−958	−607	−476	−1778	−1127
Jiangyin	8.69	2	4	313	−101	106	517	−167	175	−262	−951	−606	−432	−1569	−1000
Wujin	14.85	91	256	248	−57	96	701	−160	270	−274	−969	−621	−774	−2734	−1754
Jintan	7.10	146	197	245	81	163	331	110	220	136	−364	−114	184	−492	−154
Liyang	10.84	177	365	302	−115	94	623	−237	193	127	−386	−129	263	−796	−267
Yixing	10.34	147	289	514	−1011	−248	1011	−1987	−488	−262	−950	−606	−514	−1867	−1191
Dantu	5.07	371	357	615	−276	170	592	−266	163	141	−367	−113	136	−353	−109
Jurong	8.03	238	363	403	−8	198	614	−12	301	125	−374	−124	191	−570	−190
Danyang	9.58	47	85	416	−152	132	758	−276	241	130	−364	−117	237	−663	−213
Jiaxing	6.57	−124	−155	412	−523	−55	514	−652	−69	−11	−752	−381	−13	−938	−476
Jiashan	4.13	−103	−81	110	−631	−260	87	−494	−204	−48	−786	−417	−38	−616	−327
Pinghu	4.81	127	116	349	−466	−58	319	−425	−53	274	−923	−324	250	−843	−296
Haiyan	2.74	41	21	489	−549	−30	254	−285	−16	279	−943	−332	145	−490	−173
Haining	3.92	151	113	483	−556	−36	359	−413	−27	274	−951	−338	204	−708	−252
Tongxiang	4.42	128	107	504	−634	−65	424	−533	−55	122	−846	−362	102	−711	−304
Huzhou	6.02	−309	−353	163	−1184	−511	186	−1354	−584	92	−915	−411	106	−1046	−470
Changxing	5.62	42	45	422	−863	−221	450	−922	−236	71	−901	−415	76	−962	−443
Anji	4.18	−66	−52	178	−1025	−423	141	−814	−336	58	−873	−408	46	−694	−324
Deqing	3.11	17	10	379	−684	−152	224	−404	−90	77	−930	−426	46	−550	−252
Yuhang	5.27	−91	−92	344	−371	−13	345	−372	−13	120	−1061	−456	120	−1034	−457
Linan	3.06	−220	−128	7	−381	−187	4	−222	−109	218	−754	−268	127	−439	−156
Minhang	3.49	62	41	294	−239	28	195	−158	18	257	−974	−359	171	−645	−238
Jiading	4.29	238	194	387	−197	95	315	−161	77	281	−940	−329	229	−766	−268
Chuangsha	3.71	188	133	339	−295	22	239	−208	15	286	−925	−319	202	−653	−225
Nanhui	4.11	152	119	189	−279	−45	148	−208	−35	287	−939	−326	224	−734	−255
Qingpu	5.68	−52	−56	432	−1042	−305	467	−1125	−329	2	−756	−377	3	−816	−407
Songjiang	5.90	−287	−322	249	−1010	−381	278	−1131	−426	−60	−816	−438	−67	−914	−491
Jinshan	5.63	−77	−83	213	−1481	−634	228	−1585	−679	275	−945	−335	294	−1011	−359
Fengxian	5.87	9	10	171	−270	−49	191	−301	−55	280	−944	−332	321	−1053	−370
Baoshan	3.13	200	119	393	−92	151	234	−55	90	311	−876	−283	185	−522	−168
Chongming	3.73	195	138	283	47	165	201	34	117	165	−108	28	117	−77	20
Tai-Lake region	**232**	**34**	**1483**	**340**	**−521**	**−91**	**14987**	**−22977**	**−3994**	**46**	**−744**	**−349**	**2022**	**−32790**	**−15380**

WA = Weighted average of annual mean SOC change (kg C ha^{-1} yr^{-1}) by the area of each polygon; ASC = Average annual SOC change (kg C ha^{-1} yr^{-1}); TSC = Total SOC change (Gg C); Max = Maximum value of ASC (or TSC); Min = Minimum value of ASC (or TSC); Ave = Average of maximum and minimum ASC (or TSC).

that from P5, and the average clay content sourced from C5 for 24 counties was also lower than that from P5. Some research showed that high initial SOC content and low clay content is favorable for C losses [22,57,58].

As can be seen from the Table 5, a big number of counties where the average annual-, total SOC change modeled with the C14 and P5 differed greatly. There was only one county that gained SOC from 1982 to 2000, while other 36 counties lost SOC in C14. The SOC losses of C14 ranged from 360 kg C ha^{-1} yr^{-1}

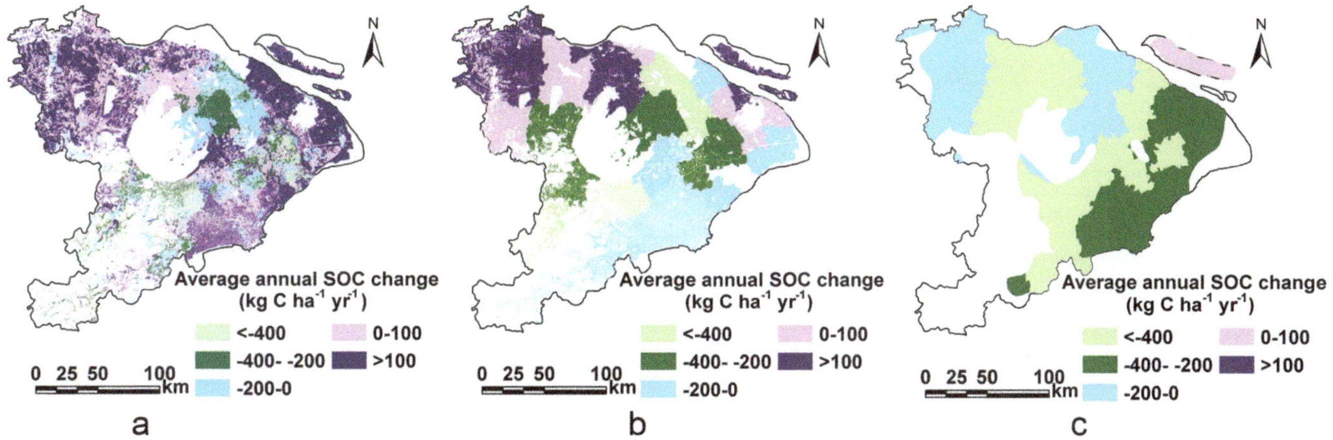

Figure 7. Spatial distribution of average annual SOC change modeled with P5, C5 and C14 in the Tai-Lake region, China (a: P5, b: c5, and c:C14).

to 620 kg C ha^{-1} yr^{-1} in most counties. With the total SOC change as modeled with P5 inputs as the baseline, the relative deviations of counties in Zhangjiagang, Taicang, Kunshan, Wuxi, Jiangyin, Changxing, Deqing, and Fengxian were more than 1000%. Only five counties (Wuxian, Huzhou, Linan, Songjiang, and Chongming) in C14 had relatively low deviation (<100%). The main reasons were that the average pH value of C14 in most counties ranged from 6.5 to 7.5, which were closer to neutral than that from C5 and C14. Moreover, the average initial SOC contents of C14 in most counties were higher than 20 g kg^{-1}, which was also much higher than that from P5 or C5. Therefore, high SOC losses occurred in C14.

The modeled data at county scale in three simulation methods indicated the underestimation with the county-based database was related to its soil data source and simulation unit resolution, especially the coarse soil maps (1:14,000,000) that missed relatively small soil patches containing low or high soil properties (i.e., initial SOC content, pH, and clay content) which were sensitive to SOC change. This would also explain why the precision of soil database plays an important role in elevating the accuracy of modeled SOC change at regional scale.

Conclusions

Using different spatial information, process-based models integrated with GIS databases can play an important role in describing C biogeochemical cycles, such as targeting mitigation efforts to the most beneficial regions. However, SOC models have often been applied to regions with high heterogeneity but limited spatially differentiated soil information and simulated unit resolution.

Simulation results indicate that total SOC change from 1982 to 2000 in the top layer (0–30 cm) of the 2.3 M ha of paddy rice fields in the Tai-Lake region was +1.48 Tg C for P5. However, discrepancies in the results existed among the three databases, because different soil data and basic simulation units were used. The total SOC changes in the Tai-Lake region were -3.99 Tg C and -15.38 Tg C for C5 (or C14), respectively, from 1982 to 2000. With the total SOC change as modeled with P5 inputs as the baseline, the relative deviation of C5 was lower than C14 due to the more precise soil data. In contrast, the relative deviation of C14 was higher than other databases due to using coarser soil data and low-resolution simulation units. In addition, with the same basic simulation unit, average annual-, total SOC change between C5 and C14 for the Tai-Lake region also had a large discrepancy due to the use of different soil data. The comparison demonstrated that the most sensitive factors (e.g., initial SOC content and pH) for modeling SOC dynamics should be given a high priority during the input data acquisition as they contribute disproportionately to the uncertainties produced during the upscaling process [20]. The results also indicate that improving the performance of the biogeochemical DNDC model is essential in creating accurate models of the soil carbon cycle.

Acknowledgments

We gratefully acknowledge support for this research from the National Natural Science Foundation of China (No. 41001126), and the National Basic Research Program of China (973 Program) (2010CB950702). Sincere thank is also given to Professor Changsheng Li (University of New Hampshire, USA) for his useful advice on DNDC model.

Author Contributions

Conceived and designed the experiments: DSY XZS SHX. Performed the experiments: LMZ SXX. Analyzed the data: LMZ SXX. Contributed reagents/materials/analysis tools: SXX YCZ. Wrote the paper: LMZ.

References

1. Eswaran H, Berg EVD, Reich P (1993) Organic carbon in soil of the world. Soil Science Society of America Journal 57:192–194.
2. Lal R (2006) World soils and greenhouse effect: An overview, in soils and global change. Encyclopedia of Soil Science. doi:10.1081/E-ESS-120042696.
3. Shi XZ, Yang RW, Weindorf DC, Wang HJ, Yu DS, et al. (2010) Simulation of organic carbon dynamics at regional scale for paddy soils in China. Climatic Change 102:579–593.
4. Li QK (1992) Paddy soil of China. Beijing: Science Press. 514 p.
5. Gong ZT (1999) Chinese soil taxonomic classification. Beijing: Science Press. 5–215 p.
6. Liu QH, Shi XZ, Weindorf DC, Yu DS, Zhao YC, et al. (2006) Soil organic carbon storage of paddy soils in China using the 1:1,000,000 soil database and their implications for C sequestration. Global Biogeochemical Cycles 20:GB3024. doi:10.1029/2006GB002731.
7. Jenkinson DS, Rayner JH (1977) The turnover of soil organic matter in some of Rothamsted classical experiments. Soil Science 125:298–305.
8. Smith P, Smith JU, Powlson DS, Arah JRM, Chertov OG, et al. (1997) A comparison of the performance of nine soil organic matter models using datasets from seven long term experiments. Geoderma 81:153–225.

9. Ardö J, Olsson L (2003) Assessment of soil organic carbon in semi-arid Sudan using GIS and the CENTURY model. Journal of Arid Environments 54:633–651.
10. Shirato Y (2005) Testing the suitability of the DNDC model for simulating long-term soil organic carbon dynamics in Japanese paddy soils. Soil Science and Plant Nutrition 51(2):183–192.
11. Tang HJ, Qiu JJ, van Ranst E, Li CS (2006) Estimations of soil organic carbon storage in cropland of China based on DNDC model. Geoderma 134:200–206.
12. Cerri CEP, Easter M, Paustian K, Killian K, Coleman K, et al. (2007) Predicted soil organic carbon stocks and changed in the Brazilian Amazon between 2000 and 2030. Agriculture, Ecosystems and Environment 122:58–72.
13. Huang Y, Yu YQ, Zhang W, Sun WJ, Liu SL, et al. (2009) Agro-C: A biogeophysical model for simulating the carbon budget of agroecosystems. Agricultural and Forest Meteorology 149:106–129.
14. Tang HJ, Qiu JJ, Wang LG, Li H, Li CS, et al. (2010) Modeling soil organic carbon storage and its dynamics in croplands of China. Agricultural Sciences in China 9(5):704–712.
15. Li CS, Frolking S, Frolking TA (1992) A model of nitrous oxide evolution from soil driven by rainfall events: I. Model structure and sensitivity. Journal of Geophysical Research 97:9759–9776.
16. Li CS, Frolking S, Frolking TA (1992) A model of nitrous oxide evolution from soil driven by rainfall events:II. Model applications. Journal of Geophysical Research 97:9777–9783.
17. Tonitto C, David MB, Li CS, Drinkwater LE (2007) Application of the DNDC model to tile-drained Illinois agroecosystems: Model comparison of conventional and diversified rotations. Nutrient Cycling in Agroecosystems 78 (1):65–81.
18. Pathak H, Li CS, Wassmann H (2005) Greenhouse gas emissions from Indian rice fields: calibration and upscaling using the DNDC model. Biogeoscience 2:113–123.
19. Neufeldt H, Schäfe M, Angenendt E, Li CS, Kaltschmitt M, et al. (2006) Disaggregated greenhouse gas emission inventories from agriculture via a coupled economic-ecosystem model. Agriculture, Ecosystems and Environment 112:233–240.
20. Zhang F, Li CS, Wang Z, Wu HB (2006) Modeling impacts of management alternatives on soil carbon storage of farmland in Northwest China. Biogeosciences 3:451–466.
21. Wang LG, Qiu JJ, Tang HJ, Li CS, van Ranst E (2008) Modelling soil organic carbon dynamics in the major agricultural regions of China. Geoderma 147:47–55.
22. Li CS, Mosier A, Wassmann R, Cai ZC, Zheng XH, et al. (2004) Modeling greenhouse gas emissions from rice-based production systems: Sensitivity and upscaling. Global Biogeochemical Cycles 18:GB1043. doi:10.1029/2003 GB002045.
23. Cai ZC, Sawamoto T, Li CS, Kang GD, Boonjawat J, et al. (2003) Field validation of the DNDC model for greenhouse gas emissions in East Asian cropping systems. Global Biogeochemical Cycles 17 (4): GB1107, doi:10.1029/2003 GB002046.
24. Xu Q, Lu YC, Liu YC, Zhu HG (1980) Paddy soil of Tai-Lake region in China. Shanghai: Science Press.
25. Shi XZ, Yu DS, Warner ED, Sun WX, Petersen GW, et al. (2006) Cross-reference system for translating between genetic soil classification of China and Soil Taxonomy. Soil Science Society of America Journal 70:78–83.
26. Soil Survey Staff in USDA (2010) Keys to Soil Taxonomy (11th Edition). Washington: USDA-Natural Resources Conservation Service.
27. Li CS, Frolking S, Harriss R (1994) Modeling carbon biogeochemistry in agricultural soils. Global Biogeochemical Cycles 8 (3):237–254.
28. Li CS, Narayanan V, Harriss R (1996) Model estimates of nitrous oxide emissions from agricultural lands in the United States. Global Biogeochemical Cycles 10 (2):297–306.
29. Li CS, Qiu JJ, Frolking S, Xiao XM, Salas W, et al. (2002) Reduced methane emissions from large-scale changes in water management in China's rice paddies during 1980–2000. Geophysical Research Letters 29 (20):1972, doi:10.1029/2002GL01 5370.
30. Li CS (2007) Quantifying greenhouse gas emissions from soils: Scientific basis and modeling approach. Soil Science and Plant Nutrition 53 (4):344-352.
31. Qiu JJ, Wang LG, Tang HJ, Li H, Li CS (2005) Studies on the situation of soil organic carbon storage in croplands in northeast of China. Agricultural Sciences in China 37 (8):1166–1171.
32. Zhao YC, Shi XZ, Weindorf DC, Yu DS, Sun WX, et al. (2006) Map scale effects on soil organic carbon stock estimation in north China. Soil Science Society of America Journal 70:1377–1386.
33. Institute of Soil Science (1986) The soil atlas of China. Beijing: Institute of Soil Science, Academia Sinica, Cartographic Publishing House.
34. National Soil Survey Office of China (1993–1997) Soils in China (Vol. 1–6). Beijing: Agricultural Publishing House.
35. Zhang LM, Yu DS, Shi XZ, Xu SX, Wang SH, et al. (2012) Simulation soil organic carbon change in China's Tai-Lake paddy soils. Soil and Tillage Research 121:1–9.
36. Zhang LM, Yu DS, Shi XZ, Xu SX, Weindorf DC, et al. (2009) Quantifying methane emissions from rice fields in the Taihu region, China by coupling a detailed soil database with biogeochemical model. Biogeosciences 6:739–749.
37. Xu SX, Zhao YC, Shi XZ, Yu DS, Li CS, et al. (2013) Map scale effects of soil databases on modeling organic carbon dynamics for paddy soils of China. Catena 104:67–76.
38. Yu DS, Yang H, Shi XZ, Warner ED, Zhang LM, et al. (2011) Effects of soil spatial resolution on quantifying CH_4 and N_2O emissions from rice fields in the Tai Lake region of China by DNDC model. Global Biogeochemical Cycles 25:GB2004. doi:10.1029/2010GB003825.
39. Yu DS, Zhang LM, Shi XZ, Warner ED, Zhang ZQ, et al. (2013) Soil assessment unit scale affects quantifying CH_4 emissions from rice fields. Soil Science Society of America Journal 77:664–672.
40. Li CS (2007) Quantifying soil organic carbon sequestration potential with modeling approach. In: Tang HJ, Van Ranst E, Qiu JJ (Eds.) Simulation of soil organic carbon storage and changes in agricultural cropland in China and its impact on food security. Beijing: China Meteorological Press. 1–14 p.
41. Gou J, Zheng XH, Wang MX, Li CS (1999) Modeling N_2O emissions from agriculture fields in Southeast China. Advances in Atmospheric Sciences 16 (4):581–592.
42. China Meteorological Administration (2011) China meteorological data daily value. China Meteorological Data Sharing Service System, Beijing, China. http://cdc.cma.gov.cn/index.jsp.
43. Lu RK, Shi TJ (1982) Agricultural chemical manual. Beijing:China Science Press. 142 p.
44. Whitmore AP, Klein-Gunnewiek H, Crocker GJ, Klír J, Körschens M, et al. (1997) Simulating trends in soil organic carbon in long-term experiments using the Verberne/MOTOR model. Geoderma 81:137–151.
45. Admassu Y, Shakoor A, Wells N (2012) Evaluating selected factors affecting the depth of undercutting in rocks subject to differential weathering. Engineering Geology 124:1–11.
46. Leech NL, Barret KKC, Morgan G (2008) SPSS for intermediate statistics. New York: Lawrence Erlbaum Associates. 270 p.
47. Dielman TE (2001) Applied regression analysis for business and economics. California: Duxbury Thomson Learning. 647 p.
48. Arnold RW (1995) Role of soil survey in obtaining a global carbon budget. In: Lal R, Kimble J, Levine E, Stewart BA (Eds.) Advances in Soil Science: Soils and Global Change. Boca Raton, FL: CRC Press. 57–263 p.
49. Zhong B, Xu YJ (2011) Scale effects of geographical soil datasets on soil carbon estimation in Louisiana, USA: a comparison of STATSGO and SSURGO. Pedosphere 21 (4):491–501.
50. Liu YH, Yu ZR, Chen J, Zhang FR, Reiner D, et al. (2006) Changes of soil organic carbon in an intensively cultivated agricultural region: A denitrification-decomposition (DNDC) modelling approach. Science of the Total Environment 372:203–214.
51. Liu Q, Sun B, Jie XL, Li ZP (2009) The spatial-temporal dynamic change and simulation of county-scale paddy soil organic carbon red soil hilly region. Acta pedologica sinica 46 (6):1059–1067.
52. Xu SX, Shi XZ, Zhao YC, Yu DS, Wang SH, et al. (2012) Spatially explicit simulation of soil organic carbon dynamics in China's paddy soils. Catena 92:113–121.
53. Studdert GA, Monterubbianesi GM, Domínguez GF (2011) Use of RothC to simulate changes of organic carbon stock in the arable layer of a Mollisol of the southeastern Pampas under continuous cropping. Soil and Tillage Research 117:191–200.
54. Yu YQ, Huang Y, Zhang W (2012) Modeling soil organic carbon change in croplands of China, 1980-2009. Global and Planetary Change 82–83:115–128.
55. Wu TY, Schoenau JJ, Li FM, Qian PY, Malhi SS (2004) Influence of cultivation and fertilization on total organic carbon and carbon fractions in soils from the Loess Plateau of China. Soil and Tillage Research 77:59–68.
56. Pacey JG, DeGier JP (1986) The factors influencing landfill gas production. In: Energy from landfill gas. Proceeding of a conference jointly sponsored by the United Kingdom Department of Energy and the United States Department of Energy (October 1986). 51–59 p.
57. Burke IC, Lauenroth WK, Conffin DP (1995) Soil organic matter recovery in semiarid grassland: implications for the conservation reserve program. Ecological Monographs 5:793–801.
58. Kay BD (1998) Soil structure and organic carbon: a review. In: Lal R, Kimble JM, Follett RF. Soil Processes and the carbon cycle. Boca Raton, FL: CRC Press.169–198 p.

20

Comparative Analysis of Normalised Difference Spectral Indices Derived from MODIS for Detecting Surface Water in Flooded Rice Cropping Systems

Mirco Boschetti[1]*, Francesco Nutini[1,2], Giacinto Manfron[1,2], Pietro Alessandro Brivio[1], Andrew Nelson[3]

1 Institute for Electromagnetic Sensing of the Environment, National Research Council, Milan, Italy, **2** Department of Agricultural and Environmental Sciences – Production, Landscape, Agroenergy, University of Milan, Milan, Italy, **3** Social Sciences Division, International Rice Research Institute, Los Baños, Philippines

Abstract

Identifying managed flooding in paddy fields is commonly used in remote sensing to detect rice. Such flooding, followed by rapid vegetation growth, is a reliable indicator to discriminate rice. Spectral indices (SIs) are often used to perform this task. However, little work has been done on determining which spectral combination in the form of Normalised Difference Spectral Indices (NDSIs) is most appropriate for surface water detection or which thresholds are most robust to separate water from other surfaces in operational contexts. To address this, we conducted analyses on satellite and field spectral data from an agronomic experiment as well as on real farming situations with different soil and plant conditions. Firstly, we review and select NDSIs proposed in the literature, including a new combination of visible and shortwave infrared bands. Secondly, we analyse spectroradiometric field data and satellite data to evaluate mixed pixel effects. Thirdly, we analyse MODIS data and Landsat data at four sites in Europe and Asia to assess NDSI performance in real-world conditions. Finally, we test the performance of the NDSIs on MODIS temporal profiles in the four sites. We also compared the NDSIs against a combined index previously used for agronomic flood detection. Analyses suggest that NDSIs using MODIS bands 4 and 7, 1 and 7, 4 and 6 or 1 and 6 perform best. A common threshold for each NDSI across all sites was more appropriate than locally adaptive thresholds. In general, NDSIs that use band 7 have a negligible increase in Commission Error over those that use band 6 but are more sensitive to water presence in mixed land cover conditions typical of moderate spatial resolution analyses. The best performing NDSI is comparable to the combined index but with less variability in performance across sites, suggesting a more succinct and robust flood detection method.

Editor: Guy J.-P. Schumann, NASA Jet Propulsion Laboratory, United States of America

Funding: Funding for this research came from RIICE (Remote Sensing based Information and Insurance for crops in emerging Economies) funded by the Swiss Agency for Development and Cooperation (http://www.sdc.admin.ch) and from GRiSP (Global Rice Science Partnership) the CGIAR funded Research Program on Rice (http://www.grisp.net). The funders had no role in study design, data collection and analysis, decision to publish, or preparation of the manuscript.

Competing Interests: The authors have declared that no competing interests exist.

* E-mail: Boschetti.m@irea.cnr.it

Introduction

Timely and accurate information on crop typology and status is required to support suitable action to better manage agricultural production and reduce food insecurity [1,2]. More specifically, spatial information on where, when and how staple crops are cultivated is an important input for spatialized crop growth models for yield forecasts and yield gap analyses [3,4]. Digital cartographic data related to crops and croplands at global/regional scales (i.e. [2,4–9]) represent a static or snapshot view of a strong dynamic process and although they are valuable and essential datasets they do not provide information to monitor and model the annual and seasonal changes in crop status and crop production which are prerequisites for monitoring food security. There is a recognised need to develop and test methods which can provide operational information for staple crops, including rice [10] and automated classification and detection techniques are vital to ensure sustainability and transparency.

We focus on rice for several reasons. Firstly, it is the world's most important staple crop; it is the second largest in terms of harvested area after wheat, but is by far the most important in terms of human consumption, especially in low- and lower-middle-income countries (FAOSTAT, 2012). Thus, reliable and timely information on the global rice crop is required to guide food security and food price policies. Secondly, it truly is a global crop cultivated across a wide range of environmental conditions. Rainfall in rice-growing areas can vary from over 5,000 mm/y along Myanmar's Arakan Coast to less than 100 mm in Al Hasa Oasis in Saudi Arabia; average temperature during the rice season can vary from highs of 33°C in Sindh, Pakistan, to lows of 17°C in northern Japan; and rice can be cultivated from sea level up to 2600 m on the terraces of Nepal and Bhutan [11]. Thirdly, rice is grown under a wide range of management conditions such as highly mechanised, irrigated, single summer cropping (i.e. Italy, Japan, the U.S., Australia, Brazil); the more marginal rainfed rice systems across Latin America, sub-Saharan Africa, and South and South-East Asia; in rotation with other crops such as the vast rice/wheat double-cropping areas of India and China; intensive, irrigated triple cropping (i.e. Indonesia, Vietnam); and various other double and triple mixed-cropping combinations across Asia, Africa and Latin America.

Distilling the results of the many studies on rice crop detection and characterisation into an operational system is clearly challenging due to the geographic extent of rice, the range of environments under which it is cultivated and the diversity of crop management methods. Rice has been well studied in the remote sensing literature, and several mapping methods have been proposed to identify rice cultivated area based on the agronomic practise of pre-season flooding which characterises the majority of rice cropping systems (namely, irrigated rice, lowland rainfed rice and deepwater rice) which account for over 90% of the 154 M ha [11] cultivated with rice each year.

The detection of this flooded condition at the start of the growing season, followed by a rapid increase in observed biomass, is the key element for discriminating a rice crop from other natural vegetated surfaces and/or other crops. Although assessments of different methods for detecting biomass and changes in biomass under different environments are abundant, there is no comparable body of research on methods for detecting agronomic flooding. A thorough investigation of methods to detect agronomic flooding would seem to be a prerequisite for developing an operational rice crop detection methodology. In this paper, we focus attention on robust and reliable normalised difference spectral indices (NDSIs) to detect agronomic flooding practises which are particular to rice.

1.1 Normalised Difference Spectral Indices in the Literature

NDSIs have been proposed as an operational tool to perform water surface detection in different contexts [12,13] and NDSIs related to water surface presence and vegetation indices that monitor rice biomass growth have been proposed as part of a paddy rice detection algorithm [14–17]. In general, water-related indices are proposed as a combination of shortwave infrared (SWIR, 1250–2500 nm), the most suitable spectral domain due to the presence of specific physical water absorption features, and near infrared (NIR, 700–1250 nm) [18] or visible (VIS, 350–700 nm) spectral regions [19–21]. However, the formulation of a water index can be found in different combinations of bands as well as with different nomenclature for the same formulation. Ji et al. [22] provide a useful review of the existing, and sometimes confusing, nomenclatures of NIR/SWIR water indices and suggest a reference terminology in relation to the portion of the SWIR region being used.

Ouma and Tateishi [23] used Landsat data to perform open water detection and tested several NDSI combinations using a Normalised Difference Water Index (NDWI) nomenclature. They identified the NIR/SWIR normalised difference as the best performing index. The same index was also proposed by Xiao et al. [16] as the Land Surface Water Index (LSWI) and the same formulation is found with other nomenclature: the Normalised Difference Infrared Index (NDII) [24], the Normalised Difference Shortwave-Infrared Index (NDSWIR) [25] and the shortwave infrared water stress index (SIWSI) [26], albeit with reversed band order in the equation.

The capability of NDWI/LSWI to detect water is based on the comparison between the SWIR (1600 nm) and NIR (800 nm) spectral bands (see Table 1) which are sensitive to liquid water thickness – both in open water bodies and vegetation - in the ground area sensed by a radiometer. Even though this NIR/SWIR formulation is useful in detecting water presence, it is theoretically not diagnostic since it cannot separate a flooded soil from plant water content in the presence of a dense vigorous canopy. For this reason, the majority of NIR/SWIR indices, such as the SIWSI and NDSWIR, are proposed for monitoring plant water content/stress or to detect fire-affected areas.

The influence of vegetation in the LSWI can be removed or at least reduced, thus leaving only the water signal component, by subtracting a vegetation NDSI such as the Normalised Difference Vegetation Index (NDVI) [27] or the Enhanced Vegetation Index (EVI) [28]. Xiao et al. [16] proposed, and applied [14,17], this concept to detect flooded paddy rice fields using MODIS data if LSWI plus some threshold value is equal to or greater than either NDVI or EVI [14]. A similar approach was adopted by Gond et al. [29] to detect small water areas and humid vegetation in semi-arid areas of Africa. Despite the good performance of this combined index (LSWI-EVI or LSWI-NDVI), it has been noted that flood detection using this approach strongly depends on a locally adapted threshold which has varied from 0.05 [14] to 0.15 (for early-season rice in China) and to 0.21 (for late-season rice in China) in the literature [15]. Calibrating this threshold for different locations, seasons and years is non-trivial, and is one reason to assess flood detection NDSIs with the aim of identifying robust detection methods with fixed thresholds.

Table 1 provides a non-exhaustive summary of the most common indices adopted for water (standing or vegetation) detection. The table indicates the reference and the original purpose of the indices and groups them in relation to the wavelength adopted in the general formulation of an NDSI.

$$NDSI = \frac{\rho_i - \rho_j}{\rho_i + \rho_j} \quad (1)$$

where i and j refer to the different spectral bands used.

In summary NIR-NIR(SWIR) indices are mainly proposed for vegetation water detection while VIS-SWIR combinations are almost all proposed for water detection. Interestingly, the traditional NDVI (VIS-NIR combination) has also been used for water mapping [30].

1.2 Objectives

Clearly many water detection NDSIs and associated thresholds are available. We posit that any operational rice detection method which relies on the interpretation of optical time series imagery must have high skill in detecting agronomic flooding with low commission and omission errors. We also posit that this has not received sufficient attention and that there is little guidance on which indices and thresholds are the most robust. To address this, our research tested the performance of a range of NDSIs for detecting water to determine which NDSIs were most suitable for agronomic flood detection across various rice agro-ecosystems. Hence, in this study we make an intercomparison of NDSIs from table 1 and additionally we compare the NDSIs against the combined LSWI-EVI which we refer to as the B2B6-EVI index.

The objective of this research was to identify and assess robust flood detection NDSIs that met three criteria. The indices must (i) provide diagnostic information on the presence of a thick water layer, (ii) have minimal confusion with plant/soil water content and (iii) be able to separate the surfaces (water, soil, vegetation) for automatic detection purposes.

To reach these objectives, we performed a number of tests and experiments using both field and satellite data. Unlike previous studies such as Ji et al. [21], we:

i. compared a range of NDSIs available in the literature, including new combinations of VIS(GREEN/RED)-SWIR spectral bands as well as the combined B2B6-EVI index;

Table 1. List of spectral vegetation indices proposed for water detection.

PANEL A

Spectral range	SI	Original purpose	Equation	MODIS bands	Reference	Code in this manuscript
NIR-NIR	Normalised difference water index	Vegetation liquid water	$NDWI = \frac{\rho_{858} - \rho_{1240}}{\rho_{858} + \rho_{1240}}$	b2, b5	[18]	$NDSI_{B2B5}$
	Normalised difference moisture index	Forest analysis and detection	$NDMI = \frac{\rho_{858} - \rho_{1240}}{\rho_{858} + \rho_{1240}}$	b2, b5	[51]	
	Shortwave infrared water stress index	Vegetation water content	$\mathrm{SIWSI} = \frac{\rho_{1240} - \rho_{858}}{\rho_{1240} + \rho_{858}}$	b5, b2	[26]	
NIR-SWIR	Normalised difference infrared index	Vegetation water content	$NDII = \frac{\rho_{858} - \rho_{1650}}{\rho_{858} + \rho_{1650}}$	b2, b6	[24]	$NDSI_{B2B6}$
	Normalised difference shortwave-infrared index	Identification of burn scar	$\mathrm{NDSWIR} = \frac{\rho_{858} - \rho_{1640}}{\rho_{858} + \rho_{1640}}$	b2, b6	[25]	
	Shortwave infrared water stress index	Vegetation water content	$\mathrm{SIWSI} = \frac{\rho_{1640} - \rho_{858}}{\rho_{1640} + \rho_{858}}$	b6, b2	[26]	
	Normalised difference water index	Change on lake shorelines	$NDWI_3 = \frac{\rho_{858} - \rho_{1640}}{\rho_{858} + \rho_{1640}}$	b2, b6	[23]	
	Normalised burn ratio	Burn severity	$NDWI = \frac{\rho_{858} - \rho_{2130}}{\rho_{858} + \rho_{2130}}$	b2, b7	[46]	$NDSI_{B2B7}$
VIS-SWIR	Normalised difference water index	Open water detection	$NDWI = \frac{\rho_{555} - \rho_{1640}}{\rho_{555} + \rho_{1640}}$	b4, b6	[21]	$NDSI_{B4B6}$
	Modified NDWI	Water detection	$\mathrm{MNDWI} = \frac{\rho_{555} - \rho_{1640}}{\rho_{555} + \rho_{1640}}$	b4, b6	[52]	
	Normalised difference pond index	Detection of small water bodies	$\mathrm{MDPI} = \frac{\rho_{1640} - \rho_{555}}{\rho_{1640} + \rho_{555}}$	b6, b4	[53]	
	Normalised difference water index	Water end member selection	$NDWI = \frac{\rho_{645} - \rho_{1640}}{\rho_{645} + \rho_{1640}}$	b1, b6	[19]	$NDSI_{B1B6}$
	Normalised difference water index	Open water detection	$NDWI = \frac{\rho_{555} - \rho_{2130}}{\rho_{555} + \rho_{2130}}$	b4, b7	[21]	$NDSI_{B4B7}$
	Normalised difference flood index_2	Flood condition	$NDFI_2 = \frac{\rho_{645} - \rho_{2130}}{\rho_{645} + \rho_{2130}}$	b1, b7	Tested in this study	$NDSI_{B1B7}$
VIS-NIR	Normalised difference water index	Open water detection	$NDWI = \frac{\rho_{555} - \rho_{858}}{\rho_{555} + \rho_{858}}$	b4, b2	[21,48]	$NDSI_{B4B2}$
	Normalised difference water index	Open water detection	$NDWI = \frac{\rho_{555} - \rho_{1240}}{\rho_{555} + \rho_{1240}}$	b4, b5	[21]	$NDSI_{B4B5}$
	Normalised difference flood index_1	Flood condition	$NDFI_1 = \frac{\rho_{645} - \rho_{1240}}{\rho_{645} + \rho_{1240}}$	b1, b5	Tested in this study	$NDSI_{B1B5}$
	Normalised difference vegetation index	Water Mapping	$NDVI = \frac{\rho_{858} - \rho_{645}}{\rho_{858} + \rho_{645}}$	b2, b1	[27]*, [30]**	$NDSI_{B1B2}$

PANEL B

Spectral combination	Original purpose	Spectral combination	MODIS bands	Reference	Code in this manuscript
Combined index	Rice flood mapping	LSWI-EVI[&]	b1, b2, b3, b6	[14,15][&]	B2B6-EVI

Table 1. Cont.

PANEL A						
Spectral range	**SI**	**Original purpose**	**Equation**	**MODIS bands**	**Reference**	**Code in this manuscript**
Combined index		Rice flood mapping/ water bodies and wetland	LSWI-NDVI[&] OR NDVI-NDWI[$]	b1, b2, b6	[14][&], [16,29][$]	B2B6-NDVI
Enhanced vegetation index		Vegetation monitoring	$EVI = \frac{(\rho_{858} - \rho_{645})}{(\rho_{858} + 6 * \rho_{645} - 7.5 * \rho_{469} + 1)}$	b1, b2, b3	[28]	EVI

Panel A lists the NDSIs tested in this work. For each NDSI is indicated the corresponding mathematical equation and the original reference. Panel B reports equations for combined index B2B6-EVI and B2B6-NDVI calculation.
(*) original formulation for vegetation monitoring;
(**) proposed for water mapping;
(&) LSWI (Land Surface Water Index) is equal to the B2B6 MODIS index;
($) NDWI derived from SPOT-VGT data is equal to B2B6 MODIS index.

ii. used real spectroradiometric measurements acquired by field and satellite data (rather than simulated data from simple linear combinations of spectral library data) to evaluate mixed pixel effects on water indices;
iii. used MODIS data in conjunction with Landsat data at four different sites to evaluate the NDSI performance in real-world conditions where land cover mixture among vegetation, bare soil and water can alter the signal; and
iv. analysed the performance of the NDSIs on MODIS time series in four study areas.

This is the first time that published NDSIs and related thresholds for water detection have been compared, calibrated, validated and ranked across different environments on real satellite MODIS data. This analysis is necessary as a preliminary study for a remote sensing-based operational rice detection and monitoring system starting from the detection of agronomic flooding in a complex and mixed land cover environment. In all stages, of the analysis we compare the NDSIs to the performance of the combined B2B6-EVI index. A further comparison against the combined B2B6-NDVI index is shown in File S1.

Materials and Methods

Remote sensing analysis of different crop growth and bare soil conditions (dry, moist, saturated or flooded) was conducted by exploiting both field and satellite data. Field measurements provided accurate data under the controlled conditions necessary to test hypotheses on NDSI performance in different physical conditions. Satellite data were used to assess the applicability of the field data findings in real-world conditions by (i) testing the accuracy of NDSI water detection in pure water pixels, (ii) analysing the impact of different mixed pixel conditions on NDSI performance and (iii) evaluating NDSI behaviour on multi-years MODIS time series. The satellite data were acquired in different locations and different rice cropping systems to permit an evaluation of the NDSIs under different environments and management conditions.

2.1 Field Data

Field spectral data were acquired with Field Spec® Full Range spectro-radiometer (Analytical Spectral Device Inc., U.S., 350–2500 spectral range at 1 nm spectral resolution) in rice fields in an agricultural district of northern Italy. This district provides most (>95%) Italian rice production and about 60% of the total European rice production [31]. Two different data sets were exploited: (i) an archive of spectral measurements acquired from two agronomic experiments, carried out on private land, in 2004 and 2006 at a site located south of Milan, Opera (45°23′ N, 9°11′ E), and (ii) a data set from 2011 under real farm conditions in Pavia province in several fields (site 1: 45°10′N, 9°6′E; site 2: 45°10′N, 9°7′E; and site 3: 45°8′N, 9°9′E) characterised by different soil conditions (dry soil, wet soil and flooding) before sowing and/or plant emergence. For indirect field data acquisition, like spectral measurements, in these locations a specific permission was not required. Five spectra were collected along a transect in each plot/site preceded and followed by a measurement on a reference panel (Spectralon®). Data were collected in nadir direction, about 1 m above the canopy, between 12:00 h and 13:00 h under stable solar conditions. The vegetation/soil reflectance was derived using the following formula:

$$R_t(\lambda) = \frac{L_{target}(\lambda)}{L_{ref}(\lambda)} * R_{ref}(\lambda) \quad (2)$$

where $R_t(\lambda)$ is the target (vegetation, soil or water) reflectance, $L_{target}(\lambda)$ is the radiance measured above vegetation/soil/water, $L_{ref}(\lambda)$ is the radiance of the reference panel and $R_{ref}(\lambda)$ is the specific calibration reflectance coefficient of the reference panel.

Table 2 reports a synthesis of the field data, including GPS coordinates of each location, and auxiliary measurements that were conducted simultaneously with the spectral measurements. Further details on the experimental design of the 2004 and 2006 field data sets as well as the measurement protocols are provided in [32,33].

2.2 Satellite Data and Study Sites

Four Landsat images, acquired over rice-producing regions in Asia and Europe, were processed to provide a spatial representation of the physical status of rice cultivated area at high resolution (30 m). The four images represent well-known rice areas: Piemonte-Lombardia regions (Italy) – the largest rice-growing areas in Europe, West Godavari (India) – a highly productive district in Andhra Pradesh, the Mekong delta (Vietnam) – the major rice-producing region of the country, and Siem Reap (Cambodia) – where rice is cultivated extensively. Each area has distinct geography, environment and crop management practices that represent most of the variation in rice systems.

Table 2. Field data characteristics and cardinality of available spectra for each data set (#).

Data set	Year	Location	Lat. (N)	Lon. (E)	Sampling area	Field condition and measure	Additional measure
Experimental data set	2004*	Opera	45°23′	9°11′	1 field, 40 plots$	Rice in flooded field (# 320)	LAI/fAPAR, Nitrogen, SPAD, picture
	2006**	Opera	45°23′	9°11′	1 field	Dry soil (# 4)	Water level, picture
					2 field	Wet soil (# 10)	Water level, picture
					3 field	Flood soil&,£ (# 24)	Water level, picture
					1 field	Rice& (# 6)	Water level, picture
Farm conditions data set	2011***	Carbonara Ticino	45°10′	9°6′	1 field§	Dry soil (# 18)	Water level, picture
		San Martino Siccomario	45°10′	9°7′	1 field§	Dry soil (# 16)	Water level, picture
					2 field§	Flood soil (5–10 cm) (# 37)	Water level, picture
		Travacò Siccomario	45°8′	9°9′	1 field§	Flood soil (15 cm) (# 18)	Water level, picture

*eight weekly measurements from 17 June to 16 August;
**28 June;
***19 May.
$2 varieties, 4 replicates, 5 treatments (40 plots);
&Different water height (3–10 cm),
£Presence of algae (# 16).
§1 transect per field, about 15 spectra for each transect.

The Italian site (ITA) is situated at the confluence of Sesia, Ticino and Po rivers where rice is cultivated once per year in irrigated and highly managed cropping systems. The crop cycle lasts for 130–150 days from the second half of April/early May (seeding period) to the end of September (harvesting) [34]. Flooding starts in April and finishes at the end of August. Fields are either flooded a few days before seeding for wet seeding or flooded three weeks after seeding for dry seeding.

The Indian site (IND) is situated in the southern state of Andhra Pradesh, where irrigated rice is mainly grown in the deltaic tracts of the Godavari river [35]. Up to three crops per year are possible and two of these may be cultivated with rice [36]. Rice sowing occurs in May–June for the *Kharif* or monsoon season, and December–January for the *Rabi* or winter season.

The other two Asian sites, Cambodia (KHM) and Vietnam (VNM), are located in the Lower Mekong river basin. The monsoon rain from May to October leads to annual flooding along the Mekong river from August to December [12]. The KHM site includes Tonle Sap Lake, where rice cultivation starts in January–February after the long monsoon flood recession [37]. The Mekong Delta site (VNM) can have up to three irrigated rice crops per year with three flooding periods, November–December (winter–spring season), April–June (summer–autumn season) and August (autumn–winter season) [38].

The US Geological Service (USGS) archive – Global Visualization Viewer (GloVis; http://glovis.usgs.gov/) – was used to select appropriate Landsat images for these sites. Acquisition dates were selected according to the above rice season information to capture flooded rice fields at the beginning of the crop season (Table 3). We selected Landsat images from April 2003 in Italy (path-row 194-28/29), January 2002 in India (path-row 142-48) and November 2000 in both Vietnam and Cambodia (path-row 125-53 and 127-51, respectively). We selected subsamples within each Landsat footprint to capture rice areas and surrounding areas with other crops, natural vegetation, bare/wet soil conditions and water bodies. These subsamples provided good representative areas for testing the skill of the NDSIs for rice flood detection.

The MODIS TERRA MOD09A1 500-m product, validated version collection V005, was selected to perform NDSI analysis. The MOD09A1 product provides an 8-day composite in seven surface reflectance spectral bands in the RED (630–690 nm; b1), NIR (780–900 nm; b2), BLUE (450–520 nm, b3), GREEN (530–610 nm, b4), NIR (1230–1250 nm; b5), SWIR (1550–1750 nm, b6) and SWIR (2090–2350 nm, b7) wavelengths. Science Data Sets provided for this product include a pixel-specific quality assessment and the day of the year (DOY) with solar view and zenith angles of the acquisition selected in the composite. All data are geocoded in the Sinusoidal projection. Composite data for the tiles that cover southern Europe (h18v04), eastern India (h28v07) and South-East Asia (h25v07) were downloaded for the four-year period (2000–2003) in order to have the same temporal and spatial extent as the Landsat images. NDSIs were calculated for each MODIS scene.

Finally, within each study site we selected locations – based on the Landsat imagery – where rice cultivation was clearly visible and we obtained four years (2000–2003) of MOD09A1 time series data from the University of Oklahoma web service (www.eomf.ou.edu/visualization/gmap/).

Figure 1 reports a synthetic view of the selected study areas and satellite data. Table 3 shows information on the MODIS and Landsat images that were analysed, indicating the geographic location (countries and coordinates) as well as image reference (tiles and pixel coordinates) and the number of rice seasons per year.

2.3 Pre-processing and Analysis of Field Data

In order to simulate the response of the MODIS sensor, the spectral reflectance collected on the ground was resampled with the specific MODIS spectral response functions. The band-equivalent MODIS reflectance (R_{MODISi}) was derived by integration over each band's spectral response function using IDL code (*EXELIS, Visual Information Solution*). MODIS reflectances (R_{MODISi}) simulated from field data were combined to calculate the NDSIs (Table 1). NDSIs are labelled using their first and second wavelengths $\lambda_1\lambda_2$, i.e. b1.b2 when MODIS Band1 (RED) and

Table 3. Study areas: satellite data available and selected location used to study the temporal series of NDSIs.

Land use		**MODIS Terra**		**Landsat 7 ETM+**			**Location for temporal series**			
Sites	*# rice seasons*	*Tile*	*date**	*path-row*	*date*	*Analysed area [km]*	*Lon. (E)*	*Lat. (N)*	*MODIS Row*	*MODIS Column*
ITA	Single	h18v04	25/04/03	194-28/29	24/04/03	80×63	8.26°	45.24°	1143	1396
VNM	Triple	h28v07	09/11/00	125-53	07/11/00	187×85	105.54°	10.14°	2367	934
IND	Double	h25v07	01/01/02	142-48	03/01/02	180×87	16.70°	81.84°	792	2014
KHM	Single	h28v07	01/11/00	127-51	05/11/00	138×60	103.71°	13.53°	1553	200

*Reference composite date.

MODIS Band2 (NIR) are used, and grouped by spectral range as VIS/NIR ($NDSI_{b4.b2}$, $NDSI_{b1.b2}$, $NDSI_{b4.b5}$, $NDSI_{b1.b5}$), VIS/SWIR ($NDSI_{b4.b6}$, $NDSI_{b4.b7}$, $NDSI_{b1.b6}$,$NDSI_{b1.b7}$), NIR/NIR ($NDSI_{b2.b5}$) and NIR/SWIR ($NDSI_{b2.b6}$, $NDSI_{b2.b7}$).

One-way analysis of variance (ANOVA) followed by the post-hoc Tukey HSD (honestly significant difference) multiple range test were used to compare NDSI responses to different targets (flooded field, wet soil, dry soil and vegetation) in order to highlight which NDSI was more sensitive and theoretically more robust in detecting and differentiating between targets. The vegetation class was divided into two sub-classes for leaf area index (LAI): less than or equal to 2.0 representing a sparse canopy (i.e. a young rice crop or shortly after transplanting) and greater than 2.0 representing full canopy cover. Statistical analysis was performed using the R software environment (R version 2.15.0, packages agricolae and stat; www.r-project.org/). Finally, in order to assess the skill of each index in separating water surfaces from other surfaces, we computed a measure of separability, *S*:

$$S=|\mu_{i,w}-\mu_{i,u}|/(\sigma_{i,w}+\sigma_{i,u}) \qquad (3)$$

where $\mu_{i,w}$ and $\sigma_{i,w}$ are, respectively, the mean and standard deviation of fully flooded conditions and $\mu_{i,u}$ and $\sigma_{i,u}$ are, respectively, the mean and standard deviation of other surfaces: pure soil, full cover vegetation or mixed conditions. The *S* statistic assesses the degree of discrimination between two groups. A low value of *S* (<1) denotes that the distribution of the classes overlap significantly and the ability to separate (or discriminate) the two

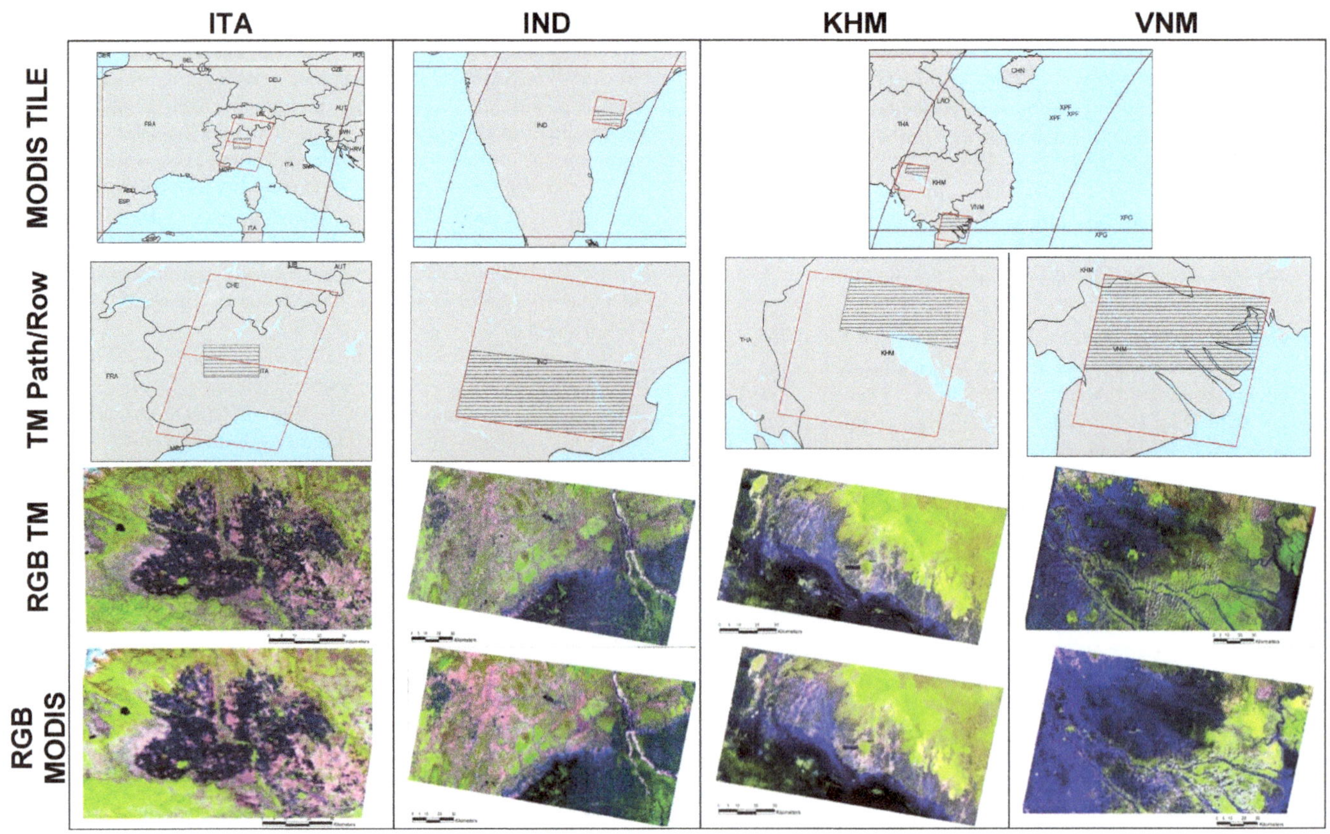

Figure 1. Study areas and satellite data analysed for the four case studies (ITA, IND KHM, VNM). Black and red boxes indicate, respectively, the MODIS tile (H18v04,H25v07 and H27v07) and Landsat (194-28/29, 125-53, 142-48 and 127-51) extent. Black lines indicate the analysed area for which a Landsat – RGB 543- and MODIS-RBG 621– colour composite is provided.

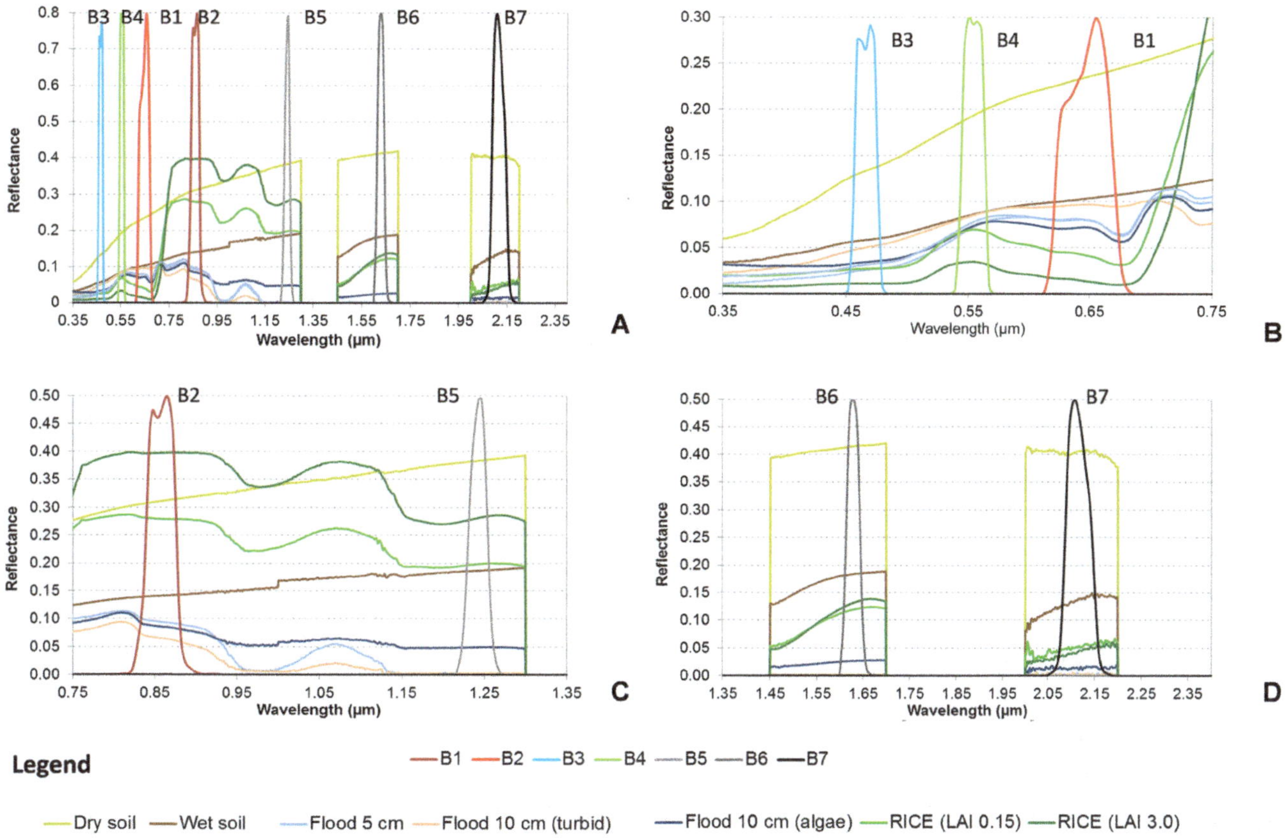

Figure 2. Comparison of different surface responses in different spectral ranges: full range 350–2500 (a), visible [0.35–0.75] (b), near infrared [0.75–1.35] (b) and shortwave infrared [1.4–2.5] (b). The position of MODIS spectral response function of bands 1–7 is overlayed on the field data.

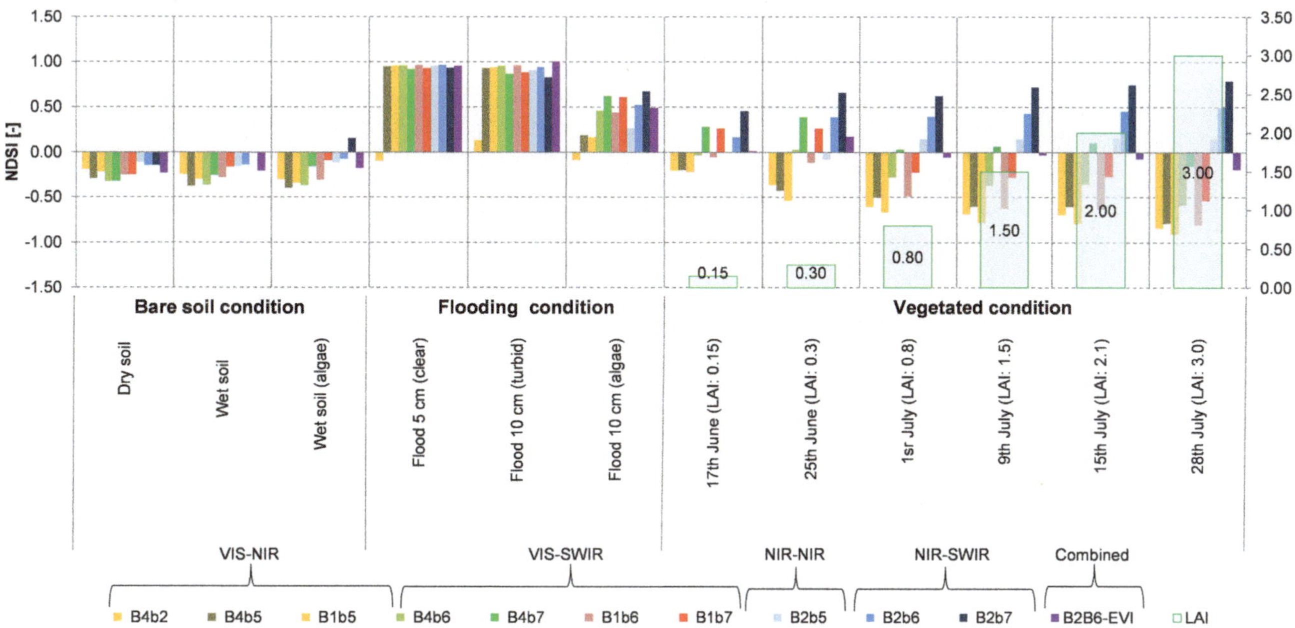

Figure 3. Behaviour of spectral indices (NDSIs) in different paddy rice conditions from bare, dry soil to dense vegetation.

Table 4. Results of ANOVA between flood condition (W) and other classes: wet soil (WS), dry soil (DS), sparse canopy (V1) and full canopy cover vegetation (V2) and post-hoc Tukey test.

		VIS/NIR						VIS/SWIR																		NIR/NIR			NIR/SWIR						COMPOSED		
		B4B2			B1B2			B4B5			B1B5			B4B6			B4B7			B1B6			B1B7			B2B5			B2B6			B2B7			B2B6-EVI		
Class	r	μ	σ	G	μ	σ	G	μ	σ	G	μ	σ	G	μ	σ	G	μ	σ	G	μ	σ	G	μ	σ	G	μ	σ	G	μ	σ	G	μ	σ	G	μ	σ	G
W	79	−0.01	0.02	**a**	0.04	0.02	**a**	0.69	0.04	**a**	0.75	0.04	**a**	0.78	0.03	**a**	0.74	0.02	**a**	0.75	0.04	**a**	0.78	0.03	**a**	0.71	0.030	**a**	0.80	0.030	**a**	0.75	0.020	**a**	0.85	0.05	**a**
WS	10	−0.35	0.05	**bc**	−0.33	0.07	**b**	−0.28	0.13	**b**	−0.04	0.14	**b**	−0.21	0.13	**b**	−0.03	0.11	**bc**	−0.04	0.10	**b**	−0.21	0.10	**b**	0.05	0.100	**b**	0.13	0.100	**cd**	0.30	0.100	**b**	−0.56	0.15	**bc**
DS	38	−0.20	0.01	**b**	−0.12	0.007	**ab**	−0.31	0.02	**b**	−0.29	0.01	**b**	−0.36	0.02	**b**	−0.37	0.02	**c**	−0.29	0.02	**b**	−0.36	0.01	**bc**	−0.12	0.006	**b**	−0.17	0.009	**d**	−0.18	0.009	**c**	−0.42	0.022	**b**
V1	35	−0.49	0.03	**c**	−0.61	0.03	**c**	−0.40	0.03	**bc**	−0.07	0.04	**bc**	−0.20	0.03	**b**	0.14	0.03	**b**	−0.07	0.04	**b**	−0.20	0.04	**b**	0.11	0.030	**b**	0.33	0.020	**bc**	0.58	0.030	**ab**	−1.8	0.025	**d**
V2	35	−0.76	0.01	**d**	−0.90	0.01	**d**	−0.70	0.02	**c**	−0.48	0.01	**c**	−0.48	0.02	**b**	−0.06	0.02	**bc**	−0.48	0.01	**c**	−0.48	0.02	**c**	0.14	0.009	**b**	0.46	0.010	**b**	0.74	0.009	**a**	−1.1	0.085	**c**

Letters indicate statistically different groups ($\alpha = 0.01$).

groups is poor. A high value of S (>>1) denotes that the means are well separated and that the two classes are relatively easy to discriminate. The numerator of the S statistic describes the difference in mean values between the two considered classes, while the denominator provides a degree of noise/dispersion of the classes. Classes with larger σ's have wider distribution, lower S values and thus will be more likely to overlap. Therefore, the S statistic can be effectively used to compare the ability of different NDSIs to discriminate between water and other classes.

2.4 NDSI Satellite Data Analysis

2.4.1 Threshold definition and evaluation on pure pixel condition. The subsample of each Landsat 7 ETM+ image, centered on a rice-growing area with evident flood condition, was selected (Figure 1 and Table 3) and classified by unsupervised ISODATA algorithm (*EXELIS, Visual Information Solution*) without performing any atmospheric correction [39]. Subsampling at the KHM site was particularly necessary to exclude strongly cloud contaminated areas while at the VNM site we analysed only the Landsat image that overlapped the MODIS tile. We characterised the four sites into three land cover (LC) classes describing the surface physical condition at the time of the satellite overpass: water (mainly representing the typical condition of agronomic flooding in rice cultivation but also river and water bodies), vegetation (natural system and established crops) and bare soil (dry or wet). We also identified a no data class corresponding to cloud cover and shadow. These high-resolution land cover (LC) maps were used to characterise the composition of each MODIS pixel and to interpret the changes in NDSI values in relation to mixture components. To perform this analysis, a random selection of more than 3500 MODIS pixels was extracted for each site, thus creating a database that recorded for each pixel the percentage of different land cover classes as provided by LC maps as well as the MODIS band response and NDSI values.

From this database, we selected 4000 pure MODIS pixels – distributed equally across each site –2000 of water and 2000 of no water (1000 of soil and 1000 of vegetation). In order to define thresholds for surface water detection with different NDSIs, this 4000 pixel sample was randomly split into two samples equally distributed across each site and across LC classes: the first group was used for calibration of NDSI thresholds and the other for validation of the NDSI classification performance.

We used a recursive partitioning technique implemented in R (rpart package) to identify the best performing threshold for each analysed NDSI. The partitioning is based on minimal error fitting on the calibration data set and it identifies the NDSI value that best discriminate pure flooded pixels from pure vegetated or pure soil pixels.

Accuracy of the classification provided by the identified thresholds was assessed by the metrics computed from the error or confusion matrix using the validation/testing set: Overall Accuracy (OA), Commission Error (CE) and Omission Error (OE) were calculated to identify the best performing NDSI. Accuracy was also expressed in terms of the kappa coefficient of agreement and a Z-test was used to evaluate the significance of differences in classification outputs [40].

2.4.2 Mixture effect analysis. To analyse the effect of different subpixel water percentages on NDSI values, we extracted a stratified random sample of 625 MODIS pixels from each of the four test sites. The 625 pixels were equally distributed across five classes of MODIS subpixel water percentage (100-80%, 80-60%, 60-40%, 40-20%, 20-0%); consequently, a data set of 2500 ($125 \times 5 \times 4$) pixels was obtained.

Table 5. Separability (S) between flood condition and other classes: wet soil (WS), dry soil (DS), sparse canopy (V1) and full canopy cover vegetation (V2).

	VIS/NIR								VIS/SWIR								NIR/NIR		NIR/SWIR				COMBINED	
	B4B2		B1B2		B4B5		B1B5		B4B6		B4B7		B1B6		B1B7		B2B5		B2B6		B2B7		B2B6-EVI	
Class	S	Rank	S	Rank	S	Rank	S	Rank	S	Rank	S	Rank	S	Rank	S	Rank	S	Rank	S	Rank	S	Rank	S	Rank
WS	1.20	8	0.93	11	1.27	6	1.22	7	**1.40**	**1**	1.30	5	1.38	3	1.39	2	1.10	10	1.14	9	0.82	12	1.35	4
DS	0.96	11	0.77	12	2.27	9	2.14	10	2.79	6	3.38	2	2.64	7	3.22	4	2.64	8	3.25	3	**3.74**	**1**	2.89	5
V1	1.63	8	1.80	6	1.99	4	**2.10**	**1**	2.01	3	1.52	9	2.08	2	1.67	7	1.39	10	1.26	11	0.47	12	1.92	5
V2	3.46	4	**4.61**	**1**	3.19	6	3.81	3	3.15	7	2.27	9	4.01	2	3.44	5	1.76	10	1.12	11	0.02	12	2.76	8
S_{AVG}	1.81		2.03		2.18		2.32		2.34		2.12		**2.53**		2.43		1.72		1.69		1.26		2.23	
$RANK_{AVG}$	7.75		7.50		6.25		5.25		4.25		6.25		**3.5**		4.5		9.5		8.5		9.25		5.5	

SAVG refers to the average separability value of each index for water with respect to all the other classes. Rank refers to the relative position (1 = best and 12 = worst) of each index separability with respect to the others and RankAVG the mean value.

Table 6. Separability (S) score between pure water MODIS pixels and other LCs (soil and vegetation) for the different NDSIs and combined index at all test sites.

	VIS/NIR								VIS/SWIR								NIR/NIR		NIR/SWIR				Combined	
	B4B2		B1B2		B4B5		B1B5		B4B6		B4B7		B1B6		B1B7		B2B5		B2B6		B2B7		B2B6-EVI	
Site	S	Rank	S	Rank	S	Rank	S	Rank	S	Rank	S	Rank	S	Rank	S	Rank	S	Rank	S	Rank	S	Rank	S	Rank
IND	2.0	11	1.3	12	3.3	6	2.6	10	4.2	3	4.5	2	3.3	5	4.0	4	3.2	7	3.1	8	3.0	9	**4.7**	**1**
ITA	1.6	10	1.3	12	2.6	7	2.1	8	3.5	.2	3.4	4	2.7	5	3.4	3	2.7	6	1.8	9	1.5	11	**4.8**	**1**
KHM	1.5	10	1.6	8	1.8	7	2.0	6	2.0	5	2.2	3	2.2	2	**2.6**	**1**	1.5	9	0.5	12	−0.2	11	2.2	4
VNM	2.5	8	2.8	7	3.6	3	**3.9**	**1**	3.5	4	3.3	6	3.8	2	3.5	5	1.8	10	1.4	11	1.2	12	2.1	9
S_{AVG}	1.90		1.80		2.80		2.70		3.30		3.30		3.00		**3.40**		2.30		1.70		1.40		**3.40**	
$RANK_{AVG}$	9.75		9.75		5.75		6.25		3.5		3.75		3.5		**3.25**		10		12		10.75		3.75	

SAVG refers to the average separability value of each index for water with respect to all the other classes at the four sites. Rank refers to the relative position (1 = best and 12 = worst) of each index separability with respect to the others and RankAVG the mean value.

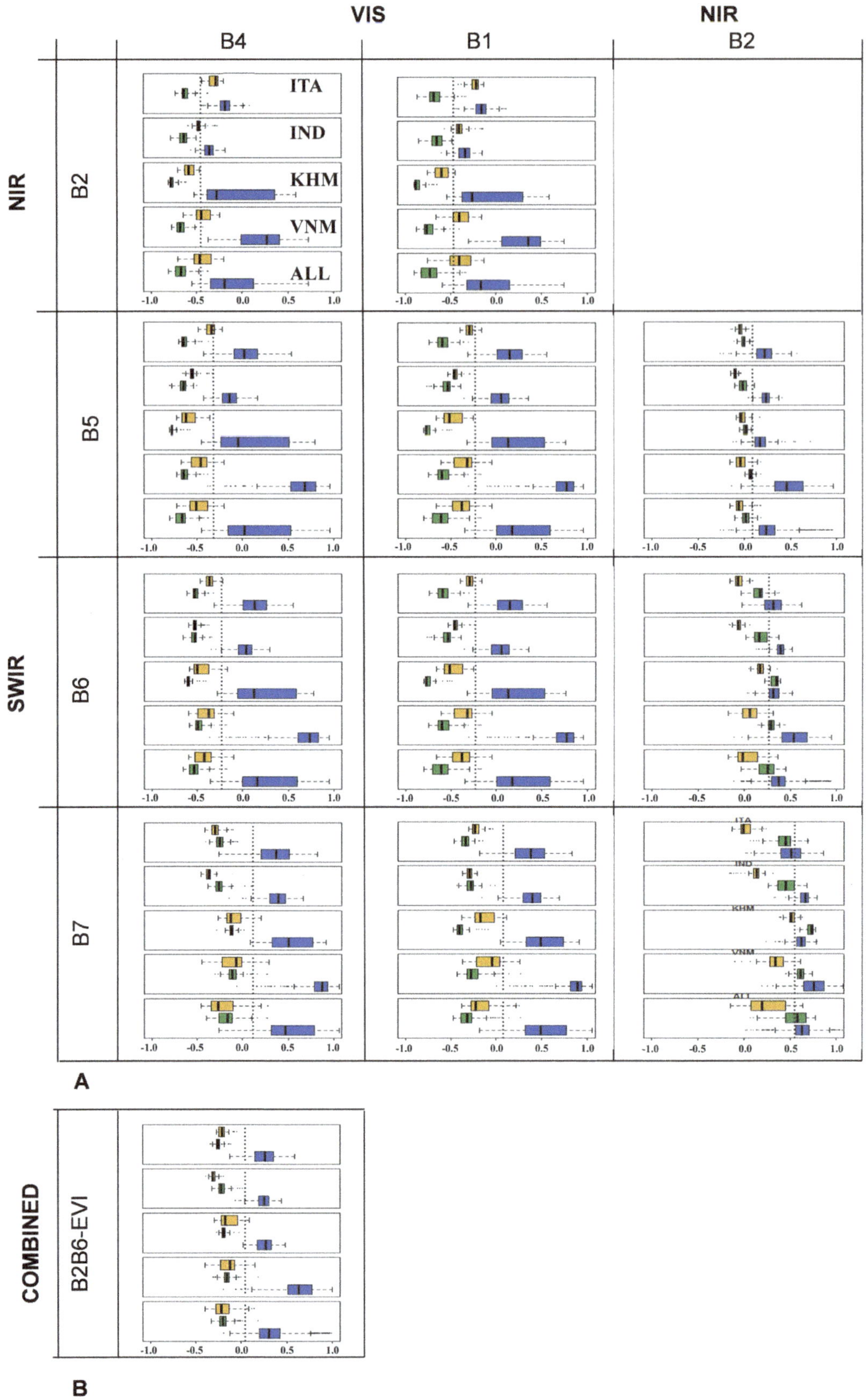

Figure 4. Boxplots of analysed indices for pure water pixels (blue), pure soil pixels (orange) and pure vegetation pixels (green). Every boxplot has the same cardinality. The vertical dotted lines show the calculated threshold. Panel (a) shows the NDSIs and Panel (b) the combined B2B6-EVI index.

Table 7. Threshold values (T) for the NDSI/combined index and detection accuracy (%) on pure pixel validation data set.

	VIS/NIR				VIS/SWIR				NIR/NIR	NIR/SWIR		COMBINED
	B4B2	**B1B2**	**B4B5**	**B1B5**	**B4B6**	**B4B7**	**B1B6**	**B1B7**	**B2B5**	**B2B6**	**B2B7**	**B2B6-EVI**
T	−0.451	−0.470	−0.312	−0.258	−0.228	0.114	−0.223	0.084	0.092	0.271	0.555	0.044
OA	87%	81%	95%	91%	**97%**	**96%**	**97%**	**96%**	93%	80%	71%	**97%**
CE	19%	26%	4%	6%	2%	2%	3%	3%	3%	20%	28%	**1%**
OE	4%	4%	7%	14%	**3%**	7%	**3%**	6%	10%	21%	29%	5%

This selection was used in a second-order polynomial regression between the NDSI values and MODIS subpixel water percentage. Noise equivalent (NE) (Eq. 4) was then computed to measure the sensitivity of each NDSI to changes in relation to water percentage:

$$NE = \frac{RMSE}{\left[\frac{d(NDSI)}{d(\%\,water)}\right]} \tag{4}$$

where RMSE is the root mean square error of the water percentage-NDSI regressive model and d(NDSI)/d(%water) is the first derivative of the NDSI with respect to water percentage. NE was chosen since it permits a direct comparison of indices with different scales and dynamic ranges [41–43]. NE provides a measure of how well the NDSI responds to water percentage across its entire range of variation. A low NE value means that an NDSI is sensitive to water presence.

2.4.3 Mapping performance in mixture conditions. Thematic cartography produced by the automatic classification of low-/medium-resolution satellite data can be affected by low-resolution bias when compared to high-resolution reference data [44]. Low-resolution bias is the inaccuracy introduced by the difference in spatial resolution between high- and low-resolution data and is not related to the performance of a classification algorithm. The bias, linked to the characteristics of the features on the ground and a function of shape, size and fragmentation of the target under analysis, can be computed for each test site using a Pareto boundary analysis [44]. This approach assesses the potential maximum accuracy that a classification method can achieve in a specific experimental condition as a function of the actual fragmentation of the environment. Thus it can be used to rank the performance of the NDSIs.

For each site we derived moderate resolution surface water maps from each NDSI using the identified thresholds (§2.4.1) to evaluate the performance of the NDSIs in real conditions where mixed pixel problems can affect classification. These moderate resolution NDSI water maps were compared with the high resolution LC map derived from Landsat for each site and their performance was quantified with respect to the Pareto boundary in the Commission Error (CE) – Omission Error (OE) space.

Results and Discussion

Figure 2 shows the field spectral data for different surfaces analysed together with the position of the seven MODIS spectral response functions in the range 350–2500 nm. Figure 2a shows the entire spectral range for which data were acquired; Figure 2b, 2c and 2d provide details for the VIS, NIR and SWIR regions, respectively.

In the visible (VIS) range (MODIS bands B1, B3 and B4), the wet soil and flooded field conditions have similar spectral responses, with a slight increase from short to long visible wavelengths. An increase in water presence, from saturated soil to a 5 to 10 cm deep water layer, reduces the reflectance, mainly in the RED region, but with a lesser effect when compared to the change observed between dry and wet soil. Dry soils show a typical increase, with the highest reflectance for RED (up to 25%). On the contrary, vegetation behaviour presents the lowest reflectance in the photosynthetically active radiation region (PAR, 350–700 nm), confirming the typical absorption features which correspond to the BLUE and RED bands due to chlorophyll action. It is interesting to note that, when water, between 5 and 10 cm deep, is present on soil, the main changes in the VIS region are due to the presence of algae or suspended matter. Algae in flooded soil show an absorption feature in RED at 670 nm, whereas turbid water behaves similarly to saturated soil and exhibits high reflectance. The B1 RED band shows the strongest differences between the three land conditions of dry soil, flood/wet soil and vegetation (Figure 2b).

In the near infrared (NIR) range (B2 and B5), the surfaces have a different albedo behaviour: B2 shows high values for dense and healthy vegetation conditions, followed by dry soil, low dense vegetation, wet soil and flooded. Band B5 shows a strong sensitivity to water content and drops are visible in vegetation and flooded soil. The presence of algae increases the reflectance in this spectral range; a flood condition shows values close to zero (Figure 2c). Indeed, for wavelengths greater than 1100 nm, the presence of a water layer produces the largest differences between the spectra.

In the shortwave infrared (SWIR) range (B6 and B7), water absorption features dominate the spectral response. The absorption coefficient increases with wavelength in an exponential way with specific high peaks at 1400 and 1925 nm [45]. Flooded areas present a signal close to zero even when algae or sediments are present. The vegetation spectral response also decreases markedly, reaching an absolute value of 5%, showing a reflectance two-thirds lower than wet soil (Figure 2d).

3.1 NDSI Analysis from Field Data

Figure 3 shows the values of the tested NDSIs calculated on the R_{MODISi} that was simulated from the field spectra acquired in 2004 and 2005 in the experimental field. The data – from left to right – represent a plausible temporal development of paddy conditions during a crop season, ranging from bare soil to flooding and then through the vegetative stages from germination to tillering prior to flowering. NDSIs – grouped by wavelength region combinations – are represented on the primary y axis in yellow, green/red and blue tones for VIS/NIR, VIS/SWIR, NIR/NIR and NIR/SWIR, respectively. The combined B2B6-EVI index is

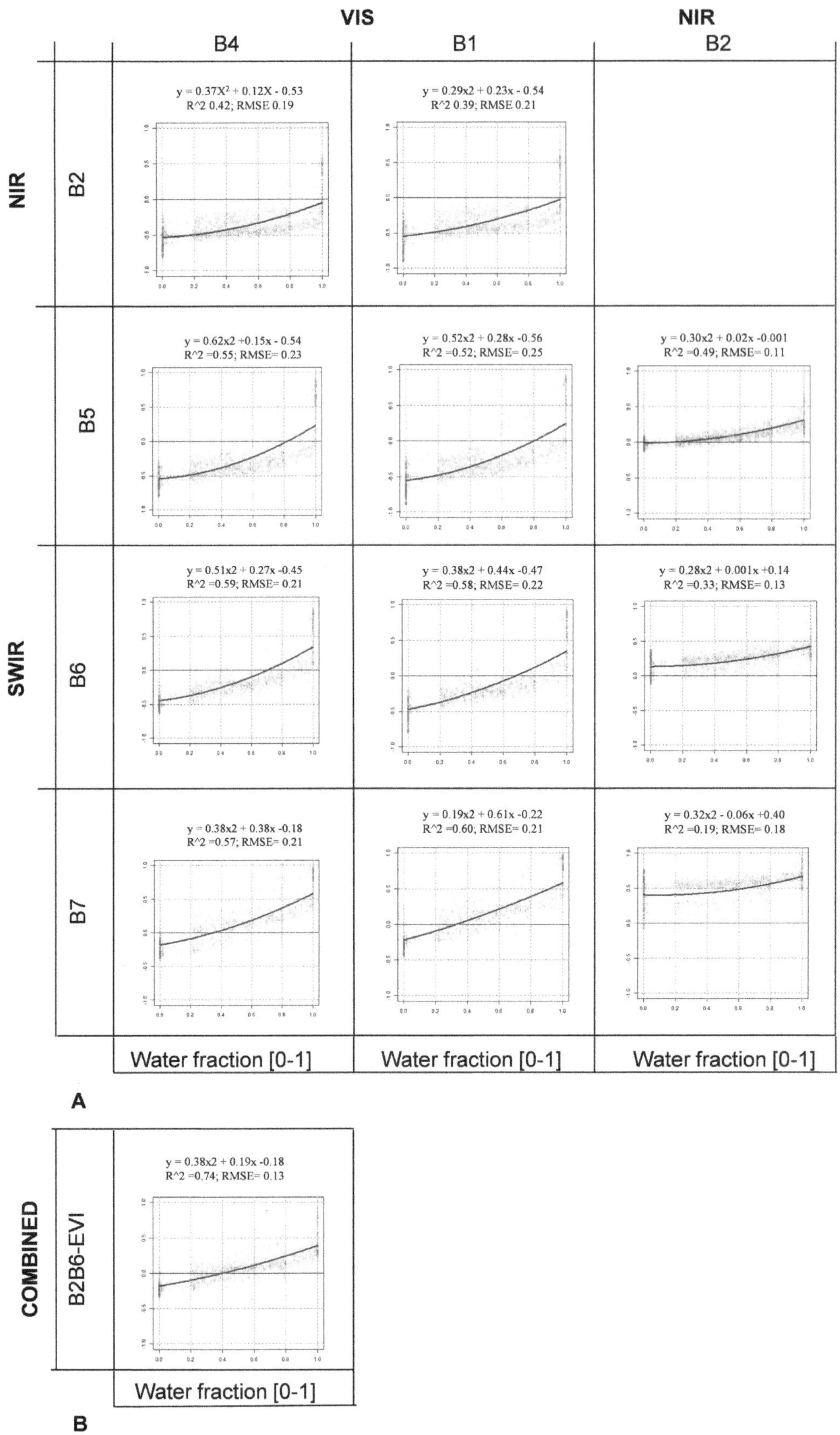

Figure 5. Correlation between the 11 NDSIs and water percentage in MODIS pixels. Panel (a) shows the NDSIs and Panel (b) the combined B2B6-EVI index.

Table 8. Regression parameters and performances (r^2 and RMSE) for the NDSI/combined index vs water fraction relation.

SITE	Coeff.	VIS/NIR				VIS/SWIR				NIR/NIR	NIR/SWIR		COMBINED
		B4B2	B1B2	B4B5	B1B5	B4B6	B4B7	B1B6	B1B7	B2B5	B2B6	B2B7	B2B6-EVI
ITA	intercept	−0.44	−0.42	−0.46	−0.44	−0.41	−0.26	−0.39	−0.25	−0.02	0.05	0.2	−0.21
	x	0.26	0.31	0.21	0.26	0.12	0.03	0.19	0.13	−0.02	−0.16	−0.26	0.13
	x^2	0.01	−0.04	0.31	0.26	0.47	0.64	0.4	0.55	0.28	0.46	0.61	0.37
	r^2	0.54	0.42	0.72	0.65	0.82	0.82	0.77	**0.83**	0.64	0.55	0.36	0.71
	RMSE	0.13	0.18	0.17	0.2	0.14	0.15	0.16	0.15	0.09	0.12	0.2	0.10
IND	Intercept	−0.52	−0.5	−0.56	−0.54	−0.49	−0.28	−0.47	−0.25	−0.06	0.05	0.28	−0.24
	x	0.2	0.13	0.36	0.29	0.54	0.81	0.46	0.73	0.26	0.38	0.45	0.65
	x^2	−0.05	0.02	0.08	0.15	0.02	−0.12	0.09	−0.05	0.05	−0.01	−0.06	−0.14
	r^2	0.58	0.36	0.87	0.78	**0.93**	**0.93**	0.88	**0.93**	0.87	0.72	0.61	0.72
	RMSE	0.08	0.12	0.09	0.12	0.08	0.10	0.10	0.09	0.06	0.10	0.15	0.10
VNM	Intercept	0.45	0.65	0.47	0.73	0.67	0.85	0.96	1.17	−0.12	−0.07	−0.09	−0.17
	x	−0.56	−0.58	−0.54	−0.57	−0.42	−0.1	−0.46	−0.15	0.03	0.20	0.5	0.43
	x^2	0.36	0.27	0.76	0.59	0.48	0.11	0.28	−0.14	0.62	0.45	0.37	0.00
	r^2	0.8	0.81	**0.91**	**0.91**	**0.91**	**0.91**	**0.91**	**0.91**	0.7	0.55	0.46	0.81
	RMSE	0.19	0.21	0.19	0.2	0.17	0.15	0.18	0.16	0.16	0.17	0.16	0.07
KHM	intercept	−0.64	−0.7	−0.64	−0.71	−0.5	−0.09	−0.6	−0.25	−0.01	0.24	0.62	−0.13
	x	−0.07	0.2	0.01	0.27	0.14	0.1	0.52	0.69	0.11	−0.06	−0.27	0.50
	x^2	0.69	0.45	0.77	0.54	0.62	0.55	0.31	0.09	0.09	0.14	0.27	0.29
	r^2	0.69	0.7	0.79	0.79	0.82	**0.87**	0.82	0.86	0.64	0.11	0.03	0.75
	RMSE	0.23	0.23	0.22	0.23	0.19	0.13	0.21	0.16	0.07	0.09	0.10	0.15

shown in purple. The secondary y axis shows field LAI in green to provide information on the status of the rice crop in terms of biomass and development.

Table 4 reports the results of the ANOVA and post-hoc Tukey test to compare NDSI responses among different targets. Classes that are significantly different ($p<0.01$) from each other in terms of NDSI values are indicated with different letters. Table 5 reports the values of the separability metric (S) computed between flooded condition (W) and each of wet soil (WS), dry soil (DS), sparse canopy cover (V1) and full canopy cover vegetation (V2) for each NDSI and the combined B2B6-EVI index. A global score for separability is provided in terms of both average separability value (S_{AVG}) and ranking position ($Rank_{AVG}$), where rank is from best (1) to worst (12) separability performance between the different surfaces.

It is interesting to note that, despite some specific differences, the NDSIs belonging to the same spectral category behave in a similar way. NDSIs in VIS-SWIR (B4B6, B4B7, B1B6 and B1B7) have significantly different values for water with respect to the other classes (see groups in ANOVA analysis, Table 4) and hence high S values (S_{AVG}: 2.34, 2.12, 2.53 and 2.43; Table 5) and first positions in separability ranking ($RANK_{AVG}$: 4.25, 6.25, 3.5 and 4.5; Table 5). These indices are clearly diagnostic of water surfaces, having positive values only for flood conditions and negative values for soil (both wet and dry) and vegetation cover. The only exception is B4B7 that has positive values for sparse vegetation but is still significantly different from water conditions.

Also, VIS/NIR NDSIs that use band 5 (B1B5 and B4B5) show a performance similar to that of the previous category in terms of average separability (2.32 and 2.18) and average ranking position (5.25, 6.25). However, we observe that B4B5, that was proposed by [21] based on their analysis of spectral library data is ranked sixth both for separability score and average rank.

The post-hoc test shows that both B2B7 and B1B2 cannot separate water from all the other classes. B2B7 fails to distinguish water from vegetation and BIB2 fails to distinguish water from dry soil, even if they show the highest separability performance with respect to dry soil (B2B7) and dense vegetation (B1B2).

B2B5 is able to separate water from other surfaces and negative values are reported only for dry soil conditions. Gao [18] proposed this index to monitor liquid water in vegetation and it is based on water absorption features at 1240 nm. The results are therefore in agreement with the theoretical assumption: higher values must be interpreted as higher water presence (W>>V2>V1>WS; Table 4). The B2B6 results can be interpreted in a similar way; however, the use of a SWIR band with longer wavelength (1600 nm), where water absorption is stronger, determines that vegetation values are no longer significantly different from those of wet soil. This index also has negative values for dry soil and can easily separate bare soil conditions from flooding ($S=3.25$; rank = 3).

Finally, B2B7, similar to B2B5 and B2B6, easily separates water from dry soil ($S=3.74$; rank = 1) but is not able to differentiate classes that present medium/high wet conditions, and dense canopy cover has values that are similar to those of water ($S=0.02$; rank = 12). The B2B7 index was originally proposed to evaluate burn severity on vegetated areas as a consequence of water loss [46]. It is important to highlight that these last three indices were developed and adopted to monitor water content, primarily in vegetation, and are therefore sensitive and strongly correlated to the total amount of water that can be remotely sensed.

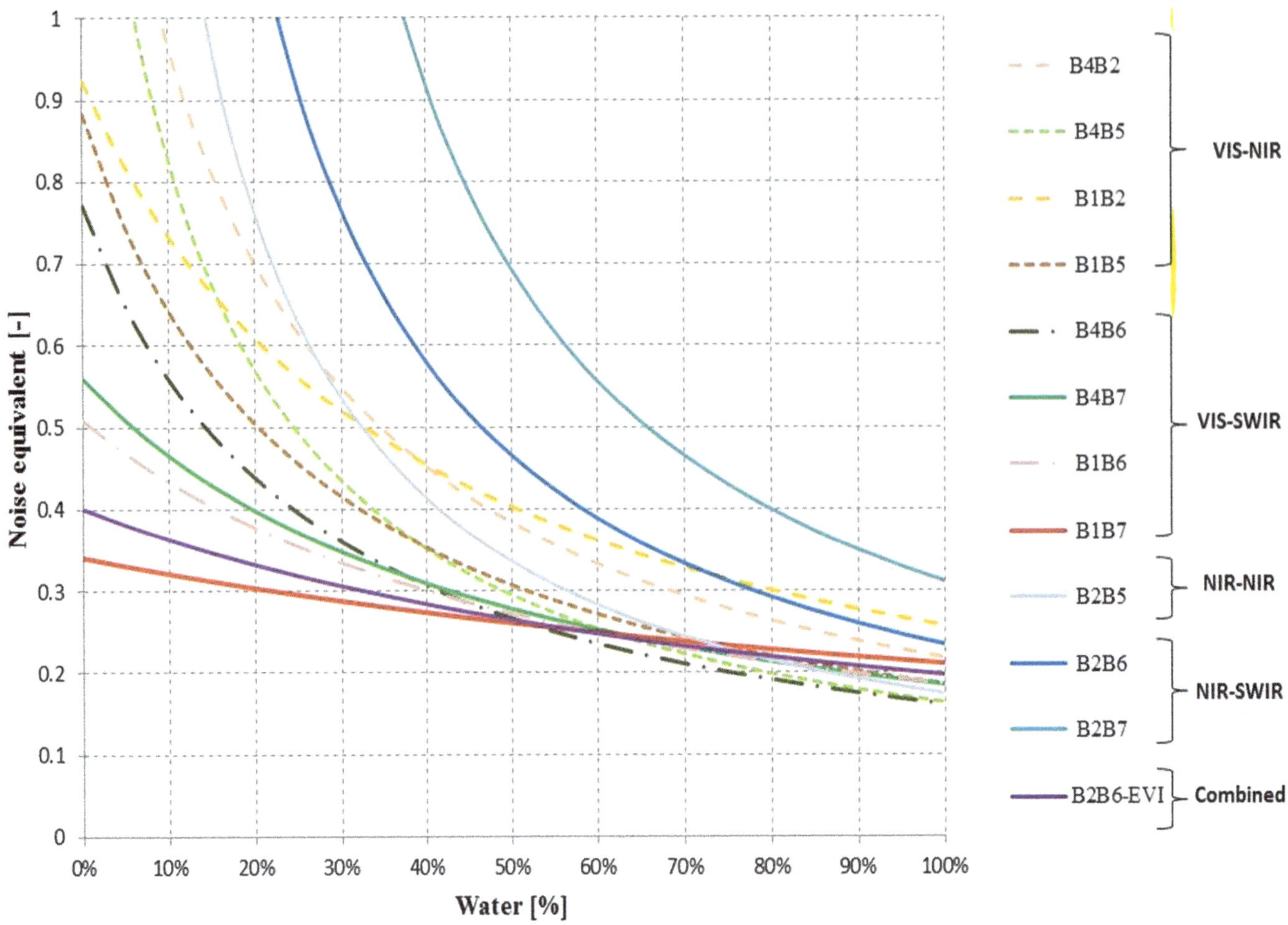

Figure 6. Noise equivalent (see Eq. 4) as a function of water percentage for the 11 NDSIs and the combined B2B6-EVI index.

The combined B2B6-EVI index is a good indicator of water condition. ANOVA demonstrates that the flood condition is statistically different from other classes and the separability analysis shows an average value that is comparable to the VIS/SWIR categories though it is placed fifth in the ranking.

3.2 NDSI Analysis from Satellite Data

The high-resolution maps, derived from Landsat (see §3.4), show agricultural landscapes dominated by the agronomic flooding of rice fields but with a certain degree of heterogeneity as a consequence of different timing in field management activities (i.e., flooded or bare soil conditions) and the presence of other crops and natural vegetation patches. It is this heterogeneity of land cover and land use that permits an analysis of MODIS NDSI performance in real-world conditions. We first report the results of the pure MODIS pixel separability analysis with threshold definition and then the mixed MODIS pixel NDSI sensitivity analysis.

3.2.1 Separability analysis on MODIS pixels. Table 6 shows the results of the separability analysis conducted on 2000 pure MODIS pixels to determine which NDSIs best discriminate pure water pixels from other LC classes. The separability and ranking are given for each test site as well as average values across sites. The NDSIs of the VIS/SWIR family (B4B6, B4B7, B1B6, and B1B7) outperform the other NDSIs consistently across sites. In particular, B1B7 and B4B6 are the best across all sites ($Rank_{AVG}$: 3.255) but these NDSIs have the highest ranking position only for KHM (B1B7), whilst B4B6, B4B7 and B1B6 are ranked second in ITA, IND and KHM/VNM respectively. Only the B1B5 NDSI, from the VIS/NIR family, outperforms the VIS/SWIR category at the VNM site (S: 3.9, Rank: 1).

These results agree with the analysis of field data: the VIS/SWIR family performs better than other NDSIs. Among them, the average separability value indicates a slightly better performance for B1B7 (S_{AVG}: 3.4) and B4B7 (S_{AVG}: 3.3), both based on longer SWIR wavelength, followed by B4B6 (S_{AVG}: 3.3) and B1B6 (S_{AVG}: 3.0).

The analysis of the combined index B2B6-EVI on real MODIS data confirmed the field data results. This index performs well with a S_{avg} value equivalent to B1B7. However, this value is a consequence of B2B6-EVI being the best ranked index for IND and ITA, whilst in KHM and in particular in VNM the index is not highly ranked. This variability in ranking across sites, which is higher than the ranking variability in any of the VIS/SWIR indices, suggests that the combined B2B6-EVI index is sensitive to local conditions and supports the previous statements on the need for locally adaptive thresholds for this index.

3.2.2 Threshold calibration and validation. According to the separability analysis, NDSIs of the VIS/SWIR family have greater skill in distinguishing water from other surfaces. However, a further comparison was made using a calibration approach to

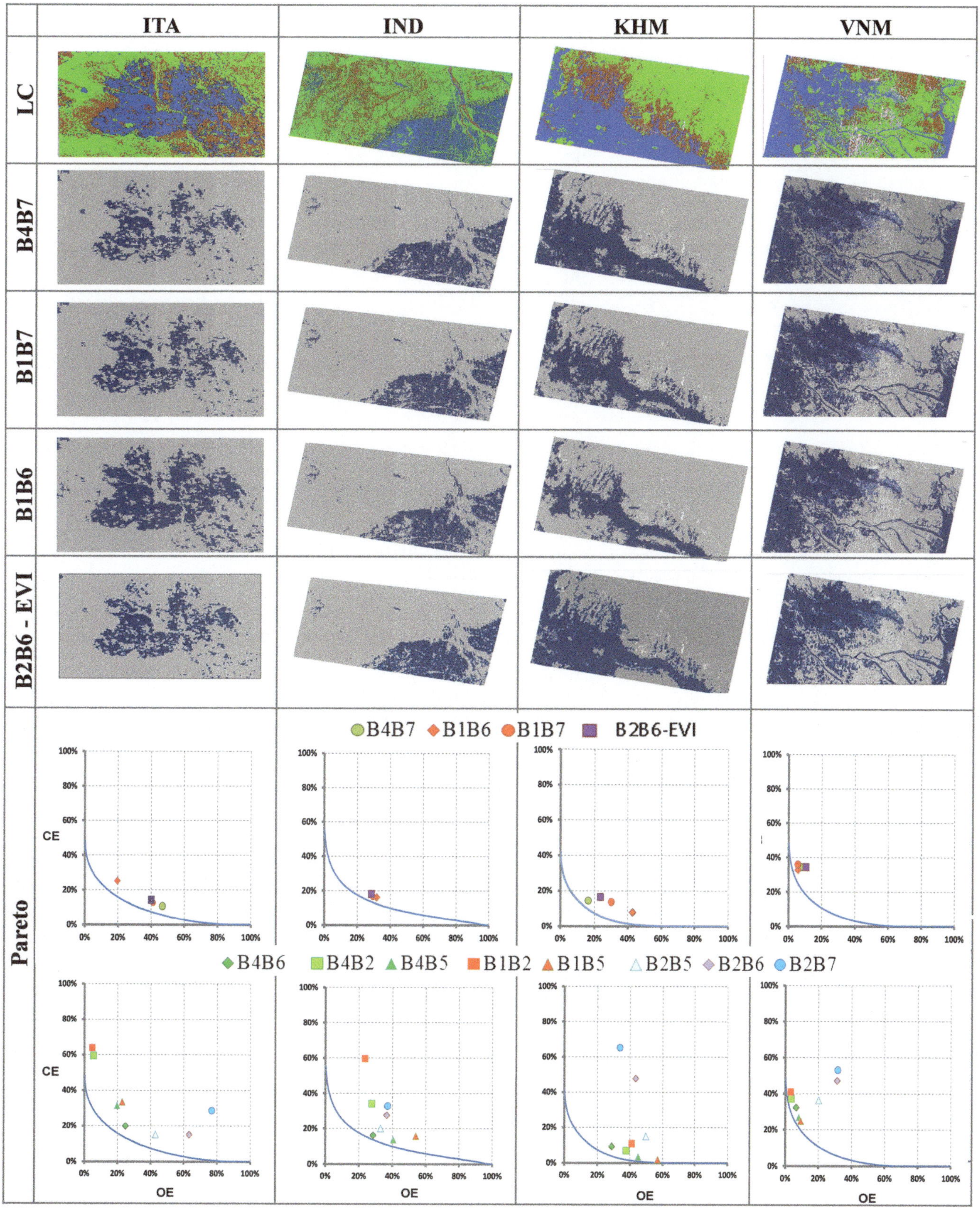

Figure 7. The upper section shows high resolution LC maps (blue: Water; green: vegetation; orange: bare soil; white: clouds) from Landsat analysis and water maps from MODIS for VIS/SVIR indices using global thresholds at the four test sites. The lower section shows the performance of VIS/SWIR, VIS/NIR, NIR/SWIR indices and the combined B2B6-EVI index in the Omission/Commission Error space. Pareto boundaries in the Omission/Commission Error space for each site are provided.

Table 9. Overall Accuracy (%) of water mapping for the 11 NDSIs/combined index at the four test sites (IND, ITA, VNM, KHM).

		VIS/NIR				VIS/SWIR				NIR/NIR	NIR/SWIR		COMBINED
	Site/ALL	**B4B2**	**B1B2**	**B4B5**	**B1B5**	**B4B6**	**B4B7**	**B1B6**	**B1B7**	**B2B5**	**B2B6**	**B2B7**	**B2B6-EVI**
OA	IND	85%	70%	89%	86%	**91%**	90%	90%	90%	89%	86%	85%	90%
	ITA	64%	57%	86%	85%	**89%**	87%	88%	88%	87%	83%	79%	88%
	VNM	78%	75%	84%	**86%**	81%	80%	81%	79%	76%	66%	61%	79%
	KHM	84%	82%	83%	79%	87%	**89%**	82%	85%	78%	65%	42%	86%
OA	Value $_{AVG}$	78%	71%	86%	84%	**87%**	**87%**	85%	86%	83%	75%	67%	86%
	Rank $_{AVG}$	8.5	10.5	5.5	6.75	**1.75**	3.75	4.5	4	7.5	9.75	11.25	4.25
OE	Value $_{AVG}$	19%	**18%**	28%	36%	22%	25%	25%	27%	36%	44%	45%	26%
	Rank $_{AVG}$	3.0	**2.5**	8	9.25	4.25	5.75	5.5	5	9.5	10	9.75	5.5
CE	Value $_{AVG}$	35%	44%	**19%**	**19%**	**19%**	**19%**	21%	20%	22%	34%	45%	21%
	Rank $_{AVG}$	8.5	10	**3.5**	**3.5**	4.25	5	4.75	5.5	7.5	8.75	10.5	6.25

Average performance ($Value_{AVG}$, %) and rank ($Rank_{AVG}$) of OA, OE and CE are provided.

define the threshold on the basis of a training data set and validating the results on a separate/independent test data set.

Figure 4 shows the boxplots of the calibration data set results in a matrix layout. The rows and columns of the matrix represent the possible NDSI band combinations, while each matrix element shows the boxplots of the respective NDSI values for water, soil and vegetation pixels for each site (500 training samples each for ITA, IND, KHM and VNM) and 'globally' across all sites (2000 training samples for ALL). Each boxplot shows the statistical distribution (minimum, lower quartile, median, upper quartile and maximum) and possible outliers for pure MODIS pixels fully covered by water (blue), soil (orange) and vegetation (green). Every boxplot has the same data cardinality (Table 3 and §2.4); hence, the extent of each box depends only on the data distribution.

In each case, water pixels have the highest NDSI values, as expected from the ANOVA of the field data (Table 4). The median NDSI values for water show that ITA, IND and KHM behave in a similar way, but that VNM has consistently different behaviour from the other three sites, with much higher NDSI. The boxplots show that B4B2 and B1B2 indices from the VIS/NIR family globally confuse water with soil, confirming the field data analysis in which both of them have the lowest separability for the dry soil (DS) condition (Rank: 11 and 12, respectively). The B1B2 index corresponds to NDVI so it is not surprising that it clearly separates vegetation from other categories and that there is overlap between soil and water. The boxplots of B2B6 and B2B7 show a confusion between water and vegetated pixels, again confirming the separability analysis conducted on field data (Rank: 11 and 12, respectively). These results are also in agreement with the post-hoc test on field spectra (Table 4) and confirm the choices of several authors in combining B2B6 with a vegetation index (EVI or NDVI) to reduce this effect. Panel B of Figure 4 provides the results for the combined B2B6-EVI index confirming the capability of this index to separate water for other surfaces. However, when compared to B1B7 we see that the water class is closest to the vegetation class and when data from all sites are grouped, the index values for these categories can overlap. The dotted line on each panel shows the NDSI threshold determined by the application of the recursive partitioning technique to distinguish water from other LC classes and values are reported in Table 7.

These thresholds were tested on the validation dataset (500 test pixels per site), and the Overall Accuracy (OA) was computed with the confusion matrices on each NDSI and the combined B2B6-EVI index. We report in Table 7 the results across all four sites. The best mapping accuracy (OA>96%) was obtained for the VIS/SWIR (B4B6, B4B7, B1B6, B1B7) indices and the combined B2B6-EVI index, which agrees with the separability analysis results. Among them, NDSIs with shorter SWIR wavelength (B1B6 and B4B6) have a slightly higher OA (97%) due to a lower Omission Error (OE~3%). Although OA values are similar (96% and 97%), the Z-test showed that the mapping accuracies obtained with NDSIs that adopt a different SWIR band are statistically different with a confidence level of >95%. Thus, the performance of B1B6 and B4B6 is not significantly different and these indices detect pure water pixels slightly better than B1B7 and B4B7. The blue boxplots in Figure 4 provide an interpretation of the results: the higher OE for B1B7 and B4B7 is mainly caused by low values of NDSI assumed from water pixels in Italy that overlap with the other targets.

The VIS/SWIR indices with a RED band (B1B7 and B1B6) have some skill in separating vegetation and soil pixels compared with the VIS/SWIR indices that use the GREEN band and the combined B2B6-EVI index. The boxplot in Figure 4 shows that the B1B7 and B1B6 indices have higher values in vegetation conditions than B4B6, B4B7 and the combined B2B6-EVI index, in which the boxplots for soil and vegetation are similar. Hence, the Commission Errors (CE) for B1B7 and B1B6 are caused mainly by soil cover, while the CE of B4B7, B4B6 and the combined B2B6-EVI index could result from a confusion across the non-water land cover classes (see also Figure S1 in File S1).

Globally, the analysis of pure MODIS pixels suggests that (i) the indices B1B7 and B4B7 have more skill in separating water pixels from other LCs (Table 6) and that (ii) B1B6 and B4B6 have slightly better performances in detecting pure water pixels (Table 7). Again, the combined B2B6-EVI index is comparable to this category of NDSIs.

3.3 The Impact of Different Pixel Mixture Conditions on NDSI Mapping Performance

The sensitivity of the NDSIs to water presence was assessed by regression analysis and Noise Equivalent analysis on a data set selected to guarantee equal distribution across sites and classes of

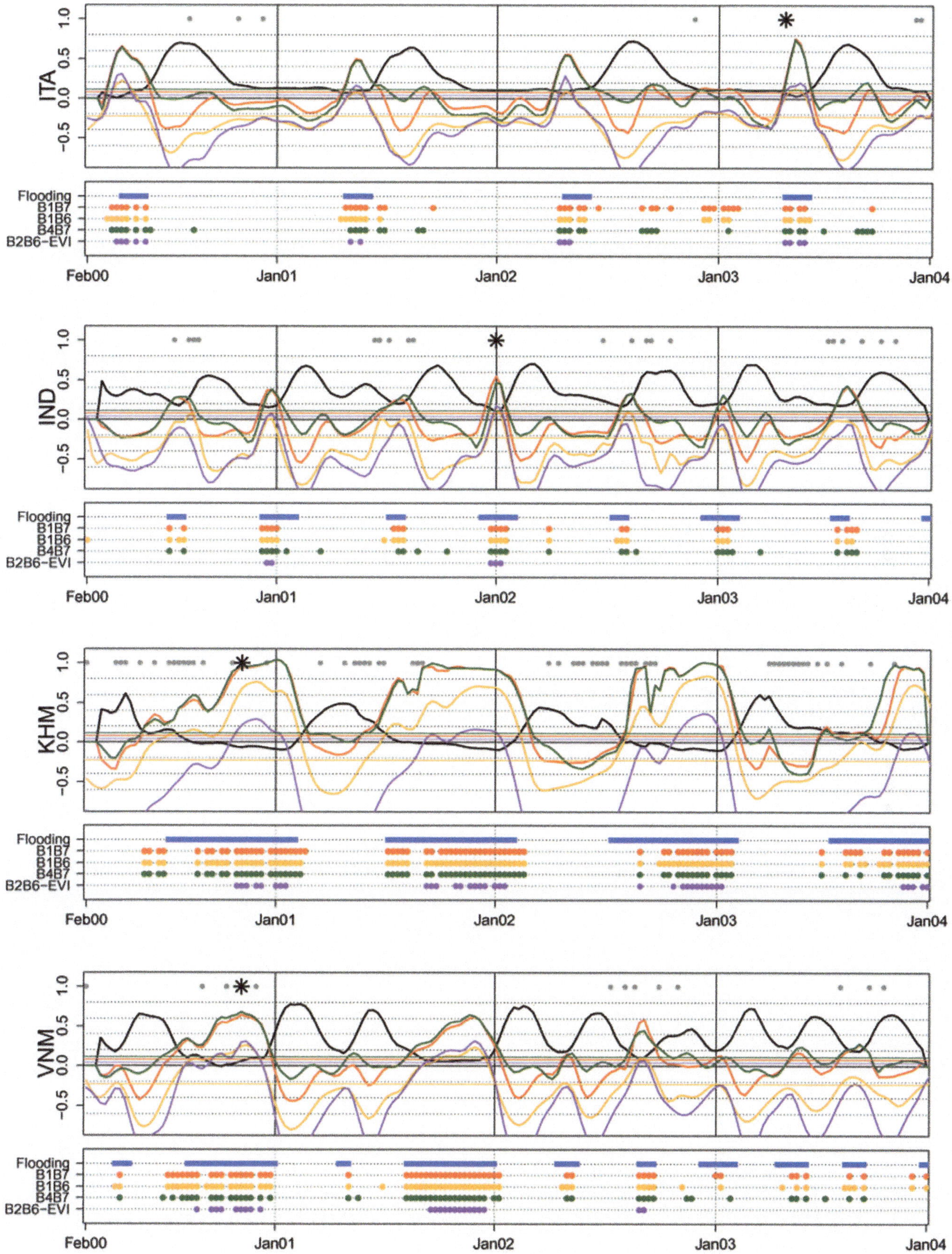
ITA
A
Flooding
B1B7
B1B6
B4B7
B2B6-EVI
IND
B
KHM
C
VNM
D
Feb00
Jan01
Jan02
Jan03
Jan04

Figure 8. NDSI and combined B2B6-EVI index temporal series for 2000–2003 for four selected rice cultivation areas in Italy (ITA), India (IND), Cambodia (KHM) and Vietnam (VNM). For each site, the top panel shows smoothed profiles of EVI, B1B7, B1B6, B4B7 and B2B6-EVI with black, red, dark green, orange and purple continuous lines respectively. Horizontal lines show the threshold for each index. Grey points indicate cloud contamination, black stars indicate dates of Landsat image acquisition. For each site, the bottom panel shows the flood detection by each index for each MODIS composite date.

water percentage (see §2.4). Figure 5 shows the correlation analysis as a scatterplot matrix, with one plot per NDSI showing the NDSI value on the x axes and the MODIS pixel water percentage on the y axes. Since NDSIs are generally related to water presence in a non-linear manner, we fitted a second-order polynomial regression through the data to better capture saturation behaviour at high water presence. The regression fit and model, across all sites, are shown on each plot and in each case the NDSI value increases more rapidly when the water presence in the MODIS pixel is greater than 70%. Globally, the combined B2B6-EVI index shows the highest correlation with water presence in MODIS pixels ($r^2 = 0.74$) followed by B1B7 ($r^2 = 0.60$). NDSIs from the family NIR/SWIR (B2B6, B2B7) have the lowest correlation with water presence, $r^2 = 0.33$ and 0.19 respectively (See also Figure S2 in File S1 for B2B6-NDVI performance).

Table 8 reports r^2 and RMSE also for each regression model calculated on data from single sites. Higher correlation is obtained for the VIS/SWIR family of NDSI at all sites, with the coefficient of determination (r^2) permanently greater than 0.65 and above 0.9 in six cases and above 0.8 in twelve cases. The indices from the VIS/SWIR family outperform the combined B2B6-EVI index in each study area and the minimum r^2 for the combined index is 0.77 while for B1B7 it is 0.83. This again could indicate that the combined B2B6-EVI index is sensitive to local conditions.

NE was computed for each NDSI and the results are shown in Figure 6. NE has low values when an NDSI has both high correlation (low RMSE) and high sensitivity (high first derivative) to water presence. Figure 6 shows that the VIS/SWIR NDSI family has the lowest NE (best sensitivity) when water covers less than 50% of the MODIS pixel. Among them, B1B7 has the highest sensitivity to water presence (NE<0.35) when water covers less than 50% of the pixel, followed by B1B6 (NE<0.5) and B4B7 (NE<0.6). The combined B2B6-EVI index is also very sensitive to water presence and is second placed (NE<0.4) after B1B7 (see also Figure S3 in File S1 for B2B6-NDVI performance). The regression and NE analyses both suggest that the B1B7 NDSI is the most sensitive to water presence in mixed land cover MODIS pixels and is comparable to the combined B2B6-EVI index.

The NE results also show that the sensitivity of all indices is comparable (NE<0.3) at high water percentages (>80%) but the NIR/NIR and NIR/SWIR NDSI families lose sensitivity when water presence in MODIS pixels falls below 60%.

3.4 Map Accuracy: NDSI Performance Comparison with Pareto Boundary

The top section of Figure 7 shows the classified Landsat maps and the results of applying the validated thresholds to the first three NDSIs of the VIS/SWIR family and the combined B2B6-EVI index on the MODIS images across the four sites. These four spectral combinations have consistently outperformed the other NDSIs in the prior analysis steps so we focus on these for reasons of clarity. The lower section of Figure 7 shows the Pareto boundary corresponding to the best possible classification solution for each NDSI in the CE-OE space for each site (see also Figure S4 in File S1 for B2B6-NDVI performance).

The VIS/SWIR family of NDSI is the closest to the Pareto boundaries and never in a region dominated by other NDSIs. In IND and VNM, the mapping accuracy of the VIS/SWIR family of NDSI is comparable (IND: OE~30% and CE~20%; VNM: OE~10% and CE~35%). At the other two sites, ITA and KHM, the VIS/SWIR indices perform in a different way. In ITA, NDSIs that use SWIR band 7, and the combined B2B6-EVI index, clearly show a higher OE (~45%) with respect to the VIS/SWIR indices that use band B6 (OE~25%). In KHM the VIS bands (RED and GREEN) seem to influence the OE: B1B7> B4B7 (~30% vs. ~16%) and B1B6> B4B6 (~43% vs. ~29%). In this case B4B7 dominates the combined B2B6-EVI index. B1B5 and B4B5 also show good results being close to the Pareto boundary; however, these indices always have the highest omission error. The Pareto boundary plots also show that the IND and ITA sites have a more fragmented landscape (OE up to 80%, CE up to 50%), and hence are more challenging to map with low-resolution data.

Table 9 provides a synthesis of the results for the mapping accuracy on the entire area covered by subsets of the Landsat images. Globally, B4B6 has the highest OA (87%) and first ranking position (1.75), followed by B4B7 (OA 87% and rank 3.75), B1B7 (OA 86% and rank 4.00), B1B6 (OA 85% and rank 4.5) and B4B5 (OA 86% and rank 5.5) but the difference in OA across these five best performing NDSIs is negligible. The combined B2B6-EVI index is again comparable to the VIS/SWIR NDSIs (OA 86% and rank 4.25).

3.5 Analysis of Temporal Profiles of NDSIs for MODIS Data

Figure 8 shows four years (2000–2003) of temporal series of NDSIs for a single MODIS pixel covering rice production areas at each of the four study sites (Table 3). The same indices used in Figure 7 have been selected to analyse the paddy fields' temporal signal: B1B7 (red line), B4B7 (dark green line), B1B6 (orange line) and the combined B2B6-EVI index (purple line). EVI (black line) is also reported to describe vegetation growth and to help interpret the agronomic dynamics of the crop. In order to facilitate the temporal interpretation of the NDSI dynamics, the data have been smoothed using a Savitzky–Golay filter weighted for cloud contamination [47]. The continuous coloured lines represent the results of the smoothing procedure which removes the high and low extremes of the original NDSI signals, thus reducing the residual noise due to potential contamination that was not completely eliminated by the MODIS composite algorithm.

Lowland rice is characterised, and can therefore be detected when vegetation growth follows a period of controlled flooding [14]. All the described situations show that agronomic flooding is clearly visible when the indices (NDSI or combined) values surpass the previously derived thresholds reported in the figure as horizontal continuous lines. Flood detection for the indices are reported for each site in the bottom panel where coloured points represents flood detection for each MODIS composite date. This water detection can be interpreted as agronomic flooding because it is followed by an increase in EVI values during the expected crop calendar. Besides this general feature, it is important to note how the MODIS NDSIs are able to identify and characterise different cropping systems. Italy and Cambodia show only one crop season, albeit with different seasonality at each site.

At the European site (ITA, Lat 45°), rice is cultivated in summer with a spring flooding (April–May). At this site, B1B7 and B4B7 identified flood conditions also during the senescent part of the crop cycle and again in December 2002–January 2003 (together with B1B6) when there is permanent water in the field in winter due to heavy rainfall. In Cambodia (KHM, Lat 13°) the season occurs much earlier in spring after a very long period of flood (October–January). The four indices perform in a similar way, detecting flooding in the same periods. The India site (IND, Lat 82°) has a double-season rice cropping system with short-duration varieties, and rice growing is preceded by a short period of flooding which is detected by all four indices. Finally, Vietnam shows triple rice cultivation with three occurrences of flooding followed by a rapid increase in and senescence of EVI, typical of the intensive cultivation of short-duration rice varieties under irrigated conditions. It is interesting to note that the Vietnam site (VNM, Lat 10°) shows a double season in 2000 and 2001 and a triple season in 2002–2003. Moreover, the winter crop is preceded by an intense flooding period whilst the middle-season flooding is less evident in the NDSI signal. This analysis of the multi-year temporal profiles of real MODIS data confirms the results obtained in mapping water on a single date: when using moderate satellite observations, VIS/SWIR indices (B1B7, B4B7, B1B6, and B4B6- data not shown) provide a diagnostic indication of water presence in flooded rice fields. In general these indices have a similar behaviour and water detection across the different indices and thresholds varies only slightly. One index can have a better performance than another in connection with specific conditions of landscape heterogeneity, water thickness and percentage of water presence in the MODIS pixel.

The four indices provide similar values in connection with the SWIR band used, although NDSIs using B7 are more sensitive to water presence. Moreover, the use of the RED band (B1) provides additional information to distinguish bare soil from vegetated conditions. The red and orange lines (Figure 8) show the highest amplitude and have strong drops in connection with maximum vegetation growth shown by the EVI black lines. A flat behaviour of the indices in bare soil conditions can be seen only at the Italian site where fields have no vegetative cover during the winter.

In general, the combined B2B6-EVI index detects flooding less frequently than any of the selected best performing NDSIs. This is particularly evident in IND and VNM where some flooding events are not detected at all. In the IND case study one season is missed in 2000, 2001 and 2002 and both seasons in 2003. Similarly in VNM one season is not detected in 2001 and 2002 and all the three seasons are missing in 2003. These results were obtained using a global threshold value (T_{global} 0.045), however when an optimum local threshold value is calculated for each case study ($T_{ITA} = -0.1255$; $T_{IND} = 0.186$, $T_{KHM} = -0.22$ and $T_{VNM} = -0.243$) all the agronomic flooding events were detected.

We also tested thresholds available in literature, Xiao et al. [14] and Peng et al [15] respectively equal to −0.05 and −0.21 for each site. These values detect all rice flooding events in ITA, IND and KHM, however in VNM case study the second season – April/May - in 2002 is missed by both and the −0.05 thresholds also misses one season in 2001 and all seasons in 2003.

Conclusions

In this study, we have analysed and compared a range of NDSIs for detecting surface water in flooded rice fields with the primary aim to identify the most robust strategy, using either the best index or a combination of several indices, that could be used as part of a rice mapping and monitoring system. From a remote sensing point of view, paddy rice flood conditions can be physically interpreted as a shallow water body where soil bottom characteristics strongly influence the reflectance in the VIS-NIR domain. For this reason, we tested NDSIs proposed in the literature specifically for water body detection (e.g. NDWI or MDPI, see Table 1) and/or developed for water content monitoring but also adopted in standing water monitoring. Finally we compared these NDSIs with a combined index (LSWI-EVI equal to B2B6-EVI) usually adopted for rice flood monitoring (see Table 1).

We first classified the proposed NDSIs into different categories in relation to the spectral band used in their calculation, and we added other possible combinations of spectral bands which used the RED and long wavelength SWIR band (2090–2350 nm). The groups showed similar behaviour in relation to the physical processes that govern the targets' reflection/absorption features in the different spectral domains considered. The NIR-SWIR (800 vs. 1640/2130 nm) and NIR-NIR (800 vs. 1240) indices are proposed in the literature to estimate and monitor plant water content, and in our experimental case they behaved coherently with those assumptions: the higher the NDSI value, the greater the water presence up to completely flooded conditions. On the contrary, the VIS-NIR (550/670 vs. 1240 nm) and VIS-SWIR (550/670 vs. 1640/2130 nm) indices are diagnostic mainly for the detection of water/non-water conditions [21,48].

The analysis of field spectral data helped to interpret the physical basis behind the band combinations for the various indices, and the ANOVA of the experimental data revealed that all the indices, except for B2B7, provide values that are significantly different for water with respect to the other classes. Specific thresholds for each NDSI were derived from a calibration data set of pure MODIS pixels (water and no water) and their performance was validated on independent data. VIS/SWIR indices outperform the other category, showing OA greater than 96%. The NE analysis showed that the B1B7 index was more sensitive than other NDSIs in low water percentage conditions, but, when water presence exceeds 60%, B4B6, B1B6 and B4B7 slightly outperform B1B7 due to a saturation effect of the B1–B7 combination at high water presence percentages. Very similar results were obtained with the combined B2B6-EVI index.

NDSI water maps at MODIS spatial resolution were validated with Landsat water/no-water data using the Pareto boundary method to account for potential maximum accuracy in relation to low-resolution bias. Results demonstrate again that VIS/SWIR indices are the best and that, in the experimental case, the indices that use B4 bands (B4B6 and B4B7) perform slightly better. The combined B2B6-EVI index also performed well.

The analysis of four years of temporal MODIS data helped visualise the NDSI response throughout the year and the consistency in detection across years. It also confirmed the flood detection results from the single date mapping. All the analysed VIS/SWIR indices detected agronomic flooding of the rice crop except for a few particular events. The combined B2B6-EVI index performed less well with three different global thresholds, and only matched the performance of the VIS/SWIR NDSI when local thresholds were used. From both an information theory view point and an operational detection system viewpoint, a single NDSI (two bands) and a global threshold is preferable to a combined NDSI (four bands when EVI used and three bands when NDVI is use) and a locally adaptive threshold. The small number of sites for the temporal analysis means that this observation should be tested further and we provide some indication of future research directions below.

In conclusion, the VIS/SWIR indices outperform the other NDSIs in water detection with low-resolution data where mixed pixel problems can strongly affect results. Results show that one combination of VIS (GREEN and RED) and SWIR (B6 and B7) bands can outperform another depending on the specific conditions of landscape heterogeneity, water thickness and percentage of water presence in the MODIS pixel. In general NDSIs that use band 7 are much more sensitive to water presence but produce slightly more Commission Error when compared with the ones that use band 6. Therefore, the choice of which VIS/SWIR index to use depends on the aim of water mapping. For rice monitoring, flood mapping is a prerequisite necessary to identify the crop. If adopting a method like the one proposed by [16,17], crop vegetation growth must follow flood detection. In this framework, an NDSI with high sensitivity to water presence (i.e. less prone to omission) is preferable because the other algorithm criteria (a rapid increase in biomass post-flood) will delete false positive detections. On the other hand, if a more conservative water map is required, NDSIs from the VIS/SWIR family that use the B6 band are suggested.

Rice-growing environments and management systems are diverse, and a robust agronomic flood detection could be developed using a combination of the best-performing NDSIs or combined ones like B2B6-EVI instead of trying to identify one superior NDSI. Stroppiana et al. [49] and Bordogna et al. [50] provided a formal framework which exploits the redundancy and complementary information provided by different spectral combinations, and this method could be adopted as part of an agronomic flood detection. The authors demonstrate the method for burn area mapping in which multiple indices are shown to provide consistent and coherent information over the target of interest as well as complementary, incoherent behaviour over confusing surfaces. Following this approach, a robust agronomic flood mapping method could first use the more conservative indices to select *seed pixels* which have a greater probability to detect only flooded fields (thus reducing false positive error) and then expand the flood detection to the neighbouring pixels with a pixel-growing algorithm on the basis of the results of less conservative indices. Further studies should be conducted to evaluate the feasibility of this kind of approach, for example, testing how positive and negative information provided by a set of NDSIs derived from operational multispectral sensors can be exploited to assess and revise the integrated evidence of agronomic flood in a rice cropping system.

Finally, we have relied on MODIS and Landsat data for our analysis, but, for an operational system, SPOT-VGT data, or the forthcoming PROBA-V with the same spectral characteristics (www.esa.int/SPECIALS/Proba/SEM9FS4PVFG_0.html, accessed September 2012), can also be exploited using a combination of RED B2 (610–680 nm) and MIR (1580–1750 nm) bands. In the future, new opportunities will be provided by the GMES-ESA Sentinel mission (www.esa.int/esaLP/SEMM4T4KXMF_LPgmes_0.html, accessed September 2012). The pair of optical Sentinel-2 satellites, with the first satellite planned for launch in 2014, will routinely deliver high-resolution optical images with VIS-NIR and SWIR bands (1600 and 2200 nm) at 10 and 20 m spatial resolution, respectively. The revisit time of 5 days at the equator and 2–3 days at mid-latitudes would permit rice monitoring in fragmented environments characterised by small fields.

Supporting Information

File S1 B2B6-NDVI analysis. Comments and figures with results for the combined B2B6-NDVI index compared to B2B6-EVI and B1B7.

Acknowledgments

The authors would like to thank all the colleagues that provided precious help in conducting the field campaign. In particular we want to acknowledge Mr Mauro Musanti for the costly LAI2000 and spectral measurements of the 2004–2005 data set. A particular thanks to Mr Fedeli, who accommodated our experiments on his farm.

Author Contributions

Conceived and designed the experiments: MB AN. Performed the experiments: MB FN GM. Analyzed the data: MB AN FN PAB. Wrote the paper: MB AN.

References

1. Fischer G (2009) World food and agriculture to 2030/50: how do climate change and bioenergy alter the long-term outlook for food, agriculture and resource availability? How to feed the World in 2050. Proceedings of a technical meeting of experts, Rome, Italy, 24–26 June 2009. FAO.
2. Pittman K, Hansen MC, Becker-Reshef I, Potapov P V., Justice CO (2010) Estimating Global Cropland Extent with Multi-year MODIS Data. Remote Sens 2: 1844–1863. doi:10.3390/rs2071844.
3. Lobell DB, Cassman KG, Field CB (2009) Crop Yield Gaps: Their Importance, Magnitudes, and Causes. Annu Rev Environ Resour 34: 179–204. doi:10.1146/annurev.environ.041008.093740.
4. Licker R, Johnston M, Foley JA, Barford C, Kucharik CJ, et al. (2010) Mind the gap: how do climate and agricultural management explain the "yield gap" of croplands around the world? Glob Ecol Biogeogr 19: 769–782. doi:10.1111/j.1466-8238.2010.00563.x.
5. You L, Wood S (2006) An entropy approach to spatial disaggregation of agricultural production. Agric Syst 90: 329–347.
6. Monfreda C, Ramankutty N, Foley JA (2008) Farming the planet: 2. Geographic distribution of crop areas, yields, physiological types, and net primary production in the year 2000. Global Biogeochem Cycles 22: 1–19. doi:10.1029/2007GB002947.
7. Ramankutty N, Evan AT, Monfreda C, Foley JA (2008) Farming the planet: 1. Geographic distribution of global agricultural lands in the year 2000. Global Biogeochem Cycles 22: 1–19. doi:10.1029/2007GB002952.
8. Portmann FT, Siebert S, Petra D (2010) MIRCA2000-Global monthly irrigated and rainfed crop areas around the year 2000: A new high-resolution data set for agricultural and hydrological modeling. Glob Biogeochem Cycles 24: GB1011. doi:10.1029/2008GB003435.
9. Sacks WJ, Deryng D, Foley JA, Ramankutty N (2010) Crop planting dates: an analysis of global patterns. Glob Ecol Biogeogr 19: 607–620. doi:10.1111/j.1466-8238.2010.00551.x.
10. G20 (2011) Cannes Summit Final Declaration – Building our common future: Renewed collective action for the benefit of all: 20.
11. Maclean JL, Dawe DC, Hardy B, Hettel GP (2002) Rice almanac. 3rd edn. Wallingford, Oxon: CABI Publishing. doi:10.1093/aob/mcg189.
12. Sakamoto T, Vannguyen N, Kotera A, Ohno H, Ishitsuka N, et al. (2007) Detecting temporal changes in the extent of annual flooding within the Cambodia and the Vietnamese Mekong Delta from MODIS time-series imagery. Remote Sens Environ 109: 295–313. doi:10.1016/j.rse.2007.01.011.
13. Baret F, Bartholomé E, Bicheron P, Borstlap G, Bydekerke L, et al. (2006) VGT4Africa user manual First edition.
14. Xiao X, Boles S, Liu J, Zhuang D, Frolking S, et al. (2005) Mapping paddy rice agriculture in southern China using multi-temporal MODIS images. Remote Sens Environ 95: 480–492. doi:10.1016/j.rse.2004.12.009.
15. Peng D, Huete AR, Huang J, Wang F, Sun H (2011) Detection and estimation of mixed paddy rice cropping patterns with MODIS data. Int J Appl Earth Obs Geoinf 13: 13–23. doi:10.1016/j.jag.2010.06.001.
16. Xiao X, Boles S, Frolking S, Salas W, Moore B, et al. (2002) Observation of flooding and rice transplanting of paddy rice fields at the site to landscape scales in China using VEGETATION sensor data. Int J Remote Sens 23: 3009–3022. doi:10.1080/01431160110107734.
17. Xiao X, Boles S, Frolking S, Li C, Babu JY, et al. (2006) Mapping paddy rice agriculture in South and Southeast Asia using multi-temporal MODIS images. Remote Sens Environ 100: 95–113.
18. Gao B (1996) NDWI-A normalized difference water index for remote sensing of vegetation liquid water from space. Remote Sens Environ 58: 257–266.
19. Rogers AS, Kearney MS (2004) Reducing signature variability in unmixing coastal marsh Thematic Mapper scenes using spectral indices. Int J Remote Sens 25: 2317–2335. doi:10.1080/01431160310001618103.
20. Uchida S (2007) Monitoring of paddy rice planting with complex cropping pattern using satellite remote sensing data -a case of west java, Indonesia. Proc. 28 th Asian Conference on Remote Sensing, Kuala Lumpur,. p. 5.

21. Ji L, Zhang L, Wylie B (2009) Analysis of Dynamic Thresholds for the Normalized Difference Water Index. Photogramm Eng Remote Sensing 75: 1307–1317.
22. Ji L, Zhang L, Wylie BK, Rover J (2011) On the terminology of the spectral vegetation index (NIR − SWIR)/(NIR+SWIR). Int J Remote Sens 32: 6901–6909. doi:10.1080/01431161.2010.510811.
23. Ouma YO, Tateishi R (2006) A water index for rapid mapping of shoreline changes of five East African Rift Valley lakes: an empirical analysis using Landsat TM and ETM+ data. Int J Remote Sens 27: 3153–3181. doi:10.1080/01431160500309934.
24. Hardisky MA, Klemas V, Smart RM (1983) The Influence of Soil Salinity, Growth Form, and Leaf Moisture on-the Spectral Radiance of ~ partina alterniflora Canopies. Photogramm Eng Remote Sensing 49: 77–83.
25. Gerard F, Plummer S, Wadsworth R, Sanfeliu AF, Iliffe L, et al. (2003) Forest Fire Scar Detection in the Boreal Forest With Multitemporal SPOT-VEGETATION Data. IEEE Trans Geosci Remote Sens 41: 2575–2585.
26. Fensholt R, Sandholt I (2003) Derivation of a shortwave infrared water stress index from MODIS near- and shortwave infrared data in a semiarid environment. Remote Sens Environ 87: 111–121. doi:10.1016/j.rse.2003.07.002.
27. Rouse JW, Hass RH, Deering DW, Sehell JA (1974) Monitoring the vernal advancement and retrogradation (Green wave effect) of natural vegetation.
28. Huete A, Didan K, Miura T, Rodriguez EP, Gao X, et al. (2002) Overview of the radiometric and biophysical performance of the MODIS vegetation indices. Remote Sens Environ 83: 195–213.
29. Gond V, Bartholomé E, Ouattara F, Nonguierma A, Bado L (2004) Surveillance et cartographie des plans d'eau et des zones humides et inondables en régions arides avec l'instrument VEGETATION embarqué sur SPOT-4. Int J Remote Sens 25: 987–1004. doi:10.1080/0143116031000139908.
30. Chipman JW, Lillesand TM (2007) Satellite-based assessment of the dynamics of new lakes in southern Egypt. Int J Remote Sens 28: 4365–4379. doi:10.1080/01431160701241787.
31. Bocchi S, Callegarin A, Baldi G (2003) Rice production system in Italy and its sustainability. In: INIA U, editor. Proceedings of the 3rd Temperate Rice Conference. Punta del Este, Uruguay. 12–22.
32. Stroppiana D, Boschetti M, Brivio PA, Bocchi S (2009) Plant nitrogen concentration in paddy rice from field canopy hyperspectral radiometry. F Crop Res 111: 119–129. doi:10.1016/j.fcr.2008.11.004.
33. Boschetti M, Bocchi S, Stroppiana D, Brivio PA (2006) Estimation of parameters describing morpho-physiological features of Mediterranean rice varieties for modelling purposes. Ital J Agron 49: 40–49.
34. Boschetti M, Stroppiana D, Brivio PA, Bocchi S (2009) Multi-year monitoring of rice crop phenology through time series analysis of MODIS images. Int J Remote Sens 30: 4643–4662. doi:10.1080/01431160802632249.
35. Gumma MK, Nelson A, Thenkabail PS, Singh AN (2011) Mapping rice areas of South Asia using MODIS multitemporal data. J Appl Remote Sens 5: 053547. doi:10.1117/1.3619838.
36. Cheralu C (2011) Status Paper on Rice in Andhra Pradesh. Rice nowledge Manag Portal: 1–36. Available: http://www.rkmp.co.in/sites/default/files/ris/rice-state-wise/Rice State Wise Andhra Pradesh_0.pdf.
37. Nesbitt HJ (1997) Rice production in Cambodia. Manila: International Rice Research Institute.
38. Bouvet A, Le Toan T (2011) Use of ENVISAT/ASAR wide-swath data for timely rice fields mapping in the Mekong River Delta. Remote Sens Environ 115: 1090–1101. doi:10.1016/j.rse.2010.12.014.
39. Song C, Woodcock CE, Seto KC, Lenney MP, Macomber SA (2001) Classification and Change Detection Using Landsat TM Data: When and How to Correct Atmospheric Effects? Remote Sens Environ 75: 230–244.
40. Rosenfield GH, Fitzpatrick-Lins K (1986) A coefficient of agreement as a measure of thematic classification accuracy. Photogramm Eng Remote Sensing 52: 223–227.
41. Govaerts YM, Verstraete MM, Pinty B, Gobron N (1999) International Journal of Designing optimal spectral indices: A feasibility and proof of concept study. Int J Remote Sens 2: 1853–1873.
42. Viña A (2005) New developments in the remote estimation of the fraction of absorbed photosynthetically active radiation in crops. Geophys Res Lett 32: L17403. doi:10.1029/2005GL023647.
43. Viña A, Gitelson AA (2011) Sensitivity to Foliar Anthocyanin Content of Vegetation Indices Using Green Reflectance. IEEE Geosci Remote Sens Lett 8: 463–467.
44. Boschetti L, Flasse SP, Brivio PA (2004) Analysis of the conflict between omission and commission in low spatial resolution dichotomic thematic products: The Pareto Boundary. Remote Sens Environ 91: 280–292. doi:10.1016/j.rse.2004.02.015.
45. Palmer KF, Williams D (1974) Optical Properties of water in the near infrared. J Opt Soc Am 64: 1107–1110.
46. Key CH, Benson NC (1999) Measuring and remote sensing of burn severity: the CBI and NBR. In: Neuenschwander LF, Ryan KC, editors. Joint Fire Science Conference and Workshop, Vol. II, Boise, ID, 15–17 June 1999. Boise, University of Idaho and International Association of Wildland Fire. p. 284.
47. Chen J, Jönsson P, Tamura M, Gu Z, Matsushita B, et al. (2004) A simple method for reconstructing a high-quality NDVI time-series data set based on the Savitzky-Golay filter. Remote Sens Environ 91: 332–344. doi:http://dx.doi.org/10.1016/j.rse.2004.03.014.
48. McFeeters SK (1996) The use of the Normalized Difference Water Index (NDWI) in the delineation of open water features. Int J Remote Sens 17: 1425–1432. doi:10.1080/01431169608948714.
49. Stroppiana D, Bordogna G, Boschetti M, Carrara P, Boschetti L, et al. (2012) Positive and Negative Information for Assessing and Revising Scores of Burn Evidence. IEEE Geosci Remote Sens Lett 9: 363–367.
50. Bordogna G, Boschetti M, Brivio PA, Carrara P, Stroppiana D, et al. (2012) Handling heterogeneous bipolar information for modelling environmental syndromes of global change. Environ Model Softw: 1–17. doi:10.1016/j.envsoft.2012.03.016.
51. Wilson EH, Sader SA (2002) Detection of forest harvest type using multiple dates of Landsat TM imagery. Remote Sens Environ 80: 385–396.
52. Xu H (2006) Modification of normalised difference water index (NDWI) to enhance open water features in remotely sensed imagery. Int J Remote Sens 27: 3025–3033. doi:10.1080/01431160600589179.
53. Lacaux JP, Tourre YM, Vignolles C, Ndione JA, Lafaye M (2007) Classification of ponds from high-spatial resolution remote sensing: Application to Rift Valley Fever epidemics in Senegal. Remote Sens Environ 106: 66–74.

21

Spatial and Temporal Variations of Crop Fertilization and Soil Fertility in the Loess Plateau in China from the 1970s to the 2000s

Xiaoying Wang[1,2], Yanan Tong[1,2]*, Yimin Gao[1], Pengcheng Gao[1], Fen Liu[1], Zuoping Zhao[1], Yan Pang[1]

1 College of Natural Resources and Environment, Northwest A&F University, Yangling, China, **2** Key Laboratory of Plant Nutrition and the Agri-environment in Northwest China, Ministry of Agriculture, Yangling, China

Abstract

Increased fertilizer input in agricultural systems during the last few decades has resulted in large yield increases, but also in environmental problems. We used data from published papers and a soil testing and fertilization project in Shaanxi province during the years 2005 to 2009 to analyze chemical fertilizer inputs and yields of wheat (*Triticum aestivum* L.) and maize (*Zea mays* L.) on the farmers' level, and soil fertility change from the 1970s to the 2000s in the Loess Plateau in China. The results showed that in different regions of the province, chemical fertilizer NPK inputs and yields of wheat and maize increased. With regard to soil nutrient balance, N and P gradually changed from deficit to surplus levels, while K deficiency became more severe. In addition, soil organic matter, total nitrogen, alkali-hydrolysis nitrogen, available phosphorus and available potassium increased during the same period. The PFP of N, NP and NPK on wheat and maize all decreased from the 1970s to the 2000s as a whole. With the increase in N fertilizer inputs, both soil total nitrogen and alkali-hydrolysis nitrogen increased; P fertilizer increased soil available phosphorus and K fertilizer increased soil available potassium. At the same time, soil organic matter, total nitrogen, alkali-hydrolysis nitrogen, available phosphorus and available potassium all had positive impacts on crop yields. In order to promote food safety and environmental protection, fertilizer requirements should be assessed at the farmers' level. In many cases, farmers should be encouraged to reduce nitrogen and phosphate fertilizer inputs significantly, but increase potassium fertilizer and organic manure on cereal crops as a whole.

Editor: Cheng-Sen Li, Institute of Botany, China

Funding: The authors thank the Special Fund for Agro-scientific Research in the Public Interest of China (201103003) and the Soil Quality Foundation of China (2012BAD05B03) for their financial support. The funders had no role in study design, data collection and analysis, decision to publish, or preparation of the manuscript.

Competing Interests: The authors have declared that no competing interests exist.

* Email: tongyanan@nwsuaf.edu.cn

Introduction

China has only 9% of the world's arable land and feeds nearly 22% of the world population [1–2]. This depends heavily on increasing grain production with the use of chemical fertilizers. Before the 1970s, farmers maintained the original agricultural practices, such as crop rotation, diversified plantation, manure application and legume crop integration, for soil fertility maintenance and pest and disease control. Since the late 1980s, the practice of applying organic manure in arable cropping systems has nearly come to an end [2–6]. From then on, almost all available organic manure has been used on vegetables and fruit trees, while the nutrients for cereal crops have been mainly in the form of chemical fertilizers. From 1970 to 2010, total annual grain production in China increased from 240 to 546 million tons (a 128% increase). However, inorganic fertilizer application increased from 3.51 to 55.62 million tons (a 1485% increase) over the same period [7].

Soil quality indicators are measurable soil properties that benefit food production or other specific functions, including physical, chemical and biological characteristics [8]. The increase or decrease in single soil index values, such as soil organic matter, total nitrogen and available nutrients, amplitude of variation and variation in time, can be used as a monitoring index for agricultural land management [9–11]. Given the spatial and temporal variation in characteristics of soil quality, it is necessary to compare or analyze two or more phase changes to understand the nature and mechanisms of soil quality [12].

Farmland fertilization is one of the most effective ways to maintain soil fertility and increase crop yields [13–15]. For this reason, information on household fertilization levels is of great value. In addition, wheat and maize are two of the most important food crops throughout the world, and they account for 51.7% of the total area for food crops and 53.5% of the total food production in 2010 in China [7]. Chemical fertilizer consumption data from official Chinese statistics do not contain information on usage for each kind of crop. It is imprecise to analyze and evaluate fertilizer efficiency using total amounts, because the distribution and application of fertilizer on specific crops are ambiguous [16].

Thus, the objectives of this study were to: (1) reveal the spatial and temporal variations of chemical fertilization and yields of wheat and maize at the farmers' level from the 1970s to the 2000s in the Loess Plateau in China; (2) reveal the spatial and temporal variations of soil fertility over the same period; and (3) reveal the relationships among fertilizer inputs, crop yields and soil fertility.

Materials and Methods

Ethics Statement

This study has been approved by the Agricultural Technology Extension Center of Shaanxi province, which is responsible for fertilization and soil fertility in Shaanxi province. All data in this study can be published and shared.

Study area

Shaanxi province (Figure 1) is located in the middle reaches of the Yellow River and the upper reaches of the Yangtze River of the eastern part of northwest China, and it falls between latitudes 31°42′ and 39°35′N, and longitudes 105°29′ and 111°15′E. The area is 2.058×10^5 km^2, extending about 880 km from north to south and 160 to 490 km from east to west. The whole province from north to south can be divided into four agro-ecological zones, which include the Loess Plateau area of northern Shaanxi, the Weibei dry plateau, the Guanzhong irrigated area and the Qin-Ba mountain area of southern Shaanxi; the previous three regions belong to the Loess Plateau and in this study they are abbreviated as North, Weibei, and Guanzhong, respectively. The Loess Plateau region in China, covers five provinces (including Shaanxi province), stretches over an area of 0.62 million km^2, and consists of typical semiarid and arid areas with rainfed farming [17–18]. Winter wheat is planted in the regions of Weibei and Guanzhong, while summer maize is planted in the Guanzhong region and spring maize in the North and Weibei regions. Main soil types and climatic conditions in the different regions are shown in Table 1.

Data sources

The data from the 1970s to the 1990s was extracted from 380 published papers reporting household fertilization and soil fertility in the study area; the screening process and results are shown in Figure 2. Data from the 2000s was collected from the project "soil testing and formulated fertilization in Shaanxi province during the years 2005 to 2009."

Statistics

The data were analyzed by EXCEL software. In this study, we used the following equations to analyze the soil nutrient balance and partial factor productivity (PFP) of fertilizer:

$$\textit{Soil nutrient balance} = \textit{nutrient input rate} - \textit{nutrient output rate} \quad \text{(Eq 1)}$$

where the nutrient input rate represents chemical fertilizer input, and the nutrient output rate represents amounts extracted in crop products and above ground biomass;

$$PFP = Y/F \quad \text{(Eq 2)}$$

where Y represents crop yields, and F represents chemical fertilizer input.

Results

Spatial and temporal variations of chemical fertilization and yields of wheat and maize at the farmers' level in different regions of Shaanxi province

The average chemical fertilizer NPK inputs for both wheat and maize at the farmers' level increased for decades in the different regions (Figure 3). In the Weibei and Guanzhong regions, chemical fertilizer N inputs for wheat in the 1970s were 45 kg ha^{-1} and 52 kg ha^{-1}, respectively, and in the 2000s they increased to 185 kg ha^{-1} and 195 kg ha^{-1}, respectively. In these two regions, chemical fertilizer P_2O_5 inputs were 45 kg ha^{-1} and 46 kg ha^{-1} in the 1970s and they increased to 112 kg ha^{-1} and 115 kg ha^{-1} in the 2000s. In the 1980s, farmers started to use the chemical fertilizer K_2O for wheat, which was increased from 0.5 kg ha^{-1} and 2.3 kg ha^{-1} to 22.8 kg ha^{-1} and 22.5 kg ha^{-1}, respectively, during the 1980s to the 2000s in the two regions. For maize in the North, Weibei and Guanzhong regions, chemical fertilizer N inputs were 48 kg ha^{-1}, 89 kg ha^{-1} and 36 kg ha^{-1} and they increased to 237 kg ha^{-1}, 223 kg ha^{-1} and 244 kg ha^{-1}, respectively, from the 1970s to the 2000s. Unlike wheat, from the 1980s onward farmers were awarded for using the chemical fertilizers P_2O_5 and K_2O for maize, and their use has increased greatly.

In accordance with increased chemical fertilizer NPK inputs (Figure 3), the average yields of wheat and maize showed increasing trends in the different regions over the four decades (Figure 4). In the Weibei and Guanzhong regions, from the 1970s to the 2000s, yields of wheat changed from 1883 kg ha^{-1} and 3377 kg ha^{-1} to 4269 kg ha^{-1} and 6437 kg ha^{-1}, with increase rates of 127% and 91%, respectively. In the North, Weibei and Guanzhong regions, yields of maize changed from 3636 kg ha^{-1}, 2519 kg ha^{-1} and 4232 kg ha^{-1} to 7867 kg ha^{-1}, 7077 kg ha^{-1} and 6886 kg ha^{-1}, with increase rates of 116%, 181% and 63%, respectively, for the same period.

Spatial and temporal variations of soil nutrient balance from the inputs and uptake on wheat and maize plots in different regions of Shaanxi province

Because the farmers tended not to use organic manure for cereal crops, especially from the 1980s onward, the soil nutrient inputs only include chemical fertilizers, and the nutrient uptakes include those extracted in crop products and above ground biomass. The nutrient balance was calculated as the difference between the average input and uptake (Eq. 1). Other losses, from leakage and gaseous loss, were not included in these calculations. In the 1970s, N was deficient on wheat and maize plots in the different regions (except for maize plots in the Weibei region). Then from the 1980s N was consistently at surplus levels, and it displayed an upward trend with time. In the 2000s, N surpluses on wheat plots were 74 kg ha^{-1} and 29 kg ha^{-1} in the Weibei and Guanzhong regions, respectively; meanwhile N surpluses on maize plots were 64 kg ha^{-1}, 67 kg ha^{-1} and 93 kg ha^{-1} in the North, Weibei and Guanzhong regions, respectively (Figure 5).

In the Weibei and Guanzhong regions, the amount of surplus P_2O_5 on wheat plots increased each year from the 1970s to the 2000s, and surplus amounts increased from 24 kg ha^{-1} and 9 kg ha^{-1} to 65 kg ha^{-1} and 44 kg ha^{-1}, respectively. In the North, Weibei and Guanzhong regions, P_2O_5 was deficient on maize plots in the 1980s; then it gradually reached surplus levels until the 2000s with the increased application of chemical fertilizer phosphorus. The balance of P_2O_5 on maize plots increased from −34 kg ha^{-1}, −7 kg ha^{-1} and −35 kg ha^{-1} to 29 kg ha^{-1}, 28 kg ha^{-1} and −11 kg ha^{-1}, respectively, in the three regions from the 1980s to the 2000s. It is worth noting, that winter wheat and summer maize were in a rotation system in the Guanzhong region, so total P_2O_5 was in surplus in this region in the 2000s and the amount was 33 kg ha^{-1} (Figure 5).

Although farmers have been awarded for using K_2O chemical fertilizer in recent years, the amount used was still small (Figure 3), and it was usually from compound fertilizers. So K_2O deficiency has become more serious (Figure 5). In the 2000s, K_2O deficiency

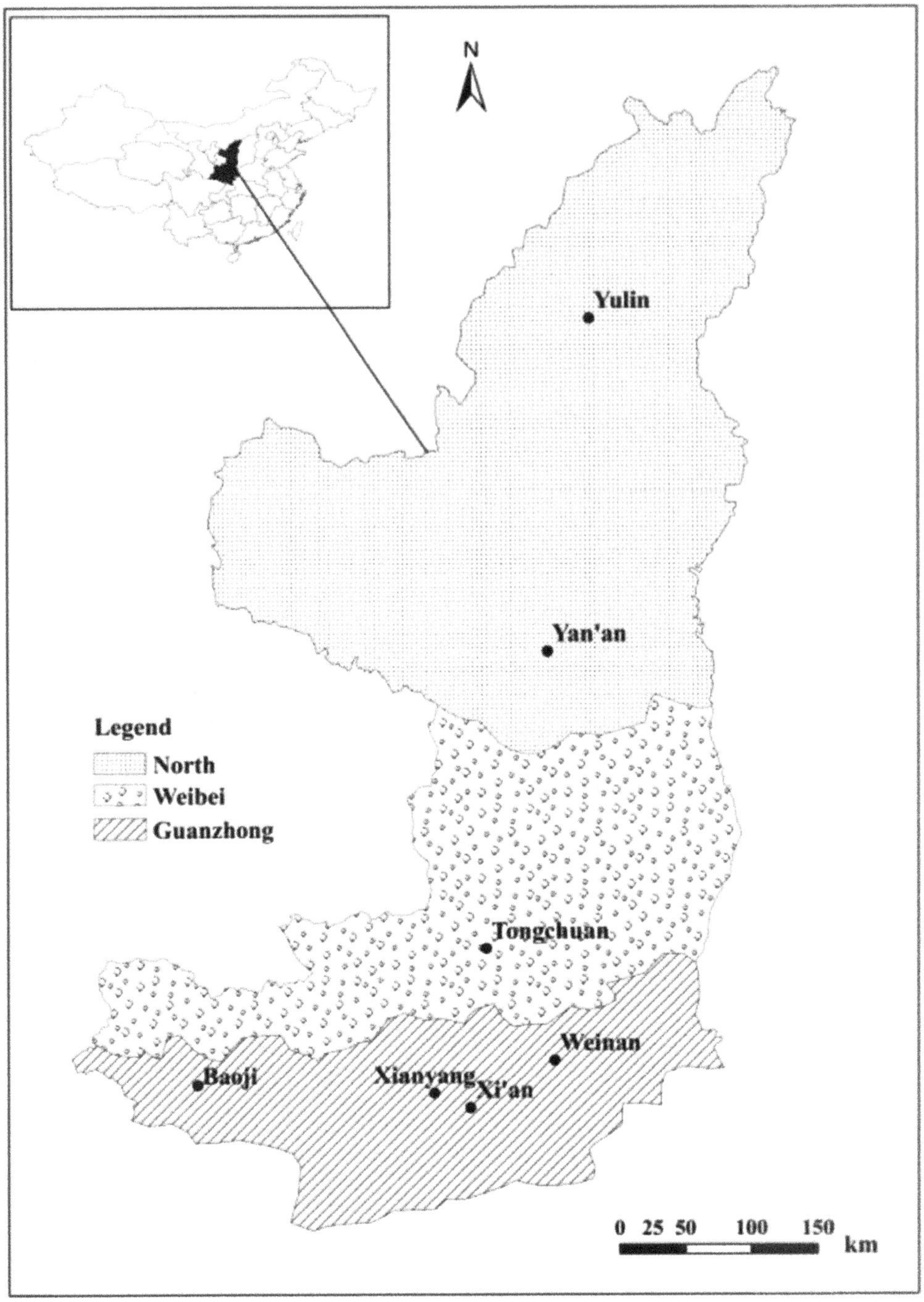

Figure 1. Map of the study area.

Table 1. Main soil types and climatic conditions in the different regions.

Region	Main soil types	Annual mean temperature (°C)	Annual precipitation (mm)
North	Castanozems, Sierozems, Loess soils	8~11	275~590
Weibei	Black loess soils, Loess soils	9~13	530~630
Guanzhong	Cinnamon soils	10~14	600~720

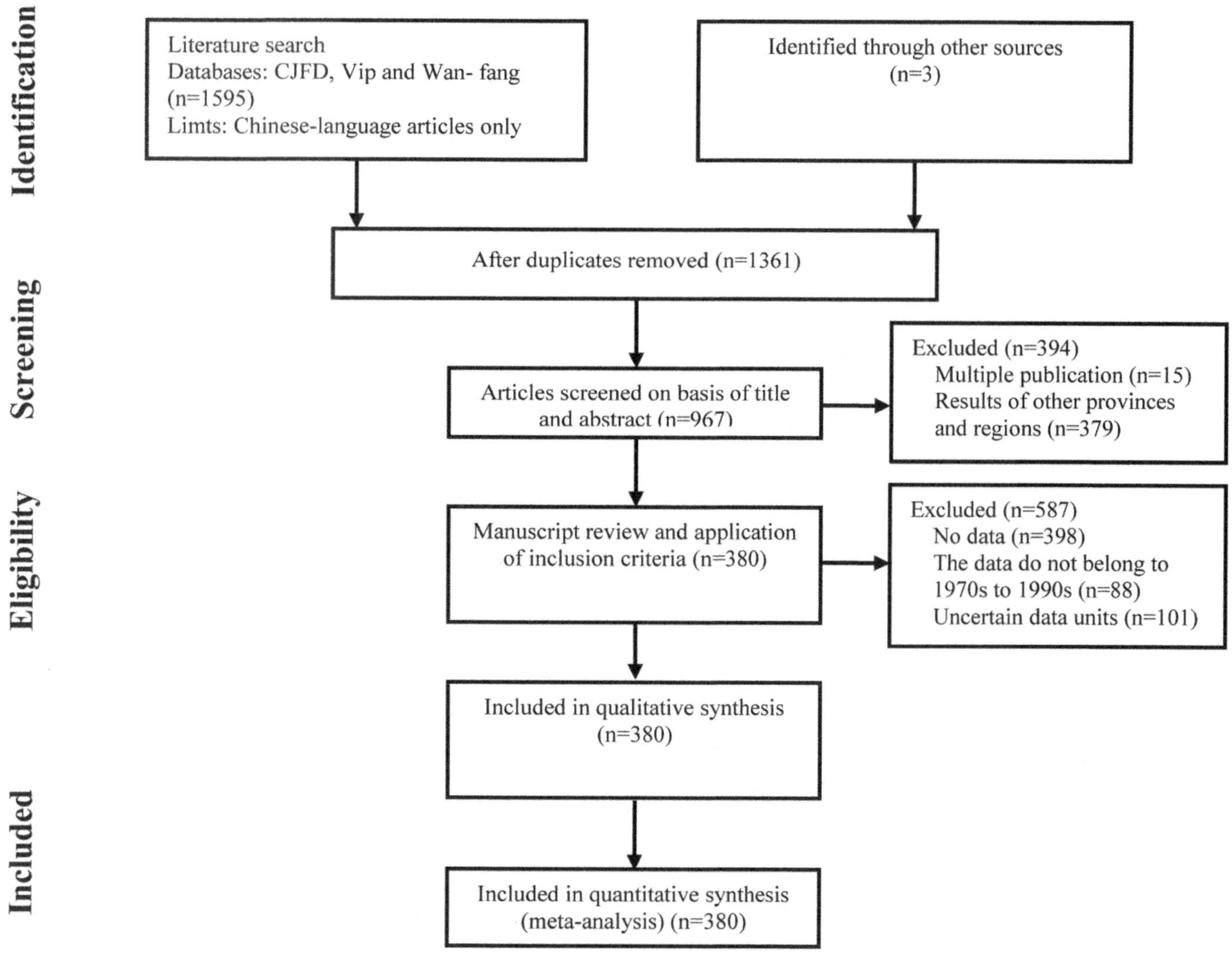

Figure 2. The screening process and results for literature from the 1970s to the 1990s.

levels on wheat plots were -102 kg ha^{-1} and -165 kg ha^{-1} in the Weibei and Guanzhong regions, respectively; meanwhile K_2O deficiency levels on maize plots were -179 kg ha^{-1}, -137 kg ha^{-1} and -147 kg ha^{-1} in the North, Weibei and Guanzhong regions, respectively.

Spatial and temporal variations of soil fertility in different regions of Shaanxi province

In the different regions of Shaanxi province, soil fertility indexes, including organic matter, total nitrogen, alkali-hydrolysis nitrogen, available phosphorus and available potassium, all increased from the 1970s to the 2000s. Simultaneously, each of these five indicators increased from the north to the south during the same period (North<Weibei<Guanzhong) (Figure 6). In the North, Weibei and Guanzhong regions from the 1970s to the 2000s, organic matter varied from 0.57%, 1.01% and 1.12% to 0.83%, 1.26% and 1.50%, with increase rates of 46%, 26% and 43%, respectively; total nitrogen varied from 0.04%, 0.07% and 0.07% to 0.05%, 0.08% and 0.09%, with increase rates of 42%, 9% and 14%, respectively; alkali-hydrolysis nitrogen varied from 29.95 mg kg^{-1}, 20.43 mg kg^{-1} and 30.81 mg kg^{-1} to 35.20 mg kg^{-1}, 58.70 mg kg^{-1} and 68.40 mg kg^{-1}, with increase rates of 18%, 187% and 122%, respectively; available phosphorus varied from 4.98 mg kg^{-1}, 7.13 mg kg^{-1} and 9.90 mg kg^{-1} to 8.10 mg kg^{-1}, 14.60 mg kg^{-1} and 26.40 mg kg^{-1}, with increase rates of 63%, 105% and 167%, respectively; available potassium varied from 85.60 mg kg^{-1}, 56.78 mg kg^{-1} and 111.75 mg kg^{-1} to 99.60 mg kg^{-1}, 160.70 mg kg^{-1} and 170.40 mg kg^{-1}, with increase rates of 16%, 183% and 52%, respectively.

Relationships among fertilizer inputs, crop yields and soil fertility in different regions of Shaanxi province

Because farmers used little P and K fertilizers in the 1970s and 1980s (Figure 3), only PFP of N, NP and NPK were calculated in the study (Eq. 2). The PFP of N, NP and NPK on wheat and maize decreased from the 1970s to the 2000s as a whole in the different regions (Table 2). The PFP of N on wheat in the Weibei and Guanzhong regions were 42 kg kg^{-1} and 65 kg kg^{-1}, respectively, in the 1970s, which decreased to 23 kg kg^{-1} and 33 kg kg^{-1}, respectively, in the 2000s. Meanwhile the PFP of N on maize in the North and Guanzhong regions were 76 kg kg^{-1} and 118 kg kg^{-1}, respectively, and they decreased to 33 kg kg^{-1} and 28 kg kg^{-1} from the 1970s to the 2000s. In the Weibei region, the PFP of N on maize changed slightly from 28 kg kg^{-1} to 32 kg kg^{-1}, which resulted from the use of high N inputs relative to the other two regions (up to 89 kg ha^{-1}) in the 1970s (Figure 3). This led to

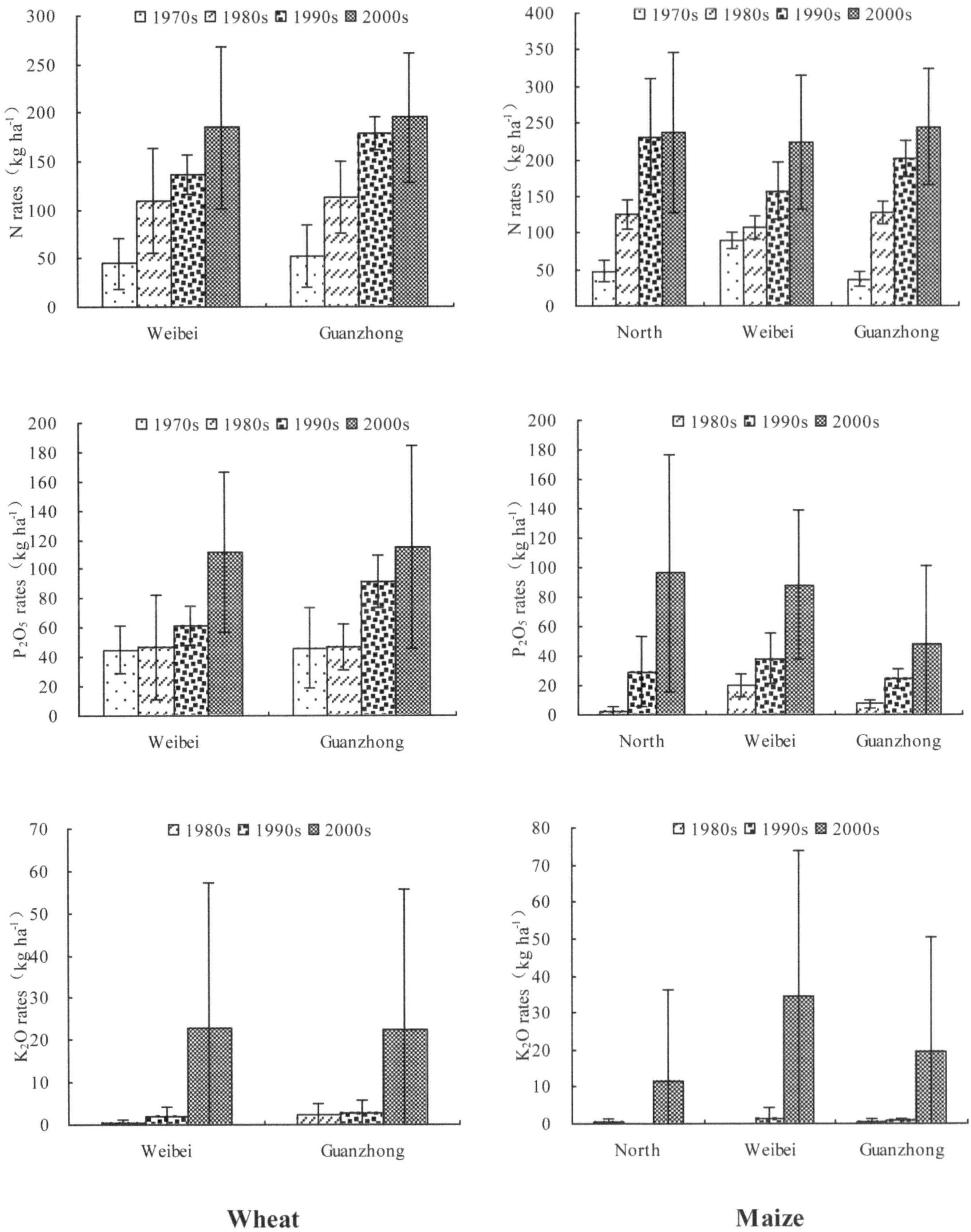

Figure 3. Variations of chemical fertilization for wheat and maize at the farmers' level in different regions of Shaanxi province (error bars show standard deviations).

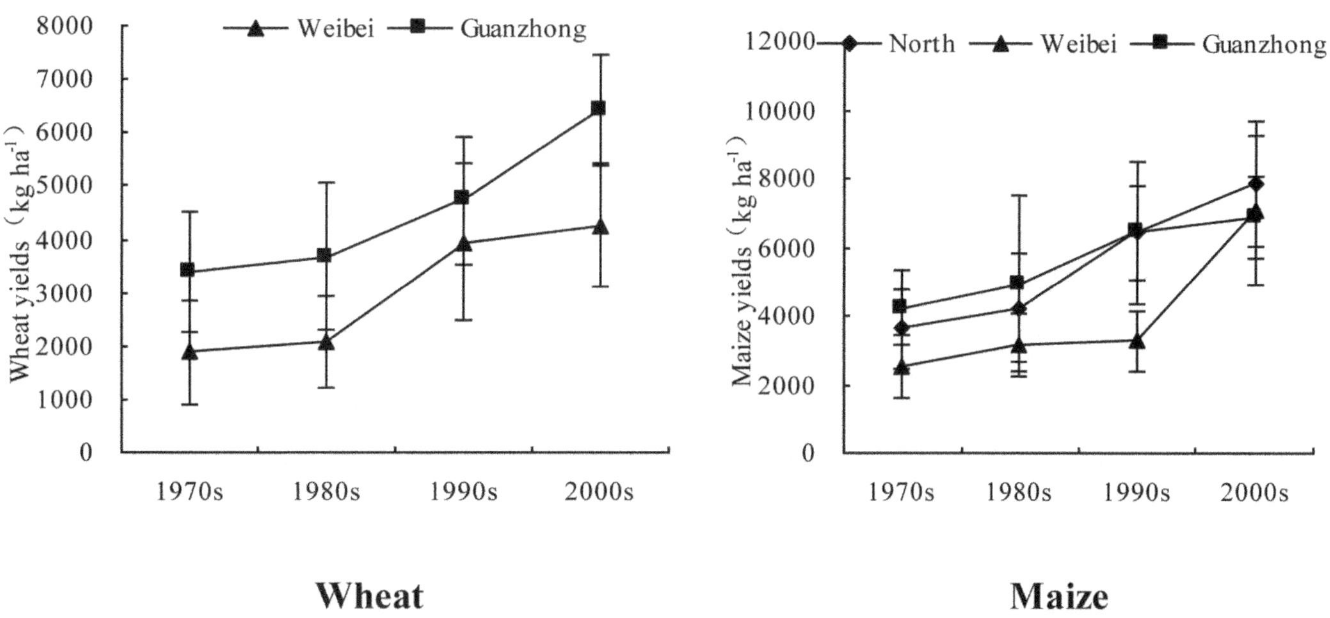

Figure 4. Variations of yields for wheat and maize at the farmers' level in different regions of Shaanxi province (error bars show standard deviations).

a low PFP of N in that period. The PFP of NP on wheat decreased to 14 kg kg^{-1} and 21 kg kg^{-1} in the Weibei and Guanzhong regions, respectively; in maize it decreased to 24 kg kg^{-1}, 23 kg kg^{-1} and 24 kg kg^{-1} in the North, Weibei and Guanzhong regions, respectively. Similar to N and NP, the PFP of NPK on wheat decreased to 13 kg kg^{-1} and 19 kg kg^{-1} in the Weibei and Guanzhong regions, respectively; in maize it decreased to 23 kg kg^{-1}, 20 kg kg^{-1} and 22 kg kg^{-1} in the North, Weibei and Guanzhong regions, respectively (Table 2).

In order to find relationships among soil fertility, crop yields and fertilizer rates, we used the Weibei region as an example. The values of fertilization, crop yields and soil fertility did not have one to one correspondence from the 1970s to the 1990s, so their mean value from each period was examined (Figures 7 and 8). Although the sample size was small and some relationships did not reach significant levels, with the increase in N fertilizer inputs, soil total nitrogen and alkali-hydrolysis nitrogen both increased. P fertilizer increased soil available phosphorus and K fertilizer increased soil available potassium significantly (Figure 7). At the same time, soil organic matter, total nitrogen, alkali-hydrolysis nitrogen, available phosphorus and available potassium all had positive impacts on wheat yields (Figure 8).

Discussion

Fertilizer use efficiency of both wheat and maize decreased from the 1970s to the 2000s as a whole in the Loess Plateau of Shaanxi (Table 2), which was consistent with national trends. Nitrogen fertilizer, phosphorus fertilizer and potassium fertilizer use efficiencies were 30–35%, 15–20% and 35–50%, respectively, from 1981 to 1983, and the average values decreased to 28%, 12% and 32% on cereal crops by 2001 to 2005 in China [19]. This suggested that the effect of chemical fertilizers on increasing grain production had diminished. The PFP of N on wheat in the Weibei and Guanzhong regions decreased to 23 kg kg^{-1} and 33 kg kg^{-1}, respectively, and the PFP of N on maize in the North, Weibei and Guanzhong regions were 33 kg kg^{-1}, 32 kg kg^{-1} and 28 kg kg^{-1}, respectively, in the 2000s (Table 2). Zhang et al. [19] reported average PFP values of N for wheat and maize of 43 kg kg^{-1} and 52 kg kg^{-1}, respectively, in China. Dobermann and Cassman [20] reported a global average PFP of N for cereals of 44 kg kg^{-1}. This indicated that nitrogen use efficiency on wheat and maize in the Loess Plateau of Shaanxi was much lower than the current national and global levels. Excessive fertilization has been the main reason for low fertilizer use efficiency in China [19]. In addition, Liu et al. [21] reported that in agro-ecosystems, surplus N increased from 1978 to 2005 throughout the country, and our findings on the Loess Plateau were consistent with this trend. For example, in the 2000s, chemical fertilizer N inputs on maize were 237 kg ha^{-1}, 223 kg ha^{-1} and 244 kg ha^{-1} in the North, Weibei and Guanzhong regions, respectively (Figure 3); meanwhile N surpluses on maize plots were 64 kg ha^{-1}, 67 kg ha^{-1} and 93 kg ha^{-1}, respectively, in the three regions (Figure 5). This indicated that excessive N fertilization was a serious problem in the Loess Plateau, and the same phenomenon has been reported many times in China, for example, in Beijing [16,22], Shandong [1,23–25], and Jiangsu [26–27]. Excessive N fertilization not only wastes resources, but also leads to many serious environmental problems [28–31] including nitrate pollution of groundwater [32–37], eutrophication of surface water [38–39], greenhouse gas emissions and other forms of air pollution [40–42], acid rain [43–46], soil acidification [36,47–50] and so on. On the other hand, a lower fertilization rate does not necessarily reduce crop yields [51]. Many studies have shown that reducing the current N application rates by 30 to 60% could increase N fertilizer efficiency, while still maintaining crop yields and substantially reducing N losses to the environment [31,52–53].

Like nitrogen, phosphate fertilizer inputs (Figure 3), P surpluses (Figure 5) and soil available phosphorus levels (Figure 6) all increased in the last 40 years on the Loess Plateau in Shaanxi. Similar results have been noted in north China and all over the country [25,54]. Yang et al. [55] reported that maintaining soil available phosphorus at a relatively high level requires a P application rate of about 80 kg ha^{-1} yr^{-1} in winter wheat/ summer maize rotation systems in the Guanzhong region. Our

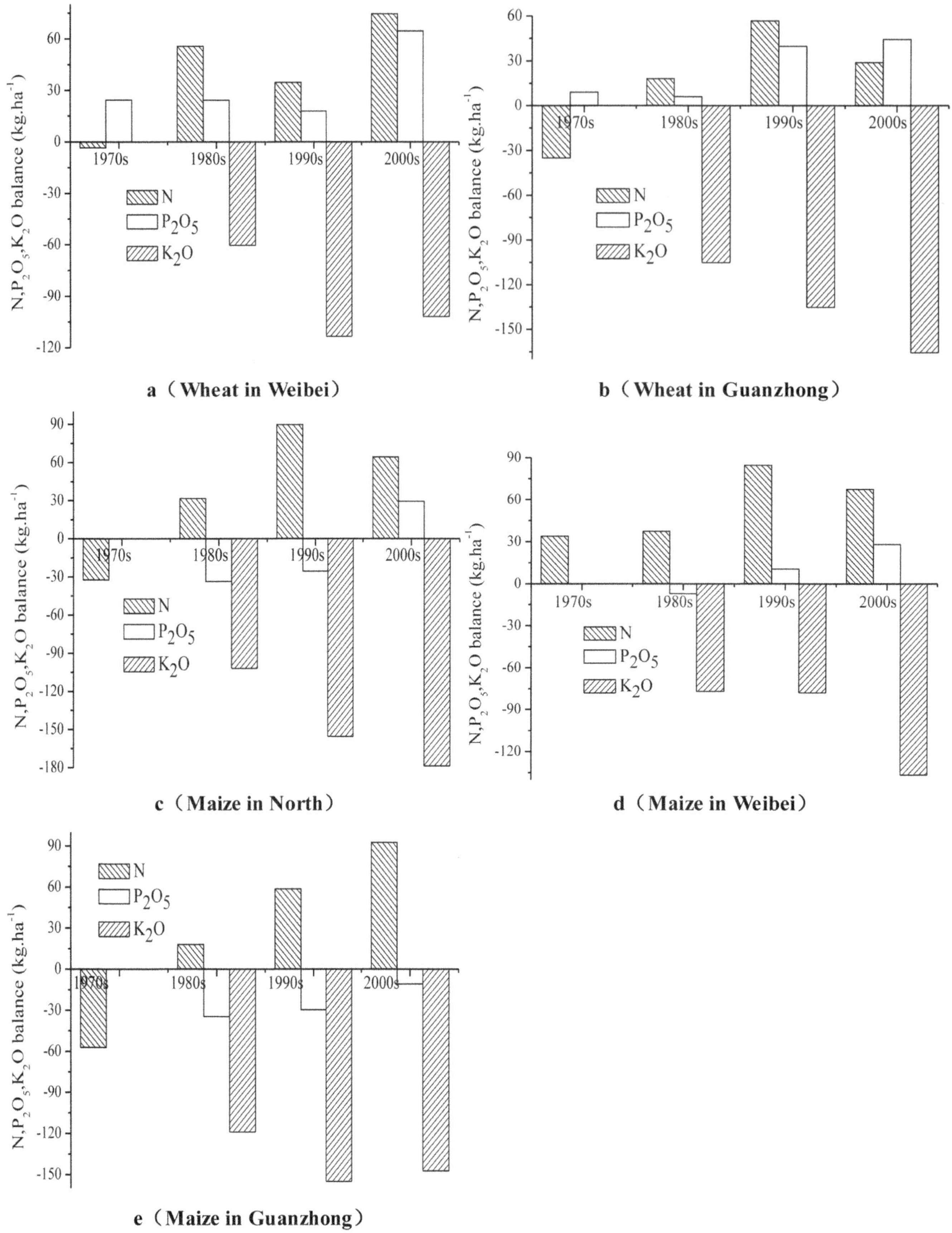

Figure 5. Variations of soil nutrient balance on wheat and maize plots in different regions of Shaanxi province.

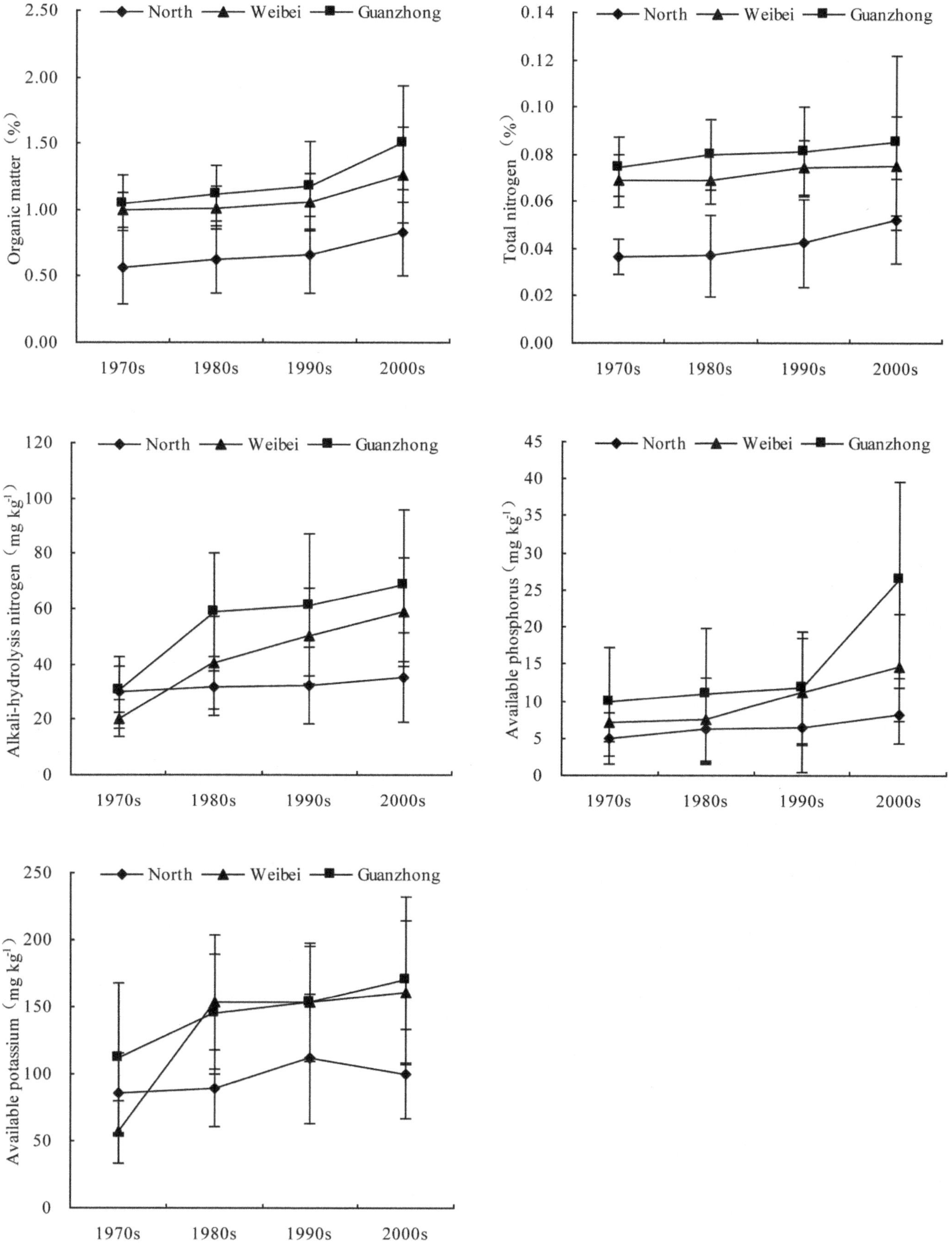
North
Weibei
Guanzhong
Organic matter (%)
2.50
2.00
1.50
1.00
0.50
0.00
Total nitrogen (%)
0.14
0.12
0.10
0.08
0.06
0.04
0.02
0.00
Alkali-hydrolysis nitrogen (mg kg-1)
120
100
80
60
40
20
0
Available phosphorus (mg kg-1)
45
40
35
30
25
20
15
10
5
0
Available potassium (mg kg-1)
250
200
150
100
50
0
1970s
1980s
1990s
2000s

Figure 6. Variations of soil organic matter, total nitrogen, alkali-hydrolysis nitrogen, available phosphorus and available potassium in different regions of Shaanxi province (error bars show standard deviations).

results showed phosphate fertilizer inputs of up to 163 kg ha^{-1} in winter wheat/summer maize rotation systems in this region in the 2000s (Figure 3). This indicated that P fertilization was also excessive, which not only wasted resources but also led to many serious environmental problems [28–31]. Phosphate fertilizer production consumes more than 80% of the phosphate rock resources [56], but phosphate rock resources are limited and high grade material is in short supply [57]. In addition, the phosphate fertilization utilization ratio of the main crops ranged from 7% to 20%. It averages 12% in China [19], which has led to phosphorus accumulation in the soil, increasing the risk of non-point source pollution from surface runoff [58]. Agricultural non-point source pollution has become an increasingly serious problem in China, primarily because it leads to eutrophication.

In spite of increased K fertilizer inputs on wheat and maize in recent years (Figure 3), the soil K balance has become increasingly negative (Figure 5) and soil available potassium has increased (Figure 6) in the last 40 years. This phenomenon was previously reported in northwest and north China [55,59–60]. Evidently, K fertilizer application was not the only source of K absorbed by crops. The primary sources of K for crops were weathering of parent materials [60–61], release of K into the soil from increased soil organic matter and changes in soil pH [61]. Yang et al. [55] found that soil organic matter content in all treatments (including those without fertilizer) significantly increased over time and soil pH dropped from the initial value of 8.65 to 8.58 from 1991 to 2010 during long-term field trials in the Guanzhong region. Our results showed that in the North, Weibei and Guanzhong regions soil organic matter increased from 0.57%, 1.01% and 1.12% to 0.83%, 1.26% and 1.50%, respectively, from the 1970s to the 2000s (Figure 6). The average soil pH has declined 0.5 units with the overuse of N fertilizer in the past two decades in China [62]. Li et al. [63] reported that the soil pH decreased from the initial value of 8.76 to 8.56 from 1992 to 2008 during long-term field trials in the North region. There may be other mechanisms involved, for example, crops might draw on K in the deeper soil layers or from the non-exchangeable pool. The contribution of K from the subsoil could be considerable [64]. Witter and Johansson [65] found that 41–47% of the K was from the subsoil for green manure crops. Many studies have shown that crops use non-exchangeable K [66–67]. Decreases in the abundance of non-exchangeable K with simultaneous increases in exchangeable and water-soluble K concentrations suggest that much of the K taken up by crops comes from non-exchangeable species via solution and exchangeable phases in a way that establishes and maintains the equilibrium between various forms of K in the soil [66].

Fertilizer rates had a large effect on soil fertility. With the increase in N fertilizer inputs, both soil total nitrogen and alkali-hydrolysis nitrogen increased; P fertilizer increased soil available phosphorus and K fertilizer increased soil available potassium significantly in the Weibei region (Figure 7). It has been reported that after 25 years of N fertilization, soil organic carbon and total nitrogen had increased by 18% and 26%, respectively, from 1984 to 2009 in the Weibei region [18]. Cai and Hao [68] also found that accumulation of soil nitrogen initially increased and then decreased with increasing nitrogen, and total nitrogen and alkali-hydrolysis nitrogen content reached the highest value or the second highest value of 135 kg ha^{-1} on wheat plots in the Weibei region, which was in accordance with findings in northwest and north China by Li et al. [63] and Lin et al. [69]. Through long-term field experimentation on the Loess Plateau in Shaanxi, Li et al. [63] and Hao et al. [70] found that with increases in P fertilizer inputs, soil available P increased significantly. Similar results have been obtained in northeast and northwest China by Geng et al. [71] and Zhao et al. [72], and also in America by Griffin et al. [73]. In addition, Li et al. [74] found that with increased K fertilizer inputs, soil available K increased significantly in a long-term field experiment on the Loess Plateau. Further-

Table 2. Variations of PFP of fertilizer on wheat and maize in the different regions (kg kg^{-1}).

Crop	Fertilizer type	Region	1970s	1980s	1990s	2000s
Wheat	N	Weibei	42	19	29	23
		Guanzhong	65	33	26	33
	$N+P_2O_5$	Weibei	21	13	20	14
		Guanzhong	34	23	17	21
	$N+P_2O_5+K_2O$	Weibei	21	13	20	13
		Guanzhong	34	23	17	19
Maize	N	North	76	34	28	33
		Weibei	28	30	21	32
		Guanzhong	118	39	32	28
	$N+P_2O_5$	North	76	33	25	24
		Weibei	28	25	17	23
		Guanzhong	118	37	29	24
	$N+P_2O_5+K_2O$	North	76	33	25	23
		Weibei	28	25	17	20
		Guanzhong	118	37	29	22

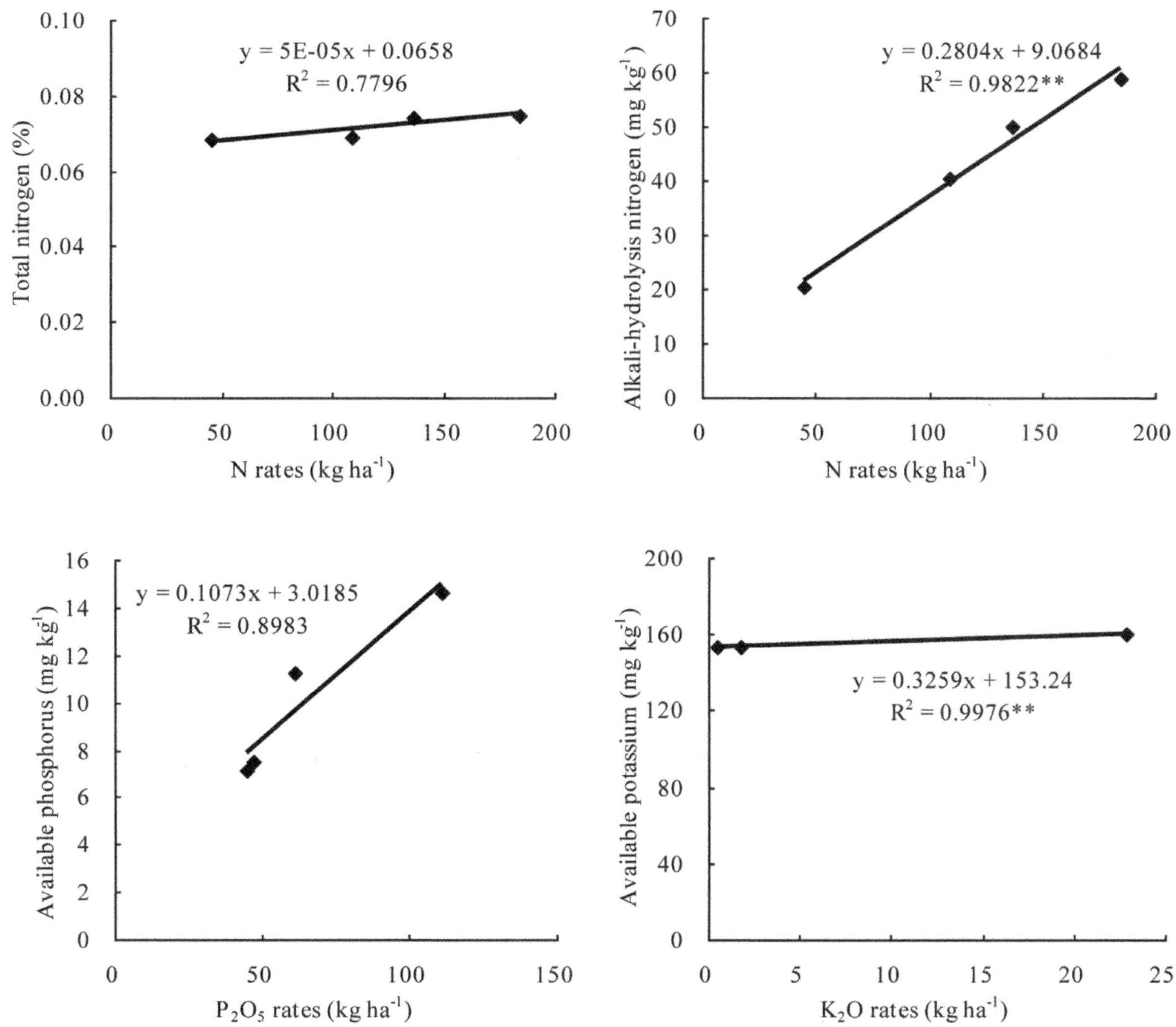

Figure 7. Relationships between N rates and total nitrogen, N rates and alkali-hydrolysis nitrogen, P_2O_5 rates and available phosphorus and K_2O rates and available potassium on wheat plots in the Weibei region of Shaanxi province. **Significance level: $P< 0.01$.

more, many studies in this area have shown that on the basis of N and P fertilizer application, long-term K fertilizer application can increase soil available K and grain yields [75–76].

Our research also found that soil fertility had a positive impact on crop yields (Figure 8). Zhou et al. [77] revealed that soil organic carbon and total nitrogen concentrations had a significant effect on crop yields in the semi-arid Loess Plateau by long-term experimentation. Higher yields without fertilizer were generally obtained in soils with higher average soil organic matter concentrations. For example, yields without fertilizer <4000 kg ha^{-1} were obtained with average soil organic matter concentrations of 1.41% for winter wheat and 1.46 for summer maize. In contrast, average soil organic matter concentrations were 1.69% for winter wheat and 1.61% for summer maize for plots with yields>6000 kg ha^{-1} without fertilizer in north China [78]. Gong et al. [79] also found that the contribution percentage of basic soil productivity to wheat yield was significantly correlated with soil organic carbon, total nitrogen, available nitrogen, available phosphorus and available potassium in long-term soil fertility experiments in north China. Similar results have been obtained in other parts of mainland China [80], indicating that inherent soil productivity contributed to the substantial increase in China's crop yields.

In addition, although the use of chemical fertilizers to supplement NPK nutrients in the soil is important, many researchers at home and abroad reported that the application of chemical fertilizer in combination with organic manure is helpful in maintaining soil fertility (especially soil organic carbon) and buffering capacity, and in reducing NO_3-N accumulation in the soil, while maintaining high soil productivity [4,81–87].

Conclusions

From the 1970s to the 2000s in the North, Weibei and Guanzhong regions of the Loess Plateau in Shaanxi province, chemical fertilizer NPK inputs and yields of wheat and maize increased at the farmers' level. In the 1970s, N was deficient on wheat and maize plots in the different regions; thereafter N was in surplus. In the same way, P gradually changed from deficit to surplus levels. In addition, soil organic matter, total nitrogen, alkali-hydrolysis nitrogen, available phosphorus and available

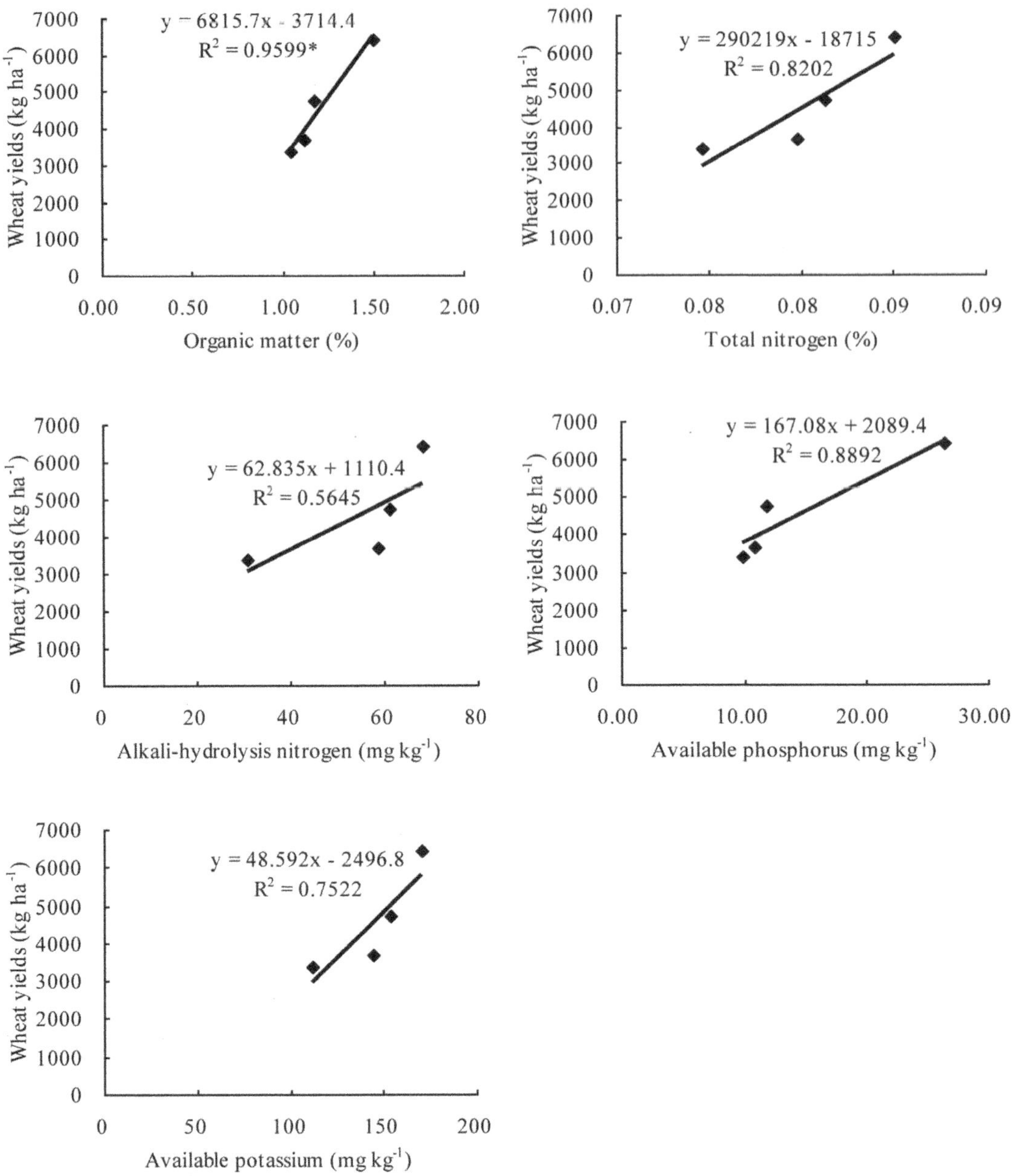

Figure 8. Relationships between wheat yield and soil organic matter, total nitrogen, alkali-hydrolysis nitrogen, available phosphorus and available potassium in the Weibei region of Shaanxi province. *Significance level: $P<0.05$.

potassium increased over the same period. However, K deficiencies became more and more severe. The PFP of N, NP and NPK on wheat and maize all decreased from the 1970s to the 2000s as a whole. With the increase in N fertilizer inputs, both soil total nitrogen and alkali-hydrolysis nitrogen increased; P fertilizer increased soil available phosphorus and K fertilizer increased soil available potassium significantly. At the same time, soil organic matter, total nitrogen, alkali-hydrolysis nitrogen, available phosphorus and available potassium all had positive impacts on crop yields. In order to promote food safety and environmental protection, farmers should be encouraged to assess their fertilizer needs carefully. Many can reduce nitrogen and phosphate fertilizer inputs significantly and increase potassium fertilizer and organic manure on cereal crops.

Acknowledgments

We are grateful to Harald Grip and Lars Lövdahl for their help in writing this paper. The authors would also like to thank the Agricultural Technology Extension Center of Shaanxi province for the help with data collection.

Author Contributions

Conceived and designed the experiments: YT PG. Analyzed the data: XW YG. Contributed reagents/materials/analysis tools: YT YG PG. Wrote the paper: XW. Collected the data: XW FL ZZ YP.

References

1. Cui ZL, Chen XP, Zhang FS (2010) Current nitrogen management status and measures to improve the intensive wheat–maize system in China. AMBIO 39: 376–384.
2. Gao C, Sun B, Zhang TL (2006) Sustainable nutrient management in Chinese agriculture: challenges and perspective. Pedosphere 16(2): 253–263.
3. Zhu ZL, Chen DL (2002) Nitrogen fertilizer use in China-Contributions to food production, impacts on the environment and best management strategies. Nutrient Cycling in Agroecosystems 63: 117–127.
4. Jiang D, Hengsdijk H, Dai TB, de Boer W, Qi J, et al. (2006) Long-term effects of manure and inorganic fertilizers on yield and soil fertility for a winter wheat-maize system in Jiangsu, China. Pedosphere 16(1): 25–32.
5. Luo SM (2007) To discover the secret of traditional agriculture and serve the modern ecoagriculture. Geographical Research 26(3): 609–615. (in Chinese).
6. Zhang F, Qiao Y, Wang F, Zhang W (2007) A perspective on organic agriculture in China: Opportunities and challenges. Proceedings of 9th German Scientific Conference on Organic Agriculture.
7. Department of Rural Surveys, National Bureau of Statistics (1971–2011) China Rural Statistical Yearbook. China Statistics Press. Beijing, China. (in Chinese).
8. Karlen DL, Mausbach MJ, Doran JW, Cline RG, Harris RF, et al. (1997) Soil quality: A concept, definition and framework for evaluation. Soil Science Society of America Journal 61: 4–10.
9. Wang XJ, Gong ZT (1998) Assessment and analysis of soil quality changes after eleven years of reclamation in subtropical China. Geoderma 81: 339–355.
10. Arshad MA, Martin S (2002) Identifying critical limits for soil quality indicators in agro-ecosystems. Agriculture, Ecosystems & Environment 88: 153–160.
11. Huang B, Sun WX, Zhao YC, Zhu J, Yang RQ, et al. (2007) Temporal and spatial variability of soil organic matter and total nitrogen in an agricultural ecosystem as affected by farming practices. Geoderma 139: 336–345.
12. Hoosbeek MR, Bryant RB (1992) Towards the quantitative modeling of pedogenesis: A review. Geoderma 55: 183–210.
13. Smil V (2001) Enriching the Earth: Fritz Haber, Carl Bosch, and the Transformation of World Food Production. MIT Press. Cambridge, UK.
14. Gong W, Yan XY, Wang JY (2011) Effect of long-term fertilization on soil fertility. Soils 43: 336–342. (in Chinese).
15. Bierman PM, Rosen CJ, Venterea RT, Lamb JA (2012) Survey of nitrogen fertilizer use on corn in Minnesota. Agricultural Systems 109: 43–52.
16. Wang SR (2002) Current status and evaluation of crop fertilization in Shaanxi province and Beijing city, Ph.D. thesis, China Agricultural University, Beijing, China. (in Chinese).
17. Liu GB (1999) Soil conservation and sustainable agriculture on the Loess Plateau: Challenges and prospects. AMBIO 28: 663–668.
18. Guo SL, Zhu HH, Dang TH, Wu JS, Liu WZ, et al. (2012) Winter wheat grain yield associated with precipitation distribution under long-term nitrogen fertilization in the semiarid Loess Plateau in China. Geoderma 189: 442–450.
19. Zhang FS, Wang JQ, Zhang WF, Cui ZL, Ma WQ, et al. (2008) Nutrient use efficiencies of major cereal crops in China and measures for improvement. Acta Pedologica Sinica 45: 915–924. (in Chinese).
20. Dobermann A, Cassman KG (2005) Cereal area and nitrogen use efficiency are drivers of future nitrogen fertilizer consumption. Science in China (Series C: Life Sciences) 48: 745–758.
21. Liu Z, Li BG, Fu J (2009) Nitrogen balance in agro-ecosystem in China from 1978 to 2005 based on DSS. Transactions of the CSAE 25(4): 168–175. (in Chinese).
22. Zhao JR, Guo Q, Guo JL, Wei DM, Wang CW, et al. (1997) The chemical fertilizer inputs and yields of grain fields in the suburbs of Beijing. Beijing Agricultural Sciences 15(2): 36–38. (in Chinese).
23. Ma WQ (1999) Current status and evaluation of crop fertilization in Shandong province, Ph.D. thesis, China Agricultural University, Beijing, China. (in Chinese).
24. Li JL, Cui DJ, Meng XX, Li XL, Zhang FS (2002) The study of fertilization condition and question in protectorate vegetable in Shouguang Shandong. Chinese Journal of Soil Science 33: 126–128. (in Chinese).
25. Zhen L, Zoebisch MA, Chen GB, Feng ZM (2006) Sustainability of farmers' soil fertility management practices: A case study in the North China Plain. Journal of Environmental Management 79: 409–419.
26. Richter J, Roelcke M (2000) The N-cycle as determined by intensive agriculture–examples from central Europe and China. Nutrient Cycling in Agroecosystems 57: 33–46.
27. Ma LH, Zhang Y, Sui B, Liu CL, Wang P, et al. (2011) The impact factors of excessive fertilization in Jiangsu province. Journal of Yangzhou University 32(2): 48–52, 80. (in Chinese).
28. Gao XZ, Ma WQ, Du S, Zhang FS, Mao DR (2001) Current status and problems of fertilization in China. Chinese Journal of Soil Science 32(6): 258–261. (in Chinese).
29. Cui ZL, Chen XP, Miao YX, Zhang FS, Sun QP, et al. (2008) On-farm evaluation of the improved soil N_{min}-based nitrogen management for summer maize in North China Plain. Agronomy Journal 100: 517–525.
30. Cui ZL, Zhang FS, Chen XP, Miao YX, Li JL, et al. (2008) On-farm evaluation of an in-season nitrogen management strategy based on soil N_{min} test. Field Crops Research 105: 48–55.
31. Ju XT, Xing GX, Chen XP, Zhang SL, Zhang LJ, et al. (2009) Reducing environmental risk by improving N management in intensive Chinese agricultural systems. Proceedings of the National Academy of Sciences 106: 3041–3046.
32. Tong YA, Emteryd O, Lu DQ, Grip H (1997) Effect of organic manure and chemical fertilizer on nitrogen uptake and nitrate leaching in a Eum-orthic anthrosols profile. Nutrient Cycling in Agroecosystems 48: 225–229.
33. Ju XT, Kou CL, Zhang FS, Christie P (2006) Nitrogen balance and groundwater nitrate contamination: Comparison among three intensive cropping systems on the North China Plain. Environmental Pollution 143: 117–125.
34. Ju XT, Liu XJ, Zhang FS, Roelcke M (2004) Nitrogen fertilization, soil nitrate accumulation, and policy recommendations in several agricultural regions of China. AMBIO 33: 300–305.
35. Yan X, Jin JY, He P, Liang MZ (2008) Recent advances on the technologies to increase fertilizer use efficiency. Agricultural Sciences in China 7(4): 469–479.
36. Guo SL, Wu JS, Dang TH, Liu WZ, Li Y, et al. (2010) Impacts of fertilizer practices on environmental risk of nitrate in semiarid farmlands in the Loess Plateau of China. Plant and Soil 330: 1–13.
37. Gao Y, Yu G, Luo C, Zhou P (2012) Groundwater nitrogen pollution and assessment of its health risks: A case study of a typical village in rural-urban continuum, China. PloS ONE 7(4): e33982.
38. Tilman D, Fargione J, Wolff B, D'Antonio C, Dobson A, et al. (2001) Forecasting agriculturally driven global environmental change. Science 292: 281–284.
39. Huang GQ, Wang XX, Qian HY, Zhang TL, Zhao QG (2004) Negative impact of inorganic fertilizer application on agricultural environment and its countermeasures. Ecology and Environment 13(4): 656–660. (in Chinese).
40. Mosier AR, Duxbury JM, Freney JR, Heinemeyer O, Minami K (1996) Nitrous oxide emissions from agricultural fields: Assessment, measurement and mitigation. Plant and Soil 181: 95–108.
41. Zhang JF, Han XG (2008) N_2O emission from the semi-arid ecosystem under mineral fertilizer (urea and superphosphate) and increased precipitation in northern China. Atmospheric Environment 42: 291–302.
42. Li H, Qiu JJ, Wang LG, Tang HJ, Li CS, et al. (2010) Modelling impacts of alternative farming management practices on greenhouse gas emissions from a winter wheat–maize rotation system in China. Agriculture, Ecosystems and Environment 135: 24–33.
43. Krusche AV, de Camargo PB, Cerri CE, Ballester MV, Lara LBLS, et al. (2003) Acid rain and nitrogen deposition in a sub-tropical watershed (Piracicaba): ecosystem consequences. Environmental Pollution 121: 389–399.
44. Menz FC, Seip HM (2004) Acid rain in Europe and the United States: an update. Environmental Science & Policy 7: 253–265.
45. Wu D, Wang SG, Shang KZ (2006) Progress in research of acid rain in China. Arid Meteorology 24(2): 70–77. (in Chinese).
46. Huang DY, Xu YG, Peng PA, Zhang HH, Lan JB (2009) Chemical composition and seasonal variation of acid deposition in Guangzhou, South China: Comparison with precipitation in other major Chinese cities. Environmental Pollution 157: 35–41.
47. Dai ZH, Liu YX, Wang XJ, Zhao DW (1998) Changes in pH, CEC, and exchangeable acidity of some forest soils in southern China during the last 32–35 years. Water, Air, and Soil Pollution 108: 377–390.
48. Zhang HM, Wang BR, Xu MG, Fan TL (2009) Crop yield and soil responses to long-term fertilization on a red soil in southern China. Pedosphere 19(2): 199–207.
49. Zhao X, Xing GX (2009) Variation in the relationship between nitrification and acidification of subtropical soils as affected by the addition of urea or ammonium sulfate. Soil Biology & Biochemistry 41: 2584–2587.
50. Huang S, Zhang WJ, Yu XC, Huang QR (2010) Effects of long-term fertilization on corn productivity and its sustainability in an Ultisol of southern China. Agriculture, Ecosystems and Environment 138: 44–50.
51. Ma WQ, István S (2008) Can sharp decrease of fertilizer input lead obvious reduction of crop yield?. Ecology and Environment 17: 1296–1301. (in Chinese).
52. Peng SB, Buresh RJ, Huang JL, Yang JC, Zou YB, et al. (2006) Strategies for overcoming low agronomic nitrogen use efficiency in irrigated rice systems in China. Field Crops Research 96: 37–47.
53. Yi Q, Zhang XZ, He P, Yang L, Xiong GY (2010) Effects of reducing N application on crop N uptake, utilization, and soil N balance in rice-wheat rotation system. Plant Nutrition and Fertilizer Science 16: 1069–1077. (in Chinese).
54. Cao N, Zhang YB, Chen XP (2009) Spatial-temporal change of phosphorus balance and the driving factors for agroecosystems in China. Chinese Agricultural Science Bulletin 25: 220–225. (in Chinese).
55. Yang XY, Sun BH, Zhang SL (2014) Trends of yield and soil fertility in a long-term wheat-maize system. Journal of Integrative Agriculture 13: 402–414.
56. Zhang WX (2011) Development and utilization trend of phosphate resources in China. Journal of Wuhan Institute of Technology 33: 1–5. (in Chinese).
57. Zhang WF, Ma WQ, Zhang FS, Ma J (2005) Comparative analysis of the superiority of China's phosphate rock and development strategies with that of the United States and Morocco. Journal of Natural Resources 20: 378–386. (in Chinese).

58. van Bochove E, Thériault G, Dechmi F, Leclerc ML, Goussard N (2007) Indicator of risk of water contamination by phosphorus: Temporal trends for the Province of Quebec from 1981 to 2001. Canadian Journal of Soil Science 87: 121–128.
59. Liu EK, Yan CR, Mei XR, He WQ, Bing SH, et al. (2010) Long-term effect of chemical fertilizer, straw, and manure on soil chemical and biological properties in northwest China. Geoderma 158: 173–180.
60. Tan DS, Jin J Y, Jiang LH, Huang SW, Liu ZH (2012) Potassium assessment of grain producing soils in North China. Agriculture, Ecosystems and Environment 148: 65–71.
61. Munson RD (1985) Potassium in Agriculture. Soil Science Society of America Madison, Wisconsin, USA.
62. Guo JH, Liu XJ, Zhang Y, Shen JL, Han WX, et al. (2010) Significant acidification in major Chinese croplands. Science 327: 1008–1010.
63. Li Q, Xu MX, Liu GB, ZhaoYG, Tuo DF (2013) Cumulative effects of a 17-year chemical fertilization on the soil quality of cropping system in the Loess Hilly Region, China. Journal of Plant Nutrition and Soil Science 176: 249–259.
64. Kautz T, Amelung W, Ewert F, Gaiser T, Horn R, et al. (2013) Nutrient acquisition from arable subsoils in temperate climates: A review. Soil Biology & Biochemistry 57: 1003–1022.
65. Witter E, Johansson G (2001) Potassium uptake from the subsoil by green manure crops. Biological Agriculture & Horticulture 19: 127–141.
66. Singh M, Singh VP, Reddy DD (2002) Potassium balance and release kinetics under continuous rice-wheat cropping system in Vertisol. Field Crops Research 77: 81–91.
67. Sharma A, Jalali VK, Arora S (2010) Non-exchangeable potassium release and its removal in foot-hill soils of North-west Himalayas. Catena 82: 112–117.
68. Cai Y, Hao MD (2013) Effects of long-term nitrogen fertilization on wheat in Loess Plateau. Journal of Triticeae Crops 33: 983–987. (in Chinese).
69. Lin ZA, Zhao BQ, Yuan L, Bing-So H (2009) Effects of organic manure and fertilizers long-term located application on soil fertility and crop yield. Scientia Agricultura Sinica 42: 2809–2819. (in Chinese).
70. Hao MD, Fan J, Wei XR, Pen LF, Lu L (2005) Effect of fertilization on soil fertility and wheat yield of dryland in the Loess Plateau. Pedosphere 15(2): 189–195.
71. Geng YH, Cao GJ, Ye Q, Qi QG, Wu P, et al. (2013) Effects of different phosphorus applications on soil available phosphorus, phosphorus absorption and yield of spring maize. Journal of South China Agricultural University 34: 470–474. (in Chinese).
72. Zhao J, Hou ZA, Li SX, Liu LP, Huang T, et al. (2014) Effects of P rate on soil available P, yield and nutrient uptake of maize. Journal of Maize Sciences 22: 123–128. (in Chinese).
73. Griffin TS, Honeycutt CW, He Z (2003) Changes in soil phosphorus from manure application. Soil Science Society of America Journal 67: 645–653.
74. Li LF, Hao MD, Li YM, Gao CQ (2009) Research on characteristics of spatial distribution and availability of soil potassium forms under long-term fertilization in the dryland of the Loess Plateau. Agricultural Research in the Arid Areas 27: 127–131, 142. (in Chinese).
75. Wang HT, Jin JY, Wang B, Zhao PP (2010) Effects of long-term potassium application and wheat straw return to cinnamon soil on wheat yields and soil potassium balance in Shanxi. Plant Nutrition and Fertilizer Science 16: 801–808. (in Chinese).
76. Zhang YL, Lu JL, Jin JY, Li ST, Chen ZQ, et al. (2012) Effects of chemical fertilizer and straw return on soil fertility and spring wheat quality. Plant Nutrition and Fertilizer Science 18: 307–314. (in Chinese).
77. Zhou ZC, Gan ZT, Shangguan ZP, Zhang FP (2013) Effects of long-term repeated mineral and organic fertilizer applications on soil organic carbon and total nitrogen in a semi-arid cropland. European Journal of Agronomy 45: 20–26.
78. Fan MS, Lai R, Cao J, Qiao L, Su YS, et al. (2013) Plant-based assessment of inherent soil productivity and contributions to China's cereal crop yield increase since 1980. PloS ONE, 8(9): e74617.
79. Gong FF, Zha Y, Wu XP, Huang SM, Xu MG, et al. (2013) Analysis on basic soil productivity change of winter wheat in fluvo-aquic soil under long-term fertilization. Transactions of the Chinese Society of Agricultural Engineering 29(12): 120–129. (in Chinese).
80. Tang YH, Huang Y (2009) Spatial distribution characteristics of the percentage of soil fertility contribution and its associated basic crop yield in mainland China. Journal of Agro-Environment Science 28: 1070–1078. (in Chinese).
81. Gami SK, Ladha JK, Pathak H, Shah MP, Pasuquin E, et al. (2001) Long-term changes in yield and soil fertility in a twenty-year rice-wheat experiment in Nepal. Biology and Fertility of Soils 34: 73–78.
82. Yang SM, Li FM, Malhi SS, Wang P, Suo DR, et al. (2004) Long-term fertilization effects on crop yield and nitrate nitrogen accumulation in soil in northwestern China. Agronomy Journal 96: 1039–1049.
83. Mando A, Ouattara B, Somado AE, Wopereis MCS, Stroosnijder L, et al. (2005) Long-term effects of fallow, tillage and manure application on soil organic matter and nitrogen fractions and on sorghum yield under Sudano-Sahelian conditions. Soil Use and Management 21: 25–31.
84. Li J, Zhao BQ, Li XY, Jiang RB, Bing SH (2008) Effects of long-term combined application of organic and mineral fertilizers on microbial biomass, soil enzyme activities and soil fertility. Agricultural Sciences in China 7(3): 336–343.
85. Banger K, Kukal SS, Toor G, Sudhir K, Hanumanthraju TH (2009) Impact of long-term additions of chemical fertilizers and farmyard manure on carbon and nitrogen sequestration under rice–cowpea cropping system in semi-arid tropics. Plant and Soil 318: 27–35.
86. Majumder B, Mandal B, Bandyopadhyay PK (2008) Soil organic carbon pools and productivity in relation to nutrient management in a 20-year-old rice-berseem agroecosystem. Biology and Fertility of Soils 44: 451–561.
87. Moharana PC, Sharma BM, Biswas DR, Dwivedi BS, Singh RV (2012) Long-term effect of nutrient management on soil fertility and soil organic carbon pools under a 6-year-old pearl millet–wheat cropping system in an Inceptisol of subtropical India. Field Crops Research 136: 32–41.

Prunus persica Crop Management Differentially Promotes Arbuscular Mycorrhizal Fungi Diversity in a Tropical Agro-Ecosystem

Maria del Mar Alguacil[1]*, Emma Torrecillas[1], Zenaida Lozano[2], Maria Pilar Torres[3], Antonio Roldán[1]

1 CSIC-Centro de Edafología y Biología Aplicada del Segura, Department of Soil and Water Conservation, Campus de Espinardo, Murcia, Spain, **2** Universidad Central de Venezuela (UCV), Facultad de Agronomía, Instituto de Edafología, El Limón, Campus Universitario, Maracay, Venezuela, **3** Departamento de Biología Aplicada, Area de Botánica, Universidad Miguel Hernández, Elche, Alicante, Spain

Abstract

Due to the important role of arbuscular mycorrhizal fungi (AMF) in ecosystem functioning, determination of the effect of management practices on the AMF diversity in agricultural soils is essential for the sustainability of these agro-ecosystems. The objective of this study was to compare the AMF diversity in *Prunus persica* roots under two types of fertilisation (inorganic, with or without manure) combined with integrated or chemical pest management in a Venezuelan agro-ecosystem. The AM fungal small-subunit (SSU) rRNA genes were subjected to PCR, cloning, sequencing and phylogenetic analyses. Twenty-one different phylotypes were identified: 15 belonged to the genus *Glomus,* one to *Claroideoglomus*, two to *Paraglomus*, one to *Acaulospora,* one to *Scutellospora* and one to *Archaeospora.* The distribution of the AMF community composition differed as a consequence of the treatment effects. The treatment combining organic and inorganic fertilisation with chemical pest control had the highest AMF richness and the treatment combining inorganic fertilisation with chemical pest had the lowest. The real causes and effects of these differences in the AMF community are very difficult to establish, since the crop management regimes tested were composed of several interacting factors. In conclusion, the crop management practices can exert a significant influence on the populations of AMF. The treatment combining organic and inorganic fertilisation with chemical pest control appears to be the most suitable agricultural management strategy with respect to improving the AMF diversity in this crop under tropical conditions, and thus for maintaining the agricultural and environmental sustainability of this agro-ecosystem.

Editor: Lee A. Newsom, The Pennsylvania State University, United States of America

Funding: This research was supported by funding from Ramon and Cajal programme (Ministerio de Educación y Ciencia, Spain). MM Alguacil was supported by the Ramon and Cajal programme (Ministerio de Educación y Ciencia, Spain). The funders had no role in study design, data collection and analysis, decision to publish, or preparation of the manuscript.

Competing Interests: The authors have declared that no competing interests exist.

* E-mail: mmalguacil@cebas.csic.es

Introduction

The soil is a complex matrix containing microorganisms that play a key role in the functioning of terrestrial ecosystems. They mediate many processes, including nutrient cycles, organic matter decomposition, soil aggregate formation and plant performance. Arbuscular mycorrhizal fungi (AMF) are among the most-important soil microorganisms, being obligate symbionts in the roots of most land plants in both natural and agricultural ecosystems, where they increase plant uptake of mineral nutrients, especially phosphorus [1]. Other beneficial effects of AMF are plant growth promotion [2], increased tolerance of drought [3], heavy metals [4] and plant protection agents [5]. In fact, in a previous study carried out at the site that is also the subject of the current work [6], it was found that galls produced in *Prunus persica* roots due to infection with *Meloidogyne incognita* were extensively colonized by AMF, whose function might be to act as protection agents against opportunistic pathogens. Furthermore, the diversity of AMF influences a number of important ecosystem processes, including plant productivity, plant diversity and soil structure [7,8,9].

Due to the important role of AMF in ecosystem functioning, knowledge of the diversity of the AMF colonising the roots of crop plants in agricultural soils is essential for sustainable management of these agro-ecosystems. Fertilisation is a common practice used to increase the nutrient availability to crops and hence their yields. Studies carried out in recent years have considered the effect of different fertilisation treatments and cropping systems on AMF diversity. Thus, a general decrease in AMF diversity has been found with the use of mineral fertilisers [10,11,12,13,14], although not in all cases [15]. Others studies showed that fertilisation with manures stimulated the AMF populations [13,16,17], but studies on the species composition of the AMF community colonising crop roots in response to other management practices are scarce [18,19,20,21].

Prunus persica (L.) Batsch. (peach) is a fruit tree, native to Asia, introduced into Venezuela. Peach production in Venezuela is an activity that generates steady employment and is aimed primarily at the domestic market [22]. At present, there are 2,500 hectares that can produce more than 15,000 metric tons of fruit. Peach production in Venezuela is a cropping system in which fertilisers and pest control are combined in order to maximise yields while maintaining a suitable soil nutrient content.

Due to the economic importance of this fruit crop, the elucidation of whether there is a fertiliser/pest management combination that can maintain or increase the AMF diversity colonising the roots is an important step towards sustainable soil use and therefore protection of biodiversity. Therefore, the objective of this study was to compare the AMF diversity in *P. persica* roots under two fertilisation treatments (inorganic, with or without manure) combined with integrated or chemical pest management.

Materials and Methods

Ethics Statement

No specific permits were required for the described field studies since these locations are not privately-owned or protected in any way. Field studies did not involve endangered or protected species.

Study site and Sampling

The study was conducted in a *P. persica* orchard located at the "Colonia Tovar" Aragua State, in the north of Venezuela (latitude 10° 29' N, longitude 67° 07' W, 1790 msl). The climate is tropical temperate (mean annual temperature of 16.8 °C, annual average rainfall of 1271 mm). The soil was classified as a sandy loam Inceptisol [23]. The soil characteristics were: pH of 5.18, 5.75% clay, 40.5% silt, 53.75% sand, 6.46 cmol kg^{-1} of cationic exchange capacity, Total N 2.7 g kg^{-1}, Available P 32 µg g^{-1}, 5.9% organic matter and bulk density 1.29 g cm^{-3}

The plants used in this survey were 13-year old peach (*Prunus persica* (L). Batsch cv. Criollo Amarillo). The experimental sampling was a randomised block design with two factors and four replication blocks (100 m^2 each) in an experimental area of approximately 1000 m^2. The first factor consisted of two types of fertilization and the second factor was two different procedures of pest control. Four treatments were established in the sampling design. The treatments were selected in order to provide more sustainable practices to producers, since the regular management practices are limited to exclusively use of inorganic fertilization and an excessive use of pesticides

T1: Combination of organic and inorganic fertilization (ComFert) and integrated pest management (IntM).

T2: Inorganic fertilization (InorgFert) and integrated pest management (IntM).

T3: Inorganic fertilization (InorgFert) and chemical pest control (ChemM).

T4: Combination of organic and inorganic fertilization (ComFert) and chemical pest control (ChemM).

-ComFert consisted of application of chicken manure (1400 kg ha^{-1}), urea (140 kg ha^{-1}), complex fertilizer (NPK) 12-12-17 (280 kg ha^{-1}), and potassium sulfate (40 kg ha^{-1}).

-InorgFert consisted of application of urea (140 kg ha^{-1}), complex fertilizer (NPK) 12-12-17 (400 kg ha^{-1}) and potassium sulfate (70 $kgha^{-1}$).

-IntM consisted of weekly applications of *Beauveria bassiana* (300 g spores ha^{-1}) for one month, subsequently weekly applications for two months of *Trichoderma harzianum* (300 g spores ha^{-1}) and lastly applications every 15 days for two more months of *Trichoderma harzianum* (300 g spores ha^{-1}). We applied these products as biocontrol agents against fungal diseases.

-ChemM consisted of applications of different chemicals from the beginning of flowering aimed at insect pests control and subsequently the incidence of foliar diseases. Thus, weekly applications for six weeks of Profenofos 0.6 kg a.i. ha^{-1} (Curacron ®) + Mancozeb 8 kg i.a. ha^{-1} (Dithane ®) were made. For control of *Oidium leucoconium*, was applied twice the mixture Urea 10 kg ha^{-1} + Flusilazol 0.4 kg a.i. ha^{-1} (Punch ®) + Mancozeb 4 kg a.i. ha^{-1} (Dithane ®), then weekly applications of Mancozeb 4 kg a.i. ha^{-1} (Dithane ®) + Profenofos 0.6 kg a.i. ha^{-1} (Curacron ®) + Endosulfuran 2.8 kg a.i. ha^{-1} (Thionil). For control of *Monilia cinerea* were performed four weekly applications of Carbendazin 2 kg a.i. ha^{-1} (Bavistin ®).

The treatments were applied for one year and sampling was conducted after fruit harvest (February 2011). Four plants (one per block) of each treatment were sampled providing 16 samples in total. The roots were sampled using three soil cores from three points/single tree/block.

Root DNA extraction and PCR

All PCR experiments were run using DNA preparations consisting of pooled roots of individual plants. DNA extractions from 16 root samples were carried out.

For each sample (total 16), total DNA was extracted from (0.1 g) fine root material using a DNeasy plant mini Kit following the manufacturer's recommendations (Qiagen). The roots samples were placed into a 2-ml screw-cap propylene tube together with two tungsten carbide balls (3 mm) and beaten (3 min, 13000 r.p.m.) using a mixer mill (MM 400, Retsch, Haan, Germany). The extracted DNA was resuspended in 20 µl of water. Several dilutions of extracted DNA (1/10, 1/50, 1/100) were prepared and 2 µl were used as template. Partial small-subunit (SSU) ribosomal RNA gene fragments were amplified using nested PCR with the universal eukaryotic primers NS1 and NS4 [24]. PCR was carried out in a final volume of 25 ml using the "ready to go" PCR beads (Amersham Pharmacia Biotech, Piscataway, N.J.), 0.2 mM dNTPs and 0.5 mM of each primer (PCR conditions: 94 °C for 3 min, then 30 cycles at 94 °C for 30 s, 40 °C for 1 min, 72 °C for 1 min, followed by a final extension period at 72 °C for 10 min). Two µl of several dilutions (1/10, 1/20, 1/50 and 1/100) from the first PCR were used as template DNA in a second PCR reaction performed using the specific primers AML1 and AML2 [44]. PCR reactions were carried out in a final volume of 25 ml using the "ready to go" PCR beads (Amersham Pharmacia Biotech, Piscataway, N.J.), 0.2 mM dNTPs and 0.5 mM of each primer (PCR conditions: 94 °C for 3 min, then 30 cycles of 1 min denaturation at 94 °C, 1 min primer annealing at 50 °C and 1 min extension at 72 °C, followed by a final extension period of 10 min at 72 °C). Positive and negative controls using PCR positive products and sterile water respectively were also included in all amplifications. All the PCR reactions were run on a Perkin Elmer Cetus DNA Thermal Cycler. Reactions yields were estimated by using a 1.2% agarose gel containing GelRed™ (Biotium).

Cloning and sequencing

The PCR products were purified using a Gel extraction Kit (Qiagen) cloned into pGEM-T Easy (Promega) and transformed into *Escherichia coli* (XL2-Blue). Thirty two positive transformants were screened in each resulting SSU rRNA gene library, using 0.8 units of RedTaq DNA polymerase (Sigma) and a re-amplification with AML1 and AML2 primers with the same conditions described above. Product quality and size were checked in agarose gels as described above. All clones having inserts of the correct size in each library were sequenced.

Clones were grown in liquid culture and the plasmid extracted using the QIAprep Spin Miniprep Kit (Qiagen). The sequencing was done by Laboratory of Sistemas Genómicos (Valencia, Spain) using the universal primers SP6 and T7. Sequence editing was done using the program Sequencher version 4.1.4 (Gene Codes

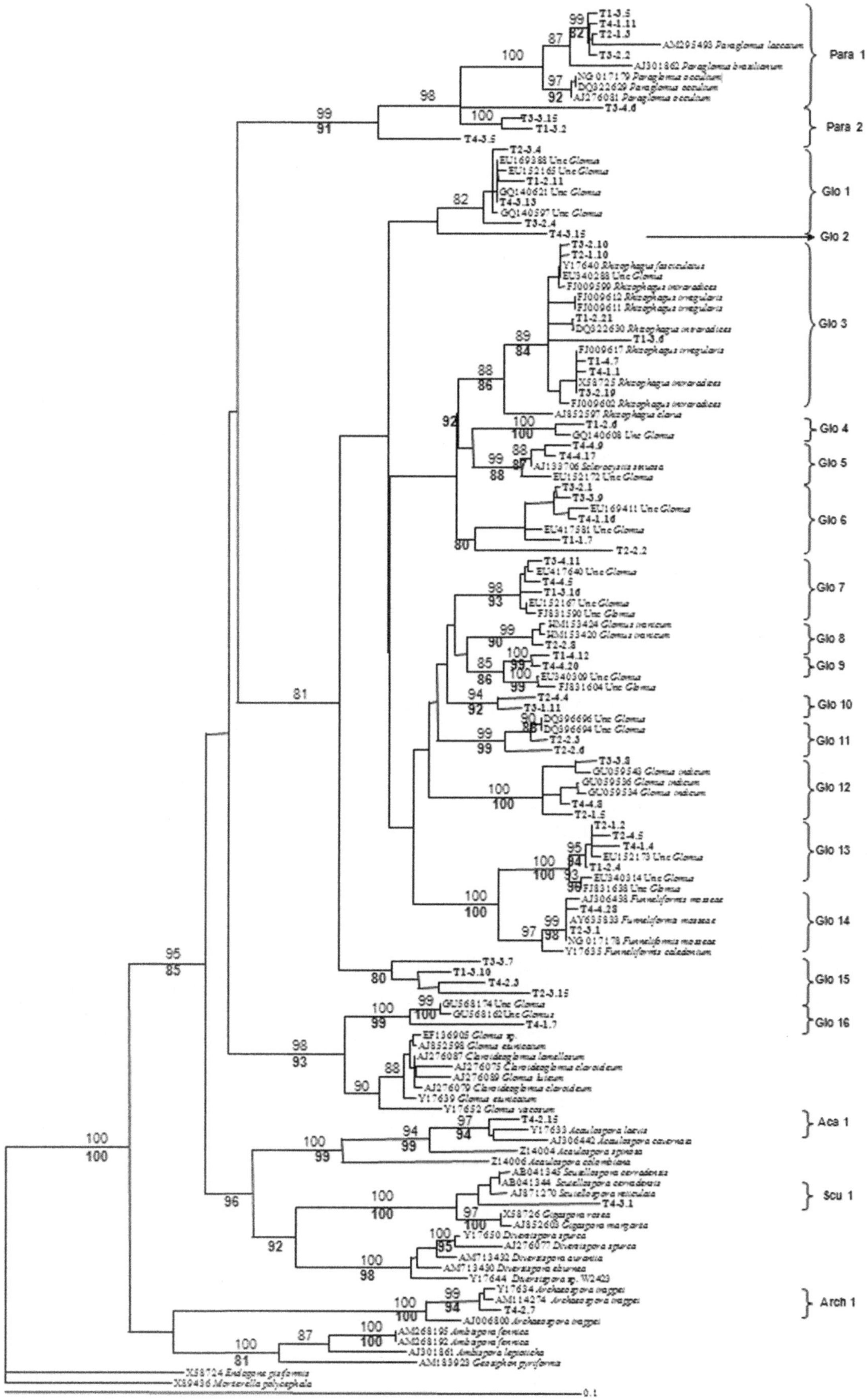
Para 1
Para 2
Glo 1
Glo 2
Glo 3
Glo 4
Glo 5
Glo 6
Glo 7
Glo 8
Glo 9
Glo 10
Glo 11
Glo 12
Glo 13
Glo 14
Glo 15
Glo 16
Aca 1
Scu 1
Arch 1

Figure 1. Phylogenetic tree of AMF sequences isolated from the *Prunus persica* roots under different treatments (T1: Combination of organic and inorganic fertilization and integrated pest management; T2: Inorganic fertilization and integrated pest management; T3: Inorganic fertilization and chemical pest control; T4: Combination of organic and inorganic fertilization and chemical pest control), reference sequences corresponding to the closest matches from GeneBank as well as sequences from cultured AMF taxa including representatives of the major taxonomical groups. Numbers above branches indicate the bootstrap values determined for Neighbour-Joining (NJ) analysis; bold numbers below branches indicate the bootstrap values of the maximum likelihood analysis. Sequences are labelled with the number of treatment from which they were obtained (T1, T2, T3, T4) and the clone identity number, Group identifiers (for example Glo 1) are AM fungal sequences types found in our study. Since identical sequences were detected, the clones producing the same sequence for each treatment were represented once in the alignment for clarity (Table S1 in File S1 material show a detailed description of the total number of clones of each AMF phylotype that were recovered from each treatment).

Corporation). Unique sequences of the clones generated in this study have been deposited at the National Centre for Biotechnology Information (NCBI) GenBank (http://www.ncbi.nlm.nih.gov) under the accession numbers HE613450 to HE613504.

Phylogenetical analysis

Sequence similarities were determined using the Basic Local Alignment Search Tool (BLASTn) sequence similarity search tool [25] provided by GenBank.

Phylogenetic analysis was carried out on the sequences obtained in this study and those corresponding to the closest matches from GenBank as well as sequences from cultured AMF taxa including representatives of the major taxonomical groups described by Schüßler et al. [26]. Sequences were aligned using the program ClustalX [27] and the alignment was adjusted manually in GeneDoc [28]. Neighbour-joining (NJ) [29] and maximum likelihood (ML) phylogenetic analyses were performed with the programs PAUP4.08b [30] and RAxML v.7.0.4 [31], respectively. The evolutionary distances for the NJ tree were computed using the maximum composite likelihood method with 1000 bootstrap replicates. For the ML analysis, a GTR-GAMMA model of evolution was used. The ML bootstrap values were calculated with 1000 replicates using the same substitution model. *Endogone pisiformis* Link and *Mortierella polycephala* Coem, were used as the out-groups.

Statistical analysis

Treatments effects on the number of phylotypes per root sample were compared using analysis of variance and comparisons among means were made using the Duncan's test calculated at $P<0.05$. The effect of two factors: types of fertilizer and pest management on AMF community composition were tested using a two-way analysis of variance, The statistical procedures were carried out with the software package SPSS 19.0 for Windows.

Canonical-correspondence analysis (CCA) with the relative abundance of clones per AMF sequence types found in *P. persica* roots under different treatments was performed. The results were summarized in an ordination diagram conducted in CANOCO for Windows v. 4.5 [32]. CCA is a multivariate statistical method that allows comparisons of AM fungal community compositions between four treatments.

The Shannon-Weaver (H') index was calculated as an additional measure of diversity, as it combines two components of diversity, i.e., species richness and evenness. It is calculated from

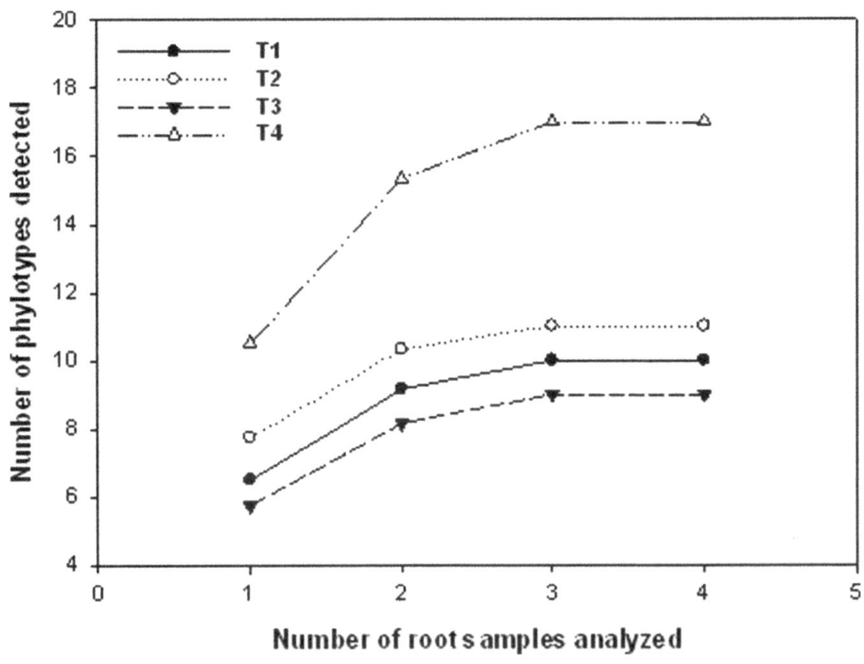

Figure 2. Sampling effort curves for *Prunus persica* roots under different treatments analysed. T1: Combination of organic and inorganic fertilization and integrated pest management; T2: Inorganic fertilization and integrated pest management; T3: Inorganic fertilization and chemical pest control; T4: Combination of organic and inorganic fertilization and chemical pest control. The number of clones for each AMF phylotypes in each root sample was used to construct the sampling effort curves (with 95% confidence intervals) using the software EstimateS 8.00 (Colwell, [33]). The sample order was randomized by 100 replications.

the equation H' = $-\sum p_i(\ln p_i)$, where p_i is the proportion of individuals found in the ith species (in a sample, the true value of p_i is unknown but is estimated as n_i/N, [here and throughout, n_i is the number of individuals in the ith species; N is the total number of individuals of all species]).

The number of clones for each AMF phylotypes in each soil sample was used to construct the sampling effort curves (with 95% confidence intervals) using the software EstimateS 8.00 [33]. The sample order was randomized by 100 replications.

Results

Sequence identity and phylogenetic analysis

In our study, 16 root samples (four repetitions per treatment) that were subjected to DNA extraction produced clonable PCR products of the expected size (about 795 bp). Overall, 512 clones from 16 clone libraries were screened by PCR; out of these, 449 contained the small-subunit rRNA gene fragment and an average of 15 clones per root sample were sequenced (in total, 242 sequences). According to the BLAST search in GenBank, sequences with a high degree of similarity (98−99%) to taxa belonging to the phylum Glomeromycota were produced by 227 clones. The remaining 15 clones produced incomplete sequences.

The 227 sequences were grouped in 21 different AMF sequence types or phylotypes, with sequence similarities varying from 97 to 100% and bootstrap values ≥ 80% (Fig. 1). These phylotypes were grouped in five families: the Glomeraceae, Paraglomeraceae, Acaulosporaceae, Gigasporaceae and Archaeosporaceae. Sixteen of these sequence groups belonged to the genus *Glomus*, two to the genus *Paraglomus*, one to the genus *Acaulospora*, one to the genus *Scutellospora* and one to the genus *Archaeospora*.

For the number of clones sequenced, the sampling effort curves showed a decreasing rate of accumulation of phylotypes, reaching the asymptote (Fig. 2). This pattern indicates that the clones analysed covered the AMF diversity colonising the *P. persica* roots under the four treatments. Therefore, no more clones were sequenced.

Nine phylotypes corresponded to morphologically-defined species: six were related to sequences from single, morphologically-described species (Para 1 to *Paraglomus laccatum*, Glo 5 to *Sclerocystis sinuosa*, Glo 8 to *Glomus iranicum*, Glo G12 to *Glomus indicum*, Glo 14 to *Funneliformis mosseae* and Arch 1 to *Archaeospora trappei*) and three were related to sequences belonging to two or three different, morphologically-described species (Glo 3 was related to a species group including *Rhizophagus intraradices/ irregularis/fasciculatus*, Aca 1 to an *Acaulospora cavernata/laevis* group and Scu 1 to a *Scutellospora cerradensis/reticulata* group). Eight phylotypes were related to uncultured glomalean species that have not been characterised morphologically (Glo 1, Glo 4, Glo 6, Glo 7, Glo 9, Glo 11, Glo 13 and Glo 16) and the remaining four phylotypes were not related to any sequences of AMF in the database (Glo 2, Glo 10, Glo 15 and Para 2) (Fig. 1).

The AMF community composition

There were significant differences in the AMF taxon richness. The ComFert+ChemM treatment harboured the highest mean number of AMF phylotypes per root sample (8.00), which was significantly different from the value for the InorgFert+ChemM treatment (4.25) according to Duncans multiple-comparison test. The mean number of AMF phylotypes detected in the tree roots receiving the ComFert+IntM or InorgFert+IntM treatments was the same (5.75) and no significant differences between these and either of the treatments mentioned above were found.

The factorial analysis showed that both factors: fertilization and pest management had a significant effect on the AMF community composition ($p = 0.002$, $F = 9.734$ and $p = 0.007$, $F = 7.412$, respectively) The interaction between these factors was not significant ($p = 0.906$, $F = 0.014$).

The AMF communities of tree roots in the InorgFert+ChemM treatment had the lowest diversity ($H = 1.78$), with the lowest total number of AMF sequence types (9). The trees from the ComFert+IntM and InorgFert+IntM treatments had similar AMF diversity (H≈2.0), while the treatment ComFert+ChemM yielded the highest number of different AMF sequence types (17) and showed the highest diversity index ($H = 2.69$). In the CCA diagram (Fig. 3), the different distributions of the AMF phylotypes, as a consequence of the treatments, can be observed. The symbols representing different treatments are distant to each other, which demonstrates that the treatments had a significant effect on the AMF community composition, with the different treatments hosting distinct phylotypes. This diagram also shows the AMF phylotypes found exclusively in each treatment.

Discussion

In this study we compared the diversity of AMF in *Prunus persica* roots under two types of fertilisation (inorganic, with or without manure) combined with integrated or chemical pest management in a tropical agro-ecosystem.

It is worth noting the high number of phylotypes found in this study (twenty-one) in comparison with other studies carried out also in agricultural soils. Daniell et al. [34] and Helgason et al. [35] found 10 phylotypes in arable soils around North Yorkshire, U.K. Toljander et al. [36] found eight phylotypes, all belonging to the

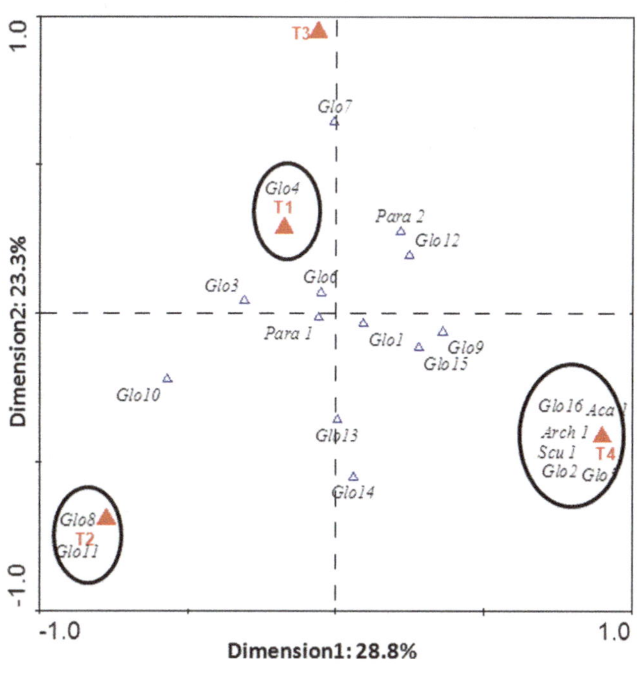

Figure 3. Canonical Correspondence analysis (CCA) of the AM fungal community composition found in the roots of *P. persica* under different treatments (T1: Combination of organic and inorganic fertilization and integrated pest management; T2: Inorganic fertilization and integrated pest management; T3: Inorganic fertilization and chemical pest control; T4: Combination of organic and inorganic fertilization and chemical pest control). Full triangles represent the treatments and the open triangles the AMF phylotypes. Open circles represent the AMF phylotypes exclusively found in individual treatments.

Table 1. Relative abundance of the different AMF sequence types observed in *Prunus persica* roots under the different treatments analysed.

	Treatments			
Phylotypes	**ComFert+IntM (n = 60)**	**InorgFert+IntM(n = 55)**	**InorgFert+ChemM (n = 49)**	**ComFert+ChemM (n = 63)**
Para 1	13.3	14.6	12.2	11.1
Para 2	3.3	0	4.1	3.2
Glo1	23.3	10.9	8.2	19.1
Glo2	0	0	0	3.2
Glo3	26.7	25.5	16.3	3.2
Glo4	3.3	0	0	0
Glo5	0	0	0	4.8
Glo6	8.3	5.5	6.1	4.8
Glo7	10.0	0	28.6	3.2
Glo8	0	3.6	0	0
Glo9	3.3	0	0	3.2
Glo10	0	9.1	4.1	0
Glo11	0	10.9	0	0
Glo12	0	3.6	16.3	11.1
Glo13	5.0	9.1	0	7.9
Glo14	0	3.6	0	3.2
Glo15	3.3	3.6	4.1	9.5
Glo16	0	0	0	3.2
Aca 1	0	0	0	3.2
Scu 1	0	0	0	3.2
Arch 1	0	0	0	3.2

ComFert: Combination of organic and inorganic fertilization; IntM: Integrated pest management; InorgFert: Inorganic fertilization; ChemM: Chemical pest control.

genus *Glomus,* in a field experiment using different organic and mineral fertilisers, while Hijri et al. [37] found 10 phylotypes in a conventional maize field in Germany. Also Verbruggen et al. [38] found that arbuscular mycorrhizal fungi richness varied from one to 11 phylotypes among organically and conventionally managed fields in the Netherlands. Alguacil et al. [39] found nine phylotypes in a study carried out in tropical savanna soils planted with leguminous forage under different doses of phosphorus fertiliser.

We observed different AMF community composition between treatments. Some studies have pointed out that soil physical-chemical properties, such as pH and nutrient content, are the factors influencing the structure of AMF communities in agricultural systems [17,36,40,41,42,43,44]. In our study, we did not find significant differences in soil characteristics among treatments (Table S2 in File S1), so the observed differences in the number of AMF phylotypes or AMF richness can be attributed to the different treatments applied. It has been reported that the AMF diversity is higher in soils amended with different organic substrates (or that the AMF diversity increases in organically-managed soils) [13,17,36,45,46]. Moreover, the addition of chemical pest-control products could have prevented the colonisation by non-AM fungi inside the roots, favouring the most-tolerant AMF species.

In contrast to some studies which indicated that species of the *Paraglomeraceae* appear to be rare or poor in agricultural soils [37,47,48], we found that Para 1 was one of the most-abundant groups in our study (12.8% of the AMF clones analysed), together with Glo 1 (15.9%) and Glo 3 (17.6%) (Table 1); Glo 6 and Glo 15 were found in all treatments, but their occurrence frequency was low (<6.2% of clones). The presence of the genus *Paraglomus* has been observed also in other agricultural management studies using group-specific primers [37,49] or the same pair of primers as ourselves [50]. In the case of Glo 3 (related to the *R. intraradices/ irregularis* species complex group), our results are in accordance with several studies carried out in agricultural soils where this AMF taxon showed the highest representation in the clone libraries [36,37,50,51,52]. On the other hand, this species also showed the highest abundance of clones in *P. persica* roots when integrated pest management treatments (ComFert+IntM and InorgFert+IntM) were applied (Table 1). Several studies have shown that the presence of *Trichoderma harzianum* significantly increases root colonisation by *Rhizophagus intraradices* in melon crops [53,54], *R. intraradices* being the only taxon that increased the *T. harzianum* populations [53]. Therefore, a synergistic relationship between *T. harzianum* and *R. intraradices* could have existed in *P. persica* roots under the ComFert+IntM and InorgFert+IntM treatments, increasing their respective abundances.

Interestingly, in contrast to Lee et al. [55] who reported that the AML1/AML2 primer pairs do not amplify sequences belonging to the family *Archaeosporaceae*, we detected the Arch 1 phylotype, which showed 99% homology with sequences related to *Archaeospora trappei.*

There were phylotypes that occurred exclusively in some treatments; for example, Glo 2, Glo 5, Glo 16, Aca 1, Scu 1 and Arch 1 seemed to be specific for trees treated with ComFert+ ChemM. The low abundance and specificity found for these phylotypes could be attributable to different colonisation strategies by these different phylogenetic groups. For example, it has been reported that mycelia of *Acaulospora* species have low root and soil colonisation levels [56]

Phylotypes Glo 8 and Glo 11 only appeared in the roots of trees receiving the InorgFert+IntM treatment and Glo 4 occurred exclusively with the ComFert+IntM treatment. The remaining phylotypes did not have a clear distribution, occurring haphazardly in two or three treatments, as in the case of Glo 14, related to sequences belonging to *Funneliformis mosseae*. Although in our study this phylotype was of low abundance (6.81%), this taxon appears to be one of the most-typical and dominant taxa in many agricultural fields [34,35,37,47,57]. Together with *R. intraradices*, these taxa are sometimes called the "typical AMF of arable lands" [47].

Conclusions

The real causes and effects of these differences in the AMF community composition observed among treatments are very difficult to establish, bearing in mind that the different crop management regimes studied consist of several influencing parameters. In fact, the ComFert+ChemM - which produced the highest diversity of AMF - included the highest number of parameters. Further investigation of AMF diversity, including analysis of each factor separately and subsequently their interactions, could help to ascertain the cause of the effects reported here.

Different crop management strategies can exert a clear influence on the populations of AMF. The treatment including a combination of organic and inorganic fertilisation together with chemical pest control appears to be the most-suitable with respect to improve of the AMF diversity in this crop under tropical conditions, thus improving the agricultural and environmental sustainability of this agro-ecosystem.

Supporting Information

File S1 Supporting tables. Table S1. AMF sequences obtained in the present study for each phylotype. **Table S2.** Chemical properties of soil in response to different treatments analysed at the time of sampling (n = 4).

Author Contributions

Conceived and designed the experiments: MMA ZL AR. Performed the experiments: MMA ZL AR. Analyzed the data: MMA ET MPT. Contributed reagents/materials/analysis tools: MMA ET AR. Wrote the paper: MMA AR.

References

1. Smith SE, Read DJ (2008) Mycorrhizal Symbiosis. New York: Academic Press. 787 p.
2. Caravaca F, Alguacil MM, Figueroa D, Barea JM, Roldán A (2003) Re-establishment of *Retama sphaerocarpa* as a target species for reclamation of soil physical and biological properties in a semi-arid Mediterranean area. Forest Ecol Manag 182: 49–58.
3. Querejeta JI, Allen MF, Alguacil MM, Roldán A (2007) Plant isotopic composition provides insight into mechanisms underlying growth stimulation by AM fungi in a semiarid environment. Funct Plant Biol 34: 683–691.
4. Hildebrandt U, Regvar M, Bothe H (2007) Arbuscular mycorrhiza and heavy metal tolerance. Phytochemistry 68: 139–146.
5. Pozo MJ, Azcón-Aguilar C (2007) Unraveling mycorrhiza-induced resistance. Curr Opin Plant Biol 10: 393–398.
6. Alguacil MM, Torrecillas E, Lozano Z, Roldán A (2011) Evidence of differences between the communities of arbuscular mycorrhizal fungi colonizing galls and roots of *Prunus persica* infected by the root-knot nematode *Meloidogyne incognita*. Appl Environ Microb 77: 8656–8661.
7. Van Der Heijden MGA, Klironomos JN, Ursic M, Moutoglis P, Streitwolf-Engel R, et al. (1998) Mycorrhizal fungal diversity determines plant biodiversity, ecosystem variability and productivity. Nature 396: 69–72.
8. Rillig MC (2004) Arbuscular mycorrhizae and terrestrial ecosystem processes. Ecol Lett 7: 740–754.
9. Wilson GWT, Rice CW, Rillig MC, Springer A, Hartnett DC (2009) Soil aggregation and carbon sequestration are tightly correlated with the abundance of arbuscular mycorrhizal fungi: Results from long-term field experiments. Ecol Lett 12: 452–461.
10. Joner EJ (2000) The effect of long-term fertilization with organic or inorganic fertilizers on mycorrhiza-mediated phosphorus uptake in subterranean clover. Biol Fertil Soils 32: 435–440.
11. Na Bhadalung N, Suwanarit A, Dell B, Nopamornbodi O, Thamchaipenet A, et al. (2005) Effects of long-term NP-fertilization on abundance and diversity of arbuscular mycorrhizal fungi under a maize cropping system. Plant Soil 270: 371–382.
12. Bradley K, Drijber RA, Knops J (2006) Increased N availability in grassland soils modifies their microbial communities and decreases the abundance of arbuscular mycorrhizal fungi. Soil Biol Biochem 38: 1583–1595.
13. Gryndler M, Larsen J, Hrselová H, Rezácová V, Gryndlerová H, et al. (2006) Organic and mineral fertilization, respectively, increase and decrease the development of external mycelium of arbuscular mycorrhizal fungi in a long-term field experiment. Mycorrhiza 16: 159–166.
14. Sasvári Z, Hornok L, Posta K (2011) The community structure of arbuscular mycorrhizal fungi in roots of maize grown in a 50-year monoculture. Biol Fertil Soils 47: 167–176.
15. Beauregard MS, Hamel C, Atul N, St-Arnaud M (2010) Long-term phosphorus fertilization impacts soil fungal and bacterial diversity but not AM fungal community in alfalfa. Microb Ecol 59: 379–389.
16. Mäder P, Edenhofer S, Boller T, Wiemken A, Niggli U (2000) Arbuscular mycorrhizae in a long-term field trial comparing low-input (organic, biological) and high-input (conventional) farming systems in a crop rotation. Biol Fertil Soils 31: 150–156.
17. Alguacil MM, Díaz-Pereira E, Caravaca F, Fernández DA, Roldán A (2009) Increased diversity of arbuscular mycorrhizal fungi in a long-term field experiment via application of organic amendments to a semiarid degraded soil. Appl Environ Microb 75: 4254–4263.
18. Alguacil MM, Lumini E, Roldán A, Salinas-García JR, Bonfante P, et al. (2008) The impact of tillage practices on arbuscular mycorrhizal fungal diversity in subtropical crops. Ecol Appl 18: 527–536.
19. Van Der Gast CJ, Gosling P, Tiwari B, Bending GD (2011) Spatial scaling of arbuscular mycorrhizal fungal diversity is affected by farming practice. Environ Microbiol 13: 241–249.
20. Borriello R, Lumini E, Girlanda M, Bonfante P, Bianciotto V (2012) Effects of different management practices on arbuscular mycorrhizal fungal diversity in maize fields by a molecular approach. Biol Fertil Soils 48: 911–922
21. Lumini E, Vallino M, Alguacil MM, Romani M, Bianciotto V (2011) Different farming and water regimes in Italian rice fields affect arbuscular mycorrhizal fungal soil communities. Ecol Appl 21: 1696–1707.
22. Avilán L, Leal F (1990) Suelos, fertilizantes y encalado para frutales.; América E, editor. Caracas, Venezuela.
23. SSS (2010) Keys to Soil Taxonomy. Washington DC: USDA. Natural Resources Conservation Service. 338 p.
24. White TJ, Bruns T, Lee S, Taylor J (1990) Amplification and direct sequencing of fungal ribosomal RNA genes for phylogenetics. In: Innis MA, Gelfand, D.H., Sninsky, J.J., and White, T.J., editor. PCR protocols A guide to methods and applications. San Diego. pp. 315–322.
25. Altschul SF, Madden TL, Schäffer AA, Zhang J, Zhang Z, et al. (1997) Gapped BLAST and PSI-BLAST: a new generation of protein database search programs. Nucleic Acids Res 25: 3389–3402.
26. Schüßler A, Schwarzott D, Walker C (2001) A new fungal phylum, the Glomeromycota: Phylogeny and evolution. Mycol Res 105: 1413–1421.
27. Thompson JD, Gibson TJ, Plewniak F, Jeanmougin F, Higgins DG (1997) The ClustalX windows interface: flexible strategies for multiple sequence alignment aided by quality analysis tools. Nucleic Acids Res 24: 4876–4882.
28. Nicholas KB, Jr. NHB (1997) GeneDoc: a tool for editing and annotating multiple sequence alignments.
29. Saitou N, Nei M (1987) The neighbor-joining method: a new method for reconstructing phylogenetic trees. Mol Biol Evol 4: 406–425.
30. Swofford DL (2002) PAUP: Phylogenetic Analysis Using Parsimony (and other methods). Version 4.08b10 for Machintosh ed. Sunderland, MA.: Sinauer Associates Inc.
31. Stamatakis A (2006) RaxML-VI-HPC: maximum likelihoodbased phylogenetic analyses with thousands of taxa and mixed models. Bioinformatics 22: 2688–2690.
32. ter Braak CFJ, Smilauer P (2004) CANOCO reference manual and CanoDraw for Windows. version 4.5 ed. Wageningen, Netherlands: Biometris. pp. Software for canonical community ordination.

33. Colwell R (2005) EstimateS: statistical estimation of species richness and shared species from samples, version 8.0.
34. Daniell TJ, Husband R, Fitter AH, Young JPW (2001) Molecular diversity of arbuscular mycorrhizal fungi colonising arable crops. FEMS Microbiol Ecol 36: 203–209.
35. Helgason T, Daniell TJ, Husband R, Fitter AH, Young JPW (1998) Ploughing up the wood-wide web? [4]. Nature 394: 431.
36. Toljander JF, Santos-González JC, Tehler A, Finlay RD (2008) Community analysis of arbuscular mycorrhizal fungi and bacteria in the maize mycorrhizosphere in a long-term fertilization trial. FEMS Microbiol Ecol 65: 323–338.
37. Hijri I, Sýkorová Z, Oehl F, Ineichen K, Mäder P, et al. (2006) Communities of arbuscular mycorrhizal fungi in arable soils are not necessarily low in diversity. Mol Ecol 15: 2277–2289.
38. Verbruggen E, Van Der Heijden MGA, Weedon JT, Kowalchuk GA, Röling WFM (2012) Community assembly, species richness and nestedness of arbuscular mycorrhizal fungi in agricultural soils. Mol Ecol 21: 2341–2353.
39. Alguacil MM, Lozano Z, Campoy M, Roldán A (2010) Phosphorus fertilisation management modifies the biodiversity of AM fungi in a tropical savanna forage system. Soil Biol Biochem 42: 1114–1122.
40. Gosling P, Hodge A, Goodlass G, Bending GD (2006) Arbuscular mycorrhizal fungi and organic farming. Agric Ecosyst Environ 113: 17–35.
41. Helgason T, Fitter AH (2009) Natural selection and the evolutionary ecology of the arbuscular mycorrhizal fungi (Phylum Glomeromycota). J Exp Bot 60: 2465–2480.
42. Schreiner RP, Mihara KL (2009) The diversity of arbuscular mycorrhizal fungi amplified from grapevine roots (*Vitis vinifera* L.) in Oregon vineyards is seasonally stable and influenced by soil and vine age. Mycologia 101: 599–611.
43. Balestrini R, Magurno F, Walker C, Lumini E, Bianciotto V (2010) Cohorts of arbuscular mycorrhizal fungi (AMF) in *Vitis vinifera*, a typical Mediterranean fruit crop. Env Microbiol Rep 2: 594–604.
44. Oehl F, Laczko E, Bogenrieder A, Stahr K, Bösch R, et al. (2010) Soil type and land use intensity determine the composition of arbuscular mycorrhizal fungal communities. Soil Biol Biochem 42: 724–738.
45. Verbruggen E, Röling WFM, Gamper HA, Kowalchuk GA, Verhoef HA, et al. (2010) Positive effects of organic farming on below-ground mutualists: Large-scale comparison of mycorrhizal fungal communities in agricultural soils. New Phytol 186: 968–979.
46. Gosling P, Ozaki A, Jones J, Turner M, Rayns F, et al. (2010) Organic management of tilled agricultural soils results in a rapid increase in colonisation potential and spore populations of arbuscular mycorrhizal fungi. Agric Ecosyst Environ 139: 273–279.
47. Oehl F, Sieverding E, Ineichen K, Mäder P, Boller T, et al. (2003) Impact of land use intensity on the species diversity of arbuscular mycorrhizal fungi in agroecosystems of Central Europe. Appl Environ Microb 69: 2816–2824.
48. Oehl F, Sieverding E, Ineichen K, Ris EA, Boller T, et al. (2005) Community structure of arbuscular mycorrhizal fungi at different soil depths in extensively and intensively managed agroecosystems. New Phytol 165: 273–283.
49. Galván GA, Parádi I, Burger K, Baar J, Kuyper TW, et al. (2009) Molecular diversity of arbuscular mycorrhizal fungi in onion roots from organic and conventional farming systems in the Netherlands. Mycorrhiza 19: 317–328.
50. Santos-González JC, Nallanchakravarthula S, Alström S, Finlay RD (2011) Soil, but not cultivar, shapes the structure of arbuscular mycorrhizal fungal assemblages associated with strawberry. Microb Ecol 62: 25–35.
51. Mathimaran N, Ruh R, Jama B, Verchot L, Frossard E, et al. (2007) Impact of agricultural management on arbuscular mycorrhizal fungal communities in Kenyan ferralsol. Agric Ecosyst Environ 119: 22–32.
52. Cesaro P, Van Tuinen D, Copetta A, Chatagnier O, Berta G, et al. (2008) Preferential colonization of *Solanum tuberosum* L. roots by the fungus *Glomus intraradices* in arable soil of a potato farming area. Appl Environ Microb 74: 5776–5783.
53. Martínez-Medina A, Pascual JA, Lloret E, Roldán A (2009) Interactions between arbuscular mycorrhizal fungi and Trichoderma harzianum and their effects on Fusarium wilt in melon plants grown in seedling nurseries. J Sci Food Agric 89: 1843–1850.
54. Martínez-Medina A, Roldán A, Pascual JA (2011) Interaction between arbuscular mycorrhizal fungi and *Trichoderma harzianum* under conventional and low input fertilization field condition in melon crops: Growth response and Fusarium wilt biocontrol. Appl Soil Ecol 47: 98–105.
55. Lee J, Lee S, Young JPW (2008) Improved PCR primers for the detection and identification of arbuscular mycorrhizal fungi. FEMS Microbiol Ecol 65: 339–349.
56. Hart MM, Reader RJ (2002) Taxonomic basis for variation in the colonization strategy of arbuscular mycorrhizal fungi. New Phytol 153: 335–344.
57. Bharadwaj DP, Lundquist PO, Alström S (2007) Impact of plant species grown as monocultures on sporulation and root colonization by native arbuscular mycorrhizal fungi in potato. Appl Soil Ecol 35: 213–225.

PERMISSIONS

All chapters in this book were first published in PLOS ONE, by The Public Library of Science; hereby published with permission under the Creative Commons Attribution License or equivalent. Every chapter published in this book has been scrutinized by our experts. Their significance has been extensively debated. The topics covered herein carry significant findings which will fuel the growth of the discipline. They may even be implemented as practical applications or may be referred to as a beginning point for another development.

The contributors of this book come from diverse backgrounds, making this book a truly international effort. This book will bring forth new frontiers with its revolutionizing research information and detailed analysis of the nascent developments around the world.

We would like to thank all the contributing authors for lending their expertise to make the book truly unique. They have played a crucial role in the development of this book. Without their invaluable contributions this book wouldn't have been possible. They have made vital efforts to compile up to date information on the varied aspects of this subject to make this book a valuable addition to the collection of many professionals and students.

This book was conceptualized with the vision of imparting up-to-date information and advanced data in this field. To ensure the same, a matchless editorial board was set up. Every individual on the board went through rigorous rounds of assessment to prove their worth. After which they invested a large part of their time researching and compiling the most relevant data for our readers.

The editorial board has been involved in producing this book since its inception. They have spent rigorous hours researching and exploring the diverse topics which have resulted in the successful publishing of this book. They have passed on their knowledge of decades through this book. To expedite this challenging task, the publisher supported the team at every step. A small team of assistant editors was also appointed to further simplify the editing procedure and attain best results for the readers.

Apart from the editorial board, the designing team has also invested a significant amount of their time in understanding the subject and creating the most relevant covers. They scrutinized every image to scout for the most suitable representation of the subject and create an appropriate cover for the book.

The publishing team has been an ardent support to the editorial, designing and production team. Their endless efforts to recruit the best for this project, has resulted in the accomplishment of this book. They are a veteran in the field of academics and their pool of knowledge is as vast as their experience in printing. Their expertise and guidance has proved useful at every step. Their uncompromising quality standards have made this book an exceptional effort. Their encouragement from time to time has been an inspiration for everyone.

The publisher and the editorial board hope that this book will prove to be a valuable piece of knowledge for researchers, students, practitioners and scholars across the globe.

LIST OF CONTRIBUTORS

Noel P. Gurwick
Smithsonian Environmental Research Center. Edgewater, Maryland, United States of America

Lisa A. Moore
Environmental Defense Fund, San Francisco, California, United States of America

Charlene Kelly
Western Carolina University, North Carolina, United States of America

Patricia Elias
Union of Concerned Scientists, Washington, District of Columbia, United States of America

Yi Liu, Jing Wang, Zhiguo Li, Yong Tao, Juan Xie and Junfeng Pan
Laboratory of Aquatic Botany and Watershed Ecology, Wuhan Botanical Garden, Chinese Academy of Sciences China, Wuhan, China

Dongbi Liu
Institute of Plant Protection and Soil Fertilizer, Hubei Academy of Agricultural Sciences, Wuhan, China

Guoshi Zhang and Fang Chen
Laboratory of Aquatic Botany and Watershed Ecology, Wuhan Botanical Garden, Chinese Academy of Sciences China, Wuhan, China
China Program, International Plant Nutrition Institute (IPNI), Wuhan, China

Judith Prommer, Wolfgang Wanek, Florian Hofhansl and Rebecca Clare Hood-Nowotny
Department of Microbiology and Ecosystem Science, University of Vienna, Vienna, Austria

Daniela Trojan, Pierre Offre, Tim Urich and Christa Schleper
Department of Ecogenomics and Systems Biology, University of Vienna, Vienna, Austria

Stefan Sassmann
Core Facility of Cell Imaging and Ultrastructure Research, University of Vienna, Vienna, Austria

Barbara Kitzler
Institute of Forest Ecology and Soil, Federal Research and Training Centre for Forests, Natural Hazards and Landscape, Vienna, Austria

Gerhard Soja
Department of Health and Environment, Austrian Institute of Technology, Tulln, Austria

Ya-Lin Hu and De-Hui Zeng
State Key Laboratory of Forest and Soil Ecology, Institute of Applied Ecology, Chinese Academy of Sciences, Shenyang, People's Republic of China

Li-Le Hu
Chinese Research Academy of Environmental Sciences, Beijing, People's Republic of China

Xin-kai Zhu and Wen-shan Guo
Key Laboratory of Crop Genetics and Physiology of Jiangsu Province, Yangzhou University, Yangzhou, China

Xiu-liang Jin and Sen-nan Song
Key Laboratory of Crop Genetics and Physiology of Jiangsu Province, Yangzhou University, Yangzhou, China Beijing Research Center for Information Technology in Agriculture, Beijing Academy of Agriculture and Forestry Sciences, Beijing, China
National Engineering Research Center for Information Technology in Agriculture Beijing, China

Hai-kuan Feng, Zhen-hai Li,Xiao-yu Song, Guijun Yang and Xin-gang Xu
Beijing Research Center for Information Technology in Agriculture, Beijing Academy of Agriculture and Forestry Sciences, Beijing, China
National Engineering Research Center for Information Technology in Agriculture Beijing, China

Guangli Wang, Xiao Chen, Wenlong Yue, Hui Zhang, Feng Li and Minghua Xiong
College of Life Sciences, Huaibei Normal University, Huaibei, People's Republic of China

Jennifer Cunniff, Sarah Wilkinson, Mark Rees and Colin P. Osborne
Department of Animal and Plant Sciences, University of Sheffield, Sheffield, United Kingdom

Michael Charles
School of Archaeology, University of Oxford, Oxford, United Kingdom

Glynis Jones
Department of Archaeology, University of Sheffield, Sheffield, United Kingdom

Douglas A. Burns
United States Geological Survey, Troy, New York, United States of America,

Laurel G. Woodruff
United States Geological Survey, Mounds View, Minnesota, United States of America

Paul M. Bradley
United States Geological Survey, Columbia, South Carolina, United States of America

William F. Cannon
United States Geological Survey, Reston, Virginia, United States of America

Ivan R. Baxter and Gregory Ziegler
United States Department of Agriculture, Agricultural Research Service, Plant Genetics Research Unit, Donald Danforth Plant Science Center, St. Louis, Missouri, United States of America

Brett Lahner, Michael V. Mickelbart and Rachel Foley
Purdue University, Department of Horticulture and Landscape Architecture, West Lafayette, Indiana, United States of America

John Danku and David E. Salt
University of Aberdeen, Institute of Biological and Environmental Science, Aberdeen, United Kingdom

Paul Armstrong
United States Department of Agriculture, Agricultural Research Service, Engineering and Wind Erosion Research Unit, Manhattan, Kansas, United States of America

Owen A. Hoekenga
United States Department of Agriculture, Agricultural Research Service, RW Holley Center for Agriculture and Health, Ithaca, New York, United States of America

Lars A. Brudvig
Department of Plant Biology, Michigan State University, East Lansing, Michigan, United States of America

John L. Orrock, Ellen I. Damschen, Philip G. Hahn, W.Brett Mattingly and Joseph W. Veldman
Department of Zoology, University of Wisconsin, Madison, Wisconsin, United States of America

Cathy D. Collins
Department of Biology, Colby College, Waterville, Maine, United States of America

Joan L. Walker
U.S. Forest Service Southern Research Station, Clemson, South Carolina, United States of America

Zhaojun Li and Yongchao Liang
Institute of Agricultural Resources and Regional Planning, Chinese Academy of Agricultural Sciences, Key Laboratory of Plant Nutrition and Fertilizer, Ministry of Agriculture, Beijing, China

Hua Yang and Jian Long
Guizhou Key Laboratory of Mountain Environment, Guizhou Normal University, Guiyang, China

Lu Lu
Graduate School of the Chinese Academy of Agricultural Sciences, Beijing, China

Brent F. Kim
Johns Hopkins Center for a Livable Future, Baltimore, Maryland, United States of America
Department of Environmental Health Sciences, Johns Hopkins Bloomberg School of Public Health, Baltimore, Maryland, United States of America

Jared D. Margulies
Johns Hopkins Center for a Livable Future, Baltimore, Maryland, United States of America
Department of Environmental Health Sciences, Johns Hopkins Bloomberg School of Public Health, Baltimore, Maryland, United States of America
Department of Geography and Environmental Systems, University of Maryland, Baltimore, Maryland, United States of America

Anne M. Palmer
Johns Hopkins Center for a Livable Future, Baltimore, Maryland, United States of America
Department of Environmental Health Sciences, Johns Hopkins Bloomberg School of Public Health, Baltimore, Maryland, United States of America
Department of Health, Behavior and Society, Johns Hopkins Bloomberg School of Public Health Baltimore, Maryland, United States of America

Keeve E. Nachman
Johns Hopkins Center for a Livable Future, Baltimore, Maryland, United States of America
Department of Environmental Health Sciences, Johns Hopkins Bloomberg School of Public Health, Baltimore, Maryland, United States of America
Department of Health Policy and Management, Johns Hopkins Bloomberg School of Public Health, Baltimore, Maryland, United States of America

Melissa N. Poulsen
CLF-Lerner Fellow, Johns Hopkins Center for a Livable Future, Baltimore, Maryland, United States of America
Department of International Health, Johns Hopkins Bloomberg School of Public Health, Baltimore, Maryland, United States of America

Katie L. Dix
Community Greening Resource Network, Parks & People Foundation, Baltimore, Maryland, United States of America

Xing Liu, Wenming Zhang and Huizhen Qiu
College of Resources and Environmental Sciences, Gansu Agricultural University, Lanzhou, Gansu Province, China
Gansu Provincial Key Laboratory of Aridland Crop Science, Gansu Agricultural University, Lanzhou, Gansu Province, China

Junlian Zhang
College of Agronomy, Gansu Agricultural University, Lanzhou, Gansu Province, China
Gansu Key Laboratory of Crop Genetic and Germplasm Enhancement, Gansu Agricultural University, Lanzhou, China

Tianyu Gu and Shixue Yin
College of Environmental Science and Engineering, Yangzhou University, Yangzhou, Jaingsu Province, China

Qirong Shen
Jiangsu Provincial Key Lab of Organic Solid Waste Utilization, Nanjing Agricultural University, Nanjing, China

Gao Peng, Geng Guangpo and Zhang Guangcan
Shandong Agricultural University, College of Forestry/ Taishan Mountain Forest Ecosystem Research Station/ Shandong Provincial Key Laboratory of Soil Erosion and Ecological Restoration, Taían, Shandong, P.R. China

Wang Bing
Research Institute of Forest Ecology Environmental Protection, Chinese Academy of Forestry, Beijing, P.R. China

Xinshu Liang, Yinan Gao, Xiaoying Zhang, Yongqiang Tian, Zhenxian Zhang and Lihong Gao
Beijing Key Laboratory of Growth and Developmental Regulation for Protected Vegetable Crops, Department of Vegetable Science, China Agricultural University, Beijing, P.R. China

Guanghua Yin, Jian Gu, Fasheng Zhang and Zuoxin Liu
Institute of Applied Ecology, Chinese Academy of Sciences, Shenyang, China

Liang Hao and Peifei Cong
University of Chinese Academy of Sciences, Beijing, China

Feng-Bao Zhang, Zhan-Li Wang and Ming-Yi Yang
State Key Laboratory of Soil Erosion and Dryland Farming on the Loess Plateau, Institute of Soil and Water Conservation, Northwest A&F University, Yangling, P. R. China
Institute of Soil and Water Conservation, Chinese Academy of Science and Ministry of Water Resources, Yangling, P. R. China

Esperanza Huerta and Salvador Hernandez-Daumas
El Colegio de la Frontera Sur, Unidad Campeche, Dpto. Agroecología, Campeche, México

Christian Kampichler
Universidad Juárez Autónoma de Tabasco, División de Ciencias Biológicas, Villahermosa, Tabasco, México
Sovon Dutch Centre for Field Ornithology, Natuurplaza (Mercator 3), Nijmegen, The Netherlands

Susana Ochoa-Gaona and Ben De Jong
El Colegio de la Frontera Sur, Unidad Campeche, Dpto. Sustainability Sciences, Campeche, México

Violette Geissen
University of Bonn - INRES, Bonn, Germany
Wageningen University and Research Center – Alterra, Wageningen, Gelderland, Netherlands

Shihe Xing
College of Resource and Environment, Fujian Agriculture and Forestry University, Fuzhou, China

Liming Zhang
College of Resource and Environment, Fujian Agriculture and Forestry University, Fuzhou, China
State Key Laboratory of Soil and Sustainable Agriculture, Institute of Soil Science, Chinese Academy of Sciences, Nanjing, China

Dongsheng Yu, Xuezheng Shi, Shengxiang Xu and Yongcong Zhao
State Key Laboratory of Soil and Sustainable Agriculture, Institute of Soil Science, Chinese Academy of Sciences, Nanjing, China

Mirco Boschetti and Pietro Alessandro Brivio
Institute for Electromagnetic Sensing of the Environment, National Research Council, Milan, Italy

Francesco Nutini and Giacinto Manfron
Institute for Electromagnetic Sensing of the Environment, National Research Council, Milan, Italy
Department of Agricultural and Environmental Sciences – Production, Landscape, Agroenergy, University of Milan, Milan, Italy

Andrew Nelson
Social Sciences Division, International Rice Research Institute, Los Baños, Philippines

Yimin Gao, Pengcheng Gao, Fen Liu, Zuoping Zhao and Yan Pang
College of Natural Resources and Environment, Northwest A&F University, Yangling, China

Xiaoying Wang and Yanan Tong
College of Natural Resources and Environment, Northwest A&F University, Yangling, China
Key Laboratory of Plant Nutrition and the Agri-environment in Northwest China, Ministry of Agriculture, Yangling, China

Maria del Mar Alguacil, Emma Torrecillas and Antonio Roldán
CSIC-Centro de Edafología y Biología Aplicada del Segura, Department of Soil and Water Conservation, Campus de Espinardo, Murcia, Spain

Zenaida Lozano
Universidad Central de Venezuela (UCV), Facultad de Agronomía, Instituto de Edafología, El Limón, Campus Universitario, Maracay, Venezuela

Maria Pilar Torres
Departamento de Biología Aplicada, Area de Botánica, Universidad Miguel Hernández, Elche, Alicante, Spain

Index